Physics and Applications of
Defects in Advanced Semiconductors

MATERIALS RESEARCH SOCIETY SYMPOSIUM PROCEEDINGS VOLUME 325

Physics and Applications of Defects in Advanced Semiconductors

Symposium held November 29-December 1, 1993, Boston, Massachusetts, U.S.A.

EDITORS:

M.O. Manasreh

Wright Laboratory
Wright-Patterson Air Force Base, Ohio, U.S.A.

H.J. von Bardeleben

University of Paris VII/CNRS
Paris, France

G.S. Pomrenke

Air Force Office of Scientific Research
Bolling Air Force Base, D.C., U.S.A.

M. Lannoo

Institut Supériure d'Electronique du Nord
Lille, France

D.N. Talwar

Indiana University of Pennsylvania
Indiana, Pennsylvania, U.S.A.

MATERIALS RESEARCH SOCIETY
Pittsburgh, Pennsylvania

This work was supported largely by the Air Force Office of Scientific Research under
Grant Number F49620-9410062.

Single article reprints from this publication are available through
University Microfilms Inc., 300 North Zeeb Road, Ann Arbor, Michigan 48106

CODEN: MRSPDH

Published by:

Materials Research Society
9800 McKnight Road
Pittsburgh, Pennsylvania 15237
Telephone (412) 367-3003
Fax (412) 367-4373

Library of Congress Cataloging in Publication Data

Physics and applications of defects in advanced semiconductors : symposium held
    November 29-December 1, 1993, Boston, Massachusetts, U.S.A. / editors,
    M.O. Manasreh, H.J. von Bardeleben, G.S. Pomrenke, M. Lannoo, D.N. Talwar
        p. cm.—(Materials Research Society symposium proceedings ; v. 325).
    Includes bibliographical references and index.
    ISBN: 1-55899-224-3
        1. Semiconductors—Defects—Congresses. 2. Quantum wells—Congresses.
    3. Superlattices as materials—Congresses. I. Manasreh, M.O. II. von Bardeleben,
    H.J. III. Pomrenke, G.S. IV. Lannoo, M. V. Talwar, D.N. VI. Series: Materials
    Research Society Symposium Proceedings ; v. 325.

QC611.6.D4P48 1994   94-1533
621.3815'2—dc20   CIP

Manufactured in the United States of America

Contents

PREFACE xi

ACKNOWLEDGMENTS xiii

MATERIALS RESEARCH SOCIETY SYMPOSIUM PROCEEDINGS xiv

PART I: DEFECTS IN QUANTUM WELLS
AND SUPERLATTICES

*DEEP LEVELS IN TYPE-II SUPERLATTICES 3
    John D. Dow, Jun Shen, Shang Yuan Ren, and William E. Packard

*OPTICAL SPECTROSCOPY OF DEFECTS IN GaAs/AlGaAs MULTIPLE
QUANTUM WELLS 19
    B. Monemar, P.O. Holtz, J.P. Bergman, Q.X. Zhao, C.I. Harris,
    A.C. Ferreira, M. Sundaram, J.L. Merz, and A.C. Gossard

IDENTIFICATION OF DIFFUSION ASSOCIATED DEFECTS AT III-V
SEMICONDUCTOR HETEROSTRUCTURES 31
    R. Enrique Viturro and Gary W. Wicks

VACANCY INJECTION ENHANCED Al-Ga INTER-DIFFUSION IN Si FIB
IMPLANTED SUPERLATTICE 37
    P. Chen and A.J. Steckl

EFFECTS OF DEFECTS IN GaAs/AlGaAs QUANTUM WELLS 43
    O.L. Russo, V. Rehn, T.W. Nee, and K.A. Dumas

*EFFECTS OF IMPURITIES ON THE OPTICAL PROPERTIES OF QUANTUM
DOTS, WIRES, AND MULTIPLE WELLS 49
    Garnett W. Bryant

INTERACTION BETWEEN DEFECTS AND TUNNELING EFFECT IN
HETEROSTRUCTURES 61
    D. Stievenard and M. Lannoo

ELECTRON BEAM INDUCED DEGRADATION OF 2DEG IN AlGaAs/GaAs
HETEROSTRUCTURE 67
    T. Wada, T. Kanayama, S. Ichimura, Y. Sugiyama, and M. Komuro

MAGNETOOPTICAL STUDIES OF ACCEPTORS CONFINED IN GaAs/AlGaAs
QUANTUM WELLS 73
    P.O. Holtz, Q.X. Zhao, B. Monemar, A. Pasquarello, M. Sundaram,
    J.L. Merz, and A.C. Gossard

D$^-$ IMPURITIES IN A QUASI-TWO-DIMENSIONAL SYSTEM: STATISTICS
AND SCREENING 79
    W.J. Li, J.L. Wang, J.-P. Cheng, S. Holmes, Y.J. Wang, B.D. McCombe,
    and W. Schaff

FABRICATION OF QUANTUM WIRES BY WET ETCHING TECHNIQUES 85
    H.W. Yang, S.F. Horng, and H.L. Hwang

DENSITY OF STATES OF HYDROGENIC IMPURITIES IN GaAs/GaAlAs
QUANTUM WIRES 93
    Salviano A. Leão, O. Hipólito, and A. Ferreira da Silva

*Invited Paper

PART II: DEFECTS AND IMPURITIES IN InP AND RELATED COMPOUNDS

*SEMI-INSULATING BEHAVIOR OF UNDOPED InP 101
K. Kainosho, O. Oda, G. Hirt, and G. Müller

RELIABLE IMPURITY IDENTIFICATION IN InP 113
M.L. Schnoes, T.D. Harris, S.J. Pearton, M.A. Di Giuseppe, R. Bhat,
and H.M. Cox

CORRELATION BETWEEN ELECTRICAL PROPERTIES AND RESIDUAL
DEFECTS IN Se$^+$-IMPLANTED InP AFTER RAPID THERMAL
ANNEALING 119
P. Müller, T. Bachmann, E. Wendler, W. Wesch, and U. Richter

DEEP LEVEL CHARACTERIZATION AND PASSIVATION IN
HETEROEPITAXIAL InP 125
B. Chatterjee, S.A. Ringel, R. Sieg, I. Weinberg, and R. Hoffman

PHOTOLUMINESCENCE FATIGUE IN LATERALLY-ORDERED $(GaP)_2/(InP)_2$
SHORT-PERIOD-SUPERLATTICES (SPS) GROWN BY MOLECULAR
BEAM EPITAXY 131
X.C. Liu, S.Q. Gu, E.E. Reuter, S.G. Bishop, A.C. Chen, A.M. Moy,
K.Y. Cheng, and K.C. Hsieh

PHOSPHORUS-VACANCY-RELATED DEEP LEVELS IN GaInP LAYERS
GROWN BY MOLECULAR BEAM EPITAXY 137
Z.C. Huang, C.R. Wie, J.A. Varriano, M.W. Koch, and G.W. Wicks

PART III: DEFECTS AND IMPURITIES IN SiGe/Si HETEROSTRUCTURES AND RELATED COMPOUNDS

*DEEP LEVEL DEFECTS, LUMINESCENCE, AND THE ELECTRO-OPTIC
PROPERTIES OF SiGe/Si HETEROSTRUCTURES 147
Pallab Bhattacharya, Shin-Hwa Li, Jinju Lee, and Steve Smith

ELECTRONIC CHARACTERIZATION OF DISLOCATIONS IN RTCVD
GERMANIUM-SILICON/SILICON GROWN BY GRADED LAYER EPITAXY 159
P.N. Grillot, S.A. Ringel, G.P. Watson, E.A. Fitzgerald, and Y.H. Xie

SPACE DISTRIBUTION OF DEEP LEVELS IN SiGe/Si
HETEROSTRUCTURE 165
Zhang Rong, Yang Kai, Gu Shulin, Shi Yi, Huang Hongbin, Wang Ronghua,
Han Ping, Hu Liqun, Zheng Youdou, and Li Qi

PHOTOLUMINESCENCE OF SiGe ALLOYS IMPLANTED WITH ERBIUM 171
H.B. Erzgräber, P. Gaworzewski, K. Schmalz, D. Krüger, T. Morgenstern,
A. Osinsky, and M. Vatnik

STRONG VISIBLE PHOTOLUMINESCENCE IN SILICON NITRIDE THIN
FILMS DEPOSITED AT HIGH RATES 177
Sadanand V. Deshpande, Erdogan Gulari, Steven W. Brown, and S.C. Rand

OCCURRENCE OF GROUND AND EXCITED-STATE IMPURITY BANDS IN
SILICON INVERSION LAYERS: ELECTRIC FIELD EFFECTS 183
O. Hipólito, Salviano A. Leão, and A. Ferreira da Silva

EFFECT OF PRESSURE ON ARSENIC DIFFUSION IN GERMANIUM 189
S. Mitha, S.D. Theiss, M.J. Aziz, D. Schiferl, and D.B. Poker

*Invited Paper

PART IV: DOPING AND DEFECTS IN III-V SEMICONDUCTOR MATERIALS FOR DEVICE APPLICATIONS

*CARBON DOPING OF InGaAs FOR DEVICE APPLICATIONS 197
   G.E. Stillman, S.A. Stockman, C.M. Colomb, A.W. Hanson, and
   M.T. Fresina

CONTACT-RELATED DEEP STAGES IN Al-GaInP/GaAs INTERFACE 209
   Z.C. Huang and C.R. Wie

STRUCTURAL DEFECTS IN PARTIALLY RELAXED InGaAs LAYERS 217
   P. Maigne, J.-M. Baribeau, A.P. Roth, and C.L. Boon

DISLOCATION DISTRIBUTION IN GRADED COMPOSITION InGaAs
LAYERS 223
   S.I. Molina, G. Gutiérrez, A. Sacedón, E. Calleja, and R. García

RAMAN SCATTERING BY LO PHONON-PLASMON COUPLED MODE IN
HEAVILY CARBON DOPED P-TYPE InGaAs 229
   Ming Qi, Jinsheng Luo, Junichi Shirakashi, Eisuke Tokumitsu,
   Shinji Nozaki, Makoto Konagai, and Kiyoshi Takahashi

DEEP LEVEL CHARACTERIZATION OF LP-MOCVD GROWN
$Al_{0.48}In_{0.52}As$ 235
   F. Ducroquet, G. Guillot, K. Hong, C.H. Hong, D. Pavlidis,
   and M. Gauneau

HYDROGEN PASSIVATED CARBON ACCEPTORS IN GaAs AND AlAs: NO
EVIDENCE FOR CARBON DONORS 241
   B.R. Davidson, R.C. Newman, R.E. Pritchard, T.J. Bullough, T.B. Joyce,
   R. Jones, and S. Öberg

*RECENT DEVELOPMENTS IN DOPING TECHNIQUES FOR COMPOUND
SEMICONDUCTORS 247
   J.E. Cunningham and W.T. Tsang

EFFECTS OF ION-IRRADIATION AND HYDROGENATION ON THE
DOPING OF InGaAlN ALLOYS 261
   S.J. Pearton, C.R. Abernathy, W.S. Hobson, and F. Ren

ANNEALING EFFECT ON PHOTOLUMINESCENCE PROPERTIES OF
Be-DOPED MBE GaAs 267
   Hajime Shibata, Yunosuke Makita, and Akimasa Yamada

ORIENTED CARBON PAIR DEFECTS STABILIZED BY HYDROGEN
IN *AS-GROWN* GaAs EPITAXIAL LAYERS 273
   Y.M. Cheng, M. Stavola, C.R. Abernathy, and S.J. Pearton

HYDROGEN PASSIVATION EFFECTS IN HETEROEPITAXIAL InSb GROWN
ON GaAs BY LPMOCVD 279
   Byueng-Su Yoo, Sang-Gi Kim, and El-Hang Lee

DIPOLE RELAXATION CURRENT IN N-TYPE $Al_xGa_{1-x}As$ 285
   Luis V.A. Scalvi, L. Oliveira, and M. Siu Li

PART V: IMPURITIES IN SEMICONDUCTORS

*OXYGEN DOPING OF GaAs DURING OMVPE CONTROLLED INTRODUCTION
OF IMPURITY COMPLEXES 293
   Y. Park, M. Skowronski, T.S. Rosseel, and M.O. Manasreh

*Invited Paper

DEEP LEVEL STRUCTURE OF SEMI-INSULATING MOVPE GaAs GROWN
BY CONTROLLED OXYGEN INCORPORATION 305
   J.W. Huang and T.F. Kuech

PROTON IRRADIATION DAMAGE IN Zn AND Cd DOPED InP 311
   George C. Rybicki and Wendell S. Williams

*THEORY OF RARE-EARTH IMPURITIES IN SEMICONDUCTORS 317
   C. Delerue and M. Lannoo

Rh: A DOPANT WITH MID-GAP LEVELS IN InP AND InGaAs AND
SUPERIOR THERMAL STABILITY 329
   H. Scheffler, B. Srocka, A. Dadgar, M. Kuttler, A. Knecht, R. Heitz,
   D. Bimberg, J.Y. Hyeon, and H. Schumann

METASTABLE DEFECTS AND STRETCHED EXPONENTIALS 335
   David Redfield and Richard Bube

HYDROGENATION OF GALLIUM NITRIDE 341
   M.S. Brandt, N.M. Johnson, R.J. Molnar, R. Singh, and T.D. Moustakas

OPTICAL SPECTROSCOPY OF A NITROGEN-HYDROGEN COMPLEX
IN ZnSe 347
   J.A. Wolk, J.W. Ager III, K.J. Duxstad, E.E. Haller, N.R. Taskar,
   D.R. Dorman, and D.J. Olego

CHARACTERIZATION OF DEFECTS IN N-TYPE 6H-SiC SINGLE
CRYSTALS BY OPTICAL ADMITTANCE SPECTROSCOPY 353
   A.O. Evwaraye, S.R. Smith, and W.C. Mitchel

PART VI: DEFECTS IN LOW TEMPERATURE
GROWN SEMICONDUCTORS

*DEFECTS IN LOW-TEMPERATURE-GROWN MBE GaAs 361
   David C. Look

STOICHIOMETRY RELATED PHENOMENA IN LOW TEMPERATURE
GROWN GaAs 371
   M. Missous and S. O'Hagan

IMPROVEMENT OF THE STRUCTURAL QUALITY OF GaAs LAYERS
GROWN ON Si WITH LT-GaAs INTERMEDIATE LAYER 377
   Zuzanna Liliental Weber, H. Fujioka, H. Sohn, and E.R. Weber

PRESSURE RATIO (P_{As}/P_{Ga}) DEPENDENCE ON LOW TEMPERATURE
GaAs BUFFER LAYERS GROWN BY MBE 383
   M. Lagadas, Z. Hatzopoulos, M. Calamiotou, M. Kayiambaki,
   and A. Christou

FEMTOSECOND PROBE-PROBE TRANSMISSION STUDIES OF LT-GROWN
GaAs NEAR THE BAND EDGE 389
   H.B. Radousky, A.F. Bello, D.J. Erskine, L.N. Dinh,
   M.J. Bennahmias, M.D. Perry, T.R. Ditmire, and R.P. Mariella Jr.

COMPOSITIONAL MODULATIONS AND VERTICAL TWO-DIMENSIONAL
ARSENIC-PRECIPITATE ARRAYS AND IN LOW TEMPERATURE GROWN
$Al_{0.3}Ga_{0.7}As$ 395
   K.Y. Hsieh, Y.L. Hwang, T. Zhang, and R.M. Kolbas

*Invited Paper

CLUSTERS AND THE NATURE OF SUPERCONDUCTIVITY IN
LTMBE-GaAs 401
    N.A. Bert, V.V. Chaldyshev, S.I. Goloshchapov, S.V. Kozyrev,
    A.E. Kunitsyn, V.V. Tret'yakov, A.I. Veinger, I.V. Ivonin,
    L.G. Lavrentieva, M.D. Vilisova, M.P. Yakubenya, D.I. Lubyshev,
    V.V. Preobrazhenskii, and B.R. Semyagin

PART VII: DEFECTS IN BULK AND
EPITAXIAL SEMICONDUCTORS

*PHOTOLUMINESCENCE IMAGING OF III-V SUBSTRATES AND
EPITAXIAL HETEROSTRUCTURES 409
    W. Jantz, M. Baeumler, Z.M. Wang, and J. Windscheif

PROPERTIES OF AN ARRAY OF DISLOCATIONS IN A STRAINED
EPITAXIAL LAYER 419
    Tong-Yi Zhang

DOMINANT MIDGAP LEVELS IN THE COMPENSATION MECHANISM
IN GaAs 425
    H. Shiraki, Y. Tokuda, E. Tohyama, K. Sassa, and N. Toyama

IDENTIFICATION OF THE 0.15 eV DONOR DEFECT IN BULK GaAs 431
    Z-Q. Fang, J.W. Hemsky, and D.C. Look

DEFECT REDUCTION BY THERMAL CYCLIC GROWTH IN GaAs GROWN
ON Si BY MOVPE 437
    W. Kürner, R. Dieter, K. Zieger, F. Goroncy, A. Dörnen,
    and F. Scholz

DEEP LEVELS INDUCED BY $SiCl_4$ REACTIVE ION ETCHING
IN GaAs 443
    N.P. Johnson, M.A. Foad, S. Murad, M.C. Holland, and
    C.D.W. Wilkinson

BAND EDGE OPTICAL ABSORPTION OF MOLECULAR BEAM EPITAXIAL
GaSb GROWN ON SEMI-INSULATING GaAs SUBSTRATE 449
    M. Shah, M.O. Manasreh, R. Kaspi, M.Y. Yen, B.A. Philips,
    M. Skowronski, and J. Shinar

A TEM STUDY OF DEFECT STRUCTURE IN GaAs FILM ON
Si SUBSTRATE 457
    Sahn Nahm, Hee-Tae Lee, Sang-Gi Kim, and Kyoung-Ik Cho

TEMPERATURE AND POLARIZATION DEPENDENCE OF THE OPTICAL
ABSORPTION IN $ZnGeP_2$ AT TWO MICROMETERS 463
    M. Shah, M.C. Ohmer, D.W. Fischer, N.C. Fernelius, M.O. Manasreh,
    P.G. Schunemann, and T.M. Pollak

PART VIII: SUPERLATTICES AND
QUANTUM WELLS

*TUNING OF ENERGY LEVELS IN A SUPERLATTICE 471
    Francois M. Peeters

OPTICAL AND ELECTRONIC PROPERTIES OF GaAs/AlAs
RANDOM SUPERLATTICES 481
    E.G. Wang, J.H. Xu, W.P. Su, and C.S. Ting

*Invited Paper

A TIGHT-BINDING THEORY OF THE ELECTRONIC STRUCTURES
FOR RHOMBOHEDRAL SEMIMETALS AND Sb/GaSb, Sb/AlSb
SUPERLATTICES 487
    J.H. Xu, E.G. Wang, C.S. Ting, and W.P. Su

THE GROWTH OF InAsSb/InGaAs STRAINED-LAYER SUPERLATTICES
BY METAL-ORGANIC CHEMICAL VAPOR DEPOSITION 493
    R.M. Biefeld, K.C. Baucom, S.R. Kurtz, and D.M. Follstaedt

OPTICAL PROPERTIES AND ELECTROLUMINESCENCE OF ORDERED
AND DISORDERED AlAs/GaAs SUPERLATTICES 499
    D.J. Arent, R.G. Alonso, G.S. Horner, A.E. Kibbler, J.M. Olson,
    X. Yin, M.C. Delong, A.J. Springthorpe, A. Majeed, D.J. Mowbray,
    and M.S. Skolnick

OPTICAL INVESTIGATION OF STRAINED-LAYER GaInAs/GaInAsP
HETEROSTRUCTURES 507
    I. Queisser, V. Härle, A. Dörnen, and F. Scholz

CATHODOLUMINESCENCE SPECTROSCOPY STUDIES OF GROWTH
INDUCED DEEP LEVELS AT GaInP 513
    R. Enrique Viturro, John D. Varriano, and Gary W. Wicks

AUTHOR INDEX 519

SUBJECT INDEX 523

Preface

This proceedings volume contains manuscripts from Symposium L, entitled "Defects in Advanced Semiconductors: Physics and Applications." It was held during the MRS Fall Meeting, November 29-December 1, 1993 in Boston, Massachusetts. Symposium L was focused on the physics and applications of defects in semiconductor quantum wells, superlattices, and bulk materials grown by various techniques. The symposium lasted three days, one evening being devoted to a poster session, and consisted of 14 invited talks and 64 contributed papers. The symposium was truly an international meeting with papers originating from 20 countries led by the United States, Germany, France, Japan, Sweden, and the U.K.

Defect characterization, identifications, and their influence on material properties and device performance are a major subject in the physics and applications of semiconductors. This volume focuses on defects in advanced semiconductors or precisely quantum wells, superlattices, and heterostructures. In Part I of this proceedings the invited and contributed papers deal with defects in type I and type II superlattices based on III-V semiconductors such as GaAs/AlGaAs multiple quantum wells (MQWs). Some of the topics include optical spectroscopy of defects in GaAs/AlGaAs MQWs, defects injections and diffusions in heterostructures and impurity effects on the electronic states in quantum wires and quantum dots. Part II deals with defects and impurities in bulk and epitaxial InP and related compounds, for example, the semi-insulating behavior of undoped InP is discussed.

Recently, SiGe/Si quantum wells and heterostructures have been the subject of an increasing interest due to their applications in electronic and opto-electric devices. Defects, dislocation distributions, and doping in these heterostructures are discussed in Part III. Doping engineering is an interesting subject in the field of electronic devices such as heterojunction bipolar transistors (HBTs). Part IV presents selected papers in this area such as doped layers of InGaAs for HBT applications. Impurity control and incorporation of such elements as O, H, Zn, Cd, and rare earth impurities and impurity complexes in III-V semiconductors are the focus of Part V.

Recently, low temperature grown (LTG) GaAs attracted the attention of many researchers in the field of defects in semiconductors. This is partly because LTG GaAs contains a large concentration of arsenic antisite-related defects. The electrical and optical properties of LTG GaAs have been the subject of various investigation in recent years. Part VI deals with defects, stoichiometry, and precipitations in this material. Defects in bulk and epitaxial semiconductors are the subject of Part VII. Related topics such as deep levels and dislocations in III-V materials and defect imaging are included. The final Part VIII addresses various topics on the optical properties and theory of superlattices and quantum well structures.

The symposium was extremely well attended with most papers presented in this volume. High standards were maintained in reviewing, selecting and accepting the papers in this volume. We believe that the main goals of Symposium L were achieved. One of the main goals was the bringing together of researchers, in one meeting, from areas of defects in semiconductors as it pertains to low dimensional systems, epitaxial layers, bulk materials, and general area of defect engineering and control. Finally, the editors hope that this volume will serve as a timely reference and hope that the collection of papers will stimulate new ideas, insight, and direction.

M.O. Manasreh
H.J. von Bardeleben
G.S. Pomrenke
M. Lannoo
D.N. Talwar

December 1993

Acknowledgments

The editors are pleased to express their appreciation to the authors, speakers, and poster presenters who presented their technical work at the meeting and composed the papers in this volume; the symposium organizers who put together the program and saw that it ran smoothly; the session chairpersons (H.J. von Bardeleben, J. Dow, G. Gumbs, M. Lannoo, M.O. Manasreh, R.C. Newman, S.J. Pearton, G.S. Pomrenke, M. Skowronski, and G.L. Witt); the organizers and contributors from Symposium M, with whom a joint session was shared; the invaluable assistance of the MRS staff in organizing the meeting and processing the proceedings; and most importantly of all the Air Force Office of Scientific Research (AFOSR), which provided the financial support for this symposium.

Symposium Support

Air Force Office of Scientific Research

MATERIALS RESEARCH SOCIETY SYMPOSIUM PROCEEDINGS

Volume 297— Amorphous Silicon Technology—1993, E.A. Schiff, M.J. Thompson, P.G. LeComber, A. Madan, K. Tanaka, 1993, ISBN: 1-55899-193-X

Volume 298— Silicon-Based Optoelectronic Materials, R.T. Collins, M.A. Tischler, G. Abstreiter, M.L. Thewalt, 1993, ISBN: 1-55899-194-8

Volume 299— Infrared Detectors—Materials, Processing, and Devices, A. Appelbaum, L.R. Dawson, 1993, ISBN: 1-55899-195-6

Volume 300— III-V Electronic and Photonic Device Fabrication and Performance, K.S. Jones, S.J. Pearton, H. Kanber, 1993, ISBN: 1-55899-196-4

Volume 301— Rare-Earth Doped Semiconductors, G.S. Pomrenke, P.B. Klein, D.W. Langer, 1993, ISBN: 1-55899-197-2

Volume 302— Semiconductors for Room-Temperature Radiation Detector Applications, R.B. James, P. Siffert, T.E. Schlesinger, L. Franks, 1993, ISBN: 1-55899-198-0

Volume 303— Rapid Thermal and Integrated Processing II, J.C. Gelpey, J.K. Elliott, J.J. Wortman, A. Ajmera, 1993, ISBN: 1-55899-199-9

Volume 304— Polymer/Inorganic Interfaces, R.L. Opila, A.W. Czanderna, F.J. Boerio, 1993, ISBN: 1-55899-200-6

Volume 305— High-Performance Polymers and Polymer Matrix Composites, R.K. Eby, R.C. Evers, D. Wilson, M.A. Meador, 1993, ISBN: 1-55899-201-4

Volume 306— Materials Aspects of X-Ray Lithography, G.K. Celler, J.R. Maldonado, 1993, ISBN: 1-55899-202-2

Volume 307— Applications of Synchrotron Radiation Techniques to Materials Science, D.L. Perry, R. Stockbauer, N. Shinn, K. D'Amico, L. Terminello, 1993, ISBN: 1-55899-203-0

Volume 308— Thin Films—Stresses and Mechanical Properties IV, P.H. Townsend, J. Sanchez, C-Y. Li, T.P. Weihs, 1993, ISBN: 1-55899-204-9

Volume 309— Materials Reliability in Microelectronics III, K. Rodbell, B. Filter, P. Ho, H. Frost, 1993, ISBN: 1-55899-205-7

Volume 310— Ferroelectric Thin Films III, E.R. Myers, B.A. Tuttle, S.B. Desu, P.K. Larsen, 1993, ISBN: 1-55899-206-5

Volume 311— Phase Transformations in Thin Films—Thermodynamics and Kinetics, M. Atzmon, J.M.E. Harper, A.L. Greer, M.R. Libera, 1993, ISBN: 1-55899-207-3

Volume 312— Common Themes and Mechanisms of Epitaxial Growth, P. Fuoss, J. Tsao, D.W. Kisker, A. Zangwill, T.F. Kuech, 1993, ISBN: 1-55899-208-1

Volume 313— Magnetic Ultrathin Films, Multilayers and Surfaces/Magnetic Interfaces— Physics and Characterization (2 Volume Set), C. Chappert, R.F.C. Farrow, B.T. Jonker, R. Clarke, P. Grünberg, K.M. Krishnan, S. Tsunashima/ E.E. Marinero, T. Egami, C. Rau, S.A. Chambers, 1993, ISBN: 1-55899-211-1

Volume 314— Joining and Adhesion of Advanced Inorganic Materials, A.H. Carim, D.S. Schwartz, R.S. Silberglitt, R.E. Loehman, 1993, ISBN: 1-55899-212-X

Volume 315— Surface Chemical Cleaning and Passivation for Semiconductor Processing, G.S. Higashi, E.A. Irene, T. Ohmi, 1993, ISBN: 1-55899-213-8

MATERIALS RESEARCH SOCIETY SYMPOSIUM PROCEEDINGS

Volume 316— Materials Synthesis and Processing Using Ion Beams, R.J. Culbertson, K.S. Jones, O.W. Holland, K. Maex, 1994, ISBN: 1-55899-215-4

Volume 317— Mechanisms of Thin Film Evolution, S.M. Yalisove, C.V. Thompson, D.J. Eaglesham, 1994, ISBN: 1-55899-216-2

Volume 318— Interface Control of Electrical, Chemical, and Mechanical Properties, S.P. Murarka, T. Ohmi, K. Rose, T. Seidel, 1994, ISBN: 1-55899-217-0

Volume 319— Defect-Interface Interactions, E.P. Kvam, A.H. King, M.J. Mills, T.D. Sands, V. Vitek, 1994, ISBN: 1-55899-218-9

Volume 320— Silicides, Germanides, and Their Interfaces, R.W. Fathauer, L. Schowalter, S. Mantl, K.N. Tu, 1994, ISBN: 1-55899-219-7

Volume 321— Crystallization and Related Phenomena in Amorphous Materials, M. Libera, T.E. Haynes, P. Cebe, J. Dickinson, 1994, ISBN: 1-55899-220-0

Volume 322— High-Temperature Silicides and Refractory Alloys, B.P. Bewlay, J.J. Petrovic, C.L. Briant, A.K. Vasudevan, H.A. Lipsitt, 1994, ISBN: 1-55899-221-9

Volume 323— Electronic Packaging Materials Science VII, R. Pollak, P. Børgesen, H. Yamada, K.F. Jensen, 1994, ISBN: 1-55899-222-7

Volume 324— Diagnostic Techniques for Semiconductor Materials Processing, O.J. Glembocki, F.H. Pollak, S.W. Pang, G. Larrabee, G.M. Crean, 1994, ISBN: 1-55899-223-5

Volume 325— Physics and Applications of Defects in Advanced Semiconductors, M.O. Manasreh, M. Lannoo, H.J. von Bardeleben, E.L. Hu, G.S. Pomrenke, D.N. Talwar, 1994, ISBN: 1-55899-224-3

Volume 326— Growth, Processing, and Characterization of Semiconductor Heterostructures, G. Gumbs, S. Luryi, B. Weiss, G.W. Wicks, 1994, ISBN: 1-55899-225-1

Volume 327— Covalent Ceramics II: Non-Oxides, A.R. Barron, G.S. Fischman, M.A. Fury, A.F. Hepp, 1994, ISBN: 1-55899-226-X

Volume 328— Electrical, Optical, and Magnetic Properties of Organic Solid State Materials, A.F. Garito, A. K-Y. Jen, C. Y-C. Lee, L.R. Dalton, 1994, ISBN: 1-55899-227-8

Volume 329— New Materials for Advanced Solid State Lasers, B.H.T. Chai, T.Y. Fan, S.A. Payne, A. Cassanho, T.H. Allik, 1994, ISBN: 1-55899-228-6

Volume 330— Biomolecular Materials By Design, H. Bayley, D. Kaplan, M. Navia, 1994, ISBN: 1-55899-229-4

Volume 331— Biomaterials for Drug and Cell Delivery, A.G. Mikos, R. Murphy, H. Bernstein, N.A. Peppas, 1994, ISBN: 1-55899-230-8

Volume 332— Determining Nanoscale Physical Properties of Materials by Microscopy and Spectroscopy, M. Sarikaya, M. Isaacson, H.K. Wickramasighe, 1994, ISBN: 1-55899-231-6

Volume 333— Scientific Basis for Nuclear Waste Management XVII, A. Barkatt, R. Van Konynenburg, 1994, ISBN: 1-55899-232-4

Volume 334— Gas-Phase and Surface Chemistry in Electronic Materials Processing, T.J. Mountziaris, P.R. Westmoreland, F.T.J. Smith, G.R. Paz-Pujalt, 1994, ISBN: 1-55899-233-2

Volume 335— Metal-Organic Chemical Vapor Deposition of Electronic Ceramics, S.B. Desu, D.B. Beach, B.W. Wessels, S. Gokoglu, 1994, ISBN: 1-55899-234-0

Prior Materials Research Society Symposium Proceedings
available by contacting Materials Research Society

PART I

Defects in Quantum Wells and Superlattices

DEEP LEVELS IN TYPE-II SUPERLATTICES

JOHN D. DOW,* JUN SHEN,† SHANG YUAN REN,* AND WILLIAM E.
PACKARD*
*Department of Physics and Astronomy, Arizona State University, Tempe,
Arizona 85287-1504, U.S.A.
†Motorola, Inc., Phoenix Corporate Research Laboratories, Tempe,
Arizona 85284, U.S.A.

ABSTRACT

Quantum confinement in superlattices affects shallow levels and band edges considerably (length scale of order 100 Å), but not deep levels (length scale of order 5 Å). Thus by band-gap engineering, one can move a band edge through a deep level, *causing the defect responsible for the level to change its doping character*. For example, the cation-on-anion-site defect in $A\ell_x Ga_{1-x}Sb$ alloys is predicted to change from a shallow acceptor to a deep acceptor-like trap as the valence band edge passes through its T_2 deep level with increasing $A\ell$ alloy content x. In a Type-II superlattice, such as $InAs/A\ell_x Ga_{1-x}Sb$ for x>0.2, where the conduction band minimum of the InAs should lie energetically below the antisite defect's T_2 level in bulk $A\ell_x Ga_{1-x}Sb$, the electrons normally trapped in this deep level (when the defect is neutral) **remotely** dope the InAs n-type in the superlattice, leaving the defect positively charged. Thus a native defect that is thought of as an acceptor can actually be a donor and control the n-type doping of InAs quantum wells. The physics of such deep levels in superlattices and in quantum wells is summarized, and related to high-speed devices.

I. INTRODUCTION

The purpose of this paper is to present the main qualitative ideas associated with deep levels in Type-II superlattices. We first review the main concepts of the modern theory of doping, and then illustrate how these ideas, when applied to superlattices in general, and to Type-II superlattices in particular, lead to new and interesting physics.

II. SUBSTITUTIONAL IMPURITIES IN BULK SEMICONDUCTORS

In 1955 Kohn and Luttinger [1] explained the main experimental facts concerning substitutional heterovalent impurities in semiconductors by introducing the envelope-function and the effective-mass approximations, and by showing that donors such as P in Si produce electrons in large-radius, small-binding-energy hydrogenic orbits around the donors. Here the ground $1s$ state Bohr radius is

$$a^* = \hbar^2 \epsilon / m^* e^2 = (0.53 \text{ Å})(\epsilon \, m_0/m^*),$$

typically of order 100 Å, and the binding energy (with respect to the conduction band minimum) is

$$E_B = (e^4 m^*/2\hbar^2 \epsilon^2) = (13.6 \text{ eV})(\epsilon^{-2} \, m^*/m_0),$$

typically of order 10 meV. (See Fig. 1 [2].) In this picture, a donor is an impurity whose nuclear charge $Z|e|$ exceeds that of the host atom it replaced. Absent from the simplest form of this picture is the central-cell potential V_c, which is necessarily strong, being at least 4 eV for P replacing Si (because the corresponding s atomic energy levels differ by that much), in contrast with the shallow impurity binding energy of order 45 meV in

3

Si. This central-cell potential accounts for the dramatically altered bonds between the impurity and its neighboring atoms. (See Fig. 1.) Also missing from the effective-mass theory is any procedure for determining the stable charge-state in the central-cell of the impurity — it is simply assumed that P is singly ionized (P^+), providing the extra positive charge with respect to Si that binds the extra P electron in a hydrogenic orbit. While this picture of shallow donors, and the corresponding picture of shallow acceptors, accounted for the main phenomena concerning conductivity in semiconductors, the incompleteness of shallow impurity effective-mass theory became apparent with the discovery of "deep" levels observed to be "bound" in the fundamental band gap both below the conduction band edge and above the valence band maximum by more than 0.1 eV. Because of their large apparent binding energies, these "deep" levels were not thermally ionizable and did not contribute to the conductivity, but they did trap carriers — hence the name "deep traps." The key to understanding deep levels was provided by data of Wolford and Streetman for O, S, Se, and N — all substituted for As and P in $GaAs_{1-x}P_x$ alloys [2]. In these substitutional crystalline alloys, the virtual crystal approximation [3] is valid, allowing one to imagine a one-electron band structure for a material with an "average" anion: $(1-x)As+xP$. This band structure has a valence band maximum at $k=0$ for all alloy compositions x, and two relative conduction band minima: one at $k=0$ that is the absolute minimum for GaAs ($x=0$), and another in the (0,0,1) and equivalent directions of the Brillouin zone, at the X-points, that are absolute minima for GaP ($x=1$). Fig. 2 shows how the band gaps (with respect to the valence band maximum) for each of these two minima vary with alloy composition x. [Whenever the $k=0$ conduction band minimum is an absolute minimum, the material is a *direct-gap* semiconductor and readily emits light, because electrons injected into the conduction band thermalize to that minimum and then recombine with holes that have thermalized to the valence band maximum, also at $k=0$. The material with composition $x=0.45$ is used in red light-emitters, because it has virtually the largest direct band gap (furthest into the visible) of the $GaAs_{1-x}P_x$ alloys. For $x>0.5$, the lowest conduction band minima are at X, the thermalized electrons are at these indirect conduction band minima, and the electrons have non-zero wavevectors, making them incapable by themselves of recombining with the zero-wavevector holes.]

$GaAs_{1-x}P_x$ ion-implanted with S and Se exhibited the expected shallow-donor levels which were "attached" to the conduction band minima and followed those minima as alloy composition x varied, with the binding energies E_B appropriate to the effective masses $m^*(k=0)$ or $m^*(k=X)$. (See Fig. 2.) Oxygen, despite being from the same Column of the Periodic Table as S and Se, did not show the expected "attached to the band-edge" behavior, but instead produced a "deep" level well within the band-gap whose energy varied almost linearly with alloy composition x. The similarity of the oxygen data to the data for N was especially striking because N is an isoelectronic defect from Column-V, having the same valence as the As or P it replaced — and so N has no long-ranged Coulomb potential, $(-e^2/\epsilon r)$, capable of producing a shallow-donor-type energy level. Yet, in GaP, the N level is closer to the conduction band minimum ("shallower") than the shallow S and Se levels; for $x\approx0.5$, the N level is more than 0.1 eV "deep" in the gap; and around $x=0.22$ the N "deep" level appears to merge with the conduction band. These data (Fig. 2) strongly suggested that S and Se, like O, each have a "deep" level, but that the S and Se deep levels lie *above* the conduction band minimum, resonant with the conduction band, where they are not normally observed, but nevertheless approximately parallel the O and N "deep" levels as alloy composition x (or pressure) is varied. Thus the data led to the notion that shallow donors such as S and Se have *both shallow and "deep" levels,* but with the deep levels outside the funda-

4

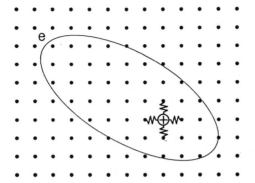

Fig. 1. Schematic illustration of a donor in a semiconductor. The dots represent host atoms. At the site labeled by a "+" there is an impurity, such as P in Si or S on a P-site in GaP. It alters the bonds with the neighbors (jagged lines). If the deep states associated with the jagged lines do not produce levels in the fundamental band gap, then the impurity is autoionized and becomes charged +|e|. This charge then binds an electron in a hydrogenic shallow-donor orbital (large ellipse).

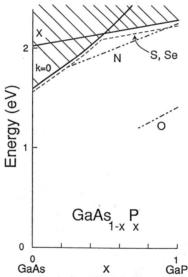

Fig. 2. Data for energy levels (in eV) of O (chained), S (dashed), Se (dashed), and N (chained) on anion sites in GaAs$_{1-x}$P$_x$ crystalline alloys, versus alloy composition x, by Wolford and Streetman [2]. The band edges (solid lines) at $k=0$ and at the X-point of the Brillouin zone ($k=(2\pi/a_L)(1,0,0)$) are also shown. The binding energies of the S and Se donors have been exaggerated a bit to facilitate displaying them and illustrating that the binding energy at the X-minimum is larger than at the $k=0$ minimum.

mental band gap and normally unobserved. Prior to that time, the distinction between shallow and deep levels was provided by the apparent binding energies with respect to a nearby band edge: shallow-level binding energies (by definition) were less than 0.1 eV, while deep levels were in the gap by more than 0.1 eV. The Wolford-Streetman data required a new definition of deep levels as those levels produced by the central-cell potential. In general, a heterovalent s- and p-bonded substitutional impurity will have both the hydrogenic series of shallow levels and four "deep" levels (one s-like and three p-like) in the vicinity of the fundamental band gap. Some or all of the "deep" levels will normally lie outside the fundamental band gap. For example, the p-like "deep" levels of O, S, Se, and N all lie well outside the gap, while the s-like deep levels of S and Se lie slightly above the conduction band minimum in $GaAs_{1-x}P_x$. The s-like "deep" level of O lies within the gap, while that of N lies within the gap for $x > 0.22$ only.

When one of the deep levels crosses a band edge or the Fermi level (as a function of host alloy composition x, for example) the ground-state and the doping character of the impurity changes. To see this, consider P in Si or S on a P site in GaP (Fig. 3), and ignore the empty p-like or T_2-symmetric "deep" state well above the conduction band minimum. The "normal" situation (Fig. 3) has the

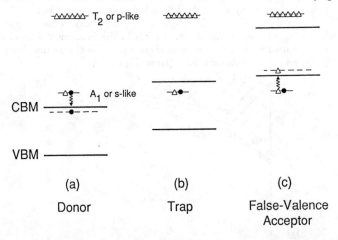

(a)   (b)   (c)

Donor   Trap   False-Valence Acceptor

Fig. 3. Energy level diagram illustrating the various ground states for a substitutional impurity whose valence is one more than the valence of the host atom it replaced. Holes are denoted by open triangles, electrons by solid circles. The valence band edge (VBM) is denoted by a heavy line, as is the conduction band edge (CBM). The T_2-symmetric p-like level of the defect contains no electrons and is irrelevant in this figure. If the A_1-symmetric or s-like level lies above the conduction band edge, its electron is autoionized, falls to the band edge and is bound by the Coulomb potential of the ionized defect (at zero temperature). Thus, in case (a) we have a shallow donor which dopes the material n-type. If, as in case (b), the s-like level lies in the gap, the neutral impurity is a deep trap for either electrons or holes and helps to make the material semi-insulating. If, as in case (c), the s-like level lies below the valence band maximum, then the impurity is an acceptor with "false valence." Here we have assumed that there is only one impurity in the solid.

6

s-like or A_1-symmetric "deep" level slightly above the conduction band minimum [4,5], occupied by one electron and one hole for the neutral impurity. Because this level is resonant with the conduction band, the electron in it is unstable, scatters out of the level, and autoionizes the impurity, thereby creating the positive charge that binds the electron in a hydrogenic shallow donor level. Since the ground state of the impurity is a shallow donor level, the impurity is termed a shallow donor, and dopes the host n-type.

However, one can imagine altering the composition of the host so that the *s*-like deep level would fall well within the gap (Fig. 3b), in which case the neutral impurity would be stable, the impurity would not be autoionized, and the *s*-like "deep" level would contain one electron and one hole — giving the impurity a "deep impurity" character. This impurity could increase or decrease its neutral charge by capturing another electron or hole, and, if the level were energetically too distant ($\sim$0.1 eV) from the nearest band edge, the carrier would not be thermalized at room temperature — and the impurity would be a "deep trap" tending to make the host semi-insulating. The physics of Fig. 3b is now generally accepted, and has been demonstrated even for P in Si [5].

One can imagine manipulating the host's band edges even more, so that the *s*-like deep level lies *below the valence band maximum*. In this case, the hole autoionizes, leaving a stable negative ion, and the Coulomb potential of this hole produces a "false-valence *acceptor*" state [6]. The reason for the terminology "false-valence" is that an impurity with one more nuclear charge than the host-atom it replaced appears, from the effective-mass viewpoint, to have one less nuclear charge than the host-atom.

We can see from these simple examples that (i) Four "deep" levels are present near the gap for all substitutional *s*- and *p*-bonded impurities, but may not lie within the gap, (ii) the relative positions of a "deep level" and the host's band edges (and Fermi energy) determine the doping character of an impurity (*e.g.*, donor, trap, acceptor), and (iii) by manipulating the energies of band edges with respect to deep levels, it is often possible to alter the doping character of an impurity.

Thus the following modern picture of doping is established: the long-ranged Coulomb potential of a heterovalent impurity produces shallow donor levels if the impurity has no "deep" levels in the gap. This Coulomb potential produces hydrogenic effective-mass states with (small) hydrogenic binding energies and (large) Bohr radii. These states have small binding energies and can be thermally ionized at room temperature, producing free carriers that account for the modest conductivities of semiconductors. The states are delocalized in real space (**r**-space) but (according to Kohn and Luttinger [1]) localized near a band extremum in **k**-space, so that they appear "attached" to a nearby band-edge and follow that edge when it is perturbed by, for example, a change in alloy composition x or applied pressure p.

In contrast, the "deep" states ψ originate from the central-cell potential V_c and the altered bonds between the impurity and the host. They are described by a Koster-Slater type theory [7]

$$(-\hbar^2/2m)\nabla^2\psi + V_c\psi = E\psi,$$

which is most conveniently solved using Green's functions [8], producing an *s*-like and a *p*-like level in the energetic vicinity of the fundamental band gap. Such levels require the sp^3 chemistry of covalent chemical bonding in semiconductors, have components from at least ten energy bands (bands made up of one *s*- and three *p*-orbitals on each of two sites per unit cell, plus at least one additional orbital such as s* [9] per site), are localized in **r**-space, are delocalized in **k**-space, and often (but not always) affect the optical properties — often acting as non-radiative traps or "killer centers" with rather large killing radii (because of the efficient exciton transport to the centers) for excitons or carriers that otherwise might radiate light. The deep levels appear *unattached* to nearby band edges, behave like the N and O levels of Fig. 2, and can have apparent binding energies with respect to the band edges that are either negative or positive, and several tenths of eV in magnitude.

A complete theory of a particular substitutional impurity in a semiconductor must account for *both* its deep levels and its shallow levels, but, in practice, it is easiest to treat one or the other. Whenever a deep level falls within a band gap, it is normally convenient to treat theoretically the Koster-Slater limit of only the central-cell potential, and to include the Coulomb effects on that level with perturbation theory [10].

The physics of an *s*-like deep level is described schematically in Fig. 4 for the example of N-doped GaP. Starting with atomic Ga and P and (for simplicity) limiting ourselves to *s*-states only (or, alternatively, to sp^9 hybrids only), we note that Ga's relevant $4s$ atomic energy level ϵ_{Ga} lies higher than P's $3s$ level ϵ_P. (Atomic energy levels go down as the effective nuclear charge increases from 3 to 5.) Pairing Ga and P atoms into a molecule leads to bonding and antibonding molecular levels (Fig. 4), with a bonding-antibonding splitting (in the extreme tight-binding limit) of approximately

$$\text{Host splitting} \sim |V_{Ga\text{-}P}|^2/(\epsilon_{Ga}-\epsilon_P).$$

The condensation of all the Ga–P molecules into a solid causes the bonding states to broaden into the valence band, while the antibonding states become the conduction band.

Replacement of one P atom by a N impurity leads to a "central-cell defect potential" $V_c(r)$ with depth of order $V_c \sim -7$ eV $\approx \epsilon_N - \epsilon_P$, and radius equal to roughly the first nearest-neighbor distance or less. The bonding-antibonding splitting for the impurity is

$$\text{Defect splitting} \sim |V_{Ga\text{-}N}|^2/(\epsilon_{Ga}-\epsilon_N),$$

much smaller than the host-splitting (because we have the nearest-neighbor transfer matrix element $V_{Ga\text{-}P}$ almost independent of the anion [11], and a much larger energy denominator for the defect), causing the N defect's *s*-like "deep level" to fall in the fundamental band gap. Note that this N deep level is not N-like, but Ga-like, and has an antibonding wavefunction similar to a Ga dangling bond. The "hyperdeep" level of Fig. 4 is N-like, but normally is far from the fundamental band gap, filled with electrons (for an electronegative impurity), and unobserved. Another way to see the host-like and Ga-like antibonding character of the deep level is to make very schematic charge-density plots (Fig. 5). The GaP host's valence band is largely P-like, while the conduction band is Ga-like. N, being very electronegative, attracts charge to its bonding hyperdeep state, which is almost totally N-like, causing the orthogonal "deep" state in the gap to be Ga-like. This picture illustrates a major point about such a deep level: it is localized in real space, having a major fraction of its charge-density, *not on the impurity, but on the nearest-neighbors*. This means that a typical deep level's *radius is of order 5 Å*, and that most deep impurities on a particular site with a specific symmetry (either *s*-like or *p*-like) have almost the same wavefunction [12-14] and almost the same energy (on the 20 eV scale of the spectral range of the sp^9 bonding).

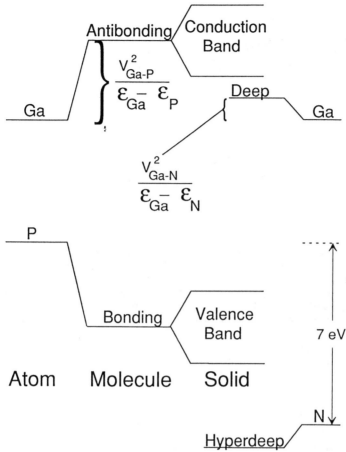

Fig. 4. Schematic illustration of the physics of deep levels. The s atomic energy levels of Ga and P produce a bonding-antibonding splitting when the atoms are brought together into a molecule. In the solid, the bonding molecular level broadens into the valence band and the antibonding level becomes the conduction band. When a N atom replaces one P atom in the solid, it exhibits a bonding-antibonding splitting, which is smaller than for Ga and P, because the N atomic energy level is $\sim$7 eV lower in energy, causing the energy denominator in the expression for the bonding-antibonding splitting to be large. As a result, the "deep" level in the gap between the valence and conduction bands falls within the gap. While this level may appear to be "bound" relative to the conduction band, it is in fact repelled upwards by the bonding levels. If the defect were oxygen instead of nitrogen, its atomic level (and hyperdeep level) would lie about $\sim$15 eV below that of P, but its deep level would only lie slightly below N's.

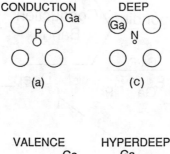

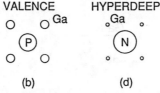

Fig. 5. Schematic charge densities for (a) the conduction band of GaP, (b) the valence band of GaP, (c) a N deep level in GaP, and (d) a N hyperdeep level in GaP. Observe that the valence band is P-like, the conduction band is Ga-like, and the highly electronegative N attracts most of the charge onto itself, in the hyperdeep level. Hence the deep level is mostly Ga-like, having very little impurity character: such deep levels have energies and wavefunctions that are almost independent of the impurity.

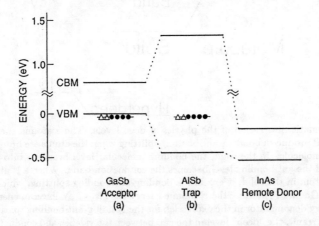

Fig. 6. Energy levels of conduction band minima (CBM) and valence band maxima (VBM) of (a) GaSb, (b) AℓSb, and (c) InAs. The energy of the T_2 or p-like antisite defect levels are also shown for Ga$_{Sb}$ and Aℓ_{Sb}. These levels cause the antisite defects to behave as shallow acceptors in GaSb, deep traps in AℓSb, and remote donors in InAs.

10

Hence, in a first approximation, *deep levels will not be affected by quantum confinement in superlattices* unless the confining width is of order 5 Å or less: one or a few atomic layers. Thus deep levels in superlattices are not very different from deep levels in bulk semiconductors.

However, *band edges in superlattices, unlike deep levels, are sensitive to quantum confinement,* and so quantum confinement can be "engineered" to change the energy of a deep level *with respect to* a nearby band edge, and hence to *change the doping character of an impurity in a superlattice.* The length scale on which band edges are affected is given by the effective-mass approximation, and is basically the shallow donor Bohr radius $a^* = \hbar^2 \epsilon / m^* e^2$, typically of order 100 Å. Thus the art of engineering the doping character of substitutional impurities in superlattices reduces to the art of engineering band edges.

III. BAND-EDGE ENGINEERING

To see how to employ these principles in order to determine the doping character of an impurity, consider the band edges of bulk GaSb, AℓSb, and InAs, plotted in Fig. 6 with the accepted band offsets appropriate for thick-layer superlattices [15].

First we define three types of superlattices: (i) Type-I superlattices such as GaSb/AℓSb, where (in the thick-layer limit) the fundamental band gap of one material (GaSb) lies entirely within the gap of the other (AℓSb); (ii) Type-II-misaligned superlattices, such as InAs/GaSb [16,17], where the conduction band minimum of one material (InAs) lies at lower energy than the valence band maximum of the other (GaSb); and (iii) Type-II-staggered superlattices, such as InAs/AℓSb [18-20], in which the energy of the valence band maximum of one material (AℓSb) lies at an energy within the fundamental gap of the other (InAs). Also shown in Fig. 6 are the predicted p-like deep levels of Ga$_{Sb}$ and Aℓ$_{Sb}$: Ga and Aℓ on an Sb site in GaSb and AℓSb, respectively. Note that this energy is virtually the same in both materials (on an absolute energy scale). This p-like level, for the neutral antisite defect, wants to contain two holes and four electrons (in otherwise undoped material), and so by our previous reasoning will be a deep trap (for up to four holes or two electrons — if we neglect Coulomb effects) in AℓSb and a (double) shallow acceptor in GaSb.

By varying the thicknesses of the layers in a superlattice fabricated from GaSb, AℓSb, and InAs, it is possible to select the *band edges of the superlattice* at almost any energy in the range defined by the bulk materials. (Here we emphasize that the superlattice is a periodic structure with its own conduction band minimum and valence band maximum. Readers accustomed to thinking in terms of quantum wells should note that the energy of the n=1 quantum-well state is the band edge of the superlattice.) For example, an AℓSb/GaSb superlattice with very thin GaSb layers would have a band gap very close to that of AℓSb. Similarly, thick InAs layers sandwiched between thin AℓSb layers would have a superlattice gap close to that of InAs, and *the electrons associate with the Aℓ$_{Sb}$ defects would spill from the AℓSb layers into the InAs layers, remotely doping the InAs n-type.* Thus, in the Type-II staggered InAs/AℓSb superlattice [18-20], for some thin AℓSb layer-thicknesses, Aℓ$_{Sb}$ is not a deep trap for both holes and electrons (making AℓSb semi-insulating) but instead is a *remote donor* controlling the number of electrons in the InAs quantum well. See Fig. 7.

The technological significance of this result is that it provides a simple explanation of why the InAs quantum wells in InAs/AℓSb systems are naturally doped n-type. These systems have very large conduction-band offsets, and so they are prime candidates for making such devices as high-speed field-effect transistors. If the processing conditions for the material always produce antisite defects Aℓ$_{Sb}$, then these defects can dope the InAs quantum wells n-type. Control of the electron concentration in the well can be achieved by employing growth conditions that produce predictable numbers of these antisite defects.

Of course, this is not the only model of process-related doping during growth of InAs wells in the InAs/AℓSb systems. Ideshita *et al.* [21] have also proposed a model which relies on there being more than one impurity. Ideshita's model has donors as well as deep acceptors in the AℓSb layers. Which model, if either, provides the correct explanation of the data remains to be determined.

IV. THEORY

The formalism that supports the physics of deep levels is very simple, having been developed by Hjalmarson *et al.* [8] for s- and p-bonded impurities in semiconductors, based on earlier ideas of Slater and Koster [7]. Applications of the formalism to deep levels both in bulk semiconductors and in superlattices have been discussed and reviewed extensively [22]. Type-II superlattices exhibit some of the most interesting phenomena in which impurities change doping character as superlattice layer thicknesses are varied [22], with the work of Shen *et al.* [16-20] providing the most details and examples of changes of doping character.

Predictions for the antisite defects Ga_{Sb} and $Aℓ_{Sb}$ are given in Fig. 8 for $N_{InAs} \times N$ [001] superlattices of InAs and either GaSb ($N=N_{GaSb}$) or AℓSb ($N=N_{AℓSb}$), following Ref. [16] and after Ref. [19]. Note how the antisite defect levels are virtually constant in energy, while the band edges move around them as the layer thicknesses are varied. Here N_{InAs} (N_{GaSb}, $N_{AℓSb}$) refers to the number of atomic bi-layers of InAs (GaSb, AℓSb) in the fundamental period of the superlattice. When the deep level, which is p-like, lies above the conduction band minimum of the superlattice, it dopes the material n-type; when it is in the gap, it promotes semi-insulating behavior, having two holes and four electrons occupying the level when the impurity is neutral. When the level lies below the valence band maximum of the superlattice, the impurity is a shallow acceptor. [Here we present results for a defect at the middle of a GaSb or AℓSb layer.]

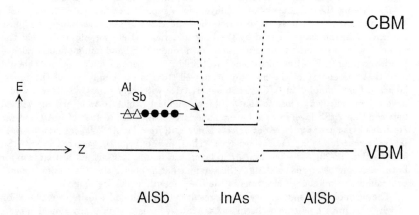

Fig. 7. Schematic illustration of an InAs quantum well sandwiched between two AℓSb layers. The Aℓ$_{Sb}$ defect is a remote donor, modulation doping the InAs n-type.

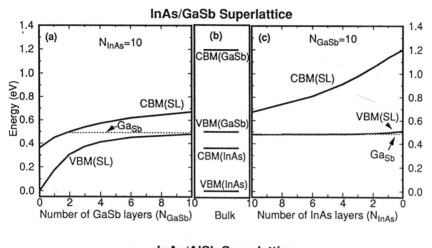

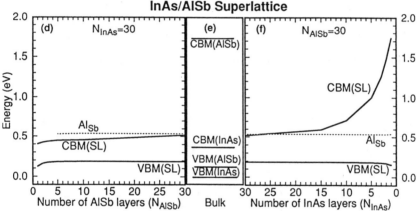

Fig. 8. Predicted energies of band edges (solid lines) and antisite deep level in (a) and (c) $N_{InAs} \times N_{GaSb}$ InAs/GaSb superlattices and in (d) and (f) $N_{InAs} \times N_{A\ell Sb}$ InAs/AℓSb superlattices, following Ref. [16] and after Ref. [19]. In panel (a) we have $N_{InAs}=10$ and N_{GaSb} variable; in (c) N_{InAs} varies while we have $N_{GaSb}=10$. In (d) we have $N_{InAs}=30$, while $N_{A\ell Sb}$ varies; in (f) we have $N_{A\ell Sb}=30$ and N_{InAs} varies. Panels (b) and (e) show the energies of the bulk band edges, for reference. The (centers of gravity of) antisite defect levels of Aℓ_{Sb} and Ga$_{Sb}$ at the layer-centers are shown as dashed lines. The splittings of these p-like levels are of order 50 meV. Here N_{InAs} denotes the number of InAs bi-atomic layers in the [001] superlattice.

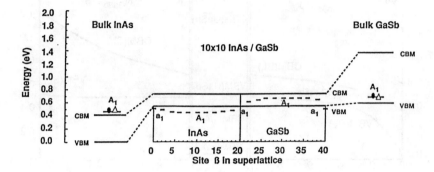

Fig. 9. Predictions [17] for A_1-like or s-like levels of C on a cation site in a 10×10 [001] InAs/GaSb superlattice. (The site $\beta = 0$ is a cation site.) Note that the deep level is almost constant in energy, moving down in energy as it approaches the center of the InAs layer, and moving up in GaSb. C wants to put one electron and one hole into this level, when the impurity is neutral. According to the theory [23], this defect is a donor in bulk InAs, and a trap (for either a hole or an electron) in bulk GaSb. In the superlattice, it is an acceptor (false-valence) in InAs layers, and a trap in GaSb layers.

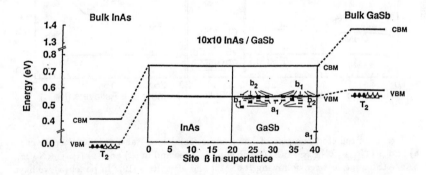

Fig. 10. Cation-site T_2-derived vacancy levels in a 10×10 InAs/GaSb [001] superlattice, after Ref. [17]. Note that the vacancy level is predicted [23] to be T_2-like or p-like, and to lie below the valence band maxima of both InAs and GaSb. In the superlattice, the theory predicts that this level moves up near the superlattice's valence band maximum in the GaSb layers, where the level is split as the defect approaches the interface. (Note that the $\beta = 40$-th site is a cation site.) The doping character of this defect changes when one or more of its sub-levels is split and ascends into the superlattice's fundamental band gap.

14

Interesting changes of doping character can occur within a superlattice, as the site of the impurity changes. For example, Fig. 9 shows how C changes from a donor in InAs to an acceptor in an InAs layer of an InAs/GaSb 10×10 superlattice, to a trap (for either holes or electrons) in a GaSb layer, to a trap in bulk GaSb, according to the theory. The s-like deep level does vary its energy somewhat as the impurity becomes closer to or farther from an interface, having higher energy in the GaSb [23].

A similar behavior is found for a p-like cation-site vacancy level in 10×10 InAs/GaSb. (See Fig. 10.) The level splits as the defect moves from the center of the GaSb layer toward an InAs/GaSb interface. In InAs, the level lies about 0.5 eV lower in energy than in GaSb, due to the electrostatic differences in the two compounds InAs and GaSb. Thus this defect is a (triple) shallow acceptor in InAs and GaSb, but can split in the superlattice, having two of its six spin-orbital levels lying in the gap. In this case, it is a single acceptor [23].

V. SUMMARY

In summary, the energies of deep levels in superlattices do not change very much in superlattices from the corresponding bulk energies. They do shift and split slightly, as shown in Figs. 9 and 10. However, a zero-order model of the effects of superlattices on deep levels is that the deep levels remain at the same energies.

The big effect of having a deep level in a superlattice is that the band-edges of the superlattice change as the layer thicknesses of the superlattice vary — and so these band-edges can rather easily pass through a (nearly constant) deep level, provided that the original level falls close to or within the fundamental band gap of either constituent of the superlattice. When this happens, the defect responsible for the deep level experiences a change of doping character — and very interesting physics. Such changes are most common in Type-II superlattices, whose band edges can be made to vary over quite a range of energies. As a result, native defects, such as the cation-on-anion site defect in InAs/Aℓ_xGa$_{1-x}$Sb superlattices, can have a range of doping characters, from acceptor, to deep trap, to remote donor. Clearly the physics of doping short-period Type-II superlattices is going to be very interesting.

Acknowledgements — We thank the U.S. Air Force Office of Scientific Research and the Office of Naval Research for their generous support of this research (Contract Nos. AFOSR-91-0418 and N00014-92-J-1425).

REFERENCES

[1] W. Kohn, in *Solid State Physics* (edited by F. Seitz and D. Turnbull, Academic Press, New York, 1957) Vol. 5, pp. 258-321; J. M. Luttinger and W. Kohn, Phys. Rev. **97**, 969 (1955).

[2] D. J. Wolford, B. G. Streetman, W. Y. Hsu, J. D. Dow, R. J. Nelson, and N. Holonyak, Jr., Phys. Rev. Letters **36**, 1400 (1976); D. J. Wolford, B. G. Streetman, W. Y. Hsu, and J. D. Dow, Proc. 13th International Conference on Physics of Semiconductors, Rome, 1976, pp. 1049; W. Y. Hsu, J. D. Dow, D. J. Wolford, and B. G. Streetman, Phys. Rev. **B 16**, 1597 (1977); D. J. Wolford, W. Y. Hsu, J. D. Dow, and B. G. Streetman, J. Lumin. **18/19**, 863 (1979); C. A. Swarts, D. L. Miller, D. R. Franceschetti, H. P. Hjalmarson, P. Vogl, J. D. Dow, D. J.

Wolford, and B. G. Streetman, Phys. Rev. **B 21**, 1708 (1980). Phys. Rev. Letters **44**, 810 (1980).

[3] S. Lee and J. D. Dow, Phys. Rev. **B 36**, 5968-5973 (1987).

[4] K. E. Newman and J. D. Dow, Solid State Commun. **50**, 587 (1984).

[5] B. A. Bunker, S. L. Hulbert, J. P. Stott, and F. C. Brown, Phys. Rev. Lett. **53**, 2157 (1984).

[6] C. S. Lent, M. A. Bowen, R. S. Allgaier, J. D. Dow, O. F. Sankey, and E. S. Ho, Solid State Commun. **61**, 83 (1987).

[7] J. C. Slater and G. F. Koster, Phys. Rev., **94**, 1498 (1954); G. F. Koster, *Solid State Physics,* Edited by F. Seitz and D. Turnbull, (Academic Press, New York, 1957) Vol. 5, p.173.

[8] H. P. Hjalmarson, P. Vogl, D. J. Wolford, and J. D. Dow, Phys. Rev. Letters **44**, 810 (1980).

[9] P. Vogl, H. P. Hjalmarson, and J. D. Dow, J. Phys. Chem. Solids **44**, 365 (1983).

[10] We normally also omit charge-state splittings of deep levels in the interest of simplifying the calculations. With each additional electron occupying it, a deep level moves up in energy about 0.1 eV. See G. Kim, J. D. Dow, and S. Lee, Arabian J. Sci. Engineer. **14**, 513 (1989); Phys. Rev. **B 40**, 7888 (1989).

[11] W. A. Harrison, *Electronic Structure and the Properties of Solids,* (Freeman, San Francisco, 1980).

[12] S. Y. Ren, Scientia Sinica **A XXVII**, p. 443 (1984); and S. Y. Ren, W. M. Hu, O. F. Sankey, and J. D. Dow, Phys. Rev. **B 26**, 951 (1982).

[13] M.-F. Li, D.-Q. Mao, and S. Y. Ren, Phys. Rev. **B 32**, 6907 (1985).

[14] G. W. Ludwig, Phys. Rev. A **137**, 1520 (1965); J. R. Niklas and J. M. Spaeth, Solid State Communications **46**, 121 (1983); J. M. Spaeth, D. M. Hofman, B. K. Meyer, Mater. Res. Soc. Symp. Proc. **46**, 185 (1985), *Microscopic Identification of Electronic Defects in Semiconductors,* ed. by N. M. Johnson, S. G. Bishop, and G. D. Watkins.

[15] S. Tiwari and D. J. Frank, Appl. Phys. Lett. **60**, 630 (1992); G. J. Gualtieri, G. P. Schwartz, R. G. Nuzzo, R. J. Malik, and J. F. Walker, J. Appl. Phys. **61**, 5337 (1987); G. J. Gualtieri, G. P. Schwartz, R. G. Nuzzo, and Q. A. Nuzzo, Appl. Phys. Lett. **49**, 1037 (1986).

[16] J. Shen, S. Y. Ren, and J. D. Dow, Phys. Rev. Letters **69**, 1089 (1992).

[17] J. Shen, S. Y. Ren, and J. D. Dow, Phys. Rev. **B 46**, 6938 (1992).

[18] J. Shen, J. D. Dow, S. Y. Ren, S. Tehrani, and H. Goronkin, J. Appl. Phys. **73**, 8313 (1993).

[19] J. D. Dow, J. Shen, and S. Y. Ren. Superlatt. Microstruct. **13**, 405 (1993).

[20] J. Shen, S. Tehrani, H. Goronkin, G. Kramer, M. Adam, and J. D. Dow. "Semi-insulating, p-type, and n-type doping in AℓSb, GaSb, and AℓSb/InAs by a single native defect," Submitted.

[21] S. Ideshita, A. Furukawa, Y. Mochizuki, and M. Mizuta, Appl. Phys. Lett. **60**, 2549 (1992).

[22] J. D. Dow, "Localized perturbations in semiconductors," In *Highlights of Condensed-Matter Theory* (Proceedings of the International School of Physics "Enrico Fermi", Course 89, Varenna, 1983), ed. by F. Bassani, F. Fumi, and M. P. Tosi (Societa Italiana di Fisica, Bologna, Italy, and North Holland, Amsterdam, 1985), pp. 465-494; R.-D. Hong, D. W. Jenkins, S. Y. Ren, and J. D. Dow, Materials Research Soc. Symp. Proc. **77**, 545-550 (1987), *Interfaces, Superlattices, and Thin Films,* ed. J. D. Dow and I. K. Schuller; J. D. Dow, S. Y. Ren, and J. Shen, NATO Advanced Science Institutes Series **B 183**: Properties of Im-

purity States in Superlattice Semiconductors, (Plenum Press, New York, 1988), pp. 175-187, edited by C. Y. Fong, I. P. Batra, and S. Ciraci; S. Y. Ren and J. D. Dow, J. Appl. Phys. **65**, 1987 (1989); S. Y. Ren and J. D. Dow, Phys. Rev. **B 39**, 7796 (1989); J. D. Dow, R.-D. Hong, S. Klemm, S. Y. Ren, M.-H. Tsai, O. F. Sankey, and R. V. Kasowski, Phys. Rev. **B 43**, 4396 (1991); J. D. Dow, J. Shen, and S. Y. Ren, in "Progress in Electronic Properties of Solids," *Physics and Chemistry of Materials with Low-dimensional Structure,* Festschrift in honor of Professor Franco Bassani, ed. by E. Doni, R. Girlanda, G. Pastori Parravicini, and A. Quattropani, (Kluwer, Dordrecht, 1989), pp. 439-449; J. Shen, J. D. Dow, and S. Y. Ren, J. Appl. Phys. **67**, 3761 (1990); S. Y. Ren, J. Shen, R.-D. Hong, S. Klemm, M.-H. Tsai, and J. D. Dow, Surf. Sci. **228**, 49 (1990); J. D. Dow, S. Y. Ren, J. Shen, R.-D. Hong, and R.-P. Wang, J. Electron. Mater. **19**, 829 (1990); J. D. Dow, S. Y. Ren, J. Shen, and M.-H. Tsai. Deep levels in superlattices. In *Impurities, Defects, and Diffusion in Semiconductors: Bulk and Layered Structures,* edited by D. J. Wolford, J. Bernholc, and E. E. Haller, Materials Research Society Symposia Proceedings No. **163**, 349 (MRS, Pittsburgh, 1990), S. Y. Ren, J. D. Dow, and J. Shen, Phys. Rev. **B 38**, 10677 (1988), and references therein; and S. Y. Ren, "Electronic Structure of Deep Impurities in Semiconductors," in *Lattice Dynamics and Semiconductor Physics, Festschrift for Professor Kun Huang,* edited by J. B. Xia, Z. Z. Gan, R. Q. Han, G. G. Qin, G. Z. Yang, H. Z. Zheng, Z. T. Zhong, and B. F. Zhu (World Scientific, Singapore, 1990), pp. 536-547.

[23] A word of caution is in order about the results of Figs. 9 and 10. We have interpreted the theoretical results as though there were zero theoretical uncertainty. In fact modest uncertainties in the theory could lead to somewhat different level positions and hence different predictions. Nevertheless our goal is to illustrate the qualitative features of the theory.

OPTICAL SPECTROSCOPY OF DEFECTS IN GaAs/AlGaAs MULTIPLE QUANTUM WELLS.

B. MONEMAR*, P. O. HOLTZ*, J. P. BERGMAN*, Q.X. ZHAO*, C.I. HARRIS*, A. C. FERREIRA*, M.SUNDARAM**, J.L.MERZ**, A.C. GOSSARD**,

* Linköping University, Department of Physics and Measurement Technology, S-581 83 Linköping, Sweden

** Center for Studies of Quantised Electronic Structures, (QUEST), University of California at Santa Barbara, CA 93016, USA

ABSTRACT

The study of electronic properties of GaAs/AlGaAs quantum wells (QWs) has traditionally been focused on intrinsic phenomena, in particular the free exciton behaviour. Defects and impurities have often been regarded as less relevant compared to the case of bulk semiconductors. Doping in QWs is important in many applications, however, and recently the knowledge about the structure of shallow donors and acceptors from optical spectroscopy has advanced to a level comparable to the situation in bulk semiconductors. A dramatic difference from the bulk case is the common occurrence of localisation effects due to interface roughness in QW structures. The recombination of bound excitons (BEs) differs drastically from bulk, BE lifetimes decrease with decreasing well thickness L_W, but increase with decreasing barrier thickness L_b (at constant L_W) below $L_b=70\text{Å}$. Exciton capture at impurities is a process which is strongly influenced by the localisation potentials from the interface roughness. The recombination process in doped QWs involves a nonradiative component, for shallow acceptors an excitonic Auger process has been identified. Deep nonradiative defects in the (MBE grown) QW as well as in the barrier material are manifested in measurements of the PL decay time vs temperature. In undoped multiple QWs the decay times vs T are consistent with thermal emission out of the well into the barrier, where nonradiative recombination via deep level defects occur. Nonradiative recombination in the well itself can be studied in electron-irradiated structures. Preliminary data also demonstrate the feasibility of hydrogen passivation of dopants as well as deep levels in the QW structures.

INTRODUCTION

Optical spectroscopy in bulk semiconductors has been a very powerful tool to investigate the electronic structure of impurities and defects in semiconductors, as well as various recombination processes for excitons or free carriers. Extensive reviews of this research area have been presented [1,2]. For quantum wells (QWs) the defects area has had less priority, probably since the dominance of intrinsic recombination processes in such structures was recognised early [3]. Doped QWs are important in many applications, however, and warrant basic studies to gain a similar detailed knowledge as for bulk semiconductors. In QW structures the properties of shallow impurities depend strongly on the position in the well with respect to the barrier interface, which makes specially grown samples with a specific narrow impurity profile necessary for detailed spectroscopic studies [4]. Results from such studies of shallow donors and acceptors in GaAs/AlGaAs multiple quantum well (MQW) samples will be briefly reviewed here with reference to recent work. Dramatic effects in optical spectra from impurities in QWs are introduced by the localisation of excitons and carriers associated with interface roughness potentials. Photoluminescence (PL) spectra and capture processes for excitons therefore differ very strongly from the bulk case, as will be shown below. Also the bound exciton (BE) properties are quite sensitive to the confinement effects in a QW. We shall discuss the effects of the variation in well width L_W as well as the variation in barrier width L_b on the radiative lifetime of BEs .

The recombination processes are both radiative and nonradiative in QW structures. At low temperatures the radiative excitonic processes dominate at not too high doping, while at higher temperatures (such as room temperature) nonradiative processes are strong, at least in MBE grown structures. At low T a nonradiative recombination channel due to an excitonic Auger effect has been identified in acceptordoped QWs. At higher temperatures recombination can partly occur via defects in the barrier, since excitons and carriers can escape from the well by thermal emission over the barrier. Nonradiative recombination via deep level defects in the well itself can conveniently be studied in samples irradiated with high energy electrons. Recent results from such experiments will be demonstrated below. Finally a brief account will be given on recent work on hydrogen passivation in QWs.

EXPERIMENTAL

The samples used in this work were typically grown by MBE at a temperature of nominally 680 °C without interruptions at the QW interfaces. The multiple QW (MQW) samples were usually prepared with 50 identical QWs, with 150 Å $Al_{0.3}Ga_{0.7}As$ barriers in between. Additional information on the samples and the experimental procedure were given in refs 4-6, and will be discussed in more detail below whenever needed.

SHALLOW DONORS AND ACCEPTORS, BOUND EXCITONS

The electronic structure of centerdoped shallow Si donors in QWs have been studied optically, both in ir transmission [7] and in PL two-electron (TET) spectra. [8] . The results for the 1s-2p and 1s-2s transition energies, respectively, are in excellent agreement with the modified effective mass (EM) model calculations in ref [9]. The BE binding energy is rather small, typically 1.6 meV for L_w = 100 Å [10], (Fig. 1 (a), and the spectra at lower L_w are strongly affected by the localisation of excitons in the QW. The internal structure of the donor BE, which is quite complicated in bulk GaAs [11], has not yet been resolved for the QW case.

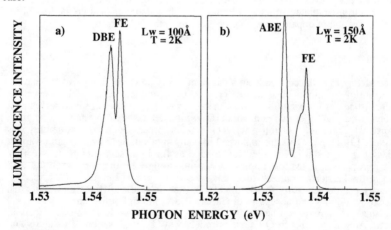

Fig. 1. PL spectra measured with above bandgap excitation at 1.5 K for a) a Be acceptor doped 150 Å wide MQW, and b) a Si donor doped 100 Å wide MQW. Both structures are center-doped to a level of 5×10^{16} cm^{-3}.

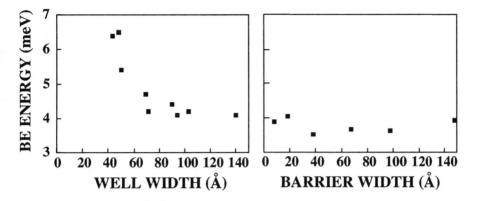

Fig. 2 The dependence of the binding energy for the central acceptor BE on a) the well width and b) the barrier width for a 100 Å wide QW.

For the case of a shallow acceptor in an MQW quite detailed investigations have recently been made. The BE has been studied in detail, for the centerdoped case (see Fig. 1 (b)), and all Be acceptor states up to 3s have been identified via two-hole transitions (THT) [12] and ir spectra [13]. The agreement with the EM model is satisfactory, even under a magnetic field perturbation [14,15]. Internal structure of the BE has also been resolved, confirming that the 5/2 component is lowest for the Be acceptor, just like in bulk GaAs [16]. The energy splittings between the sublevels in the BE case are much larger, however, about an order of magnitude larger for a 100 Å QW compared to bulk GaAs [16].

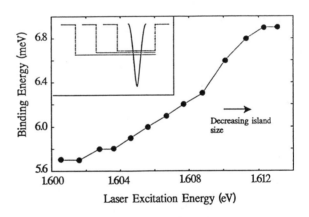

Fig. 3. Dispersion of bound exciton binding energy versus excitation photon energy for a 50 Å Be acceptor doped MQW.

Off-center acceptors have also been studied, and the binding energies for the acceptor ground state (1S-2S transition energy) as well as the BE binding energy have been measured as function of position of the acceptor in the well [4]. The agreement with available theoretical calculations in this case is only modest, however [4].

The BE binding energies vs L_w has been studied in detail, and a monotonic increase vs decreasing L_w is found (Fig 2 (a)), consistent with increased confinement of the two hole state of the bound exciton complex [5,6].

The analogous dependence of Be acceptor binding energies has been observed. Interestingly the dependence of BE binding energy (with reference to the FE energy) does not vary with L_b for 100 Å QWs until $L_w < 10$ Å (Fig 2 (b)) [17]. This proves the direct relevance of the two-hole wavefunction for the BE binding energy.

Interface localisation strongly influences the behaviour of BE spectra in narrow QWs. A trivial example of this is the observation that clear BE spectra comparable in intensity to the FE peak in QWs grown under uninterrupted interface growth conditions are only observed at doping levels about 10^{10} cm^{-2}, while in bulk GaAs BE spectra typically dominate FE spectra down to very high purity (10^{13} cm^{-3}). The interaction of the interface localisation potentials and the acceptor BE potentials has been studied in detail for $L_w = 50$ Å (see Fig 3). [18,19]. It is obvious that the interpretation of BE binding energies for narrow QWs has to be done with great care to obtain meaningful results.

The BE decay times obtained from transient data with resonant excitation reveal the oscillator strengths as a function of the parameters of the QW design. The L_w dependence shows a monotonic decrease in lifetime from about 450 ps for $L_w = 150$ Å to about 250 ps for $L_w = 50$ Å. ($L_b = 150$ Å) [20] (See fig. 4 (a)). This reflects the strong effect of confinement on the BE lifetime, which is particularly sensitive to the electron wavefunction in the BE complex [17]. With varying L_b a very dramatic effect is seen when the 2D character gets lost for low values of L_b, see Fig. 4 (b)) [17].

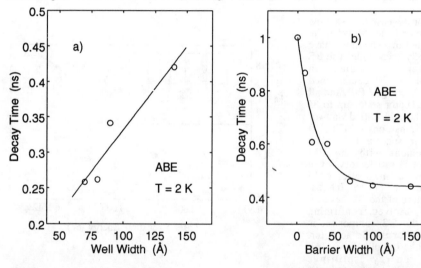

Fig. 4 (a). Decay time for the acceptor BE as a function of QW width at 2 K.

Fig. 4 (b). Decay time of the acceptor BE in a 100 Å QW versus the barrier thickness L_b, measured at 2 K.

The decay time observed at low temperatures in undoped QWs is supposed to be related to the radiative lifetime of free excitons, which is strongly affected by localisation in typical GaAs/AlGaAs QWs [21]. The very fast intrinsic lifetime of FEs is observed initially within the first 50 ps after resonant excitation with a short pulse [22]. In doped QWs excitons are quickly captured at dopants, such as the neutral acceptors, leading to BE recombination [23]. In this case the apparent decay time of the FE is usually completely dominated by the capture to the BE state [23,24]. This capture process is quite different in a QW compared to bulk, due to the strong influence of the interface localisation potentials in the QW. Localisation in these potentials compete strongly with impurity capture of excitons at low temperature, a process that is completely masking the expected power law temperature dependence of exciton capture [25,26].In fact the temperature dependence is slow at low T, and typically shows a peak below 15 K (Fig 5 (a,b)).

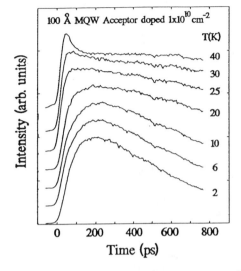

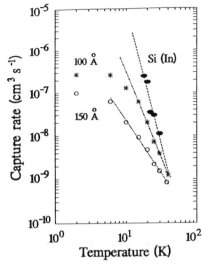

Fig. 5 (a) Temperature dependence of the onset in bound exciton emission for a 100Å acceptor doped quantum well.

Fig. 5 (b). Temperature dependence of the time averaged exciton capture rate for 100 Å and 150 Å acceptor doped QWs. Values for bulk silicon are shown for comparison.

NONRADIATIVE RECOMBINATION PROCESSES.

In intentionally doped QWs the BE recombination is normally a dominantly radiative process, like in bulk GaAs, but nonradiative processes have clearly been detected [27]. An important process is the excitonic Auger process, which acting on an ionised acceptor produces free electrons, which are spectroscopically detected via the radiative free to bound processes at acceptors. Therefore even below bandgap excitation, if resonant with free or bound exciton states, will produce free carriers in a QW [28] (Fig 6). In highly doped QWs nonradiative centres connected with dopant atoms are observed [29]. These observations are similar to, but perhaps less severe than the corresponding situation in bulk GaAs [30].

In undoped QWs the radiative recombination of the FE is the dominating process at low temperature. Non-radiative recombination processes for carriers in undoped QWs have been discussed by several authors [31-40], but no detailed study of their importance has been performed. The reason for discussing the possibility for a non-radiative recombination as a competing process to the radiative recombination is twofold. Researchers with a previous experience from double heterostructures, i.e. a GaAs layer with a thickness in the order of μm surrounded by AlGaAs barriers, are used to describe any deviation from the expected radiative recombination in terms of an interface recombination [33,37-39]. It is then assumed that excited carriers in the GaAs region, assumed to be of high quality where only radiative recombination is possible, diffuses towards the interface and there recombines nonradiatively. The non-radiative recombination is characterised in terms of a surface recombination rate.

This model has also been used in the case of QWs to motivate the importance of non-radiative recombination in these structures. In e.g. a multi QW structure the number of interfaces in relation to the GaAs region are much larger than in the case of a heterostructure, which should increase the importance of the non-radiative surface recombination. The habit of expressing the non-radiative recombination in terms of a surface recombination rate has also been kept, even if the concept has little relevance in MQWs. The origin of the interface recombination is usually not discussed in detail but is most likely to involve capture of carriers by either a multiphonon emission or an Auger process into a deep level defect. It is well known that defects are accumulated at the surface during epitaxial growth, and therefore are expected to be present with higher concentration at or close to the epitaxial interface.

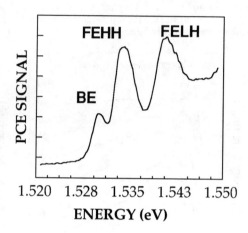

Fig. 6 Photocurrent excitation (PCE) spectrum for a Be-doped 150 Å wide QW at 1.5 K.

The second main argument for non-radiative recombination is to explain the large variations in measured PL decay time in QWs. Different studies have reported values ranging from 200 ps to 600 ps for almost comparable samples. The argument is to explain this by a sample dependent non-radiative recombination.

In our opinion none of these arguments is valid at low temperatures. There is no experimental evidence for a non-radiative recombination at low temperatures that is comparable to the radiative recombination rate. It should here be noted that the excitonic Auger process [27,28] previously discussed requires the presence of impurity atoms and are only observed in highly intentionally doped QWs. It is unlikely that this process has any importance in normal undoped structures. The large variations of the published values for the free exciton decay time is instead satisfactorily explained with the concept of localised excitons. Potential fluctuations occurs in the ideal QW due to interface roughness and compositional fluctuations in the barrier material. In a 2-dimensional system any such fluctuation leads to a localised state, in which free excitons are more or less weakly trapped. The recombination rate for a localised exciton will be more determined by the localisation potential than by the radiative property of the 2-dimensional free exciton. This will hence explain the large variations of the observed decay times.

The main argument for the negligible role of non-radiative recombination at low temperatures is however the temperature dependence of the observed decay time. All published experimental results [31-36] show an increase of the measured decay time in the temperature range from below 2 K and up to at least hundred Kelvin. This is expected for a radiative recombination process, due to the thermal redistribution of excitons, leaving a smaller number of excitons in the region with K=0, where the radiative recombination of the free exciton is allowed. This almost linear increase of the decay time is not expected for dominating non-radiative Auger processes or multiphonon capture to deep defects. From this we conclude that non-radiative recombination processes are typically of negligible importance in QWs at lower temperatures.

However, at higher temperatures the increase of the decay time is broken and at temperatures above about 200 K a decrease of the decay time is typically observed in MBE-grown structures. Figure 7 shows the measured decay time as a function of temperature for four different GaAs/AlGaAs MQW samples, with well widths L_w of 150 Å, 100 Å, 70 Å and 55 Å, respectively.

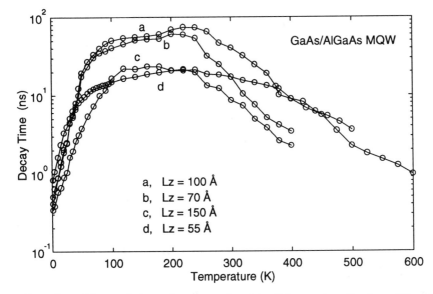

Figure 7. Measured decay times as a function of temperature for four different GaAs/AlGaAs MQW samples. The solid line is intended as a guide to the eye for each sample.

This trend has been observed in all published results of the temperature dependence of the decay time from both single QWs and multiple QWs, of various quality. This decrease of the decay time is clearly due to an onset of a non-radiative recombination mechanism, which at higher temperatures dominates over the radiative recombination. This is concluded from several experimental observations. Firstly, the decrease of the decay time is directly related to a decrease in total PL intensity. Secondly, we have measured the intensity dependence of the total PL intensity at different temperatures, as is shown in Fig. 8. At 77 K we obtain an almost linear relation between excitation intensity and PL intensity, indicating the dominance of a radiative recombination process. This interpretation is valid independently of whether the observed luminescence is due to free exciton or free carrier recombination, and only related to the absence of a non-radiative recombination channel. At 300 K we observe a relation of I_{PL} = $I_{Exc}^{1.42}$ indicating the presence of a non-radiative channel in addition to the radiative recombination.

Finally, we observe a strong difference in the intensity dependence of the decay curve measured at different temperatures. At 77 K we observe an increasingly strong non-exponential decay curve with increasing excitation intensity, reflecting the competition between an exponential exciton recombination and the non-exponential free carrier recombination. Both of these recombination processes are radiative, but the recombination of free carriers is due to its quadratic rate dominating at higher carrier concentrations, obtained at higher excitation intensities. The decay curve is slightly non-exponential even at the lowest possible excitation intensity. On the other hand, at 300 K we observe an almost perfect exponential decay for all except the highest excitation intensities. This is consistent with the presence of a dominating non-radiative recombination channel.

We would like to point out that our measurements, as well as other previously published measurements, show that the optically detected recombination from an MBE grown QW system at room temperature is dominated by non-radiative processes. This has until recently

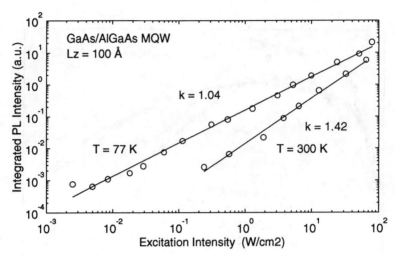

Figure 8. Spectrally integrated PL intensity as a function of cw laser excitation measured at two different temperatures, 77 K and 300 K.

been neglected and should be of importance for applications involving QW structures. This nonradiative process may probably be saturable at very high excitation intensities, however.

The process responsible for this non-radiative recombination has not been clearly identified, but two main suggestions have during the last years been proposed by several groups, based on different experimental results. The first one is recombination through defects at the interface between the well and barrier material [33,37-39], as discussed above. Even if we have excluded this process to be of importance at low temperatures, it can still be present at higher temperatures where it may be thermally activated.

The second proposal is a recombination via deep level defects in the barrier material. These will act as a recombination channel for particles in the QW through thermally activated carrier escape out of the well, followed by scattering or diffusion away from the QW and finally a non-radiative recombination in the barrier material [34-36,40]. This model has been used to obtain excellent agreement with experimental results in $Ga_{1-y}In_yP/(Al_xGa_{1-x})_{0.5}In_{0.5}P$ [34] and $In_xGa_{1-x}As/GaAs$ [36] QW's, in which case both the barrier and well recombination processes are partly radiative. The activation energy for the onset of the non-radiative recombination was in good agreement with half of the total confinement energy, supporting the model of simultaneous carrier escape (ambipolar) of electrons and holes (or excitons) out of the well.

The same clear identification has not been made for the $GaAs/Al_xGa_{1-x}As$ system. Partly because the barrier material exhibits very low PL at high temperatures and the correlation between the decay time in the well and barrier is more difficult. By calculating the activation energy for the non-radiative recombination from our experimental data we obtain values of 147 meV, 167 meV and 212 meV, for well widths of 50 Å, 70 Å, and 100 Å respectively. This is in rough agreement with the expected activation energy for carrier escape, equal to half the confinement energy, which is 167 meV, 180 meV, and 195 meV for the corresponding well widths. There are however some uncertainties with this kind of calculation. The expected decay time of the radiative recombination is not easily determined, due to the interaction between excitons and free carriers, and their combined contribution to the radiative recombination at higher temperatures. The same rough agreement has also been reported for single QW samples [35].

It should be noted that the same general trend of the decay time versus temperature, and the value for the calculated activation energy, is observed in single QWs as well as in MQWs. In the latter we can exclude the diffusion of carriers, which are thermally excited up into the barriers away from the well, as an important mechanism. These carriers are instead recaptured into an adjacent QW, and the only non-radiative recombination channel is a recombination via deep level defects in the barrier material. We therefore expect and do indeed observe a longer decay time in our MQW structures than in all published results on single QWs [32].

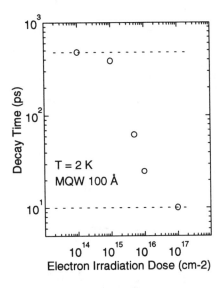

Figure 9. Measured decay time as a function of electron irradiation dose for an MQW sample with well width 100 Å. The upper dotted line is the decay time observed in the reference sample, while the lower line represents the time resolution.

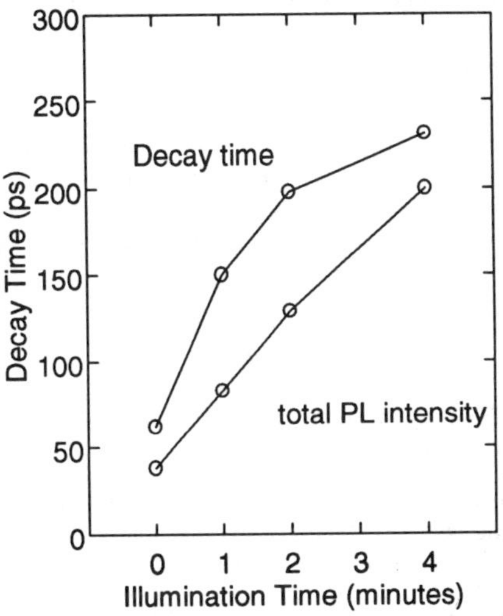

Figure 10. Measured decay time in a 100 Å MQW sample electron irradiated with a dose of 5×10^{15} cm^{-2} as a function of time (minutes) with intense optical illumination. Also shown are the corresponding total PL intensity in arbitrary units.

NON-RADIATIVE RECOMBINATION IN IRRADIATED SAMPLES

One way to study non-radiative recombination processes is to intentionally introduce a high concentration of deep level defects in the QW system. We have made this by 2 MeV electron irradiation of an originally undoped MQW sample with a QW width of $L_w = 100$ Å.

The measured decay time at low temperatures as a function of electron irradiation dose is shown in Fig. 9. The upper horizontal dotted line corresponds to the decay time in the reference sample, while the lower represents the time resolution of our decay time measurement. The total PL intensity is consequently drastically reduced with increasing irradiation dose. The decrease of the measured decay time with increasing irradiation dose is expected and shows that we have created an efficient non-radiative recombination channel.

The temperature dependence of the decay time in these samples also shows the presence of a dominating non-radiative recombination. The general trend as discussed before is more and more weakened with increasing electron irradiation dose. In samples irradiated above 5×10^{15} cm^{-2} we do not observe any change of the decay time with increasing temperature. In these cases the non-radiative recombination is completely dominating over the radiative recombination.

We have furthermore observed that the investigated non-radiative recombination channel in this sample can be reduced with optical excitation. Figure 10 shows the measured decay time at low temperature in the sample irradiated with an electron dose of 5×10^{15} cm^{-2}, as a function of time with exposure of a high intensity of laser excitation, about 100 W/cm^2. The measurements of the decay time itself were obtained with a decreased laser excitation intensity (1%). The observed increase of the decay time is also directly related to an increase of the total PL intensity. From this we conclude that we have blocked the non-radiative recombination channel. Furthermore, this condition seems to be stable during our experimental timescale (20 minutes) even when the high excitation intensity is removed. This behaviour is observed in the samples irradiated with 5×10^{15} and 1×10^{16} cm^{-2}. In the highest irradiated sample, 1×10^{17} cm^{-2}, the total intensity is too low to obtain reliable results.

HYDROGEN PASSIVATION

Hydrogen passivation is an area of extensive interest in connection with doping and defect problems in bulk semiconductors [41]. Recently low dimensional structures have also been explored, and preliminary results appear promising for further work [42]. We have studied passivation of doped QWs , both donor and acceptor doped samples. For the case of donors in QWs a set of samples with varying degree of Si doping in the well were studied, using a remote plasma reactor with a sample temperature of about 175 °C [43]. Passivation of the donors was evidenced by the strong changes in the PL spectra from the samples (disappearing of the BE line, reappearance of the FE component in strongly doped samples, see Fig. 11). This is interpreted as the result of a combination of two effects: passivation of Si donors in the QW and passivation of surface states causing the strong surface potential. The passivation procedure also causes a strongly enhanced localisation of the FEs, manifested in a strong broadening of the FE peak. The dependence of localisation on the strong surface potential implies that non-uniform surface passivation will result in a distribution of localisation potentials in the QW. The random fluctuation in surface potential brings about localisation and a broadening of the FE transition.

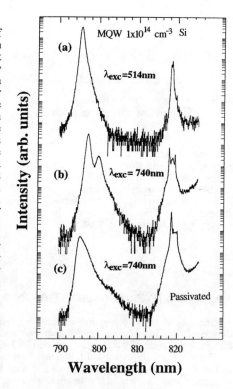

Fig. 11. Comparison of PL spectra before and after hydrogenation for a Si-doped QW.

Passivation of acceptordoped MQWs were also studied, with low energy H implantation with a Kaufmann gun, using sample temperatures of about 300 °C [44]. The Be acceptors were clearly partially compensated by this technique, as evidenced by a considerable reduction of the BE line PL intensity compared to the FE line. The effect was weaker than in the underlying GaAs buffer layer however, and there was evidence for a long term instability in the passivation process: after one year of room temperature storage the PL spectrum from the sample was nearly the same as for the virgin condition before passivation. Some deep level passivation also occurred, leading to an increase in the measured maximum PL lifetime at elevated temperatures. (see Fig 12). A too high H dose leads to a production of additional nonradiative defects, though. It should be mentioned that the process conditions were far from optimised in these preliminary experiments.

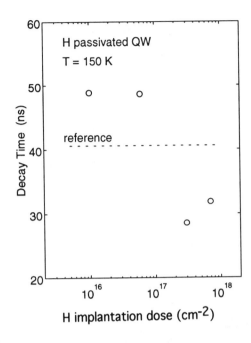

Fig 12. Variation of the decay time at 150 K with the dose of H implantation in Be-doped MQWs.

REFERENCES

1. P J Dean and D C Herbert, in Topics of Current Physics, Vol 14. Excitons, K Cho, Ed., Springer Verlag, Berlin, 1979, p 55
2. B Monemar, CRC Critical Reviews in Solid State and Materials Sciences, Vol 15, p 111, 1988
3. C Weisbuch, R C Miller, R Dingle, A C Gossard, and W Wiegmann, Solid State Commun. 37, 219 (1981)
4. G C Rune, P O Holtz, B Monemar, M Sundaram, J L Merz, and A C Gossard, Phys Rev B 44, 4010 (1991)
5. P O Holtz, M Sundaram, R Simes, J L Merz, A C Gossard, and J H English, Phys Rev B 39, 13293 (1989)
6. P O Holtz, M Sundaram, K Doughty, J L Merz, and A C Gossard, Phys Rev B 40, 12338 (1989)
7. N C Jarosik, B D Mc Combe, B V Shanabrook, J Comas, J Ralston, and G Wicks, Phys Rev Lett 54, 1283 (1985)
8. P O Holtz, B Monemar, M Sundaram, J L Merz, and A C Gossard, Phys Rev 47, 10596 (1993)
9. S Fraizzoli, F Bassani, and R Buczko, Phys Rev B 41, 5096 (1990)
!0. D C Reynolds, K K Bajaj, C W Litton, P W Yu, W T Masselink, R Fisher, and H Morkoc, Phys Rev B 29, 7038 (1984)
11. W Rühle and W Klingenstein, Phys Rev B 18, 7011 (1978)

12. P O Holtz, Q X Zhao, A C Ferreira, B Monemar, M Sundaram, J L Merz, and A C Gossard, Phys Rev B 48, 8872 (1993)
13. A A Reeder, J M Mercy, and B D McCombe, IEEE J Quantum Electron 24, 1690 (1988)
14. Q X Zhao, P O Holtz, A Pasquarello, B Monemar, A C Ferreira, M Sundaram, J L Merz, and A C Gossard, manuscript 1993
15. P O Holtz, Q X Zhao, A C Ferreira, B Monemar, A Pasquarello, M Sundaram, J L Merz, and A C Gossard, manuscript, 1993
16. P O Holtz, Q X Zhao, B Monemar, M Sundaram, J L Merz, and A C Gossard, Phys Rev B47, 15675 (1993)
17. Q X Zhao, P O Holtz, C I Harris, B Monemar, and E Veje, manuscript 1993.
18. B Monemar, P O Holtz, P Bergman, C I Harris, H Kalt, M Sundaram, J L Merz, and A C Gossard, Surface Science 263, 556 (1992)
19. C I Harris, B Monemar, P O Holtz, M Sundaram, J L Merz, and A C Gossard, Materials Science Forum Vol 117-118, 285 (1993)
20. J P Bergman, P O Holtz, B Monemar, M Sundaram, J L Merz, and A C Gossard, Phys Rev B 43, 4765 (1991)
21. D S Citrin, Solid State Commun 84, 281 (1992)
22. B Deveaud, F Clérot, N Roy, K Satzke, B Sermage, and D S Katzer, Phys Rev Letters 67, 2355 (1991)
23. C I Harris, H Kalt, B Monemar, P O Holtz, J P Bergman, M Sundaram, J L Merz and A C Gossard, Materials Science Forum, 83-87, 1363 (1992)
24. C I Harris, B Monemar, H Kalt, M Sundaram, J L Merz, and A C Gossard, manuscript 1993
25. C I Harris, B Monemar, P O Holtz, H Kalt, M Sundaram, J L Merz, and A C Gossard, Proc 10th Int Conf on the Electronic Properties in 2D Systems, (EP2DS-10), Newport, USA, 1993, Surface Science, in press.
26. C I Harris, B Monemar, P O Holtz, H Kalt, M Sundaram, J L Merz, and A C Gossard, Proc Int Conf on Excitons in Confined Systems, Montpellier, France, 1993, to be published.
27. P O Holtz, M Sundaram, J L Merz, and A C Gossard, Phys Rev B41, 1489 (1990)
28. A Ferreira, P O Holtz, B Monemar, M Sundaram, J L Merz, and A C Gossard, manuscript 1993
29. C I Harris, B Monemar, H Kalt, and K Köhler, Phys Rev B 48, 4687 (1993)
30. Jiang De-Sheng, Y Makita, K Ploog, and H J Queisser, J Appl Phys 53, 999 (1982)
31. M Gurioli, A Vinattieri, M Colocci, C Deparis, J Massies, G Neu, A Bosacchi, and S Franchi,Phys. Rev. B 44, 3115 (1991).
32. J P Bergman, P O Holtz, B Monemar, M Sundaram, J L Merz, and A C Gossard, Inst. Phys. Conf. Ser. No 123, 1992, p 73.
33. H Hillmer, A Forchel, R Sauer, and C W Tu, Phys Rev B 42, 3220 (1990)
34. P Michler, A Hangleiter, M Moser, M Geiger, and F Scholz, Phys Rev B 46, 7280 (1992)
35. M Gurioli, J Martinez-Pastor, M Colocci, C Deparis, B Chastaingt, and J Massies, Phys Rev B 46, 6922 (1992)
36. G Bacher, C Hartmann, H Schweitzer, T Held, G Mahler, and H Nickel, Phys Rev B 47, 9545, (1993)
37. H Hillmer, A Forchel, T Kuhn, G Mahler, and H P Maier, Phys Rev B 43, 13992 (1991)
38. B Sermage, F Alexandre, J Beerens, and P Tronc, Superlattices and Microstructures 6, 373 (1989)
39. M Krahl, D Bimberg, R K Bauer, D E Mars, and J N Miller, J Appl Phys 67, 434 (1990)
40. B Sermage, F Mollot, F Alexandre, and Y Gao, Inst. Phys. Conf. Ser. No 106, 1990, p 423. (Presented at Int. Symp. GaAs and Related Compounds, Karuizawa, Japan 1989).
41. S J Pearton, J W Corbett, and T S Shi, Appl Phys A 43, 153 (1987)
42. M Capizzi, C Coluzza, P Frankl, A Frova, M Colocci, A Vinattieri, and R N Sacks, Physica B 170, 561 (1991)
43. C I Harris, M Stutzmann, and K Köhler, Materials Science Forum 117-118, 339 (1993)
44. C I Harris, P O Holtz, P Bergman, B Monemar, M Capizzi, A Frova, M Sundaram, J L Merz, and A C Gossard, unpublished

IDENTIFICATION OF DIFFUSION ASSOCIATED DEFECTS AT III-V SEMICONDUCTOR HETEROSTRUCTURES

R. ENRIQUE VITURRO* and GARY W. WICKS**
*Xerox Webster Research Center, 114-41D, 800 Phillips Rd., Webster, NY 14580.
**Institute of Optics, University of Rochester, Rochester, NY 14620.

ABSTRACT

Cathodoluminescence spectroscopy is used to identify diffusion-associated III-V semiconductor defects and establish their role in AlGaAs/GaAs intrinsic and n-type impurity induced interdiffusion (Si, Ge, S, and Se) for various ambient conditions, As– and Ga–rich. These identifications involves the study of the temperature and composition dependence of these deep levels and their correlation with theoretical calculations. Our results reveal Column III vacancies and their complexes as the sole mediators of diffusion.

INTRODUCTION

Although it is widely accepted that diffusion phenomena are mediated by material defects, either present or externally induced by physical processing,[1,2] the detection and identification of these defects have until recently eluded researchers.[3,4] Results from atomic depth profiles measurements have allowed researchers to postulate several detailed mechanisms of diffusion.[1,2] However, these mechanisms are controversial and they fail to explain and predict general diffusion phenomena.[5] Here, we report a low energy cathodoluminescence spectroscopy (CLS) study of interdiffusion in a model system: AlGaAs/GaAs superlattices (SLs). These studies provide the most direct observation and identification of diffusion-associated deep levels thus far, including a first experimental indication of proposed[1,2] diffusion–associated (Si_{Ga}–V_{Ga}) complexes in Si–induced interdiffusion. We correlate these CLS results with those of Secondary Ion Mass Spectroscopy (SIMS) and C-V depth profiling (C-V).[7] Experimental details can be found elsewhere.[3,7] This methodology allows us to address the role that various defects play in the diffusion process, with and without dopants, determine defect energies, emission intensities and defect spatial distribution, and correlate them with atomic and dopant concentration profiles.

Most widespread observations of intrinsic and impurity induced layer disordering (IILD) of AlGaAs/GaAs heterostructures are[1]: (a) slow rate and high activation energies: 3–6 eV for Ga– and 4–10 eV for As– self–diffusions, (b) strongly dependent on surface annealing conditions; i.e., As overpressure increases Ga and reduces As self-diffusion rates, and (c) strong dopant effects; diffusion rates increase by orders of magnitude in doped materials. The response to surface conditions suggests diffusion mechanisms mediated by native defects whose concentrations are functions of the stoichiometry of the semiconductor.[1] It has been assumed that Al-Ga interdiffusion is enhanced on the As–rich (Ga–rich) side of the stoichiometry due to an increase in Column III vacancy (interstitial) concentration.[1] The impurity enhancement has been interpreted as a consequence of doping level and type, and not some other detailed atomistic nature of the dopant species.[2] The main factor is the presence of the dopant, which increases [V_{Ga}] equilibrium concentrations, and not its diffusion.[2] Our results partially disagree with these interpretations. We show that Al–Ga intrinsic interdiffusion is governed solely by V_{Ga}. Moreover, we show that there are contrasting differences in interdiffusion rates and patterns induced by different n–type dopants, and prove that the nature of the impurity plays a unique role, besides the dopant effect.

31

RESULTS AND DISCUSSION

Figure 1(a) shows CL spectra of $Al_{0.8}Ga_{0.2}As$/GaAs SLs (13 nm periods) annealed for 2 h at 855 °C and several As_4 pressures in evacuated spectrosil quartz ampoules. As expected, this processing causes the contamination of the outmost surface region of the samples with O, C and Si. SIMS measurements showed no other impurities present in the annealed specimens. Contaminant levels are usually below the intrinsic electron concentration at 855 °C. Previous SIMS measurements showed large variations in SL intermixing among these samples,[7] with Al–Ga interdiffusion being promoted from the surface and its rate scaling with As_4 pressure. The atomic interdiffusion correlates with changes in the electronic structure of the superlattice with anneals under different As_4 pressures. These changes are shown in Figure 1(a) by the increase in band gap (BG) energy and deep level emission with increase As_4 pressure. From the band gap energy shift the Al–Ga interdiffusion coefficient can be extracted.[7] In particular, for samples annealed under Ga–rich conditions (bottom spectrum) we observe no change in band gap energy and no deep level emissions; the observed 1.46 eV emission corresponds to band gap recombination in unaltered GaAs wells. Accordingly, SIMS measurements showed that there was no Al–Ga interdiffusion. These results show that both Ga antisites and interstitials, the point defects favored to be formed under Ga rich conditions,[6] do not mediate interdiffusion in AlGaAs/GaAs heterostructures.

Figure 1(b) illustrates in detail depth resolved CL of the $Al_{0.8}Ga_{0.2}As$/GaAs SL processed under 0.6 atm of As_4. There are intense deep level emissions at the 1.2–1.45 eV spectral region. These bands broaden and their relative intensity decreases with increasing excitation depth. The band gap emission has shifted to 1.8 eV, and it changes with excitation depth by about 30 meV. There is a weak, surface related,[3] deep level emission at 0.82 eV. All these features are common to all intrinsic processed samples, but they differ in the magnitude of band gap shift and in the relative intensity of deep level emissions. The electronic structure changes with depth correlate

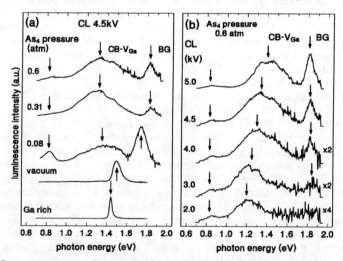

Figure 1. Room temperature CL spectra of $Al_{0.8}Ga_{0.2}As$/GaAs SLs annealed for 2 h at 855 °C. (a) at constant excitation depth for several As_4 pressures, and (b) as function of excitation depth for 0.6 atm As_4 pressure. For 1–5 keV electron beam energies excitation depths are about 30–300 nm; due to carrier mobility, recombination depths are larger by an estimated factor of 2 [8].

with the chemical composition changes shown by SIMS Al and Ga depth profiles of the processed SLs.[7] We associate the dominant 1.2–1.45 eV bands with transitions between the conduction band (CB) and Ga vacancy levels, CB–V_{Ga} (see below).[9] It appears that the optically detected Ga vacancies, the point defects favored to form under As rich conditions,[6] are the sole mediators of interdiffusion in AlGaAs/GaAs heterostructures.

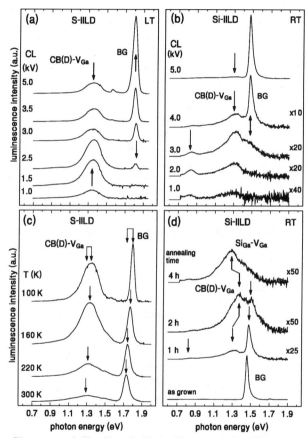

Figure 2. CL spectra of $Al_{0.8}Ga_{0.2}As$/GaAs SLs annealed at 855 °C and 1.2 atm As_4 pressure. (a) 0.5 h annealed S–IILD, and (b) 1 h annealed Si–IILD as function of excitation energy. CL at 4.0 keV: (c) as function of sample temperature for S–IILD, and (d) of Si–IILD as function of annealing time.

Figure 2(a) depicts CL spectra for S–IILD of $Al_{0.8}Ga_{0.2}As$/GaAs SL (10 nm periods) annealed for 0.5 h at 855 °C and 1.2 atm. of As_4 pressure. It shows band gap emission at 1.8 eV and deep level emission at 1.34 eV, assigned to a CB(Donor)–V_{Ga} transition, which shift with increasing excitation depth by –50 meV. These variations in the electronic structure of the processed material correlate with atomic and dopant inhomogeneities in depth observed by SIMS and C–V,[7] and shown in Figure 3(a). This figure shows good agreement between sulfur and electron concentrations at around 550 nm. This indicates a high percentage of sulfur atoms at electrically active substitutional As sites.[10]   On the other hand, the donor compensation observed at the near surface region is explained by an increase in V_{Ga} (acceptor) concentration, as shown by the 1.34 eV intensity profile in Figure 2(a).

Figure 2(b) depicts CL spectra for Si–IILD of $Al_{0.8}Ga_{0.2}As$/GaAs SLs (13 nm periods) annealed for 1 h at 855 °C and 1.2 atm. of As_4 pressure. The CL spectra of Si–IILD samples show dramatic changes with excitation depth. The bottom CL spectra, 1–3 keV, show deep level emissions at 1.3 eV and 0.82 eV, and an extremely weak emission at about 1.88 eV. This value is close to the band gap of the intermixed SL. With increasing excitation depth, the band gap emission of the unaltered superlattice is observed, whereas the deep level emission intensities decrease. For correlation purposes we also shows the SIMS data.[7] Figure 3 shows that for both Si– and Ge–IILD the interdiffusion profiles present sharp boundaries between the totally intermixed region and the unaltered region of the superlattice (S–IILD shows no sharp boundaries).[7] Note that the maximum doping level is about 3 times lower for Ge– than for Si–IILD, but the interdiffusion rate is about 3 times faster (this contradicts the predictions of the Fermi–level effect model).[2] The CL results indicate a dominant role for the defect associated with the 1.3 eV emission as a mediator for n-type IILD at AlGaAs/GaAs heterostructures.

We now characterize 1.15-1.45 eV emission bands to transitions involving CB(donors) and V_{Ga} states by studying the temperature and composition dependence of the deep level emission energies. Figure 2(c) shows CL spectra of S–IILD of AlGaAs/GaAs SLs carried out in the range 300–100 K. These measurements indicate that both band gap and 1.3 eV emissions shift from their 300 K values by approximately +60 meV with decreasing temperature. The identification of deep level transitions requires the knowledge of the "absolute" energy change of deep levels (DL), valence band (VB) and CB states (Γ band) with temperature or pressure. We use previous theoretical results for GaAs and AlAs deformation potentials,[11] which show small (large) variations for the VB (CB). Similar values are obtained for shallow donor and acceptor states, where the impurity potential is dominated by the long-range (Coulomb) term.[12,13] Deep levels are less sensitive to changes in temperature or pressure than the semiconductor bands,[13] because deep levels originate from the short-range central-cell defect potential, which is very strong within the effective radius of the defect.[13] Based on these theoretical results,[11,12,13] in Figure 4(a) we plot our calculations of the temperature dependence of the transitions. Both Γ–VB and

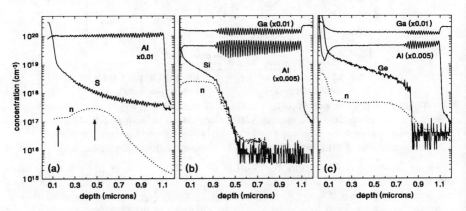

Figure 3: SIMS and C-V depth profiles of (a) S–, (b) Si– and (c) Ge-IILD of $Al_{0.8}Ga_{0.2}As$/GaAs heterostructures (all SLs 10 nm periods). Processing conditions: 0.5 h at 855 °C and 1.2 atm As_4 pressure.[7]

Γ–DL transitions have similar shifts with sample temperature. Thus, by correlating these with the experimental results (Figure 2(c)), we assign the 1.3 eV band to a transition between the CB(donor) and DL(acceptor) states, V_{Ga}, the defect favored to be formed under As rich conditions and n-type doping.

Figure 2(d) shows CL spectra of Si-IILD SLs annealed for increasing times. The deep level emission (and the SL intermixing) evolves with annealing time from 1.3 eV (1h, 1/3 intermixed) to 1.35 eV (2h, 2/3 intermixed) to 1.28 eV (4h, total intermixing). Concurrently, the concentration of in–diffused Si reaches a mid 10^{-19} cm^{-3} level across the superlattice. Figure 4(b) depicts our calculations of the change in transition energies caused by the change in CB, VB, and deep level energies (Si$_{Ga}$ and V$_{Ga}$) with increasing Al content. For Si doped GaAs the CB(D)–V$_{Ga}$ transition energy is 1.18 eV,[14] and the V$_{Ga}$

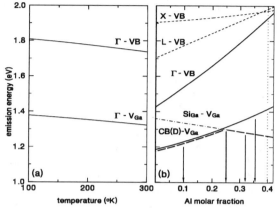

Figure 4. Schematic diagram of emission energies for Al$_x$Ga$_{1-x}$As as function of (a) temperature (x = 0.25), and (b) Al composition.

energy level lies 0.25 eV above the VB maximum. Using this value as reference and theoretical and experimental trends of the bands and V$_{Ga}$ energy positions with varying Al content,[11,13] we estimate a 0.25 eV range for the CB(D)–V$_{Ga}$.transition. Thus, the increase in CB–V$_{Ga}$ transition energy with increasing Al composition explains the observed deep level emission shift to higher energies with increasing annealing time. Also, the range of emission energies explains the previous assignment of the 1.15-1.45 eV emissions to CB–V$_{Ga}$ transitions for the intrinsic interdiffusion case.The CL spectrum of the Si–IILD 4 h annealed sample, top spectrum of Figure 2(d), shows the dominant deep level emission broadened and shifted to lower energies. A most likely explanation is the additional occurrence of transitions involving deep levels Si$_{Ga}$ and V$_{Ga}$ for Al indiffused wells and high Si concentrations, see Figure 4(b). With increasing Al composition, i.e., larger than 0.22, a large fraction of Si$_{Ga}$ (donors) becomes deep donor centers.[15] For Si and V$_{Ga}$ concentrations in the 10^{19} cm^{-3} range the average distance between randomly distributed centers is about 3 nm. Coulomb attraction between Si$_{Ga}$ donors and V$_{Ga}$ acceptors can make that distance shorter, making probable radiative transitions between these localized states.[14] These CL results arise as a first experimental indication of diffusion–associated (Si$_{Ga}$–V$_{Ga}$) complexes in Si–induced interdiffusion, and they strongly support the proposed[1,2] and recently theoretically assessed[6] role of donor–vacancy complexes as mediator of interdiffusion in III–V semiconductors under As–rich conditions.

For the range of electrochemical potentials addressed in this report the primary mediator of Al–Ga interdiffusion is V$_{Ga}$ and its complexes. These defects, which are created on the surface via a heterogeneous chemical reactions,[1] in–diffuse and create a concentration profile of available hopping sites for Al-Ga interdiffusion. This process is enhanced by n-type dopants, but this Fermi level effect does not fully explain the differences among different impurities.[16]

We propose that the relative electronegativity, χ, of the impurity with respect to the anion host determines the in–diffusion path of the impurity and subsequently, the particular Al–Ga interdiffusion pattern. Impurities with χ larger than the anion (χ_S, $\chi_{Se} > \chi_{As}$, Pauling's scale) will preferentially substitute for the anion and will in–diffuse through the As-sublattice, i.e., sulfur was shown to be substitutional solely in the As sublattice of GaAs.[10] As the Fermi level moves upwards the [V_{Ga}] increases, but impurity in–diffusion is limited by the availability of V_{As}.[7] This makes sulfur inefficient for IILD. An extreme case is that of carbon, which inhibits IILD.[17] With $\chi_C = 2.5$, carbon substitutes for As and is a p–type dopant in GaAs. Those impurities with χ smaller than the anion (χ_{Si}, $\chi_{Ge} < \chi_{As}$) in–diffuse through the Ga sublattice. In this case the Fermi level effect creates more sites for impurity in-diffusion, further increasing the dopant level, the [V_{Ga}], and favoring the formation of pair defect complexes which can fast in–diffuse.[1,2,6] The result is the total intermixing of the superlattice in regions where the impurities in–diffused, producing sharp boundaries between altered and unaltered regions of the SL.

To understand the differences in IILD within each particular dopant group, i.e., Si and Ge, further properties should be considered. For instance, the local bond energy is the main contribution to the activation energy for diffusion. Because of the openness of the zincblende structure there is no large resistance to hopping through, although, differences in atomic size may be of importance in the local tensile stress introduced to the lattice by the impurity atom. To clarify these points, theoretical comparative calculations of X–V_{Ga} complexes are necessary.

In conclusion, we have detected the diffusion–associated defects responsible for interdiffusion at AlGaAs/GaAs SLs. Our methodology strongly suggests these defects as V_{Ga} and its complexes, i.e., Si_{Ga}–V_{Ga}. The correlation of CL spectra with atomic interdiffusion and doping provides the most direct evidence of point defects associated with diffusion phenomena yet reported. We formulate a simple qualitative model for IILD which goes beyond the Fermi level effect by taking into account the properties of the in–diffusing impurities.

We thanks B. Olmsted and S. Houde-Walter for sample processing and chemical analysis.

REFERENCES

[1]   G.Deppe and N. Holonyak, Jr., J.Appl.Phys. 64, R93 (1988) and references therein.
[2]   Tan, U. Gosele, and S. Yu, Crit. Rev. Solid State Mater. Sci. 17, 47 (1991).
[3]   R. E. Viturro, B. L. Olmsted, S. N. Houde-Walter, and G. W. Wicks, J. Vac. Sci. Technol. B9, 2244 (1991) and references therein.
[4]   J–L Rouviere et al., Phys. Rev. Lett. 68, 2798 (1992).
[5]   P. Mei et al., J. Appl. Phys. 65, 2165 (1989).
[6]   J. E. Northrup and S. B. Zhang, Phys. Rev B 47, 6791 (1993).
[7]   Olmsted, S. N. Houde-Walter, and R. E. Viturro, Mat. Res. Soc. Symp. Proc. Vol. 240, 721 (1991), and Vol. 262, 867 (1992).
[8]   R. E. Viturro, to be published.
[9]   Chiang and G. L. Pearson, J. Appl. Phys. 46, 2986 (1975).
[10]   Battacharya, P. P. Pronko, ans S. C. Ling, Appl. Phys. Lett. 42, 880 (1983).
[11]   C. G. Van de Walle, Phys. Rev. B 39, 1871 (1989).
[12]   Hjalmarson, P. Vogl, D. J. Wolford, and J. D. Dow, Phys. Rev. Lett. 44, 810 (1980).
[13]   John. D. Dow et al. J. Electronic Materials 19, 829 (1990) and references therein
[14]   J. I. Pankove, Optical Processes in Semiconductors (Dover, New York, 1975) pp.141.
[15]   P. M. Mooney, J. Appl. Phys. 67, R1 (1990), D. J. Chadi and S. B. Zhang, J. Electronic Materials 20, 55 (1991).
[16]   We also observe the cubic dependence of the interdiffusion coefficient on free carrier concentration [1,2,4] for Si dopant levels up to roughly 10^{18} cm^{-3}, as shown by data extracted from Si and n diffusion front, Figure 4(b). The issue in question involves the whole range of impurity concentrations.
[17]   L.J. Guido et al., J. Appl. Phys. 67, 2179 (1990).

Vacancy Injection Enhanced Al-Ga Inter-diffusion in Si FIB Implanted Superlattice

P. Chen and A.J. Steckl
Nanoelectronics Laboratory
University of Cincinnati, Cincinnati, OH
45221-0030

Abstract

The Al-Ga inter-diffusion induced by Si FIB implantation and subsequent RTA were investigated in an $Al_{0.3}Ga_{0.7}As$/GaAs superlattice with equal 3.5 nm barrier and well widths. Si^{++} was accelerated to 50 kV as well as 100kV and implanted parallel to sample normal at doses ranging from 10^{13} to 10^{15}/cm^2. The effect of rapid thermal anneal of 10s at 950°C was characterized by SIMS technique. In the implanted region, the inter-diffusion as well as compositional mixing were significantly enhanced. An ion dose as low as 1×10^{14}/cm^2 results in a two-order of magnitude increase in the inter-diffusion coefficient, to a value of 4.5×10^{-14} cm^2/sec, in contrast to 1.3×10^{-16} cm^2/sec from RTA-only. This produces a mixing parameter of ~90%. A strong depth dependence of the mixing process was observed at implantation energy of 100keV with a pinch-off region being formed at certain depth. It is noticed that the depth where this enhancement occurred is not associated with either the maximum concentration of Si ions or of vacancies. Instead, it represents the positive maximum of the second derivative of the vacancy profile, which in turn represents the vacancy injection generated by presence of a transient vacancy concentration gradient. Based on this, a theoretical model was developed using vacancy injection as responsible for mixing.

Introduction

The compositional mixing in $Al_{1-x}Ga_xAs$/$Al_{1-y}Ga_yAs$ superlattice (SL) structures can be significantly enhanced by implantation of certain species such as Si [1-4]. In the mixed region, the energy band gap as well as the index of refraction are changed. Therefore, novel optical devices such as DFB or DBR lasers, channel waveguides and quantum wires can be fabricated by locally inducing the mixing in SL structure. Focused ion beam (FIB) implantation has been especially attractive in these applications since it provides a maskless and resistless process.

Many mixing applications require a small lateral extent. Therefore, the minimization of lateral spreading of the mixed region has to be seriously considered. In general, the lateral resolution is limited by the resolution of pattern transfer technology, the lateral profile and straggling of the ions in the solid and lateral diffusion during the post-implantation annealing. High spatial resolution can be achieved by: (a) employing short-period superlattices, which are more readily mixed; (b) low dose implantation, which has relatively smaller lateral spread; (c) rapid thermal annealing with a minimal thermal budget. On the other hand, by precisely controlling the depth of the mixed regions, one can fabricate a gain-coupled DFB superlattice laser which has a number of advantages over conventional AR-coated DFB laser [6] . Understanding of the mixing mechanism is essential to precisely control the mixing process in both horizontal and vertical directions.

The compositional mixing takes place during Al-Ga inter-diffusion between AlGaAs and GaAs layers. The inter-diffusion is known to proceed through point defects. For n-type dopant Si, the Fermi-level effect theory is widely accepted under thermal equilibrium condition. In this case, the thermal equilibrium concentration of the charged point defect is enhanced by doping and consequently the presence of the dopant is of importance for enhancing mixing [7]. The diffusion of Si impurities was also suggested to be responsible for the enhancement[2,3]. However, in a nonequilibrium process, such as post-implantation rapid thermal annealing, neither the Fermi-

37

level effect nor the impurity diffusion model is effective[8-12]. This is due to the fact that thermal equilibrium is not attained and the Si diffusion is negligible in the time frame of RTA. Meanwhile, since a vacancy gradient is introduced by ion implantation, the vacancies diffuse significantly during rapid thermal process. The motion of these nonequilibrium point defects was found to play a key role in inter-diffusion enhancement.

Experimental Procedure

The superlattice samples used in this study were grown by molecular beam epitaxy . The substrate was (100) GaAs. A GaAs buffer layer was grown first, followed by a $1\mu m$ $Al_{0.3}Ga_{0.7}As$ cladding layer , a single 30nm GaAs quantum well, a superlattice stack of 29 periods consisting of 3.5 nm GaAs wells and 3.5 nm $Al_{0.3}Ga_{0.7}As$ barriers , and a 50nm $Al_{0.3}Ga_{0.7}As$ cap.

FIB implantation was performed with a system capable of 150kV acceleration[13]. $280\mu m$ x $280\mu m$ square patterns were implanted. Si^{++} ions were accelerated to 50 and 100kV and implanted parallel to sample normal at doses ranging from 10^{13} to 3×10^{14} cm^{-2}. The Si^{++} focused ion beam had a current of 25pA and a beam diameter of ~80nm. The subsequent rapid thermal annealing was carried out at 950°C for 10 seconds, which was previously established[1] to provide a suitable thermal process for mixing while preserving the structure of the unimplanted superlattice. Secondary ion mass spectroscopy (SIMS) was employed to precisely measure the atomic composition of Al and Si as a function of depth.

Results and Discussion

SIMS depth profiles of the superlattice after 100keV implantation and subsequent annealing are shown in Fig.1. The mixing generated by 100keV implantation is enhanced toward a point roughly in the middle of superlattice stack. On the other hand, 200keV Si implantation, for which the R_p is deeper than the end of superlattice stack, induced uniform mixing throughout the entire superlattice depth. The degree of mixing is represented by the Al peak-valley ratios and the inter-diffusion coefficient can be calculated from this ratio using error function model. For these uniformly mixed samples, we have previously presented a quantitative analysis of the SIMS depth profiles[8]. An ion dose as low as 3×10^{13}cm^{-2} is capable of producing ~70% mixing. A dose of 1×10^{14}/cm^2 yields a mixing parameter of ~90% and also results in a two-order of magnitude increase in the diffusion coefficient, to a value of ~4.5×10^{-14} cm^2/sec, in contrast to 1.3×10^{-16} cm^2/sec from RTA-only.

For 100keV implanted samples (shown in Fig.1), the average degree of mixing over entire superlattice is still proportional to dose. However, the mixing at depth from ~150nm to 220nm is much more effective than elsewhere in the superlattice. With various doses, the peak-valley ratio in this region is significantly decreased and eventually a complete pinched-off is formed at depth of 170nm(from Fig.1(b)). In the following discussion, we will refer this mostly mixed region (150nm to 220nm) as the "pinch-off" region. By comparing the Si depth profiles with and without RTA, negligible Si impurity diffusion was observed, indicating that impurity diffusion is insignificant in SL mixing in the RTA time frame, unlike the case of furnace annealing. It is important to note that the maximum mixing (pinch-off point and region) takes place at a depth beyond the projected range of 100kV Si ions (R_p~100nm) (see Fig.2). This fact suggests that the degree of mixing is no longer dominated by either impurity concentration or the vacancy concentration as described in the Fermi-level effect model.

To explore the mechanism of the enhanced mixing in the pinch-off region, the depth distribution of Si and vacancy concentrations were simulated using TRIM[14]. Si depth profiles computed by TRIM were in good agreement with those measured by SIMS. The depth profile of the vacancy concentration is shown in Fig.2. As with the Si concentration, the position of maximum vacancy concentration is much shallower than the pinch-off region. The inter-diffusion coefficient as a function of depth was calculated and is shown Fig.4.

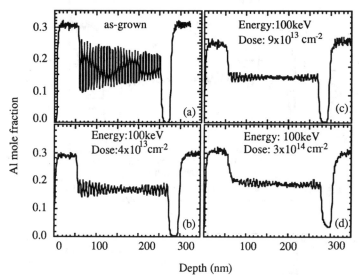

Fig.1 SIMS depth profiles of Al in superlattices implanted by 50kV Si $^{++}$ at different doses followed by rapid thermal annealing at 950°C for 10 seconds.

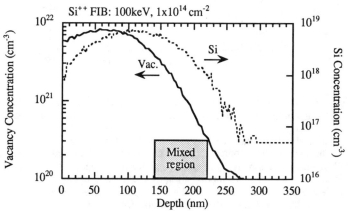

Fig.2 Depth distribution of Si and vacancy concentration after Si implantation

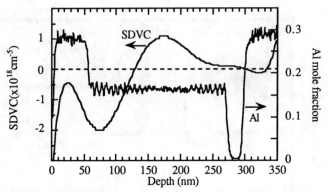

Fig.3 Depth profiles of Al mole fraction of superlattice and second
derivative of vacancy concentration (SDVC)

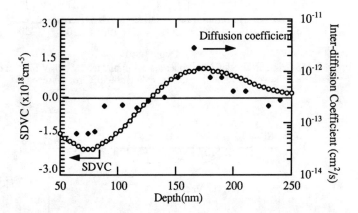

Fig.4 Inter-diffusion and second derivative of vacancy concentration (SDVC) as a
function of depth. Si ion energy: 100keV, Dose: $1 \times 10^{14} cm^{-2}$, RTA: 950°C,10s.

A similar depth-dependence was reported [9-12] for different SL structures and RTA
conditions. The contrast in Al composition between mixed and unmixed region is usually more
dramatic SL features longer period. For example, in the result reported by Lee *et al.*[9] which is
reproduced in Fig.5, the on-set of mixing abruptly occurs in a region which is deeper than the
R_p. A transient-enhanced interdiffusion theory was established by Kahen, *et al.*[9-12] for
explaining the mechanism of inter-diffusion in the time frame of RTA. In this theory, the vacancy
diffusion was calculated and the assumption of $D_{Al} = D_{(Al,eq)} (C_v / C_{v,eq})$ presumes that excess
vacancies are responsible for the enhancement in inter-diffusion. D_{Al}, $D_{(Al,eq)}$, C_v and $C_{(v,eq)}$ are
the transient Al diffusion coefficient, thermal equilibrium Al diffusion coefficient, transient
vacancy concentration and thermal equilibrium vacancy concentration, respectively. However,
according to this assumption, the maximum inter-diffusion should coincide with the peak of the
vacancy concentration, which contradicts the experimental results reported by Lee *et al.*[9] as well
as those reported in this paper.

The calculation of the second derivative of the vacancy concentration, SDVC=$(\partial^2 C_v(x,t)/\partial x^2)$, was performed. The SDVC depth distribution and inter-diffusion coefficient are shown in Fig.3 and 4 for a dose of 1×10^{14}cm^{-2}. From Fig.3, one can clearly see that the pinch-off mixing only takes place in the region where SDVC is positive and that the pinch-off point coincides to the maximum SDVC value. In another aspect, Fig.4 shows that inter-diffusion is greatly enhanced by a positive SDVC. According to Fick's second law of diffusion, the vacancy concentration time- and space-dependence are interrelated:

$$\frac{\partial C_v(x,t)}{\partial t} = D_v \frac{\partial^2 C_v(x,t)}{\partial x^2}. \tag{1}$$

A positive SDVC results in a positive $\partial C_v(x,t)/\partial t$ which indicates that vacancies are injected into the region. Therefore, we conclude that the Al-Ga inter-diffusion is enhanced by the vacancy injection. Through curve fitting, an exponential form of the inter-diffusion coefficient is derived as a function of SDVC. Hence, in the time frame of RTA, we suggest that the Al-Ga inter-diffusion coefficient can be written as:

$$D_{Al} = D_{(Al,eq)} \cdot \beta \cdot e^{[\alpha \cdot (\partial^2 C_V / \partial x^2)]}, \tag{2}$$

where α and β are constants and the exponential term represents the enhanced inter-diffusion effect of vacancy injection. Equation (2) suggests that in the time scale of RTA, the Al-Ga inter-diffusion is greatly affected by the motion of vacancies that occurs during the early period of annealing due to the implantation-induced gradient. In this period, nonequilibrium processes take place, the inter-diffusion proceeds through charged vacancies as usual and is further enhanced by vacancy injection in the region. The vacancy injection greatly enhances the Al-Ga inter-diffusion and becomes a decisive factor in the short period of annealing. However, this vacancy injection process is saturated in a few seconds[9] because the vacancy thermal equilibrium is established. Therefore, the mixing usually does not increase appreciably as the annealing time is increased beyond this point [5,9]. When the annealing time is long enough (as in furnace annealing), the mixing process will be dominated by impurity concentration as well as impurity diffusion as described in the Fermi-level effect and impurity diffusion models.

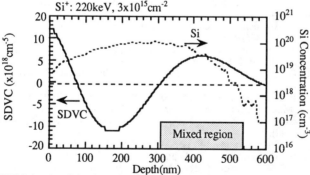

Fig.5 TRIM simulated depth profiles of Si concentration and SDVC for the Si+ implantation with 220keV at dose of 3×10^{15}cm^{-2} in GaAs/AlGaAs superlattice(period: 20nm+20nm). The thickness of SL stack is 1.6μm and complete mixing occurs in the marked region. The experiment and SIMS profiles were published by Lee et al. [9].

As a verification, we have calculated the impurity and vacancy distribution of a different superlattice structure and implantation condition, in which a strongly depth-dependent mixing was also observed[5, 9-12]. The results are shown in Fig.5. Since the period of superlattice used here was quite thick (40nm), the required dose for mixing was considerably higher. The energy of implanted Si+ was 220keV, which corresponds to a R_p of 280nm. Rapid thermal annealing was performed subsequently. Similarly to the results we reported above, the mixing only occurred from depth of ~320nm-520nm, in other words beyond the peak of both Si and vacancy concentrations. The position of the mixed region also coincides with the positive SDVC. This result strongly supports the vacancy injection mechanism we are proposing.

Summary and Acknowledgment

In summary, we have studied the Si++ FIB-induced mixing of an $Al_{0.3}Ga_{0.7}As$/GaAs superlattice structure. A fairly complete mixing can be achieved with relatively low dose at a certain depth. This result can be utilized to minimize lateral spread, reduce damage and achieve depth control. The mechanism for the depth-dependence and pinch-off mixing effect were discussed. Nonequilibrium vacancy injection has been suggested to be responsible for the high degree of mixing in the pinch-off region. Based on experimental results, a theoretical model has been proposed to describe the inter-diffusion as well as the mixing process in the time scale of RTA.

The authors would like to thank R. Kolbas for the MBE growth, S. Novak for SIMS measurement and A.G. Choo, J.T. Boyd and H.E. Jackson for many useful discussion. The partial support for this work from the National Science Foundation and the Wright-Patterson Air Force Base is acknowledged.

References

[1]   A. Steckl, P. Chen, A. Choo, H.Jackson, J. Boyd, P. Pronko, A. Ezis and R. Kolbas, Mat. Res. Soc. Symp. Proc. **240B**, 703(1992).

[2]   J. Kobayashi, M. Nakajima, Y. Bamba, T. Fukunaga, K. Matsui, K. Ishida, H. Nakashima and Koichi Ishida, Jpn J. Appl. Phys. **25**, L385(1986).

[3]   K. Matsui, J. Kobayashi, T. Fukunaga, Koichi Ishida and H. Nakashima, Jpn J. Appl. Phys. **25**, 651(1986).

[4]   S. Lee, G. Braunstein, P. Fellinger, K.B. Kahen and G. Rajeswaran, Appl. Phys. Lett. **53**, 2531(1988).

[5]   K. Ishida, E. Miyauchi, T. Morita, T. Takamori, T. Fukunaga, H. Hashimoto, Jpn. J. Appl. Phys. **26**, 285(1987).

[6]   W. Tsang, F. Choa, M. Wu, Y. Chen, R. Logan, S. Chu, A. Sergent, and C. Burrus Appl. Phys. Lett. **60**, 2580(1992).

[7]   T. Tan, U. Gosele and S. Yu, Critical Reviews in Solid State and Materials Sciences **17**, 47(1991).

[8]   A. Steckl, P. Chen, A. Choo, H. Jackson, J. Boyd, A. Ezis, P. Pronko, S. Novak and R. Kolbas, Mat. Res. Soc. Symp. Proc. **281**, 319 (1993)

[9].   S. Lee, G. Braunstein, P. Fellinger, K. Kahen and G. Rajeswaran, Appl. Phys. Lett. **53**, 2531(1988).

[10].   K. Kahen, G. Rajeswaran and S. Lee, Appl. Phys. Lett. **53**, 1635(1988)

[11].   K. Kahen and G. Rajeswaran, J. Appl. Phys. **66**, 545(1989).

[12].   S. Schwarz, T. Venkatesan, R. Bhat, M. Koza, H. Yoon, Y. Arakawa and P. Mei, Mat. Res. Soc. Symp. Proc. **56**, 321(1986).

[13]   N. Parker, W. Robinson and J. Snyder, SPIE **632**, 76(1986).

[14]   J. Ziegler, J. Biersack and U. Littmark, "The stopping and range of ions in solids", Pergamon Press (1985).

EFFECTS OF DEFECTS IN GaAs/AlGaAs QUANTUM WELLS

O.L.RUSSO[*], V.REHN[**], T,W.NEE[**] AND K.A.DUMAS[***]
* New Jersey Institute of Technology, Department of
Physics,Newark, NJ 07102
** Naval Air Warfare Center, Michelson Laboratory, China Lake,
CA. 93555
*** Jet Propulsion Laboratory, California Institute of
Technology, Pasadena, CA. 91109

ABSTRACT

High Resolution transmission electron microscopy (HRTEM) and
electroreflectance (ER) were used to explain the role of point
defects in the molecular beam epitaxy (MBE) grown PIN structure
containing five coupled (50Å/28Å) GaAs/Al$_x$Ga$_{1-x}$As quantum wells
with x = 0.25. The ER data were taken at 300K and 77K for
energies from 1.4 to 2.1 eV from which sub-band energy
transitions were determined. Data at 300K showed three
transitions whereas four were readily resolved at 77K. HRTEM data
determined the uniformity of both the wells and barriers to be
within ± 2Å, which neither caused appreciable broadening nor a
decrease in the transition probability. However, the data at
different temperatures suggest that point defects may be
responsible for the decrease in the transition probability.

INTRODUCTION

The bulk optical properties of compound semiconductors such as
GaAs and AlGaAs have received enormous attention and are well
known. Many hetero and quantum well structures, however, are
designed with a major focus on minimizing size. This emphasis on
scaling results in dimensions small enough so that quantum
effects begin to play an important role. The quantum confinement
of carriers evident in superlattices and quantum wells creates
sub band energies in both the valence and conduction bands not
observed in larger scaled structures [1-4]. Excitonic transitions
are enhanced and observed more readily even at room temperature,
largely due to the constraints of the exciton imposed by the well
size [5]. While the exciton levels and the sub band energies for
an ideal quantum or series of quantum wells can be determined
theoretically rather simply, real models add complexity to the
calculations. In particular, defects when introduced into the
hamiltonian play a significant role in establishing whether
certain energy states are realized.
We have studied, and report here, the sub-band energies in a
finite (five period) communicating quantum well structure. The
GaAs/Al$_x$Ga$_{1-x}$As molecular beam epitaxy (MBE) grown device, a PIN
configuration with a nominal value of x = 0.25, results in a
conduction barrier height of 198 meV and 132 meV for the hole
barrier height, using the 60-40 percent rule for the band offset
[6]. The electron effective mass ratio of 0.067 and heavy and
light hole effective masses of 0.45 and 0.088 respectively for
GaAs were used for the analytical calculations. The barrier
heights limit the allowed bound transitions. We have identified
only three of these transitions at 300K and four at 77K and
attribute the inconsistency to the introduction of growth defects

43

such as interfacial roughness or point defects. Our data suggest a possible qualitative explanation of the influence of interfacial roughness and point defects on the bound exciton energy levels.

The data used to determine the direct optical transitions were obtained by using electroreflectance (ER) [7-9] in the spectral region of photon energies, E, between 1.4 to 2.1eV. The spectra are given in terms of the normalized change in reflectance with energy, $\delta R/R$, where the reflectance $R = R(E)$. The transition energies are determined from the $\delta R/R$ spectra by using curve fitting of a modified Lorentz line shape given by [10,11]

$$\delta R/R = Re \ [Ce^{i\theta} \ (E - E' + i\Gamma)^k \qquad (1)$$

where Re is the real part of the function, C is a constant which determines the amplitude, θ is the phase angle, E', the transition energy between the conduction and valence sub-bands, and Γ is the lifetime broadening parameter. For an excitonic transition, the value of k = 2 [10].

The quantum mechanical solution of a particle in a finite isolated well differs from the multiple quantum well solution when the barrier widths become small. The values of the transitions we present are based on a calculation of five communicating quantum wells. The fundamental structure consisted of five 50.9 Å GaAs wells, four 28.3 Å barriers of $Al_{0.25}Ga_{0.75}As$ with the first and fifth wells bounded by 150 Å of $Al_{0.25}Ga_{0.75}As$. Solutions indicate that only energy levels corresponding to n = 1 are allowed. Because of the built-in field and the applied field of ER, the selection rules $\delta n = 0$ are relaxed, allowing transitions from the n = 2 valence to the n = 1 conduction states. However, our allocation of the 60-40 rule to the band offset instead of the 85-15 rule [4] precludes n = 2 states as allowed states. In addition, the intensity of the transition is determined by the value of the overlap integral [12] in which the wave functions are those of the valence and conduction sub-bands. Application of the overlap integral indicates that five transitions with $\delta n = 0$ are most probable.

The presence of extended defects such as surface roughnes and irregularities have been analyzed by using high resolution transmission electron micrographs (HRTEM). Variations in the well and barrier widths are measured and then used to determine the new energy levels

DISCUSSION AND RESULTS

The results of ER are interpreted on the basis of variations in the well and barrier widths as determined by HRTEM. The usual assignment of the energy level transitions from the $\delta R/R$ spectra is by curve fitting of eq. (1) near the transition energy E_i'. The spectra at 300K for photon energies from 1.4 to 2.1 eV are shown in figure 1. Calculations show that only photon energies up to 1570 meV are allowed, which agrees with the experimental data. The spectra beyond this energy are caused by bulk excitations which are broad in nature when compared with the consistently narrower excitons in the well which are about 10 meV. The spectra in a smaller range but expanded so as to illustrate the features of the transitions are shown in figure 2 and this is compared with figure 3 which shows the spectra at 77K.

A comparison of figures 2 and 3 shows several apparent features: (1) The dominant structures in the middle of the range are to one another and each has a width of about 10 meV regardless of the temperature. (2) The exciton (peak) at the GaAs edge has moved from 1.430 eV at 300K to 1.505 meV at 77K. (3) The exciton at 77K is more pronounced than that at 300K (the change in sign is due to a phase change in the angle θ of equation (1)). (4) The exciton at the GaAs edge is wider than the excitons within the well, and the transitions increase in width as the energy gets larger (i.e. for E > 1.57 eV at 300K and for about E > 1.65 eV at 77K).

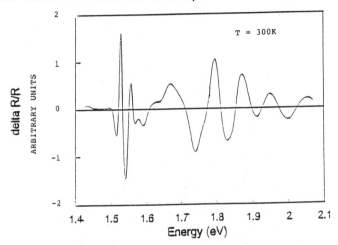

Fig.1   Electroreflectance (ER) spectra for the PIN five well GaAs/AlGaAs structure at 300K.

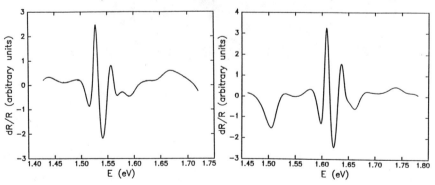

Fig.2   ER spectra emphasizing exciton energy region at 300K.

Fig.3   ER spectra emphasizing exciton energy region at 77K.

A fit of equation (1) to the curves of figures 2 and 3 yields transition energies which are given in Table 1. Although four of the transitions are resolved at 77K, only three are determined with confidence at 300K. The transition at 1.50 eV for 300K is weak and broad so that a good fit is unlikely, but the lowest transition at 77K is easily resolved and found to be 1576 meV.

Extended defects such as surface irregularities, roughness or possibly point defects could lead to boadening of the lines or change the wavefunction which would affect the transition probability. To determine the quality of the quantum well structure, high resolution transmission electron microscopy (HRTEM) was used. The micrographs are shown in figures 4 and 5. The darker broader regions, the GaAs, and the lighter, the AlGaAs are both seen to be well defined and uniform. Measurements made by counting the interference fringes, each of 2.8 Å, over an extended range, resulted in an average value of 30 ± 2 Å for the barriers and 50 ± 2 Å for the wells. To show that some of the localized dark spots are artifacts of the sample preparation, the sample, grown in the [100] direction, was cleaved along the (110) plane and viewed along the [010] direction for confirmation. An HRTEM of the cleaved sample is not shown. Calculations for the five wells show that the broadening of the energy levels is not significant for the values obtained using the HRTEM data.

Fig. 4   High Resolution TEM for a five well GaAs/Al$_{0.25}$Ga$_{0.75}$As structure.

The values of the transition energies at 77K and 300K as determined from figures 2 and 3 using equation (1) are given in TABLE 1. The results show that the transitions which are strong are well resolved for both 77K and 300K. The fact that the lowest transition at 77K is resolved and that at 300K is not suggests that the cause may be due to point defects which have a lower density at the lower temperature. The effect is indeed small, but, large enough to preclude an accurate assignment to the transition.

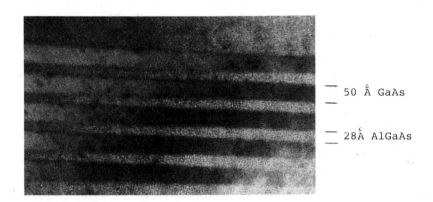

— 50 Å GaAs

— 28Å AlGaAs

Fig. 5 HRTEM microgragh for the five MBE GaAs/$Al_{0.25}Ga_{0.75}As$ quantum well structure.

TABLE 1

MEASURED ENERGIES FROM FIGURES 2 AND 3 USING EQUATION (1)

E_i' (meV)

	E_1'	E_2'	E_3'	E_4'	E_g *
300K		1526	1540	1557	1430
77K	1576	1608	1625	1640	1505

* E_g is the energy of the GaAs band edge.

The calculated values of E_i' were in the range from 1461 to 1558 meV at 300K and from 1548 to 1638 meV at 77K.

CONCLUSION

The measured values of the E_i' transitions for both 77K and 300K agreed within reason with those calculated for five communicating wells. The 60-40 % allocation to the band offset conforms with the spectrum which shows that the transitions beyond bound states are broad and not characteristic of quantum confined excitons. Although the measured values are reasonable, there are a number of ways in which errors can be introduced in the calculated energy levels. The value of x in $Al_xGa_{1-x}As$ affects the barrier heights in both the valence and conduction sub-bands so should be known accurately. The 60-40 % allocation assumed, if not a sufficiently precise model, can cause large variations in the energies when the barriers or wells are narrow. There is a significant degree of uncertainty, in particular in the value of the heavy hole mass [13].

The results of both the HRTEM and the ER studies show that neither the energy levels nor the broadening are significantly affected by the ± 2Å variation in the well or the barrier. The results at 77K shows a transition at 71 meV from the GaAs band edge, while the corresponding 300K transition, which should be at the same energy from the band edge, is not observed. Because the exciton line width is not temperature dependent, it's energy level is readily identifiable, provided the overlap integral defining the strength of the transition is not too small. One of the factors affecting the overlap integral is the temperature-dependent point defect density because it directly affects the wave function. On this basis, it appears that there is a suggestion at least that a high point defect density could obscure an allowed weak transition and not significantly affect transitions which have a large overlap integral.

REFERENCES

1. L.Esaki and R.Tsu, IBM J. Res. Dev. 14, 61 (1970).
2. M.W.Cole, Rev. Mod. Phys. 46, 451 (1974).
3. R.Dingle, W.Wiegmann, and C.H.Henry,Phys.Rev Lett.24, 593 (1974).
4. R.Dingle, Festkorperprobleme, Advances in Solid State Physics, edited by H J. Queisser (Pergamon-Vieweg) XV, 21 (1980).
5. D.A.B.Miller,D.S.Chemla, T.C.Damen, A.C.Gossard, W.Wiegmann, T.H.Wood and C.A.Burrus, Phys. Rev. Lett. 53, 2173 (1984).
6. R.C.Miller, D.A.Kleinman and A.C.Gossard, Phys. Rev.B 29, 7085 (1984).
7. M.Cardona, Solid State Phys, Suppl. 11, ed. by F.Seitz D.Turnbull and H.Ehrenreich (Academic, New York, 1969).
8. Semiconductors and Semimetals, Vol. 9, edited by R.L. Willardson and A.C.Beer (Academic, New York, 1972).
9. D.E.Aspnes, in Handbook on Semiconductors, Vol.2, Optical Properties of Solids, edited by M. Balkanski (North Holland, New York, 1980).
10 D.E.Aspnes, Surf. Sci. 37, 418 (1973).
11 D.E.Aspnes and J.E.Rowe, Phys. Rev. Lett. 27, 188 (1971)
12 See for example, E. Merzbacher, Quantum Mechanics, 2nd Ed. (John Wiley & Sons, New York, 1961). p.79.
13 M.S.Skolnick, A.K.Jain, R.A.Stradling,J.Leotin, J.C.Ousset and S.Askenazy, J.Phys.C: Sol. State Phys. 9, 2809, (1976).

EFFECTS OF IMPURITIES ON THE OPTICAL PROPERTIES
OF QUANTUM DOTS, WIRES, AND MULTIPLE WELLS

GARNETT W. BRYANT
Microphotonic Devices Branch, US Army Research Laboratory, Adelphi, MD 20783

ABSTRACT

Identifying and understanding the effects of impurities and defects in quantum dots, wires, and multiple wells is important for the development of nanostructures with good optical properties. Simple model calculations are presented to show when and how shallow impurities affect the radiative recombination of confined electron-hole pairs. Results for nanostructures are compared with results for bulk systems. Qualitative differences between bulk and confined systems are described.

INTRODUCTION

The optics of excitons in confined semiconductor systems has been studied intensively in recent years to determine how the optical properties of these systems can be tailored and enhanced [1]. In wells and wires, confinement enhances the electron-hole binding, increasing the pair overlap and oscillator strength. In dots, confinement concentrates the oscillator strength at discrete energies. In coupled multiple-well systems, switching is achieved by controlling the interwell tunneling. Identifying and understanding the effects of impurities and defects in quantum dots, wires and multiple-well systems is important for the development of nanostructures with good optical properties. Impurities and defects, especially those near the boundaries of dots and wires, provide nonradiative traps that can completely degrade the optical response of these systems.

To understand the effects of impurities on the optical properties of quantum nanostructures, one must understand exciton states in these systems. In the limit of *complete* quantum confinement, electron-hole correlation in confined dimensions is quenched, so the electron and hole occupy independent single-particle states of the confinement potential [2]. In the bulk limit, the exciton has free center-of-mass motion and internal excitations, determined by the pair reduced mass. A simple understanding is possible in each limit because the exciton behaves as a composite of two *independent* particles. Neither simple picture provides a good understanding of excitons confined in systems such as thin films [3], wide quantum wells, finite superlattices and multiple coupled wells [4], microfabricated quantum wires and dots [5,6], and large nanocrystallites and clusters [7,8]. These systems are typically too large to be quantum confined in every confined dimension but too small to be bulk systems. In these intermediate-dimensional systems, confinement strongly couples the center-of-mass and relative motion and correlation strongly couples the electron and hole motion. The confined, intermediate-dimensional exciton is a composite of two *strongly coupled* particles that exhibits nonmonotonic, nonuniversal dependences on electron and hole masses and on electron-hole interaction, which are not expected from intuition about quantum-confined or bulk excitons [9].

Confinement acts as a large defect that localizes the electron and hole globally. Pair interaction correlates the pair motion. Impurities or defects trap the electron and hole locally. All three effects compete to determine exciton states and must be accounted for. In this paper, I present simple model calculations to show qualitatively when and how shallow impurities affect the radiative recombination of confined electron-hole pairs. A simple model is used so that insight can be obtained that is applicable in general to confined nanostructures. The model is described in the next section. Results are first presented for confined intermediate-dimensional excitons as a function of confinement energies and interaction strengths to demonstrate the nonuniversal behavior of confined excitons. Results are then presented for the effects of impurities on confined intermediate-dimensional excitons. I use these results to clearly identify the effects that must be included in realistic models for impurity effects on confined excitons.

Mat. Res. Soc. Symp. Proc. Vol. 325. ©1994 Materials Research Society

THE MODEL

Hubbard models are often used to describe interacting many-body systems on a lattice of sites. In this paper I use a Hubbard model to describe a single confined exciton interacting with an impurity. The Hubbard Hamiltonian for a single interacting electron-hole pair on a lattice of sites with an impurity at one of the sites is

$$H = \sum_{i,j}^{N} \left(U_e \, c_i^* \, c_j + U_h \, d_i^* \, d_j \right) + \sum_{i}^{N} E_x \, c_i^* \, c_i \, d_i^* \, d_i + V_e \, c_n^* \, c_n + V_h \, d_n^* \, d_n . \tag{1}$$

The system has N sites. Here *site* is used loosely. A site could be a well or dot in a finite superlattice of wells or dots [4,10], an atomic site in a polymer chain, or a fictitious site defined by a local basis used to represent states in a dot or nanocrystallite. Only nearest-neighbor electron (hole) hopping, U_e (U_h), is included. The electron (hole) annihilation operator at site i is c_i (d_i). The on-site electron-hole interaction is E_x. Longer range electron-hole interactions and electron-electron and hole-hole interaction in multiexciton systems can also be included [9]. The impurity interaction is assumed to be on-site. The impurity is located at site n. For simplicity, linear chains are considered here. The results for finite two-dimensional arrays of sites are qualitatively similar [9]. Confined systems are modeled by systems with terminated ends. Systems modeled with periodic boundary conditions are also considered. As shown below, excitons in systems with periodic boundary conditions exhibit the same universal behavior as bulk excitons. One-exciton wavefunctions and energies are determined by diagonalization of Eq. (1). The nonuniversal, nonmonotonic behavior of confined excitons is most clearly revealed by properties that depend on the exciton wave function. To understand this behavior, electron and hole distributions, electron-hole correlation functions, electron-hole separation, and oscillator strengths for transitions from the no-pair state to lth one-exciton state,

$$O_l = | < l \,|\, \sum_{i}^{N} c_i^* \, d_i^* \,|\, 0 > |^2 , \tag{2}$$

are determined.

Several special cases can be treated analytically to provide key insight [9]. Here I present the final results. To understand why excitons in systems with periodic boundary conditions exhibit bulk universal behavior, consider, for simplicity, an N-site linear chain with periodic boundary conditions (a ring). For a ring, Eq. (1) can be rewritten in an electron and hole plane wave basis. By transforming the plane wave basis to the center-of-mass momentum and relative momentum basis ($K = (k_e + k_h) / 2$ and $k = (k_e - k_h) / 2$), I obtain the $K = 0$ Hamiltonian, which describes the ground state exciton,

$$H_{K=0} = \sum_{k} 2 \, (U_e + U_h) \, \hat{\cos}(k) \, \alpha_k^* \, \alpha_k + \sum_{k_1,k_2} (E_x/N) \, \alpha_{k_1}^* \, \alpha_{k_2} , \tag{3}$$

where $\alpha_k = d_{-k} \, c_k$ is the annihilation operator for a pair with $k_e = -k_h = k$. The properties of the $K = 0$ exciton in a ring scale with $E_x / (U_e + U_h)$. For a bulk exciton, $U_e + U_h$ is proportional to the exciton inverse reduced mass, so the $K = 0$ ring exciton has the same scaling properties as a bulk exciton. The properties of bulk excitons depend only on one ratio, $E_x / (U_e + U_h)$. The properties of confined excitons depend separately on the electron and hole hopping rates and the interaction strength.

The behavior of intermediate-dimensional excitons in the limit of strong interaction can be determined by the use of second-order degenerate perturbation theory. The Hamiltonian for an N-site, terminated chain to second order in the hopping is

$$H = \sum_{i,j}^{N} U_{ex}\, \alpha_i^* \alpha_j + \sum_i^N V_i\, \alpha_i^* \alpha_i , \qquad (4)$$

where $\alpha_i = c_i\, d_i$ is the on-site pair annihilation operator, and only nearest neighbor hopping is allowed. U_{ex} is the nearest-neighbor pair hopping energy, $U_{ex} = 2U_e U_h / E_x$. V_i is the correlation energy at site i due to single-particle hopping to adjacent sites. For internal sites of the chain ($i = 2,..., N-1$), $V_i = V = 2(U_e{}^2 + U_h{}^2)/E_x$. Hopping can occur in only one direction away from chain ends, so $V_1 = V_N = V / 2$. Suppression of the correlation at the chain ends induces a potential V_i, which creates a dead layer for a tightly bound pair at the chain ends. Exclusion from the ends becomes complete if $U_{ex} = 0$.

INTERMEDIATE-DIMENSIONAL EXCITONS

To characterize the effects of confinement and electron-hole interaction on intermediate dimensional excitons, I discuss the energies, oscillator strengths, and charge distributions for the one-exciton ground states of 10-site chains and rings. Results are presented for equal electron and hole hopping ($U_h / U_e = 1$), intermediate asymmetry in the electron and hole hopping ($U_h / U_e = 0.1$) and large asymmetry in the electron and hole hopping ($U_h / U_e = 0.001$). Results are presented as a function of E_x / U_e to show the transition from noninteracting, independent-particle pair states ($E_x / U_e = 0$) to strongly correlated Frenkel-like excitons ($|E_x / U_e| \gg 1$).

Exciton ground-state energies for 10-site rings and terminated chains are nearly identical. To see the *qualitative* differences between rings and terminated chains, one must consider exciton properties that depend directly on the exciton wavefunction, such as the oscillator strengths and the charge distributions. Exciton ground-state oscillator strengths, O_{ex}, for terminated chains and rings are shown in Fig. 1. For rings, O_{ex} increases monotonically as the pair interaction strength, $|E_x / U_e|$, increases and as U_h / U_e decreases. For $E_x = 0$ (see Fig. 2), $O_{ex} = 1$, as expected for a noninteracting-pair state. For large interaction strength, $O_{ex} = N$ in rings, as expected for a tightly bound electron-hole pair with equal amplitude for being at each site. The three curves for rings become the same universal curve when they are plotted as a function of $E_x / (U_e + U_h)$.

O_{ex} for terminated chains is *qualitatively* different from O_{ex} in rings. For terminated chains, O_{ex} is a *nonmonotonic* and *nonuniversal* function of E_x, U_h, and U_e. For small E_x, O_{ex} initially increases as $|E_x|$ increases from zero for all U_h (this is not visible in Fig. 2 for $U_h < 0.01$). For similar electron and hole hopping, O_{ex} continues increasing monotonically with increasing interaction. For $U_h \ll U_e$, O_{ex} increases, decreases, and then increases with increasing interaction. For small E_x, the confined exciton oscillator strength varies *nonmonotonically* with U_h / U_e. For large E_x, the confined exciton oscillator strength decreases monotonically with decreasing U_h, *opposite* to the behavior in a ring. The large E_x limit for O_{ex} is significantly less than N for a confined system.

The initial increase of O_{ex} from the noninteracting pair limit as $|E_x|$ increases can be understood by first-order perturbation theory for small $|E_x|$ [9]. When $E_x = 0$, the exciton ground state is the product of electron and hole single-particle ground states ($n_e = n_h = 1$). Interactions mix in higher energy pair states. In first-order perturbation theory, mixing in pair states with $n_e = n_h > 1$ enhances O_{ex}. In second-order perturbation theory, the normalization of the exciton wavefunction amplitude is modified by all the pair states that are coupled by the interaction, including states with n_e not equal to n_h that do not contribute to O_{ex}. When the electron and hole hopping is similar, pair states with $n_e = n_h$ are preferentially included, and O_{ex} increases monotonically with increasing interaction. When $U_h < 0.2U_e$, pair states with n_e not equal to n_h, primarily states with $n_e = 1$ and $n_h > 1$, are preferentially included and O_{ex} is reduced by the renormalization of the wavefunction amplitude. This second-order reduction is strong enough for weak hole hopping that O_{ex} decreases for a range of increasing interaction and can be reduced below the noninteracting-pair limit. In this range of parameters, interaction suppresses the exciton oscillator strength because the interaction mixes in many configurations that have no oscillator strength.

51

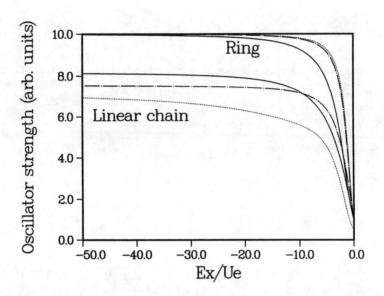

Fig. 1. Exciton ground-state oscillator strength for chains with terminated ends and for rings. $N = 10$. $U_h/U_e = 1$ (solid curves), 0.1 (dot-dash curves), and 0.001 (dotted curves).

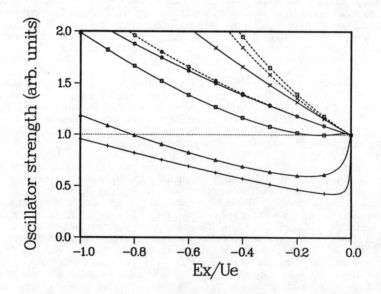

Fig. 2. Exciton ground-state oscillator strength for chains (solid curves) and rings (dashed curves) for weak interaction. $N = 10$. $U_h/U_e = 1$ (circles), 0.1 (crosses), 0.01 (squares), 0.001 (triangles), and 0.0001 (pluses).

The average electron-hole separations in the ground-state exciton of rings and terminated chains are nearly identical, decreasing monotonically with increasing interaction and decreasing U_h/U_e. The electron and hole are distributed uniformly on a ring, so that decreasing the electron-hole separation increases the electron-hole on-site correlation uniformly on the ring and increases O_{ex}. This does not happen on terminated chains. To distinguish the effects of confinement, one must consider the electron and hole distributions separately. For terminated chains, the electron and hole root mean square displacements from the chain center are shown in Fig. 3. For weak pair interaction, the electron (hole) contracts toward the chain center in the attractive potential defined by the hole (electron) distribution. This self-localization of the confined intermediate-dimensional exciton explains, for example, the 3D-2D crossover [4] that occurs for excitons in superlattices of shallow quantum wells as the barrier height is decreased ($|E_x/U_e|$ and $|E_x/U_h|$ increase). On-site correlation at the center of the chain is enhanced but pair occupation of other sites is suppressed for small $|E_x|$ and $U_h/U_e \ll 1$ because the hole contracts more than the electron. O_{ex} is suppressed, even though the electron-hole separation is smaller, because fewer sites are occupied. O_{ex} is enhanced at larger $|E_x|$ because the electron continues to contract while the hole begins to expand to increase on-site correlation away from the chain center.

For large interaction, the electron and hole distributions expand, becoming nearly equal with the pair tightly bound. The exciton center of mass has the same distribution as the electron and hole. The electron, hole, and exciton are contracted relative to electron and hole distributions for no interaction, indicating that there is a dead layer at the chain ends where the exciton is excluded. The dead layer can be understood in the large interaction limit by use of the second-order degenerate perturbation theory described in the previous section. The dead layer increases monotonically with decreasing U_h/U_e, because U_{ex}, the pair hopping rate for tunneling into the barrier that determines the dead layer, decreases with decreasing U_h/U_e. Because of the dead layer, O_{ex} is less than N. Exclusion from the ends is incomplete when the pair can hop to the ends. Exclusion from the ends becomes complete if $U_h = 0$ or $U_e = 0$. Thus, O_{ex} decreases with decreasing U_h/U_e for large $|E_x|$ in terminated chains.

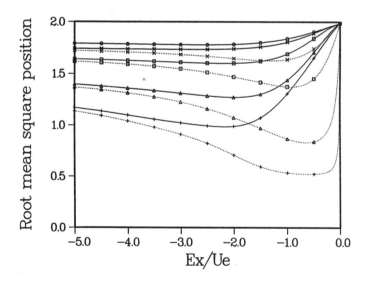

Fig. 3. Root mean square displacement of the electron (solid curves) and hole (dotted curves) in the ground-state exciton from the chain center ($N = 10$). $U_h/U_e = 1$ (circles), 0.3 (crosses), 0.1 (squares), 0.01 (triangles), and 0.001 (pluses).

The following picture (see Fig. 4) arises from these results. Two possibilities occur depending on whether the electron and hole have similar masses (in the context of the Hubbard model, if the electron and hole have similar nearest neighbor hopping rates) or the electron and hole have very different masses (there is a large asymmetry in the hopping rates). In the limit of weak interaction (the quantum-confinement limit with $E_x \ll U_h, U_e$), the electron and hole occupy the uncorrelated single-particle states defined by the confining potential. As the electron-hole interaction increases, the electron and hole contract in the potential defined by the distribution of the other particle. If the electron and hole masses are similar, the both contract similarly as E_x increases and the pair becomes strongly correlated with the exciton center-of-mass confined by the boundary of the system. If the masses are very different then the heavier particle contracts more than the lighter particle and the electron-hole overlap can be reduced, by the asymmetry in the electron and hole distributions, even as their motion becomes more strongly correlated. In the limit of very large electron-hole interaction, the exciton moves freely inside a confined system with dead layers near the boundaries. The dead layer arises because the exciton can not fully correlate near a surface. The effect of this dead layer is determined by how strongly the exciton can tunnel into the layer. Nonuniversal, nonmonotonic behavior arises as a result of the dead layer and the effects of asymmetry in the hopping.

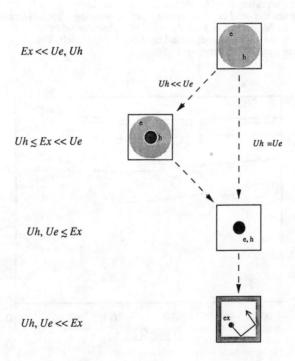

Fig. 4. Physical picture for confined intermediate-dimensional excitons.

EFFECT OF IMPURITIES ON CONFINED INTERMEDIATE-DIMENSIONAL EXCITONS

To demonstrate the effects of an impurity on a confined intermediate-dimensional exciton, I consider in this paper a single impurity interacting with an electron-hole pair which has a large asymmetry in the electron-hole hopping ($U_h/U_e = 0.001$). It is in the limit of large asymmetry in the hopping, that the nonmonotonic, nonuniversal behavior of the confined excitons, the effect of dead layers, and the asymmetric localization of the electron and hole are most pronounced. I present results for 10-site terminated chains. Qualitatively similar results are found for other chain lengths. The effects of impurities interacting with excitons on rings are similar to the effects of impurities at the middle of terminated chains. I assume that the electron-impurity and hole-impurity interaction strengths have the same magnitude and that the interaction is attractive for one particle and repulsive for the other ($V_e = -V_h$). Both the donor bound ($V_e = -0.5U_e$) and the acceptor bound ($V_h = -0.5U_e$) cases are considered. The magnitude of the impurity interaction is chosen to be large relative to the hopping for one particle and small relative to the hopping for the other particle. Results are present as a function of the pair interaction strength and the location of the impurity on the chain.

The pair binding energy to an impurity (the difference between the ground state exciton energy with and without an impurity present) is shown in Figs. 5 and 6 for a donor and acceptor bound exciton. The pair binding energy to a donor varies nonmonotonically with the pair interaction strength. This variation is different for donors at bulk and end sites. For donors at end sites, pair binding initially weakens with increasing pair interaction strength, with the pair

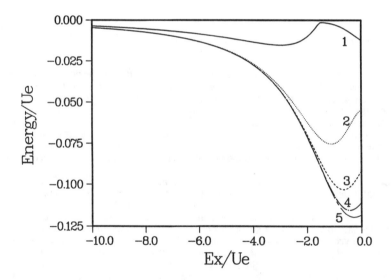

Fig. 5. Binding energy of an exciton to a donor at sites 1 (end), 2, 3, 4, and 5 (middle) of a 10-site terminated chain.

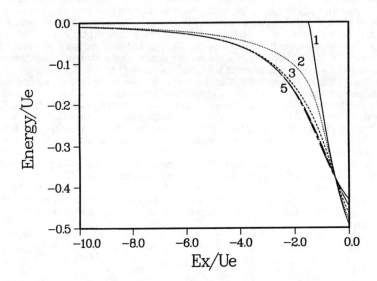

Fig. 6. Binding energy of an exciton to an acceptor at site 1 (end), 2, 3, 4, and 5 (middle) of a 10-site terminated chain.

becoming bound to the center of the confinement potential rather than to the donor. For large pair interaction strength, the pair is strongly bound together and not strongly localized by the confinement potential, so the pair can bind to an end-site donor. Binding to donors at bulk sites initially increases for increasing interaction strength because the pair interaction overcomes the hole-donor repulsion and helps localize the hole to the impurity. The binding energy of acceptor bound pairs decreases monotonically with increasing pair interaction strength. For large pair interaction strength, an end acceptor cannot bind a pair. Instead the electron becomes bound in the potential defined by the confinement and the hole is localized at the center of the electron charge distribution.

The electron charge densities at the sites on the chain are shown in Fig. 7 (8) for a pair interacting with a donor (acceptor) at the end of the chain. For a donor at the end of a chain and for weak pair interaction, the electron localizes toward the center of the chain as it would if no donor were present because the hole tries to remain localized away from the donor. For large pair interaction, the pair localizes to the site adjacent to the donor to maximize the electron-donor binding without maximizing the hole-donor repulsion. For an acceptor at the end of the chain, the hole is localized at the acceptor and the electron is localized in the confinement potential of the chain for weak pair interaction. As the interaction increases, the electron moves toward the acceptor. For large pair interaction, the pair becomes delocalized from the acceptor with the hole density (not shown) concentrated at the center of the electron density. In the limit of large pair interaction, the effect of the dead layer is apparent since the pair is localized away from the chain ends.

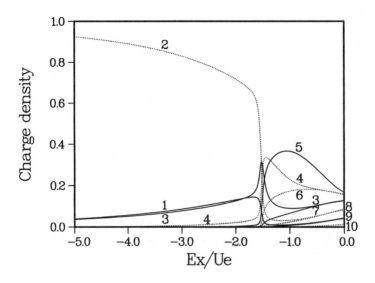

Fig. 7. Electron charge density at each site in a 10-site, terminated chain for a pair interacting with a donor at site 1. The labels indicate the site for each charge density.

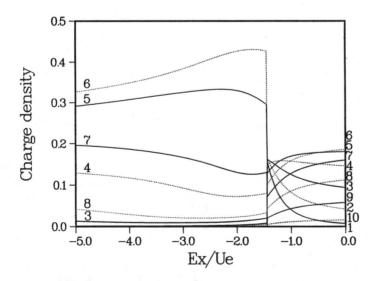

Fig. 8. Electron charge density at each site in the chain for a pair interacting with an acceptor at site 1.

The effect of impurities on the pair oscillator strength is shown in Fig. 9 for donors and acceptors at the end and at the middle of the chain. The oscillator strength for a pair interacting with a donor (acceptor) at the chain end clearly shows a large change when the pair becomes bound (unbound) to the donor (acceptor) with increasing pair interaction strength. For large pair interaction strength, the oscillator strength for a pair bound to an acceptor at any bulk site is similar to the oscillator strength for a pair bound to an acceptor at the middle of the chain (as shown in Fig. 9). The oscillator strength for pairs bound to donors depends on where the donor is located. For donors near the middle of the chain (sites 4 and 5 for the case considered here), the large pair-interaction limit corresponds to a pair localized to the sites on either side of the donor. For donors at sites near the chain end (sites 2 and 3 in this case), the exciton dead layer prevents localization of the pair to the site which is adjacent to the donor and close to the end. In this latter case the oscillator strength is reduced by half because the pair can localize to only one side of the donor rather than to both sides of the donor.

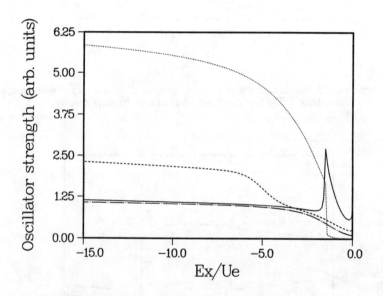

Fig. 9. Pair oscillator strength: for donors at the end (solid curve) and at the middle of the chain (dashed curve) and for acceptors at the end (dotted curve) and at the middle of the chain (dash-dot curve).

CONCLUSIONS

A simple Hubbard model has been used to gain general insight about how impurities affect the optical properties of confined nanostructures such as quantum dots, wires and coupled wells. In these systems, intermediate-dimensional excitons exhibit nonmonotonic, nonuniversal behavior with dead layers and asymmetry in the charge localization. The competing effects of confinement and pair interaction induce separate length and interaction scales for the localization of the electron and hole in the confined interacting pair. When an impurity interacts with a pair, new length and energy scales for the electron-impurity interaction and the hole-interaction must be introduced. Important modifications of exciton states arise as a result of the pair-impurity interaction.

Intermediate-dimensional excitons bound to donors or acceptors at or near system boundaries can be very different from excitons bound to bulk impurities. Excitons are bound to bulk impurities for all pair interaction strengths. In contrast, excitons bound to donors at a boundary become unbound for weak pair interaction. Excitons bound to acceptors at a boundary become unbound for strong pair interaction. For impurities at the boundary, oscillator strengths can be large or small, depending on whether the impurity binds the exciton. The oscillator strength for excitons bound to donors near a boundary is reduced because the dead layer induced by confinement can affect a bound exciton. Any effort to improve the optical properties of confined nanostructures by controlling defects and impurities must take into account that trap energies of defects and impurities and bound exciton oscillator strengths depend sensitively on the impurity or defect position in the nanostructure and on the geometry of the nanostructure. Any effort to realistically and quantitatively model impurity effects on confined excitons must be done with care to ensure that these competing effects of confinement, pair correlation, and impurity interaction are fully accounted for.

REFERENCES

[1] *Optics of Excitons in Confined Systems*, edited by A. D'Andrea, R. Del Sole, R. Girlanda, and A. Quattropani (The Institute of Physics, Bristol, 1992).
[2] G. W. Bryant, Phys. Rev. B **37**, 8763 (1988).
[3] Z. K. Tang, A. Yanase, T. Yasui, Y. Segawa, and K. Cho, Phys. Rev. Lett. **71**, 1431 (1993).
[4] I. Brener, W. H. Knox, K. W. Goosen, and J. E. Cunningham, Phys. Rev. Lett. **70**, 319 (1993) and the references therein to earlier work.
[5] K. Kash, D. D. Mahoney, B. P. Van der Gaag, A. S. Gozdz, J. P. Harbison, and L. T. Florez, J. Vac. Sci. Technol. B **10**, 2030 (1992).
[6] K. Brunner, U. Bockelmann, G. Abstreiter, M. Walther, G. Bohm, G. Trankle, and G. Weimann, Phys. Rev. Lett. **69**, 3216 (1992).
[7] M. G. Bawendi, W. L. Wilson, L. Rothberg, P. J. Carroll, T. M. Jedju, M. L. Steigerwald, and L. E. Brus, Phys. Rev. Lett. **65**, 1623 (1990).
[8] J. Wormer, M. Joppien, G. Zimmerer, and T. Moller, Phys. Rev. Lett. **67**, 2053 (1991).
[9] G. W. Bryant (unpublished).
[10] G. W. Bryant, Phys. Rev. B **47**, 1683 (1993).

INTERACTION BETWEEN DEFECTS AND TUNNELING EFFECT IN HETEROSTRUCTURES

D.STIEVENARD and M.LANNOO
IEMN-Département ISEN,UMR 9929 (CNRS), 41 Bd Vauban, 59046 Lille Cédex, France

ABSTRACT

Due to the reduced size of quantum devices, defects play a dominant role in their electrical behaviour. We theoretically show that defects can change the electrical behaviour of a single barrier or a quantum well. For the barrier and the well, we have performed a quantum calculation of a defect assisted tunneling effect based on the WKB theory to describe the wave functions and on the Oppenheimer approach for the determination of the tunneling probability. The main effects are : appearance of a negative differential resistance (NDR) in the current-voltage characteristics (I-V) of the single barrier and an apparent lowering of the band offset of the quantum well. In the latter case, we have experimentally measured this effect on an AlInAs-GaInAs-AlInAs well, using admittance spectroscopy. The apparent lowering is associated with a native defect located in the barrier, a defect studied using the DLTS technique.

INTRODUCTION

Since Esaki and Tsu [1] first proposed fabricating one-dimensional multiquantum well structures, improvements in the growth processing have led to increased interest and studies of quantum well structures. However, in many cases, "abnormal" current behaviours have been detected such as the appearance of peaks in the photocurrent characteristics of multiquantum well based diodes [2], negative differential resistance (NDR) in the current-voltage characteristics of ultrathin oxide - silicon interfaces, an effect detected via a scanning tunneling microscopy [3] or NDR detected in the I-V characteristics of a quantum well [4]. In every case, a defect-assisted resonant tunneling effect has been proposed to explain qualitatively the observed phenomenon. However, no quantitative model has been proposed.

The aim of this paper is to propose a theoretical model of defect-assisted resonant tunneling, an effect which can occur between the electronic states of a well or of the continuum of states of a material and the energy level associated with a defect located in an adjacent barrier. A quantum analysis leads to an analytical formula which is applied in the case of a single barrier or a single quantum well. A NDR is predicted in the I-V characteristics of the barrier whereas an *apparent* lowering of the conduction band offset is predicted for the well and verified experimentally for an AlInAs-GaInAs-AlInAs quantum well.

I. CASE OF A SINGLE BARRIER

Figure 1 gives the representation of the conduction band associated with a barrier. The indices L and R are used respectively for left and right. V_L and V_R are the voltages applied to the barrier. E is the total energy, ε is the energy along the z direction and $\vec{k}$ is the wave vector ($\vec{k} = \vec{k}_z + \vec{k}_{//}$). The defect is located at z_0 and the width of the barrier is W.

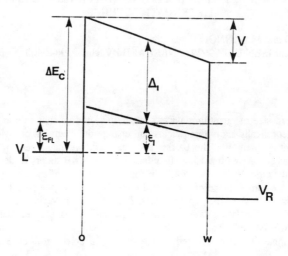

Figure 1 : Conduction band of a barrier with an energy level located in the barrier

The defect-assisted tunneling effect will occur when $E_k = E_I(z_0)$, the energy level associated with the defect located in the barrier. The flux dJ_L of particles per unit time and per unit area, going from the left to the energy level $E_I(z_0)$, is proportional to the number of defects located in a slice of width dz and is given by [5] :

$$dJ_L(zo) = \frac{1}{S}(N_T Sdz)\sum \frac{2\pi}{\hbar}|V_{ko}|^2 \delta(E_k - E_I(z_o))F_L(E_k)$$ (1)

where S is the surface area of the structure, V_{ko} = <ψ_k/V_1/ψ_0> with ψ_0 the wave function associated with the defect and ψ_k the wave function in the barrier. N_T is the defect concentration assumed uniform. $F_L(E_k) = f_L(E_k)(1 - g(E_I(z_0))Z_v - (1 - f_L(E_k))g(E_I(z_0))Z_p$, where f and g are the occupancy functions of the energy level E_k and of the energy level $E_I(z_0)$ respectively, and Z_v and Z_p the degeneracy factors of the defect when empty or filled. (In the following, to simplify the notation, we will assume that Z_v and Z_p are both unity). V_1 is the potential associated with the defect, assumed to be spherical with a radius of the order of the atomic radius $R_{at} \sim 1.25$ Å . To calculate V_{k0} , we have used the WKB theory [6] in order to describe the wave functions, as well as the Oppenheimer approach [7] for the determination of the tunneling probability.

The details of the calculation are given in [8]. Taking into account the contribution to the current of particles going from the left to the right and vice versa, the final expression of the defect assisted tunneling current is given by :

$$J \propto \frac{f_L - f_R}{\dfrac{1}{D_L} + \dfrac{1}{D_R}}$$ (2)

where f_L and f_R are the occupancy functions of the energy level E_k and of the energy level $E_I(z_0)$. As the tunneling effect is measured at low temperature (to avoid the contribution of the thermionic effect), we assume that the occupancy functions are step functions, i.e. we take $f_L = 1$ and $f_R = 0$. Thus, the behaviour of J will depend of the behaviour of D_L and D_L . Due to the mathematical form of the relation (2), one expects that J will exhibit a maximum because

D_R decreases when D_L increases, with $D_L + D_L \sim$ constant. A simulation of such a behaviour is given in Figure 2 where we have taken the following parameters : $\Delta E_C = 0.32$ eV, W=120 Å, Δ_I=0.23 eV, N_T=1x10^{18} cm^{-3}, ε_{FI}=40 meV and the temperature equals 80 K.

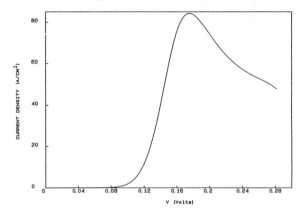

Figure 2 : Simulation of the current voltage characteristics of a single barrier, taking into account a defect assisted tunneling effect.

As shown in figure 2, a NDR occurs in the current-voltage characteristic of a single barrier. This model allows an interpretation of the "abnormal" NDR observed in the I-V curve of a single barrier. Experimental studies of this effect are in progress for on a single GaAs-GaAlAs-GaAs barrier, where the defects are introduced with electron irradiations.

II. CASE OF A SINGLE WELL

We consider a well and we assume there is a defect located in the associated barrier. Now, we consider the following phenomenon : the electrons localized in the well are not emitted through a thermionic process but via a double tunneling process. First, they tunnel to the energy level of a defect located in the barrier. Second, once captured by the defects, they tunnel towards the conduction band of the barrier. As we shall see later, the activation energy associated with this process corresponds to the band offset decreased by 1.66 times the ionization energy associated with the defect. We propose a theoretical model and an experimental study on a well, using the admittance spectroscopy technique.

II.1 The model

Figure 3a gives the representation of the conduction band associated with a well and with a defect located in the barrier, with the associated energy level $E_I(z)$ located at the energy Δ_0 from the bottom of the conduction band of the barrier. The well has a width L and its potential depth is V_0 or ΔE_C (the associated band offset). The defect is located at a distance z_0 from the edge of the well and its associated potential well is taken to be spherical. Due to tunneling through the triangular barrier associated with both the conduction band of the well and the barrier, there is a flux of electrons which tunnel from the well to the defect (and vice versa) and an other flux from the defect to the conduction band of the barrier (and vice versa). At equilibrium, the net flux is equal to zero. An electrical excitation (as in the case of the admittance spectroscopy technique), induces a change in the occupation functions of the states in the well and in the conduction band of the barrier (with respect to equilibrium). In that case, a conductivity is associated with the electrical excitation. The behaviour of this conductance is directly connected to the tunneling effect of the electrons from the well to the conduction band of the barrier. As shown in Figure 3a, the structure is symmetric, i.e. there is a tunneling effect from the conduction band of the barrier located on the left of the well to the well and from the well to the conduction band located on the right of the well. The equivalent electrical circuit is given in Figure 3b. Under a polarisation δV applied on the structure, one diode is forward-biased and the other one is reverse-biased. The total associated current will be determined by the saturation current of the reverse-biased diode. The important fact is that the two components of the current (from the left and from the right) have the same activation energy associated with the tunneling process.

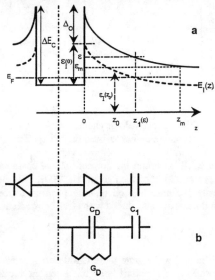

Figure 3 : Schematic conduction band of a well at equilibrium (3a), together with the electrical equivalent circuit (3b).

The important fact to note is that the main part of the current corresponds to a small value of z_0. For small values of z, we assume that the potential is linear. Then, we obtain an analytical expression for the tunneling current given by [9] :

$$J = \sigma_o \exp(-\frac{\overline{\Delta}_o}{kT})(1 - \exp(-\frac{\delta V}{kT})) \tag{3}$$

where $\overline{\Delta}_o = \Delta E_c - 1.66\,\Delta_0$. The factor 1.66 comes from the integration over z. The associated small signal conductivity σ is obtained by taking the derivative of J with respect to δV in the limit $\delta V = 0$. We obtain :

$$\sigma = \frac{\sigma_o}{kT}\exp(-\frac{\overline{\Delta}_0}{kT}) \tag{4}$$

One observes that the activation energy associated with the phenomena described just before corresponds to the band offset value decreased by $1.66\,\Delta_0$, giving an *effective* lowering of this band offset.

II.2 Experimental evidence

We have studied structures grown by molecular beam epitaxy and uniformly doped with Si, at typical concentration of 2×10^{16} cm^{-3} . They each consist of Ga$_{1-x}$In$_x$As quantum well (x=0.53 and x=0.60, with a width L=25 nm), sandwiched between two Al$_{0.52}$In$_{0.48}$As layers (0.35 μm thick). For admittance measurements, Ti-Au Schottky diodes have been deposited on the top layer. In order to distinguish the well response from the defects located in the AlInAs barrier, Deep Level Transient Spectroscopy, has been used as a complementary technique. The band bending diagram of the structure is given in Figure 3, together with the equivalent electrical circuit.

Such a structure, under an excitation of frequency ω, has a resonance defined by the condition :

$$G_D = (C_D + C1)\omega \qquad (5)$$

The conductance G_D can be written as :

$$G_D = \frac{\sigma S}{L_s} \qquad (6)$$

where L_S is the total length of the structure. Then, combining (4), (5) and (6), we obtain a new condition for the resonance :

$$kTf = \frac{\sigma_0 S}{2\pi(C_1 + C_D)L_s} \exp - \frac{\overline{\Delta}_o}{kT} \qquad (7)$$

Taking a linear variation of the Fermi level in the temperature range of measurement, equal to $\mu T + E_{Fo}^*$, we can write :

$$Log\left(\frac{1}{kTf}\right) = \frac{(\Delta E_C - 1.66\Delta_0) + E_{Fo}^*}{kT} + \mu - Log\left(\frac{\sigma_0 TS}{2\pi(C_1 + C_2)L_S}\right) \qquad (8)$$

An Arrhenius plot (see figure 4) allows one to determine the effective activation energy associated with the process, i.e. $(\Delta E_C - 1.66\Delta_0)$.

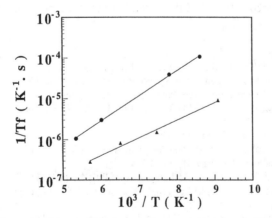

Figure 4 : Arrhenius plot of 1/Tf versus 1000/T for two indium fractions x=0.53 (•) and x=0.60 (Δ).

Figure 4 shows that at low temperature, the emission assisted by the defect is dominant and the corresponding Arrhenius plot , i.e. $(kTf)^{-1}$ versus the inverse of the temperature, for two samples (x=0.53 and 0.60 in $Ga_{1-x}In_xAs$), gives $\overline{\Delta}_0$.

Taking into account the Fermi level variation with the temperature and the conduction band discontinuity, we can deduce the energy Δ_0 of the trap responsible for the band offset lowering (see Table I). By comparing this value and the apparent barrier ($\Delta_0 + E_B$) deduced from DLTS measurements on the same AlInAs layers, we can deduce the energy barrier E_B associated with the capture. Within the precision of the measurements, this value agrees with that of the level C lying at 0.18 ± 0.02 eV below the conduction band in the AlInAs layers [10].

x	0.53	0.60
ΔE_c (eV)	0.51	0.54
$\overline{\Delta}_0 = \Delta E_C - 1.66\Delta_0 (eV)$	0.37±0.015	0.34±0.015
Δ_0 (eV)	0.085±0.015	0.12±0.015
E_B (eV)	0.085±0.015	0.09±0.015

TABLE I : Δ_0 and E_B deduced from $\overline{\Delta}_0$ and (Δ_0+E_B) for two GaInAs alloys .

CONCLUSION

We have presented two illustrations of defect-assisted tunneling, for a single barrier and a single well. In both cases, we have shown that such a mechanism leads to important changes in the electrical behaviour of the structures. Due to the diminution of the size of devices, this effect should become increasing important in the future.

ACKNOWLEDGEMENTS
We thank S.Ababou, (L.P.M. URA 358 (CNRS), INSA de Lyon) for the admittance spectroscopy measurements.

REFERENCES
1. L.Esaki and R.Tsu, IBM J.Res.Dev.**14**, 61, (1970)
2. F.Capasso, K.Mohammed and A.Y.Cho, Phys.Rev.Lett.**57**, 2303, (1986)
3. M.Tabe and M.Tanimoto, Appl.Phys.Lett.**58**, 2105, (1991)
4. M.W.Dellow, P.H.Beton, C.J.G.M.Langerak, T.J.Foster, P.C.Main and L.Eaves, Phys.Rev.Lett.**68**, 1754, (1992)
5. J.Bardeen, Phys.Rev.Lett.**6**, 57, (1961)
6. See, for example, A.Messiah, Mécanique quantique, (Dunod, Paris, 1959), p194
7. J.R.Oppenheimer, Phys.Rev.**31**, 66, (1928)
8. D.Stiévenard, X.Letartre and M.Lannoo, Appl.Phys.Lett.**61**(13), 1582, (1992)
9. D.Stiévenard, X.Letartre, M.Lannoo, S.Ababou and G.Guillot, Phys.Rev.B, (to be published)
10. A.Georgakilass, A.Christou, G.Halkias, K.Zekentes, N.Kornilios, A.Dimoulas, F.Peiro, S.Ababou, A.Tabata and G.Guillot, Journal of Electrochemical Society, **92-20**, 141,(1992)

ELECTRON BEAM INDUCED DEGRADATION OF 2DEG
IN AlGaAs/GaAs HETEROSTRUCTURE

T. WADA*, T. KANAYAMA**, S. ICHIMURA*, Y. SUGIYAMA* and M. KOMURO*
*Electrotechnical Laboratory, 1-1-4 Umezono,Tsukuba-shi, Ibaraki 305, Japan
**National Institute for Advanced Interdisciplinary Research, 1-1-4 Umezono, Tsukuba-shi,
Ibaraki 305, Japan

ABSTRACT

The effects of low-energy electron irradiation on the two-dimensional electron gases (2DEG's) in AlGaAs/GaAs heterostructures have been investigated. Not only the electron mobility of the 2DEG's but also the two-dimensional (2D) carriers are found to be reduced by the electron irradiation with the incident energies between 3.5 k and 8 keV and the electron dose of $1x10^{16}$ and $1x10^{17}$ /cm^2. The degraded mobility and the removed carriers by the low-energy electron irradiation are shown to recover by isochronal annealing to some extent, but not completely below 400℃. It is also found that considerable amount of scatterers which are created by an electron irradiation at room temperature are also created by an irradiation at 90 K. Comparing the experimental results with the Monte Carlo simulation, we speculate that the mobility degradation and the 2D carrier compensation are partly caused by the formation of complex defects in the GaAs buffer layer which are due to the excitations of core electrons of As, and that the mobility is further degraded by the formation of short-range scatterers in the heterointerface.

INTRODUCTION

Electron beam lithography plays the leading part in the fabrication of ultra-fine-structure devices in which the quantum effects dominate the device operations. In the quantum effect devices made of III-V compound semiconductors, two-dimensional electron gas (2DEG) in AlGaAs/GaAs heterostructures with an extremely high electron mobility plays an important role. If such low-energy electrons as are commonly utilized in the electron beam lithography cause any damage to the 2DEG's, the effects on the characteristics of the quantum effect devices may be serious. In fact, damage seems to be induced in 2DEG's by the irradiation of low-energy electrons.[1,2,3] The irradiation effects on bulk GaAs by high-energy electrons, that is, by 1-MeV electrons, have been studied intensively, and the mechanisms of defect formation have been considerably clarified.[4,5] On the other hand, research on the irradiation effects exerted by the relatively low-energy electrons is still limited as far as we know.[1,2,3,6]

This paper describes how the quality of 2DEG's are degraded by the irradiation of relatively low-energy electrons and the results are compared with the simulated results by the Monte Carlo method. The effects of a low temperature irradiation and isochronal anneal on the degraded 2DEG's are also investigated.

EXPERIMENTAL

The selectively doped n-AlGaAs/GaAs heterostructures with high mobility two-dimensional (2D) electrons were grown by molecular beam epitaxy. Figure 1 illustrates the structure of the samples examined in this work. The structure consists of a 1 μm-thick undoped GaAs buffer layer on semi-insulating (100) GaAs, a 20 nm undoped $Al_{0.3}Ga_{0.7}As$ spacer layer, a 50 nm Si-doped $Al_{0.3}Ga_{0.7}As$ carrier-supplying layer, and a 10 nm Si-doped GaAs cap layer. In our samples the 2DEG's lie at a depth of 80 nm from the surfaces. The electron Hall mobility of the 2DEG's is around $2x10^5$ cm^2/Vs at 14 K and the sheet carrier concentration is about $3.4x10^{11}$ /cm^2. The van der Pauw Hall elements were fabricated by conventional photolithography and the

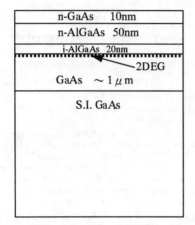

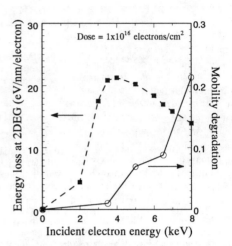

Fig.1. The schematic structure of the MBE-grown AlGaAs/GaAs heterostructure used in this study.

Fig.2. The incident electron energy dependences of the mobility degradation and the electron energy loss at the 2DEG region.

AuGe/Ni/Au metal system was used to form ohmic contacts. Van der Pauw Hall measurements were performed mostly at 14 K outside the electron irradiation chamber after the sample was irradiated perpendicularly by electrons with the energies ranging from 3.5 to 8 keV at room temperature in 2×10^{-8} Torr. The diameter of the electron beam at the sample position was about 3 cm and the fluctuation of the current density was not larger than 20 %. The current density of the electron beam was about 2 $\mu A/cm^2$ and the irradiation doses were 1×10^{16} or $1 \times 10^{17}/cm^2$. One of our samples was irradiated by keeping its temperature at 90 K and *in situ* Hall measurements were performed at the same temperature just before and after the irradiation.[7]

RESULTS

Figure 2 shows the degree of the mobility degradation as a function of the incident electron energy, when the electron dose is $1 \times 10^{16}/cm^2$. The mobility degradation is defined by

$$[\text{mobility degradation}] = [\mu_B - \mu_{irrad}]/\mu_B , \qquad (1)$$

where μ_B is the mobility before irradiation and μ_{irrad} is the mobility after irradiation. Even at the incident energy of 3.5 keV, the mobility degrades slightly and the degree of mobility degradation increases roughly superlinearly with an increase in the incident energy at least up to 8 keV. At the 8 keV irradiation, the mobility degrades by more than 20 %, but the 2D carriers decreases only by a small amount. When a sample is irradiated by 8-keV electrons with the dose of $1 \times 10^{17}/cm^2$, the mobility degrades by 95 % and the 2D carriers decreases by 40 % . On the other hand, in the case of the sample which is irradiated by keeping its temperature at 90 K, the rate of mobility degradation decreases by 15 % and the rate of 2D carrier removal diminishes by 30 % as compared with those for the sample irradiated at room temperature.

The results obtained from the 2DEG isochronally annealed for the time periods of 15 min are illustrated in Fig. 3, where the annealed sample was irradiated at room temperature by the 1×10^{17} electrons/cm^2 with the incident energy of 8 keV. The degree of mobility and 2D carrier recovery is estimated by the fraction not annealed f_{NA} defined by

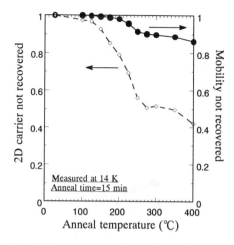

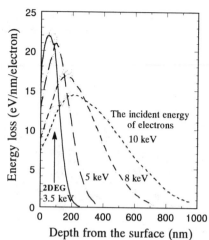

Fig.3. The temperature dependences of the recovery of mobility and 2D carriers by the isochronal anneal.

Fig.4. The depth profiles of the electron energy loss in an AlGaAs/GaAs heterostructure obtained by the Monte Carlo simulation.

$$f_{NA}=[m_B-m(T)]/\Delta m, \qquad (2)$$

where m_B is the mobility or the sheet carrier concentration before irradiation, Δm is the change in mobility or sheet carrier concentration due to the irradiation, and $m(T)$ is the mobility or the sheet carrier concentration after annealing at a temperature T. The degraded quality of the 2DEG by the electron irradiation recovers gradually between 100 and 250 °C but the recovery proceeds only slightly above 250 °C. The degraded mobility and the removed 2D carriers do not recover completely below 400 °C. The situation is quite different from the results obtained from the 1-MeV-electron-irradiated bulk GaAs,[4] where the induced degradation recovers completely by the annealing below 400 °C.

DISCUSSION

In order to examine the correlation between the incident energy dependence of the mobility degradation and the electron energy loss in an AlGaAs/GaAs heterostructure, the distributions of electron energy losses were simulated by the Monte Carlo method. The calculation procedures are essentially the same as those published elsewhere.[3,8,9] Figure 4 shows the simulated depth profiles of electron energy losses in the structure shown in Fig. 1. In this simulation, the scattering processes of 1×10^4 incident electrons were traced. When the lower energy (3.5, 5 keV) electrons are irradiated, most of energies are lost in the shallower region, that is, around the 2DEG region; on the other hand, when the higher energy (8, 10 keV) electrons are irradiated, electrons lose their energies much more broadly and the peaks of energy loss shift to the deeper region. The peak value of the energy loss profile also becomes smaller when the incident energy becomes higher. The amounts of electron energy loss at the 2DEG region are plotted against the incident electron energy in Fig. 2. In the sample with the 2DEG depth of 80 nm, the energy loss at the 2DEG region reaches maximum when the incident electron energy is around 4 keV; on the other hand, the observed degradation of electron mobility increases at least up to 8 keV in our experiments. As far as we can judge from the depth profiles of the whole electron energy losses, the mobility degradation seems to be caused not by the scatterers formed in the 2DEG region due

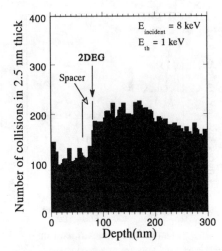

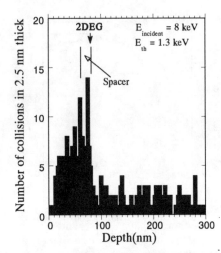

Fig. 5. The number of collisions by which electrons lose energies larger than 1 keV.

Fig. 6. The number of collisions by which electrons lose energies larger than 1.3 keV.

to the electron irradiation but by the charged defects formed in the remote region, if the created defects move very little at room temperature during and/or after electron irradiation, partly affirming Fink et al.'s conjectures.[2] However, the distance of around 80 nm from the heterointerface, where electrons with the incident energy of 8 keV deposit energies most frequently, seems to be rather too far to cause the mobility degradation effectively.

In the above discussion, we considered the amounts of the energy loss due to electron irradiation in order to compare them with the incident energy dependence of the mobility degradation. However, in such a case, where we must consider the electron collisions by which defects are created, it is not the amount of energy loss but the number of collisions by which electrons deposit energies larger than a certain threshold energy that has physical meaning. Fig. 5 illustrates one example of histograms of the collisions by which electrons deposit energies larger than 1 keV, where 1×10^6 electrons with the incident energy of 8 keV were traced by the Monte Carlo simulation. 100 times of collisions in the 2.5 nm-thick layer correspond to the defect density of 4×10^{18} /cm^3 by the incident electrons of 1×10^{16} /cm^2, if we suppose that all collisions create defects and that they do not diffuse. Such too frequent collisions with energy deposition larger than 1 keV correspond to the excitations of L-shell electrons of Ga atoms, since most of these collisions deposit energies between 1 keV and 1.3 keV according to the simulated results by the Monte Carlo simulation, and the mean electron binding energies are 1.56 keV for the Al K-shell electron, 1.38 keV for the As L-shell electron, and 1.17 keV for the Ga L-shell electron. Figure 6 illustrates the histogram of the collisions by which electrons deposit energies larger than 1.3 keV as a function of depth, for the incident electron energy of 8 keV. As shown in this figure, when we choose E_{th} of 1.3 keV, which is in between the binding energies of the As L-shell electron and Ga L-shell electron, electrons are shown to deposit energies larger than E_{th} with relatively reasonable frequency.

It has been suggested that the low-temperature mobilities of 2DEG's in AlGaAs/GaAs heterostructures are primarily determined by the scattering by ionized impurities.[10] The scattering rate at the heterointerface due to an ionized defect that resides at a distance r from the interface is approximately inversely proportional to r^2. Accordingly, we estimate the effects of collisions on the electron mobility of a 2DEG using the effective collision factor defined by

$$F = \Sigma N_{col}(r)/r^2, \qquad (3)$$

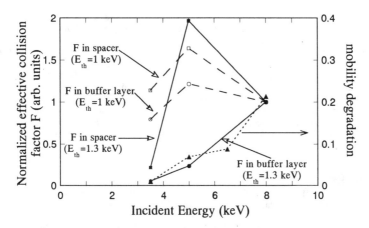

Fig.7. The incident electron energy dependence of the normalized effective collision factor which is defined in the text.

where $N_{col}(r)$ represents the number of collisions at a distance r from the heterointerface. The summations were done separately in the AlGaAs spacer layer and in the GaAs buffer layer using the obtained results by the Monte Carlo simulation with the incident electrons of 1×10^6 and the threshold energies of 1 k and 1.3 keV. Figure 7 shows the incident energy dependence of the effective collision factors normalized by the factors at the incident energy of 8 keV. Although an increase of ionized centers in the spacer layer was shown to decrease the 2DEG mobility effectively by the theoretical mobility calculation,[3] the factor F in the spacer layer becomes maximum when the incident energy is 5 keV. The situation conflicts with the experimentally observed energy dependence of the mobility degradation. On the other hand, the factor F that is obtained in the buffer layer with the threshold energy of 1.3 keV increases with an increase in the incident energy. The tendency is the same as found in the experiment. Because most of the collisions in the AlGaAs spacer layer deposit energies larger than 1.5 keV, such collisions excite Al K-shell electrons. Therefore, the excitations of the Al K-shell electron seem to be only slightly related to the formation of ionized centers. Collisions which deposit energies between 1.3 keV and 1.5 keV in the buffer layer can be attributed to the As L-shell electron excitations.

Therefore, we can speculate on the analogy of the bond-breaking mechanism of Si-O by synchrotron radiation[11] as follows. The irradiated electrons excite the core electrons of As in the GaAs buffer layer. The excited electrons and interatomic Auger transitions break bonds between As and Ga, creating complex defects which act as ionized centers. These defects, thus, become scattering centers against 2D carriers in the AlGaAs/GaAs heterostructure, resulting in a decrease in the mobility and 2D carriers. This kind of defect is thought to be formed even at 90 K. In fact, the mobility degrades considerably by the 90 K irradiation. However, the rate of the mobility degradation by the room temperature irradiation increases by 15 % as compared with the rate for the 90 K irradiation. This result suggests that another kinds of scatterers must be created probably with the assistance of diffusion at room temperature. The results obtained by the isochronal anneal experiments also suggest that at least two kinds of defects are created: one kind of defect is easily annealed below 400°C and the other one is rather hard to anneal thermally. Because large energies are frequently deposited around the AlGaAs/GaAs heterointerface as shown in Fig. 6, such short-range scatterers as are due to the precipitions of AlAs could also be formed in the heterointerface. These scatterers will cause further decrease in the mobility.

CONCLUSIONS

The effects of the low-energy electron irradiation on the 2DEG's in AlGaAs/GaAs heterostructures have been studied. Not only the electron mobility of the 2DEG's but also the 2D carriers were found to be reduced by the electron irradiation with the incident energies between 3.5 k and 8 keV and the electron dose of 1×10^{16} and 1×10^{17} /cm^2. The degraded mobility and the removed carriers by the low-energy electron irradiation were shown to recover by isochronal annealing to some extent, but not completely below 400°C. It was also found that considerable amount of scatterers which are created by an electron irradiation at room temperature are also created by an irradiation at 90 K. Comparing the experimental results with the Monte Carlo simulation, we speculate that the mobility degradation and the 2D carrier compensation are partly caused by the formation of complex defects in the GaAs buffer layer which are due to the excitations of core electrons of As, and that the mobility is further degraded by the formation of short-range scatterers in the heterointerface.

REFERENCES

1. D.C. Tsui, A.C. Gossard and G.J. Dolan, Appl. Phys. Lett. **42**, 180 (1983).
2. T. Fink, D.D. Smith and W.D. Braddock, IEEE Trans. Electron Devices **37**, 1422 (1990).
3. T. Wada, T. Kanayama, S. Ichimura, Y. Sugiyama, S. Misawa and M. Komuro, to be published in Jpn. J. Appl. Phys. **32**, Part I(1993).
4. L. W. Aukerman and R. D. Graft, Phys. Rev. **127**, 1576 (1962).
5. See, for example, D. Stievenard, X. Boddaert, J. C. Bourgoin and H. J. von Bardeleben, Phys. Rev. B **41**, 5271 (1990).
6. N. Tanaka, H. Kawanishi and T. Ishikawa, Jpn. J. Appl. Phys. **32**, 540 (1993).
7. T. Kanayama, Y. Takeuchi, Y. Sugiyama and M. Takano, Appl. Phys. Lett. **61**, 1402(1992).
8. S. Ichimura and R. Shimizu, Surf. Sci. **112**, 386 (1981).
9. S. Ichimura, Z.-J. Ding and R. Shimizu, Surf. & Interface Anal. **13**, 149 (1988).
10. K. Hirakawa and H. Sakaki, Phys. Rev. B **33**, 8291(1986).
11. Y. Sugita, Y. Nara, N. Nakayama and T. Ito, Proc. 1991 Int. MicroProcess Conf. (Jpn. J. Appl. Phys., Tokyo, 1991) JJAP Series 5 p.281.

MAGNETOOPTICAL STUDIES OF ACCEPTORS CONFINED IN GaAs/AlGaAs QUANTUM WELLS

P.O. HOLTZ*, Q.X. ZHAO*, B. MONEMAR*, A. PASQUARELLO**, M. SUNDARAM***, J.L. MERZ*** AND A.C. GOSSARD***

*   Department of Physics and Measurement Technology, Linköping University, S-581 83 Linköping, SWEDEN
**   Institut Romand de Recherche Numérique en Physique des Matériaux (IRRMA), PHB-Ecublens, CH-1015 Lausanne, Switzerland
***   Center for Quantized Electronic Structures (QUEST), University of California at Santa Barbara, Santa Barbara, CA 93106, USA

ABSTRACT

Magnetooptical studies have been performed on the shallow Be acceptor confined in the central region of narrow GaAs/AlGaAs quantum wells (QWs) with the magnetic field along the growth direction. The magnetic field dependence of the acceptor transition between the $1S(\Gamma_6)$ hh-like ground state and the excited hh-like $2S(\Gamma_6)$ state has been investigated by means of two independent techniques: Two-hole transitions of the acceptor bound exciton (BE) and resonant Raman scattering. The $1S(\Gamma_6)$ - $2S(\Gamma_6)$ transition energy as a function of the magnetic field has been measured for central acceptors in QWs of widths in the range 50 - 150 Å. The energy levels for the 1S ground states and 2S excited states of the confined acceptor with a magnetic field as a perturbation have also been calculated. These calculations predict a larger splitting between the $m_j=+3/2$ and $m_j=-3/2$ components of the acceptor $1S(\Gamma_6)$ ground state in comparison with the corresponding splitting of the excited $2S(\Gamma_6)$ state. The experimental results are in good agreement with the theoretical predictions derived without any fitting parameters. Furthermore, the Zeeman splitting of the acceptor BE emission has been measured and it is concluded that the $J = 5/2$ BE state is lowest in energy, similar to shallow acceptor BEs in bulk GaAs.

INTRODUCTION

The energy level spectrum of a hydrogenic potential, when going from 3D to 2D is a topic of fundamental interest. Both acceptors and donors have been extensively studied in the case of bulk GaAs. While states close to the conduction band edge, e.g. donor states, can be well described with a single effective-mass approximation (EMA), the corresponding description of states close to the valence band edge, e.g. acceptor states, is more demanding because of the degenerate valence band edge. A multi-band EMA model has to be employed [1] for a proper treatment of acceptors.

If we now turn to an acceptor state in the quantum well (QW) case, the fourfold degenerate Γ_8 level splits into a twofold degenerate heavy hole (hh)-like state with Γ_6 symmetry and a twofold degenerate light hole (lh)-like state with Γ_7 symmetry. However, there is an appreciable coupling between these states, which has to be taken into account in a proper description of these acceptor states. Also, the Coulomb coupling between different subbands should be included, since the acceptor energy is larger than the intersubband separation. The first calculations coping with the complicated valence band structure were presented by Masselink et al [2, 3]. Later on, extensive calculations involving the multi-band EMA [4 - 6] or the effective band-orbital model [7] have been performed.

We present in this work, a combined experimental and theoretical investigation of the dependence of the confined acceptor states on the applied magnetic field. The

73

computational method used is an extension of the calculations previously reported by Fraizzoli and Pasquarello [5, 6] for the zero field case. The experimental results are derived from spectroscopic studies of acceptor doped QWs involving partly two-hole transitions (THTs) of the acceptor bound exciton (BE) observed in selective photoluminescence (SPL) and partly resonant Raman scattering (RRS) in the presence of a magnetic field.

THEORY

The dependence of the 1S ground state together with the 2S excited states for a confined acceptor on a magnetic field applied parallel to the growth direction has been computed by extending the comprehensive theory given by S. Fraizzoli and A. Pasquarello for zero field conditions [5, 6]. A vector potential $\mathbf{A}$ including the magnetic field strength $\mathbf{B}$ according to

$$\mathbf{A} = \frac{1}{2} \mathbf{B} \times \mathbf{r} = \frac{1}{2} (-yB, xB, 0) \tag{1}$$

is introduced into the Hamiltonian for an acceptor in a single QW. Considering a single quantum well, grown in the [001] direction, which we take as the quantization axis z, the acceptor Hamiltonian then is given by a 4x4 matrix operator

$$H = H^{kin} + H^{qw} + H^c \tag{2}$$

Here H^{kin} represents the kinetic energy of the holes, H^{qw} the confinement potential originating from the valance-band discontinuity, and H^c corresponds to the potential of the acceptor and of the image-charges due to the mismatch of the dielectric constant. The kinetic-energy term H^{kin}, quadratic in $\mathbf{k} = -i\nabla + |e|\mathbf{A}/(\hbar c)$, describes the dispersion of

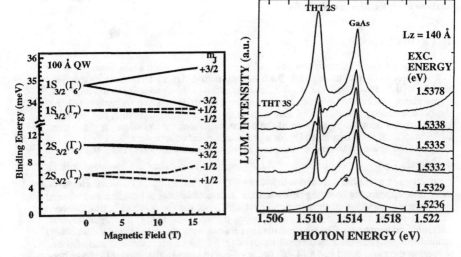

Fig. 1 The predicted development of the energy spectra for 1S and 2S states of a central acceptor in a 100 Å wide QW with an applied magnetic field.

Fig. 2 SPL spectra at zero field conditions with the excitation close to the FE (at 1.5378 eV) and the BE (at 1.5335 eV).

74

the Γ_8 valence band with the applied magnetic field $\mathbf{B} = \nabla \times \mathbf{A}$, and is given by the Luttinger-Kohn Hamiltonian [8]

$$H^{kin} = \begin{bmatrix} P+Q & L & M & 0 \\ L^+ & P-Q & 0 & M \\ M^+ & 0 & P-Q & -L \\ 0 & M^+ & -L^+ & P+Q \end{bmatrix} + 2\mu_B \kappa B J_z + q\mu_B B J_z^3 \qquad (3)$$

where

$$P = \frac{\gamma_1 \hbar^2}{2m_0} k^2 \qquad (4a)$$

$$Q = \frac{\gamma_2 \hbar^2}{2m_0} (k_x^2 + k_y^2 - 2k_z^2) \qquad (4b)$$

$$L = -i \frac{\sqrt{3}\gamma_3 \hbar^2}{2m_0} (k_x - ik_y)k_z \qquad (4c)$$

$$M = \frac{\sqrt{3}\hbar^2}{4m_0} (\gamma_2 + \gamma_3)(k_x - ik_y)^2 \qquad (4d)$$

The γ_1, γ_2, γ_3, κ and q are the Luttinger parameters describing the Γ_8 valence band. The same Luttinger Hamiltonian has earlier been successfully used to calculate the zero magnetic field case [5, 6].

If we let the Hamiltonian given in (2) act on the acceptor wave function written as a four-component envelope function, we derive the energy spectrum for the acceptor. Fig. 1 shows an example on the predicted dependence of the energy spectrum on the magnetic field including the 1S ground states and the corresponding 2S excited states for an acceptor in the center of a 100 Å wide well. The fourfold degeneracy of the 1S(Γ_8) ground state in the 3D case is lifted and splits into the hh-like 1S(Γ_6) and lh-like 1S(Γ_7) states already at zero magnetic field for the 2D case. The twofold degeneracies of the 1S(Γ_6) ($m_j = \pm 3/2$) and 1S(Γ_7) ($m_j = \pm 1/2$) Kramers doublets are in turn lifted in the presence of a magnetic field. It should be noted that a similar splitting occurs for the 2S(Γ_6) and 2S(Γ_7) excited states, but the magnitudes of the splittings are very different (Fig. 1). The Zeeman splitting of the 1S(Γ_6) ground state is significantly larger than the corresponding splitting for the 2S(Γ_6) excited state, while the reverse situation applies for the lh-like Γ_7 states.

EXPERIMENTAL

Be doped QW structures grown by MBE have been used in this study. Multiple (n= 50) QW structures with well widths in the range 50 to 150 Å sandwiched between 150 Å wide $Al_{0.30}Ga_{0.70}As$ barriers were grown. The samples were selectively doped with Be in the central 20% of the QW with a concentration ranging from 3×10^{16} cm^{-3} for the wide wells to 1×10^{17} cm^{-3} for the more narrow QWs. For the SPL and PL excitation (PLE) measurements an Ar+ ion laser pumped Titanium doped Sapphire solid-state laser was employed. The magnetooptical measurements were performed in magnetic fields up to 16 T applied perpendicular to the QW layers. The same optical fiber was used for

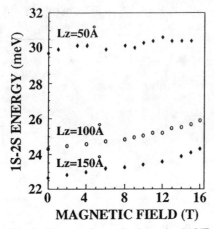

Fig. 3 The experimentally determined 1S[Γ_6] - 2S[Γ_6] transition energy as a function of an applied magnetic field for three different QW widths

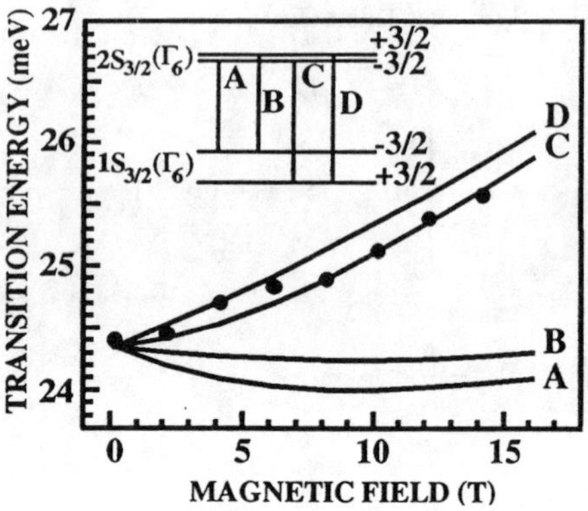

Fig. 4 Calculated dependence of the 1S[Γ_6] - 2S[Γ_6] energy on the applied magnetic field. The splitting of the 1S(Γ_6) and 2S(Γ_6) states in a magnetic field is schematically illustrated in the insert together with the four possible transitions.

coupling the excitation light from the laser onto the sample and for collecting the emission light from the sample onto the monochromator.

ACCEPTOR STATES IN THE PRESENCE OF A MAGNETIC FIELD

The quantity measured in our experiments is the energy separation 1S(Γ_6) - 2S(Γ_6) with an applied magnetic field. This quantity is achieved from the energy separation between the principle BE peak related to the 1S(Γ_6) ground state and the 2S(Γ_6) related satellite state. The former is derived directly from the PL spectrum, while the latter is obtained via two different optical techniques: THTs of the acceptor BE observed in SPL or RRS. The THT satellite is found to dominate when the excitation is close to or resonant with the FE, while the RRS satellite is as strongest with excitation close to or resonant with the BE as illustrated in Fig. 2 for a 150 Å wide QW for the zero-field case.

The dependence of the acceptor 1S[Γ_6] - 2S[Γ_6] transition energy on the applied magnetic field up to 16T for three different QW widths is shown in Fig. 3. In the presence of an applied magnetic field, both the acceptor BEs and the satellite peaks are blueshifted, but at a different rate. Accordingly, the energy separation between the acceptor BEs and the satellites, i.e. the 1S[Γ_6] - 2S[Γ_6] energy separation, increases with increasing magnetic field. The same trend is found for all QW widths investigated, although at a different rate as shown in Fig. 3.

From our calculations, we have already concluded that the predicted magnetic field splitting between the $m_j = \pm 3/2$ components of the Kramers' doublet states is significantly larger for the 1S(Γ_6) ground state than for the excited 2S(Γ_6) state as schematically illustrated in the insert in Fig. 4. There are four possible transitions between between the acceptor 1S(Γ_6) ground state and the excited 2S(Γ_6) state, denoted A, B, C and D in the insert in Fig. 4, of which two transitions, m_j = +/-3/2 ↔ m_j = -/+3/2, involve spin-flip [9]. The predicted transition energies for the four transitions between the 1S(Γ_6) and 2S(Γ_6) states are given in Fig. 4 for the case of acceptors in the center of a 100 Å wide QW. Two major branches (related to m_j = +3/2 and m_j = -3/2, respectively, of the 1S(Γ_6) ground state) of predicted transition energies are derived. The resolution in our experiment will not allow us to draw any conclusions whether the spin-conserving, m_j = +/-3/2 ↔ m_j = +/-3/2, or the spin-flip, m_j = +/-3/2 ↔ m_j = -/+3/2, transitions dominate. Also, only transitions corresponding to the high energy branch are observed in our experiments. Possible explanations for the absence or low intensity of the low energy branch could be a low oscillator strength for RRS or THT transitions related to this branch or a limited thermal population of the initial BE state.

Next, we will compare the experimentally derived data for the high energy branch for the acceptor 1S[Γ_6] - 2S[Γ_6] transition energy as a function of the applied magnetic field with the theoretically predicted results derived from the computations described above. It should be emphasized that no adjustable input parameter is used in our calculations except for the Luttinger parameter. From our computations we conclude that the experimentally determined 1S[Γ_6] - 2S[Γ_6] acceptor transition energies are very sensitive for the Luttinger parameter and we derive an accurate determination of κ = 1.2 for the hole of the confined acceptor. It should be pointed out that even for bulk GaAs, the reported values on the Luttinger parameter, κ, diverge significantly, from 0.8 up to 1.9 [10].

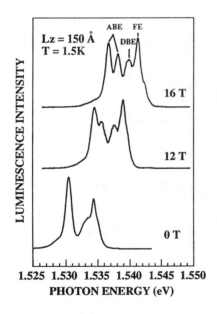

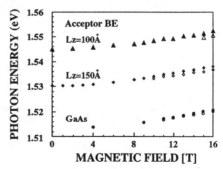

Fig. 5 Zeeman spectra for an acceptor doped 150 Å wide QW.

Fig. 6 A summary of the acceptor BE peak positions as a function of the magnetic field for two QWs in comparison with bulk GaAs.

ZEEMAN MEASUREMENTS

The development of the PL spectra with increasing magnetic field have been investigated up to 16T. The acceptor BE splits into two resolved components for higher fields as shown in Fig. 5. From this linear Zeeman splitting, ΔE, of the BE observed in PL, an effective g-value, g_{eff}, is derived from

$$\Delta E = g_{eff} \; B \; \mu_B \tag{5}$$

where B is the applied magnetic field and μ_B is the Bohr magneton. The evaluated effective g-value, g_{eff}, from the observed splitting in PL is found to be strongly dependent on the degree of confinement: For the QWs used, an effective g-value, g_{eff}, of $g_{eff} = 1.52$ is estimated for the 150 Å wide QW, which is further increased to $g_{eff} = 1.89$ in a 100 Å wide QW to be compared with a g-value of $g_{eff} = 0.66$ for the shallow acceptor in bulk GaAs (Fig. 6).

The splitting observed in PL is a combination of the splitting in the initial state, the BE state, and the splitting in the final state of the emission, i.e. between the $m_J = \pm 3/2$ states of the acceptor. The splitting in the initial state is confirmed by Zeeman measurements performed at elevated temperatures, in which the two components of the acceptor BE are strongly thermalized. The intensity of the low energy component increases with increasing temperatures. This fact is not consistent with a $J = 1/2$ BE state at lowest energy assuming that the electron g-value, $g_e < 0$, i.e. the $g_e = -0.4$ in bulk GaAs remains negative also in QWs. The observed thermalization behavior implies instead that the $J = 5/2$ BE state is lowest, similarly to what is concluded for shallow acceptors in GaAs.

SUMMARY

The magnetooptical properties of the acceptor in varying degree of confinement has been studied. The experimentally determined dependence of the acceptor $1S[\Gamma_6] - 2S[\Gamma_6]$ transition energy on the magnetic field is compared with theoretical predictions. An accurate determination of the Luttinger parameters for the bound hole is derived from these calculations. Also the behavior of the acceptor BE in the presence of a magnetic field is studied. The effective g-factor is found to be strikingly dependent on the degree of confinement. We conclude that the J=5/2 BE state is lowest in energy.

REFERENCES

1. The pioneering work on the multi-band EMA model, the k·p approximation was introduced by J.M. Luttinger and W.Kohn in Phys. Rev. 97, 869 (1955)
2. W.T. Masselink, Y.C. Chang, and H. Morkoç, Phys. Rev. B28, 7373 (1983)
3. W.T. Masselink, Y.C. Chang, and H. Morkoç, Phys. Rev. B32, 5190 (1985)
4. A. Pasquarello, L.C. Andreani, and R. Buczko, Phys. Rev. B40, 5602 (1989)
5. S. Fraizzoli and A. Pasquarello, Phys. Rev. B42, 5349 (1990)
6. S. Fraizzoli and A. Pasquarello, Phys. Rev. B44, 1118 (1991)
7. G.T. Einevoll and Y.C. Chang, Phys. Rev. B41, 1447 (1990)
8. J.M. Luttinger, Phys. Rev. 102, 1030 (1956)
9. V.F. Sapega, M. Cardona, K. Ploog, E.L. Ivchenko, and D.N. Mirlin, Phys. Rev. B45, 4320 (1992)
10. M. Reine, R.L. Agarwal, B. Lax, C.M. Wolfe, Phys. Rev. 2, 458 (1969)

D- IMPURITIES IN A QUASI-TWO-DIMENSIONAL SYSTEM: STATISTICS and SCREENING

W. J. LI,[*] J. L. WANG,[*] J.-P. CHENG,[#] S. HOLMES,[+] Y. J. WANG,[##] B. D. McCOMBE,[*] AND W. SCHAFF[**]
[*]SUNY at Buffalo, Buffalo, NY 14260
[#]Francis Bitter National Magnet Lab. MIT Cambridge, MA 02139
[+]Imperial College, London, UK
[**]Cornell University, Ithaca, NY 14853
[##]Present address: NHMEL, Florida State University, Tallahassee, FL 32306

ABSTRACT

Far infrared spectroscopic studies of electron-density- and magnetic-field-dependence of many-electron effects on silicon donor impurities confined in GaAs quantum wells are presented. At low excess electron densities, transitions from D$^-$ singlet and triplet states are observed. Temperature- and polarization-dependence measurements show that the relative absorption strengths of various spectroscopic features (D^0, D$^-$, and CR) are in qualitative agreement with a statistical calculation in thermal equilibrium in high magnetic fields. At large excess electron densities, the "D$^-$" transition energy shifts to higher energy when electron density is increased, and the magnetic-field dependence of the transition energy exhibits discontinuities in slope at integer filling factors. The spectroscopic features and their relation to excess free carriers and the role of screening and correlation are presented and discussed.

INTRODUCTION

Shallow impurities in quasi-2D semiconductor structures are important from both scientific and technological perspectives. The binding energies of shallow impurities are strongly affected by confinement; the binding energies of neutral donors can be easily increased by a factor of about 3 for donors located in the centers of GaAs wells in GaAs/AlGaAs quantum-well (QW) structures. Recently, it has been shown that in addition to the neutral donor state, a long-lived negative donor ion (Si donor binding two electrons) can be metastably created in the GaAs wells of GaAs/AlGaAs multiple-QW (MQW) structures [1,2]. For a number of years there has also been interest in the effects of a quasi-2D electron gas in screening the bound states of neutral donors in such systems [3]. MQW structures doped in both the wells and barriers with shallow donors provide a convenient testing ground for investigating these phenomena. From the (essentially classical) screening point of view, since screening by a quasi-2D electron gas is an oscillatory function of applied magnetic field (a maximum at fields corresponding to half-filled Landau levels and a minimum at filled Landau levels) [4], the impurity binding and transition energies should also oscillate. On the other hand, an alternative quantum mechanical point of view, more appropriate for large excess quasi-2D electron densities, is to consider the effects of a positive charge (a donor ion) on the states of a 2DEG in a large magnetic field. The single particle energy spectrum is modified by the positive charge, and, in addition, since the Coulomb interaction between an electron and the impurity ion destroys the translational invariance in the plane of the 2DEG, the effects of correlation among the electrons on the many electron states can be detected optically [5]. Anomalies in the cyclotron resonance (CR) spectrum of quasi-2DEG systems have been reported [6].

Mat. Res. Soc. Symp. Proc. Vol. 325. ©1994 Materials Research Society

We present systematic electron-density and magnetic-field dependence studies of many-electron effects on quasi-2D shallow-donor impurities. In the low electron density region, temperature-dependence study clearly show the existence of the two-electron (D^-) impurity bound state, and the evolution of the electron occupancy process from the one-electron (D^0) and the two-electron (D^-) bound states to free electron Landau states. In the high electron density region, "D^-" transition energy increases with increasing electron density, and whose slope exhibits discontinuities versus magnetic field near the integer LL filling factors [7]. Recent calculation of static-screening of a non-interacting electron gas can only explain part of the experimental observation [8]. The direct and exchange interactions have to be included to account for the energy shift of the impurity states [9].

EXPERIMENTAL

The samples (grown by molecular beam epitaxy) consist of 20 wells, 200Å GaAs separated by 600Å $Al_{0.3}Ga_{0.7}As$ barriers with silicon donors planar-doped at both the well and barrier centers. The important sample parameters are given in Table I. The wide barriers are intended

TABLE I. Summary of sample parameters. All parameters are nominal dimensions or densities, except the barrier-doping densities for samples 4 and 5, which were determined from quantum Hall measurements of free electrons in the wells.

Sample	Well-doping N_W 10^{10}/cm^2	Barrier-doping N_B 10^{10}/cm^2
1	2	3.5
2	2	4
3	2	8
4	2	15
5	2	28

to prevent electrons in the well from binding to their parent donors in the barriers. The free electron densities for samples 4 and 5 were determined from quantum Hall effect measurements and Shubnikov-de Haas oscillations; these are the values listed in the table. Other parameters listed in the table are nominal.

Far infrared (FIR) magneto-transmission measurements were carried out with Fourier transform spectrometers and a CO_2-pumped FIR laser in conjunction with 9T and 17T superconducting magnets.

RESULTS AND ANALYSIS

Occupancy Process of D^- Impurities

Previous photon-dose experiments [2] have clearly shown that at low temperature excess electrons introduced from the barrier regions will first bind to the neutral donors and form negatively charged D^- ions before they populate the free electron states. As a result, one can not observe screened neutral donors at low temperature.

A unique and fundamental property of D^- impurities is the existence of both singlet and triplet bound states and the associated optical transitions in the presence of a magnetic field. Theoretical calculations [10,11] for strict 2D D^- impurity in the high field limit indicate that the triplet ground state is depopulated at low temperature because it is higher than the singlet ground state energy by the amount $\Delta E = 0.1464\epsilon(0)$, where $\epsilon(0)$ is the magnetic-field-dependent binding energy of D^0 in the 2D limit. For 200Å quantum wells at 8T, $\epsilon(0)$ is approximately 15 meV [12]. If it is assumed that all transition energies can be scaled by $\epsilon(0)$ as the system deviates from the strictly 2D condition, then ΔE is approximately 2.2 mev or $\sim 25K$.

To search for the triplet state transitions, we have performed magneto-transmission experiments at various sample temperatures. In Fig 1, transmission spectra for sample 2 (upper) and sample 4 (lower) are shown. These data were taken with a FIR laser at 118.8 μm and a circular polarizer placed in front of the sample. The laser beam after the polarizer was 90% circular polarized as determined from electron CR in the two different senses of circular polarization. For sample 2, strong CR and D^- singlet transition are observed at 4.2K; at 11K, a new feature (labeled A) appears at 7.1T, which is $\sim 0.9T$ higher than CR absorption line. Since the magnetic field dependence of feature A is similar to CR, i.e. 13.2 cm^{-1}/T, feature A is approximately 12 cm^{-1} below CR. From the tilted field measurement, this feature is confinement related, and is only active in the electron CR active sense of circular polarization. While the D^- line diminishes rapidly with increasing temperature, feature A grows with temperature up to 20K, and then decreases in strength at higher temperature. This behavior is indicative of a population process in which electrons are thermally excited to a higher bound state. For a barrier doping density at $\sim 1.5 \times 10^{11}$ cm^{-2}, a feature in the same field position as A is clearly seen on the shoulder of strongly broadened CR absorption profile at 4.2K. The

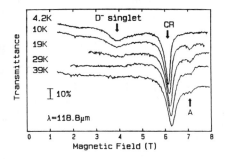

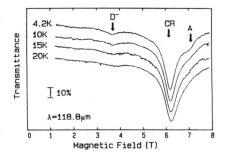

FIG. 1. Magneto-transmission spectra for sample 2 (upper): $N_B = 4 \times 10^{10}$ cm^{-2}, and sample 4 (lower): $N_B = 1.5 \times 10^{11}$ cm^{-2} at laser frequency of 118.8μm. Data were taken in Faraday geometry.

intensity of this line decreases rapidly with increasing temperature.

In strong magnetic field, the states of the two-electron D⁻ impurity can be described as the symmetric (singlet) and anti-symmetric (triplet) combinations of the two single-particle states, where the single-particle states are denoted by the Landau quantum number, l, and the azimuthal angular momentum quantum number, m, for electron 1 and 2 as $(l_1\, m_1,\, l_2\, m_2)$. In this notation, the singlet ground state is (00, 00) because electrons are most tightly bound with zero angular momentum; the triplet ground state is (00, 0-1) because it is a state with the lowest angular momentum among all available states associated with the lowest Landau level. The state (00, 0-1) is a bound state only in the presence of a magnetic field [10,11,13]. The dipole-allowed transitions are those that conserve the total angular momentum, $M = m_1 + m_2$ (a good quantum number of the system). Since the circular polarizer only selects out transitions with $\Delta M = +1$, an electron in the M=0 singlet

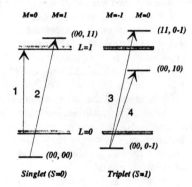

FIG. 2. Schematic energy level diagram (of the "outer" electron) for the D⁻ impurity states and the associated allowed transitions for $\Delta M = +1$. States are labeled with the single-particle notation, $(l_1 m_1,\, l_2 m_2)$, and $L = l_1 + l_2$, $M = m_1 + m_2$.

ground state (00, 00) can make transitions only to the M=1 singlet state (00, 11), and an electron in M=-1 triplet ground state (00, 0-1) can make transitions only to the M=0 triplet states (00, 10) and (11, 0-1). Note that there are two final states for triplet transitions, both are associated with L=1 $(L = l_1 + l_2)$ Landau level. Since higher angular momenta are less likely to bind in the Coulomb potential, (11, 0-1) is expected to have higher energy than (00, 10). Analytical solutions for the strictly 2D system in the high field limit [10,11] show that the singlet transition (00, 00)→(00, 11) occurs at $\hbar\omega_c + \epsilon(0)/2$ ($\hbar\omega_c$ is the CR energy); one of the triplet transition, (00, 0-1)→(11, 1-1), also occur at $\hbar\omega_c + \epsilon(0)/2$; the other triplet transition, (00, 0-1)→(00, 10), takes place at $\hbar\omega_c - \epsilon(0)/4$, which is below CR absorption line. The energy levels and the associated transitions with $\Delta M = +1$ are shown schematically in Fig. 2. From the above discussion, it is likely that feature A is, in fact, the triplet transition 4 in the figure. At high excess electron densities, this simple description is not adequate, and a many electron state perturbed by the potential of an isolated positive charge is more appropriate [9,14].

To explore further the statistical population of the two electrons in various impurity states in thermal equilibrium, we have studied the temperature dependence of the absorption intensity of CR, D⁰ and D⁻ transitions. Data were taken with a Fourier transform spectrometer. To extract information about the number carriers for each transitions, the absorption profiles were fitted with Lorentzian lines. The area of each absorption profile (proportional to the oscillator strength, f, and the electron density, N(T)) is ratioed to the corresponding area at 4.2K ($\propto$ fN(4.2K)). Since f is not a function of temperature, the ratio gives the relative density at T with respect to the density at 4.2K. The ratio N(T)/N(4.2K) vs. temperature is plotted in Fig. 3. It can been seen that the initial increase in temperature depopulates the D⁻ state rapidly. While one of the two electrons in the D⁻ site is thermally excited to the conduction band, the remaining electron that still binds to the D⁺ forms a D⁰. The population of D⁰ drops and that of LL increases as temperature is further increased. Shown in the insert of Fig. 3 are results

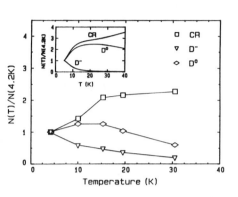

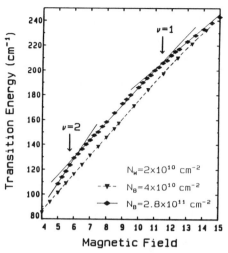

FIG. 3. The relative occupancy of D^- and D^0 impurities, and free electrons in Landau levels as a function of temperature. Solid lines are guide to the eye. Numerical results from a statistical calculation is shown in the insert.

FIG. 4. Magnetic field dependence of "D^-" singlet transition energies for sample 2 and 5. The position of integer filling factor are marked by arrows. Solid and dashed lines are guide to the eye.

from a statistical calculation in which a donor impurity was modeled as having only three possible states: no occupancy, single occupancy and double occupancy. The number densities of these states were obtained numerically from the conservation of total number of electrons. The binding energies at 8T were taken as 14 meV for D^0 [12] and 4 meV for D^- [15]. The calculated results are in qualitative agreement with the measurements.

Electron-Density and Filling Factor Dependence

Previous experiments [7] have shown that the crossover value of excess electron density for the present set of samples can be defined to be roughly $n_c = 3.5 \sim 4 \times 10^{10}$ cm^{-2}. Below n_c the added electrons, n_{ex}, will not substantially contribute to many-electron effects. Plotted in Fig. 4 is the singlet transition energy for sample 2 ($N_B = 4 \times 10^{10}$ cm^{-2}) and sample 5 ($N_B = 2.8 \times 10^{11}$ cm^{-2}). The position of the filling factors $\nu = 1$ and $\nu = 2$ were determined from quantum-Hall-resistance and Shubnikov-de Haas-effect measurements. There are two distinct effects: (1) the transition energy is blue-shifted for sample 5; the shift is ~ 13.5 cm^{-1} at $\nu = 2$ and ~ 10 cm^{-1} at $\nu = 1$; (2) the transition energy vs. magnetic field changes slope at magnetic fields corresponding to integer filling factors. A discontinuity of slope is also seen at $\nu = 1$ from sample 4 ($N_B = 1.5 \times 10^{11}$ cm^{-2}). For other low-density samples, there is no discernable slope change in this field region, since they remain within the quantum limit ($\nu < 1$).

The transition energies of D^0 and D^- including effects of linear static screening by free electrons in disordered LLs have been calculated recently [8]. The screening was taken into account with a simple static dielectric function for a non-interacting electron gas: $\epsilon(q) = 1 + q_s/q$, where the screening parameter q_s, being proportional to the density of states in a magnetic field, is maxima when LL is half-filled, and minimum when LL is filled. The calculated change of the slope of D^- transition energy vs magnetic field at even filling factor is in qualitative agreement with experimental observation. However, the transition energy of the screened D^- is always lower (red shifted) compared to that of the unscreened D^-, in contradiction to the data. It is clear that this simple picture is inadequate and that correlation effects among electrons and possibly dynamical screening are important in determining the energy levels and transition energies of the system [9,14]. Hawrylak has performed a calculation of the excitation energy of the donor-bound inter-LL excitons from a filled LL ($\nu=2$) in strong magnetic field limit with all electrons treated on the same footing [9]. A blue shift of approximately $0.15\epsilon(0)$ from the bare D^- singlet transition has been found, which comes predominately from the inter-LL and intra-LL exchange interactions among electrons. The estimated shift ΔE at $\nu=2$ is approximately 16 cm^{-1}, higher than the experimental value at 12 cm^{-1}. The agreement is expected to improve when LL mixing at finite field is taken into account.

SUMMARY AND CONCLUSIONS

We have carried out temperature and electron density dependence measurements of magneto-transmission for D^- impurities in QWs. A new transition line with energy less than CR line has been identified as transition between triplet states. The relative occupancy of D^- ions, D^0 neutral donors and Landau levels as a function of temperature has revealed the process of creating both free electrons and D^0 impurities from D^- centers. The blue shift of the "D^-" singlet transition energy with increasing excess electron density and the discontinuous slope changes at integer filling factors have demonstrated the importance of many-electron effects.

We are grateful to P. Hawrylak for helpful discussions and preprints of work prior to publication. This work was supported by ONR under Grant No. N00014-89-J-1673, and ONR/MFEL under Grant No. N00014-91-J1939.

1. S. Huant, S. P. Najda, and B. Etienne, Phys. Rev. lett. **65**, 1486 (1990); R. Mueller, D. M. Larsen, J. Waldman, and W. D. Goodhue, Phys. Rev. Lett. **68**, 2204 (1992);
2. S. Holmes, J.-P. Cheng, and B. D. McCombe, Phys. Rev. Lett. **69**, 2571 (1992).
3. See e.g.,T. Ando, A. B. Fowler, and F. Stern, Rev. Modern Phys. **54**, 437 (1982); J. A. Brum, G. Bastard, And C. Guillemot, Phys. Rev. B **30**, 905 (1984); C. Guilemot, Phys. Rev. B **31**, 1428 (1985); A. Gold, A. Phys. B **74**, 53 (1988); U. Wulf, V. Gudmundsson, and R. R. Gerhardts, Phys. Rev. B **38**, 4218 (1988).
4. T. Ando and Y. Uemura, J. Phys. Soc. Jpn. **37**, 1044 (1974).
5. W. Kohn, Phys. Rev. **123**, 1242 (1961).
6. See e.g., Z. Schlesinger, S. J. Allen, J. C. Hwang, P. M. Platzman, and N. Tzoar, Phys. Rev. B **30**, 435 (1984); J.-P. Cheng and B. D. McCombe, Phys. Rev. lett. **64**, 3177 (1990).
7. J.-P. Cheng, Y. J. Wang, B. D. McCombe, and W. Schaff, Phys. Rev. Lett. **70**, 489 (1993).
8. P. Hawrylak, Solid State Commun., submitted, 1993.
9. P. Hawrylak, preprint, 1993.
10. A. H. MacDonald, Solid State Commun. **84**, 109 (1992).
11. A. B. Dzyubenko, Phys. Lett. **A165**, 357 (1992).
12. R. L. Greene, and K. K. Bajaj, Phys. Rev. B **31**, 913 (1985).
13. D. M. Larsen, Phys. Rev. B, **20**, 5217 (1979).
14. A. B. Dzyubenko, JETP Lett. **57**, 507 (1993).
15. T. Pang, and S. G. Louie, Phys. rev. lett. **65**, 1635 (1991).

FABRICATION OF QUANTUM WIRES BY WET ETCHING TECHNIQUES

H. W. Yang, S. F. Horng, and H. L. Hwang
Department of Electrical Engineering, National Tsing Hua University Hsin-Chu, Taiwan 30043, R.O.C.

ABSTRACT

Quantum wires structures were fabricated by patterning quantum well samples with electron beam lithography and various wet chemical etching procedures. Wire structures with 800Å wire width were achieved by wet etching in NH_4OH / H_2O_2 / H_2O (20:7:973). These samples were characterized by scanning electron microscopy (SEM), photoluminescence (PL), and polarization-dependent photoluminescence excitation (PLE) measurements. The PL spectra show significantly stronger peaks than that taken from an en-etched quantum well sample. A wire width of 400Å was estimated from the blue shift of PL peaks. A 22% anisotropy was observed from polarization-dependent PLE spectra, further corroborating the existence of two-dimensional quantum confinement.

INTRODUCTION

The quantum confinement of carriers in a reduced dimensional structure such as quantum wires (QWR's) and quantum dots (QD's) inherit some interesting electrical and optical phenomena [1]. Theoretical calculations predict attractive properties of these nanostructures for optoelectronic applications [2-8]. With the reduction in dimensionality, one expects for example the narrowing of the gain spectrum of lasers as well as an increase in the exitonic oscillator strength [9,10]. These quantum confined structures are therefore highly favorable in making high-performance semiconductor lasers.

Fabrication of high-quality QWR and QD semiconductor heterostructures is still a major challenge for the current nano-fabrication and crystal growth technology. Quantum confinement along the growth direction can be easily achieved with growth techniques such as molecular beam epitaxy (MBE) or metal-organic chemical vapor deposition (MOCVD). Lateral confinement are, however, much more difficult to control. The most commonly used approach is firstly to define a mask by electron beam lithography on quantum well sample; the masked sample are then patterned by dry etching techniques such as reactive ion etching (RIE). Alternatively, focused ion beam (FIB) can be used to directly pattern the quantum well sample.

Dry etching techniques can provide better controlled aspect ratios and finer line definition for nanostructures; however, the damages result from ion bombardment usually lead to reduced luminescence efficiency and degraded electrical properties [11-13]. The alternative approach to fabricate nanostructures by wet chemical etching is therefore worth exploiting. In this paper we report our results of fabricating quantum wires with wet etching techniques.

EXPERIMENTAL TECHNIQUES

The quantum well samples were grown in a Varian GEN II MBE system at National Tsing Hua University. Semi-insulating (100) GaAs substrates obtained from Sumitomo Electronics Co. were used. These substrates were degreased in trichlroethylene, acetone, and methanol baths for 15 minutes each and rinsed in DI water for 20 minutes without surface etch. They were blown dry with filtered nitrogen gas, In-bonded to the Mo blocks, and immediately loaded into the load-lock chamber, where they were outgassed at 150°C for 2 hours. The samples were then transferred into the buffer chamber and were further outgassed at 400°C for 20 minutes. After the outgassing, they were transferred to the MBE growth chamber for the growth of the single quantum well. To remove the thin oxide on the substrates, the substrates were heated at 600°C for 20 minutes in an As_4 beam flux. During the growth, an As/Ga beam equivalent pressure ratio of 20 was chosen. The epitaxial layers consist of a 3000Å of GaAs buffer layer, a 80Å GaAs quantum well sandwiched between two $Al_{0.25}Ga_{0.75}As$ layers of 500Å followed by a 50Å GaAs capping layer.

After the growth, the samples were removed from the MBE system and were mounted to Si wafers for handling. They were baked to 170°C for 20 minutes to desorb H_2O. 500Å photoresis layer of polymethylmethacrylate (PMMA) with a OBER-100 to thinner ratio of 1:2 was spin coated on the samples. The coated samples were then baked at 170°C for 20 minutes before the electron beam exposure. The PMMA resist layers were exposed directly by electron beam with various combinations of electron dosage (90~200μC/cm^2) and beam energy (25/50KeV) to search for the optimal condition. In order to prevent the floating of PMMA layer, which would degrade the line resolution, the exposed samples were kept in a decicator for more than 16 hours prior to development without postbaking. Parts of the exposed samples were then examined by SEM to study the effects of different exposure conditions. The other samples were then subject to wet chemical etching.

Two different wet chemical etching processes were employed. For the first group of masked samples, the etching process is the two-step etching procedure proposed by Katoh et. al. [14], which consist of etching in, (1). H_2SO_4 / H_2O_2 / H_2O (8:1:120) solution at 3.5°C for 20 seconds, and (2). a Br_2/CH_3OH (1:1000) solution at 17°C for 4 seconds. The second group of samples were etched in NH_4OH / H_2O_2 / H_2O (20:7:973) for various time duration (2-5 seconds). The pH value measured for the latter solution is 9.3.

After the etching, the samples were subject to SEM observation to study the surface structures, to PL and polarization-dependent PLE characterization to investigate the electronic structures. The SEM micrographs were recorded with a JXA-840A electron probe microanalyer made by JEOL Co. All the SEM observation was done on bare sample surfaces without conductive layer coating in order to reveal the real structures on the surface. PL spectra and PLE spectra were taken from samples with different etching conditions. These samples were mounted on a cold finger of a closed-cycle He cryostat, and were cooled down to 19K (9K) for PL (PLE) characterization. For the PLE investigation, both polarization of incident light perpendicular and parallel to the wire structures were employed to determine the optical anisotropy.

RESULTS AND DISCUSSION

The effects of various combination of electron dosage and beam energy on the exposure and development of the PMMA layers were examined by SEM. We found a combination of electron dosage of 170μC/cm^2 and electron energy of 25KeV is satisfactory to expose the 500Å PMMA layer, sufficient to achieve a line resolution of 4000Å (fig. 1). Furthermore, the dosage required for 50KeV electron energy is about three times of that required for 25KeV electron beam to achieve the same exposure. However, to avoid the undesirable damages induced by high energy electron beam, 25KeV electron beam energy was chosen for all the samples used in later experiments.

With the two etching processes employed, we found that the etching rate of Br_2/Methonal system is much too high that it lead to overetch in seconds (fig. 2).

	quantum well	WIRE1	WIRE2	WIRE3	WIRE4
peak(Å)	7748	7730	7734	7734	7728
intensity(arb. unit)	0.981	1.794	2.110	2.117	2.087
FWHM(meV)	13.7	12.9	12.9	12.9	12.9

Table 1 Summary of PL characterization

$\dfrac{I_{1elh}}{I_{1ehh}}$	quantum well	WIRE1	WIRE2	WIRE3	WIRE4
parallel	0.40	0.42	0.42	0.43	0.37
perpendicular	0.40	0.50	0.52	0.54	0.58
optical anisotropy	0%	8.9%	11%	11%	22%

Table 2 Summary of PLE characterization

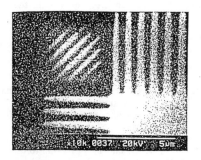

Fig. 1: SEM micrograph of a
PMMA layer exposed
with 25KeV,170μC/cm ^{2}.

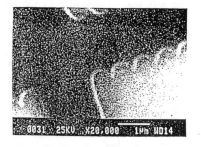

Fig. 2: SEM micrograph of a
GaAs sample overetched
in Br_2/CH_3OH.

Therefore the two-step etching procedure was not found to produce wire structures in our experiments. We believe that this difficulty might be overcome by cooling the etching solution to lower temperature. On the other hand, the etching rate of NH_4OH / H_2O_2 / H_2O is slower and wire array of 0.08um wire width can be achieved (fig. 3). However, with finer line resolution, the adhesion problem becomes more and more severe. We found that slight stirring of the the samples in the solution could lift the PMMA resist mask. The etching must therefore be done in stagnant solution.

The results of PL investigation on the wire array achieved with NH_4OH / H_2O_2 / H_2O etching described above for different duration (2,3,4,5 seconds, denoted as WIRE1, 2, 3 and 4) and an un-etch controlled sample are summarized in table 1. A typical spectrum from WIRE3 is also shown in fig. 4. Single symmetric peaks were observed in all these PL spectra. Compared to the un-etched quantum well, the PL peaks of the etched ones exhibit slight blue shifts (about 4meV) and are significantly stronger with increasing etching time investigated. Calculation with a simple particle-in-box (PIB) model, the energy shift with a 800Å lateral confinement, as observed from SEM micrographs, should be only 1.0meV or roughly 8Å. The fact that we have larger blue shift might be explained by the depletion from the wire edge which leads to narrower linewidth than that expected from SEM observation. Fitting the energy shift of 4meV with the simple PIB model, our wire width should be around 400Å. This narrowing corresponds to a depletion width of 200Å at each side, which is not an unreasonable estimate to our knowledge. The FWHM peakwidth decreases slightly from 13.7meV for the unetched quantum well to 12.9meV for all the four etched wires. The fact that the PL spectra for the etched samples show stronger and narrower peaks suggests that the damage problem, usually encountered in fabrication processes involving dry etching, is less severe in wet chemical etching.

Polarization-dependent PLE was employed to further characterize these wire structures. The results of PLE characterization is summarized in table 2 and typical spectra for WIRE3 are shown in fig. 4. Two peaks were observed with in the PLE spectra. These two peaks for the etched samples were about 88meV and 115meV higher than the bandgap of bulk GaAs at 9K, which is 1.5188eV. Assume that all the electron and hole effective masses remain isotropic, the energy shifts due to quantum confinement for transitions from heavy (light) hole to conduction band should be inversely proportional to the reduced masses of heavey (light) hole and electron. Taking the bulk values, the ratio energy shift for transitions involving heavy and light hole should be about 1.6; however, because the transition which involves light hole will increase more than that involves heavy hole, fluctuation in wire width will change this ratio of energy shifts. Our measured value is around 1.3, which is 18% less. However, because the ratio of these two peaks exhibits the expected optical anisotropy, as will be described below, we tentatively attribute these two peaks to the n=1 heavy hole and light hole to conduction band transition (denoted as 1ehh and 1elh). The fact that we have a ratio 18% less than expected indicates that we may have large fluctuation of the wire width; the actual reason is not clear to us at this point. Comparing the peak positions of the PL and PLE spectra and taking into account the shift of bandgap due to change in temperature, we observed a small Stoke shift of about 3meV, indicating the presence of damages or imperfections. Because of different laser power used in characterizing these samples, we have normalized the intensity of 1elh to 1ehh. As is the characteristic of quantum wires, the normalized 1elh should increase as the incident light changes from horizontal to vertical polarization [15]. This is indeed the case with our measured data. For a quantative comparison, we calculated the degree of anisotropy, which we tentatively define as $(I_{perpendicular} - I_{parallel})$ / $(I_{perpendicular} + I_{parallel})$, where I is the normalized transition intensity of 1elh. As expected, we found no anisotropy for the un-etched quantum well sample. On the other hand, this anisotropy increases with the etching time, indicating a better defined line structures; and it reaches a maximum value of 22% for sample 4. This 22% anisotropy may roughly be compared to the comparable linear polarization obtained by Mirin et. al [16] with quantum wires grown by migration-enhanced epitaxy regrowth on patterned substrates. The

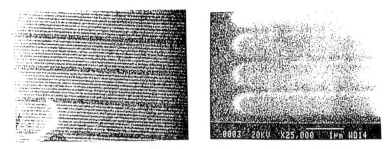

Fig. 3:   SEM micrographs of wire array with wire width 0.08μm.

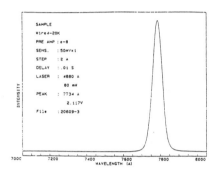

Fig. 4:   PL spectrum taken from sample WIRE3.

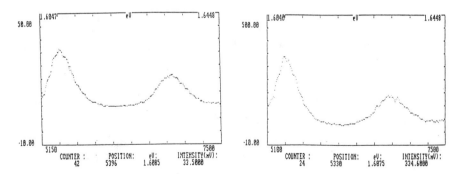

Fig. 5:   PLE spectra taken from sample WIRE3.   (left) vertical polarization; (right) horizontal polarization.

existence of anisotropy evidently corroborate the existence of two-dimensional quantum confinement.

CONCLUSION

We have investigated the fabrication of quantum wires by patterning quantum well samples with electron beam lithography and wet chemical etching. The optimal parameters for electron beam lithography were found to be $170\mu C/cm^2$ and 25KeV. Array structures with wire width of 800Å were obtained with wet etching in NH_4OH / H_2O_2 / H_2O (20:7:973). These samples were then characterized by PL, and polarization-dependent PLE measurements. The PL spectra show significantly stronger peaks than that taken from an en-etched quantum well sample, indicating less severe damage problems, as compared to wires obtained from dry etching techniques. From the blue shift of PL peaks, the actual wire width of these wires were estimated to be 400Å. A 22% optical anisotropy was observed from polarization-dependent PLE spectra, further corroborating the existence of two-dimensional quantum confinement.

ACKNOWLEDGEMENT

We are grateful to the Material Center at National Tsing Hua University for access to the GEN II MBE system and the JXA-840A SEM; and to Professor Y. F. Chen in Department of Physic at National Taiwan University for measuring the polarization-dependent PLE spectra. This work was supported by ROC National Science Council (NSC 82-0208-N-007-146).

REFERENCES

1. J. A. Brum, Phys. Rev., B43, 12082 (1991)

2. H. Sakaki, Jpn. J. Appl. Phys., 19, L735 (1980)

3. P. M. Petroff, A. C. Gossard, R. A. Logan, and W. Wiegmann, Appl. Phys. Lett., 41, 635 (1982)

4. T. Fukui and H. Saito, Appl. Phys. Lett., 50, 824 (1987)

5. H. E. G. Arnot, M. Watt, C. M. Sotomayor-Torres, P. Glew, R. Cusco, J. Bates, and S. P. Beaumont, Superlatt. Microstruct., 5, 459 (1989)

6. M. B. Stern, H. G. Craighead, P. F. Liao, and Mankiewich, Appl. Phys. Lett., 45, 410 (1984)

7. J. C. Bean, G. E. Becher, P. M. Petroff, and T. E. Seider, J. Appl. Phys. 48, 907 (1977)

8. Y. Arakawa, K. Vahala, A. Yariv, and K. Lau, Appl. Phys. Lett., 48, 384 (1986)

9. M. Kohl, D. Heitmann, P. Grambow, and K. Ploog, Phys. Rev. Lett., 63, 2124 (1989)

10. Y. Arakawa, K. Vahala, A. Yariv, and K. Lau, Appl. Phys. Lett., 47, 1142 (1985)

11. D. Gershoni, H. Temkin, G. J. Dolan, S. N. G. Chu, and M. B. Panish, Appl. Phys. Lett., 53, 995 (1988)

12. G. P. Li, L. Guo, T. Katoh, Y. Nagamune, S. Fukatsu, Y. Shiraki, and R. Ito, Jpn. J. Appl. Phys. 29, L1213 (1990)

13. S. W. Pang, W. D. Goodhue, T. M. Lyszczarz, D. J. Ehrlich, R. B. Goodman, and G. D. Johnson, J. Vac. Sci. Technol., B 6, 1916 (1988)

14. T. Katoh, Y. Nagamune, G. P. Li, S. Fukatsu, Y. Shiraki, and R. Ito, Appl. Phys. Lett, 57, 1212, (1990)

15. see, for example, G. Bastard, *Wave Mechanics Applied To Semiconductor Heterostructures*, Halsted Press, New York, 1988.

16. R. P. Mirin, I. Tan, H. Weman, M. Leonard, T. Yasuda, J. E. Bowers, and E. L. Hu, J. Vac. Sci. Technol. A 10(4), 697, 1992.

DENSITY OF STATES OF HYDROGENIC IMPURITIES IN GaAs/GaAlAs QUANTUM WIRES

SALVIANO A. LEÃO* , O. HIPÓLITO* AND A. FERREIRA DA SILVA**
* Departamento de Física e Ciência dos Materiais, Instituto de Física e Química de São Carlos, Universidade de São Paulo, 13560-250 São Carlos, SP, Brazil
** Laboratório Associado de Sensores e Materiais, Instituto Nacional de Pesquisas Espaciais, Caixa Postal 515, 12225 São José dos Campos, SP, Brazil, and Instituto de Física - UFBa, 40210 Salvador, Ba, Brazil

ABSTRACT

We investigate the occurrence of impurity density of states of hydrogenic impurities placed on the axis of a cylinder quantum-well wire of GaAs/GaAlAs structure. The effects of disorder are taken into account in the calculation. It is shown that for a specific radius of the wire, the peak energy of the density of states as a function of the impurity concentration increases very fast enlarging the bandwidth. The impurity bands are considered for the observed binding energy and radius as well as for the impurity concentration of experimental interest.

Theoretical investigations of the impurity density of states in quantum-well wire (QWW) of GaAs/GaAlAs are important for the design and interpretation of optical and transport measurements [1-9]. Following recent works [6-9], we present a though analysis of the density of states of on-axis hydrogenic impurities in these structures with circular cross section.

The system is described by the tight-binding Hamiltonian [6,7] ,

$$H = \sum_i \varepsilon_i |i><i| + \sum_{i \pm j} V_{ij} |i><j| \tag{1}$$

where $< \vec{r} \mid i >= \psi(\vec{r} - \vec{R}_i)$ is the ground-state wave-function of an electron bound to an impurity at position $\vec{R}_i$. The values of ε_i presents a small variation, and will be consider as a constant. V_{ij} is the hopping integral given by

$$V_{ij} = \left\langle i | \frac{e^2}{\varepsilon |\vec{r} - \vec{R}_i|} |j \right\rangle \tag{2}$$

93

where ϵ is the GaAs background dielectric constant. We assume all impurities laying on the z axis and the structure of circular cross-section wire with radius R as shown in Fig. 1. We calculate the density of states from the Green's functions

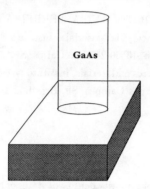

Figure 1: Schematic of a quantum-well wire with circular cross-section of radius R. The impurities are randomly distributed on the z axis.

$$D(E) = -\frac{1}{\pi} Im < G_{ij}(\varepsilon + i0^+) >,$$

$$(3)$$

where averaged diagonal elements of the propagator $< \dots >$ are given by

$$< G_{ii}(\varepsilon + i\theta^+) > = \xi(\omega)/\omega,$$

$$(4)$$

with $\xi(\omega)$ obtained from [6,7]

$$\eta(\omega) = \frac{N\xi(\omega)}{2\pi\omega^2} \int \frac{V^2(\vec{K})d\vec{K}}{1 - N\xi(\omega)/\omega V(\vec{K})}.$$

$$(5)$$

where N is the impurity concentration per unit length and $V(\vec{K})$ is the Fourier transform of V_{ij}.

The variational solutiom for a single point charge donor impurity is given by [8]

$$E = < \psi| - \nabla^2 - \frac{2}{(\rho^2 + z^2)^{1/2}} + V(\rho)|\psi >,$$

$$(6)$$

where $V(\rho)$ is the confining potential, i.e., $V(\rho) = \infty$ for $\rho > R$ and zero for $\rho < R$. Here $\rho = (x^2 + y^2)^{1/2}$ and z the direction along the axis of the wire. We use the wave functions for the cylindrical confining potential [8,9] ,

$$\psi(\rho, z) = N_0 J_0(K_{10}\rho) \ exp \ [-\lambda(\rho^2 + z^2)]^{1/2}, \rho < R$$

$$= 0 \ , \ \rho > R \qquad (7)$$

where N_0 is the normalization constant and λ the variational parameter. In order to solve Eq.(5) we need $V(\vec{K})$, which is obtained as

$$V(\vec{K}) = -(E_B + K^2)\psi(K)^2 a_0^* R_y^*. \qquad (8)$$

The value of the binding energy E_B is obtained from Eq.(6) as a function of λ and the radius of the wire in units of the effective Bohr radius R/a_0^*. We adopt $a_0^* = 100\mathring{A}$ and the effective Rydberg $R_y^* = 5.3$ meV.

The Fourier transform of $\psi(\vec{K})$ is given by

$$\psi(\vec{K}) = \sqrt{\frac{2}{\pi}} \frac{\lambda R}{\sqrt{K^2 + \lambda^2}} K_1(R\sqrt{K^2 + \lambda^2}) \qquad (9)$$

where K_1 is the modified Bessel function of the second kind.

In Fig. 2 we show the Fourier transform $V(\vec{K})$ for $E_B = 2.8R_y^*$ and $R = 1.132a_0^*$. As we can see it has a localized behavior.

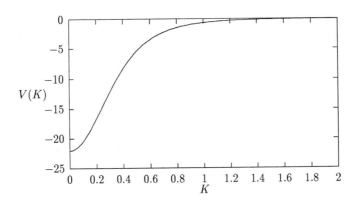

Figure 2: Fourier transform of the hopping energy for a cylindrical quantum-well wire.

With the same values for E_B and R we calculated the density of states for different concentrations and show in Fig. 3. We choose $R = 1.13a_0^*$ because it corresponds approximately to a QWW of square cross-section wire of $L = 200 \mathring{A}$ which is the experimental interest [1,2] and also for this radius it does not depend of the magnetic field, as shown by Branis et al. [9]. In Fig. 3 we show that the peak of the bands increases with the concentration broadening the impurity bandwidth. These effects will be of importance, for instance, in the understanding of optical and transport measurements.

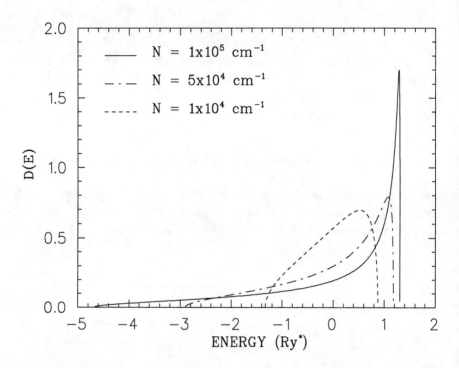

Figure 3: The calculated density of states for a cylindrical quantum-well wire of GaAs/GaAlAs structure. The origin energy is that of the bound electrons.

References

[1] P.M. Petroff, A.C. Gossard, R.A. Logan and W. Wiegman, Appl. Phys. Lett. **41**, 635 (1982).

[2] J. Cibert, P.M. Petroff, G.J. Dolan, S.J. Pearton, A.C. Gossard and J.H. English, Appl. Phys. Lett. **49**, 1275 (1986).

[3] P.C. Sercel and K.J. Vahala, Phys. Rev. **B44**, 5681 (1991).

[4] H. Sakaki, Jpn. J. Appl. Phys. **19**, L735 (1980).

[5] G. Fishman, Phys.Rev. **B34**, 2394 (1986).

[6] E.A. de Andrada e Silva, I.C. da Cunha Lima and A. Ferreira da Silva, Phys. Rev. **B37**, 8537 (1988).

[7] A. Ferreira da Silva, Phys. Rev. **B41**, 1684 (1990).

[8] J.W. Brown and H.N. Spector, J. Appl. Phys. **59**, 1179 (1986).

[9] S.V. Branis, G. Li and K.K. Bayay, Phys. Rev. **B47**, 1316 (1993).

PART II

Defects and Impurities in InP and Related Compounds

SEMI-INSULATING BEHAVIOR OF UNDOPED InP

K. KAINOSHO*, O. ODA* , G. HIRT** and G. MÜLLER**
*Nikko Kyodo Co., Ltd, Electronic Materials and Components Laboratories
3-17-35, Niizo-Minami, Toda, Saitama, 335 Japan
**Universität Erlangen-Nürnberg, Institut für Werkstoffwissenschaften 6
Martensstraße 7, 91058 Erlangen, Germany

ABSTRACT

Recently, it was found that semi-insulating behavior of undoped InP can be realized by high pressure annealing of undoped high purity InP. In the present work, studies related with the achievement of the semi-insulating state are reviewed. Purification of raw materials, effect of native defects, effect of high pressure annealing, contamination of Fe are discussed. The semi-insulation mechanism is explained by the Shockley diagram. The semi-insulating state is supposed to be achieved by the annihilation of shallow donors (presumably phosphorus vacancies) and the compensation of the residual donors with a small amount of Fe deep acceptor.

INTRODUCTION

Semi-insulating InP is a key material for high-frequency devices and OEICs which are indispensable for high speed computers, satellite communications and fiber communications. Semi-insulating InP is usually produced by doping Fe atoms, which act as deep acceptors, up to concentrations above $1 \times 10^{16} cm^{-3}$. This high Fe concentration is however undesirable since it has a possibility of deteriorating the device performances. In fact, the disadvantages of Fe-doped InP have been reported. The decrease of activation efficiency of ion implantation [1], the out-diffusion of Fe to the epitaxial layers [2], the formation of slip line defects in epitaxial layers due to Fe in substrates [3] are these disadvantages. It seems that Fe doping is a large obstacle for the rapid development of InP devices. The realization of undoped semi insulating InP is therefore strongly desired.

Regarding the possibility of semi-insulation of undoped InP, various studies have been reported. Klein et al have first reported that undoped high resistive InP (resistivity of about 10^5 Ωcm) was prepared by bulk annealing under phosphorus vapor pressure [4]. Hofmann et al have found that undoped semi-insulating InP can be obtained when high purity undoped InP is annealed [5]. Hirano et al showed that the resistivity is increased when Zn-doped InP wafers are annealed [6]. Kainosho et al reported that undoped 2 inch diameter semi-insulating InP with uniform electrical properties across the wafer can be obtained when high purity undoped InP is annealed at high temperatures and under high phosphorus pressures [7].

In the present paper, the related studies leading to the semi-insulation of InP are reviewed and the mechanism of semi-insulating behavior of InP is discussed.

PURITY OF UNDOPED InP AND ELECTRICAL PROPERTIES

Polycrystals and Single crystals

Since Fe atoms which act as deep acceptors are doped for compensating shallow donors, it was believed to be important to reduce the content of shallow donor impurities in InP material. In the case of InP single crystal growth, polycrystals are synthesized by the HB method in advance for single crystal growth. This is because direct synthesis which is common for GaAs is extremely difficult for InP due to high decomposition pressure of phosphorus. The purification of InP polycrystals was extensively studied in the beginning of InP material developments. Fig. 1 shows the relationship between the carrier concentration and the mobility of various InP polycrystals [1]. Polycrystals synthesized by the HB method using SiO_2 boats show the carrier concentrations of the level of $10^{16}cm^{-3}$. When pBN boats were used instead of SiO_2 boats, the carrier concentration was reduced to the level of $10^{15}cm^{-3}$. From these facts, it was concluded that the main shallow donor impurity during polycrystal synthesis was silicon. In fact, there was a good linear relationship between the carrier concentration and the Si concentration [8]. After having improved the synthesis conditions, the best carrier concentration of the level of $5x10^{14}$ cm^{-3} could be obtained. This level was comparable to the best purity reported for liquid phase epitaxial growth [9] and the photoluminescence has first shown free exciton peaks for bulk InP materials [10].

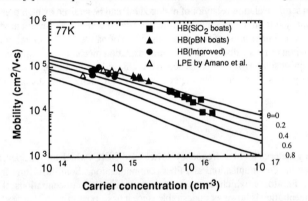

Fig. 1 Mobility and carrier concentration of various polycrystals

By using these polycrystals with various carrier concentrations, undoped single crystals have been grown by the LEC technique. The carrier concentration of undoped LEC InP was approximately $3x10^{15}cm^{-3}$ having no relationship with the purity of the polycrystalline starting material as shown in Fig. 2. This fact was very reproducible and the same phenomena have been reported by Morioka et al [11]. The reason why the carrier concentration is decreased from a level of $10^{16}cm^{-3}$ to the level of $3x10^{15}cm^{-3}$ after crystal growth can be explained by assuming that residual silicon in the polycrystals is reduced by the reaction of silicon in the melt with the encapsulant of B_2O_3. It is well known that B_2O_3 has a strong effect on purification due to the gettering effect. In fact, the chemical analysis supported this speculation. What was surprising was that the carrier concentration of the pure polycrystals is increased from $10^{14}cm^{-3}$ to $3x10^{15}cm^{-3}$ after crystal growth. This was intuitively explained by the contamination during crystal growth but the results of chemical analysis show no significant increase of the impurity content, a fact which was observed in other studies below [12]. It is also known from the crystal growth of GaAs that such a level of silicon contamination

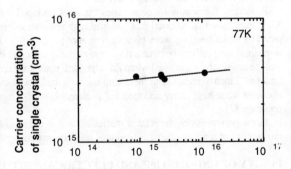

Fig. 2 Carrier concentrations of undoped InP single crystals grown by using polycrystals with various carrier concentrations

does not occur during LEC crystal growth. The fact that the carrier concentration of all single crystals show the same value is therefore speculated as being due to not only impurities but also native defects.

Wafer Annealing

In order to clarify this open question, Inoue et al have annealed undoped InP wafers with the carrier concentration of 3×10^{15}cm^{-3}, which are prepared from high purity polycrystals with the carrier concentrations of 5×10^{14}cm^{-3} under phosphorus vapor pressure of 0.5 atm in the temperature range between 520°C and 820°C [13]. It was found that the carrier concentration and the mobility are drastically changed when they are annealed at temperatures higher than 620°C. The carrier concentration is decreased to $2-3\times10^{14}$ cm^{-3}, one order of magnitude lower than that of the as-grown crystals. At temperatures higher than 720°C, the carrier concentration is slightly increased and the mobility is slightly decreased. These data have however been obtained only for crystals grown from high purity polycrystals. Single crystal wafers prepared from low purity polycrystals do not show this behavior as shown in Table I.

Table I. Electrical Properties of Various InP

Polycrystalline Starting Material		LEC Single Crystal As-grown		LEC Single Crystal Wafer Annealed	
Carrier Concentration (cm^{-3})	Mobility (cm^2/Vs)	Carrier Concentration (cm^{-3})	Mobility (cm^2/Vs)	Carrier Concentration (cm^{-3})	Mobility (cm^2/Vs)
2.3×10^{15}	49000	3.4×10^{15}	38900	3.8×10^{14}	86400
1.1×10^{16}	16300	3.6×10^{15}	37100	1.6×10^{15}	54400

These phenomena could be explained by a change of the concentration of phosphorus vacancies which act as shallow donors [14]. Single crystals grown by the LEC technique probably contain a significant concentration of phosphorus vacancies in the as-grown state. By wafer annealing under phosphorus overpressure, the concentration of phosphorus vacancies will be decreased. Under a phosphorus overpressure of 0.5atm, the phosphorus vacancies have minimum concentrations at 620°C. When the annealing temperature is increased under a constant phosphorus overpressure, the phosphorus vacancy concentration is increased and the carrier concentration is increased. This increase may be due to the constant phosphorus overpressure even for high temperatures. Then an appropriate increase of the phosphorus overpressure would suppress the increase of the carrier concentration. The decrease of the carrier concentration was not observed for single crystals prepared from low purity polycrystals as seen in Table I. The reason may be due to the residual silicon concentrations which can not be changed by annealing.

SEMI-INSULATION OF UNDOPED InP BY HIGH-PRESSURE ANNEALING

Semi-Insulating Behavior of Undoped InP

Table II shows the research activities for undoped semi-insulating InP by high-pressure annealing. Klein et al have first reported that the resistivity of undoped InP could be increased to 10^5 Ωcm by annealing at 900-940°C for about 3 weeks under a phosphorus vapor pressure of 6bars [4]. Hofmann et al have reported that undoped semi-insulating InP with the resistivity higher than 10^7 Ωcm can be obtained by annealing at 900°C for about 80h under a phosphorus vapor pressure of 5bars [5]. The mobilities of the semi-insulating undoped InP annealed under these conditions are very high (> 4000cm²/Vs) and can be compared to undoped conducting InP [15]. They also found that this semi-insulation takes place only for high purity undoped InP materials as shown in Fig. 3 [16]. Kainosho et al performed high-pressure annealing at 900°C for 20-40h under a higher phosphorus vapor pressure of 15 atm for 2 inch diameter undoped InP wafers [7]. This annealing produced semi-insulating InP wafers with the resistivity higher than $10^7\Omega$cm and the mobility greater than 4000cm²/Vs and with good uniformity as shown in Fig. 4.

Table II. Activities for undoped semi-insulating InP

Resistivity (Ωcm)	Annealing Conditions	Sample Size	Reference
3.6×10^5	900-940 °C 3 weeks $P_{P_4}=6$bar		[4]
2×10^7	900-950 °C 80 hr $P_{P_4}=5$bar	15×15mm²	[5]
1.3×10^7	900 °C 20 hr $P_{P_4}=15$atm	2 inch wafer	[7]

Effect of Stoichiometry

Table III shows the results of stoichiometry measurements of as-grown and annealed InP wafers by coulometric titration for In concentration. High-pressure annealing was found to be

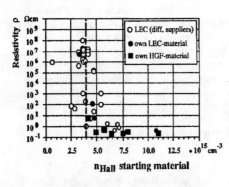

Fig. 3. Resistivity ρ of annealed undoped InP samples of different origin versus the net carrier concentration n_{Hall} of the starting material.

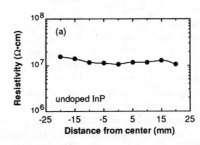

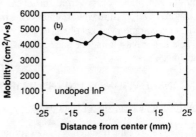

Fig. 4 Resistivity and mobility of 2 inch undoped semi-insulating InP.

effective to convert In-rich crystals to stoichiometric crystals but the conversion of crystal composition did not depend on phosphorus vapor pressure ranging from 7.5 atm to 20 atm. On the other hand, the phosphorus vapor pressure ranging from 15 atm to 25 atm did not affect the crystal composition of stoichiometric crystals within the detection limit of the coulometric titration method. The composition of crystals grown by the HB method under different phosphorus vapor pressure [16] did not vary within the detection limit neither [17]. These facts are not coincident with the case of GaAs wafer annealing [18]. The present results indicate that the phosphorus vapor pressure

Table III. Stoichiometry measurement results of InP by coulometric titration for In content

Sample	Phosphorus vapor pressur (atm)	In/InP as atomic fraction	Standard deviation (n=4)
HB polycrystals grown under various phosphorus vapor pressure			
A(25atm)	-	0.50003	±0.00005
B(17atm)	-	0.49998	±0.00003
C(40atm)	-	0.49995	±0.00004
Single crystals grown from In-rich melt			
as-grown	-	0.50020	±0.00009
annealed	7.5	0.50002	±0.00005
annealed	1 5	0.50002	±0.00007
annealed	2 0	0.50005	±0.00003
Single crystals grown from stoichiometric melt			
as-grown	-	0.49999	±0.00003
annealed	1 5	0.50004	±0.00004
annealed	2 5	0.50002	±0.00004

has not a large effect on changing the composition within the detection limit. This suggests that the nonstoichiometric region in the phase diagram of InP is very narrow compared with the case of GaAs in which the nonstoichiometric region is sufficiently large to be studied by the coulometric titration analysis. This however does not exclude the possibility that the phosphorus vapor pressure is related with intrinsic defect concentrations. It is concluded that the nonstoichiometric region of InP is too narrow to detect it by the titration analysis.

Contribution of Residual Fe

Fig. 5 shows the relationship between the Fe concentrations of undoped and Fe-doped InP wafers before and after high-pressure annealing. For InP wafers with low Fe concentrations before annealing, a Fe contamination with a level of $5 \times 10^{14} cm^{-3}$ after annealing is clearly seen. This Fe contamination level is much lower than the doping level of conventional Fe-doped semi-insulating InP which is above $1 \times 10^{16} cm^{-3}$. Hirt et al have reported similar results concerning the Fe contamination of undoped InP, which was annealed at 900°C for 80hrs under a phosphorus vapor pressure of up to 8atm. A total Fe-incorporation of $0.8-3.5 \times 10^{15} cm^{-3}$ was found by chemical trace analysis [16] and the

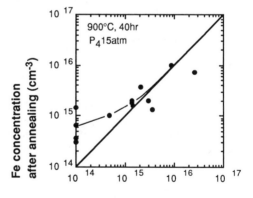

Fe concentration before annealing (cm^{-3})

Fig. 5 Relationship between the Fe concentrations of undoped and Fe-doped InP wafers before and after annealing.

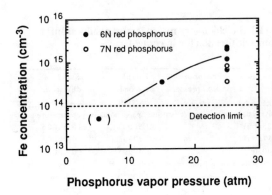

Fig. 6 Fe concentrations of undoped conductive InP

concentration of electrically active Fe could be proven to be in the range of $1 \times 10^{15} cm^{-3}$ by optical and electrical spectroscopy [16, 19].

Fig. 6 shows the Fe concentration of undoped conductive InP wafers after high-pressure annealing. It is seen that the level of Fe contamination is dependent on the phosphorus vapor pressure and the purity of red phosphorus. This result suggests that the origin of the Fe contamination may be due to the vapor source of red phosphorus. It is assumed that Fe atoms are contained as phosphides in red phosphorus, and they are transported from the vapor source to InP wafers, since the vapor pressure of Fe phosphide is known to be sufficiently high [20]. There is however the other possibility that the contamination stems from ampoule materials. The purity of ampoul material may be also a point to be considered for preventing the Fe contamination.

Fig. 7 shows the relationship between the resistivity and the Fe concentration after high-pressure annealing of undoped conductive InP and low Fe-doped InP. The resistivity of Fe-doped as-grown crystals is also shown for comparison. It is clearly seen that semi-insulating behavior of undoped InP wafers after annealing is due to slight Fe contamination. The minimum Fe concen-

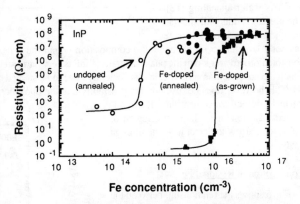

Fig. 7 Resistivity of InP as a function of Fe concentrations

tration for realizing semi-insulation of undoped InP by annealing was found to be about $5 \times 10^{14} cm^{-3}$, which is much lower than that of conventional Fe-doped semi-insulating InP.

Contribution of Shallow Donors

Table IV shows the electrical properties of undoped conductive InP wafers after high-pressure annealing. When undoped conductive InP wafers were not contaminated by Fe during annealing, they remained conductive but the carrier concentration was surprisingly lowered to below $2 \times 10^{13} cm^{-3}$ (samplesNo. 8, 9). This very low carrier concentration is not due to an increase of acceptors and substantial compensation of residual donors. It was found that the photoluminescence spectra of undoped semi-insulating InP were highly resolved with increased intensities, and the emission of excitons bound to shallow residual neutral donor in the rotational states are clearly seen (Fig. 8) [21, 22]. These photoluminescence spectra are comparable to the spectra of the purest MOCVD epitaxial layers [23]

Table VI Electrical properties of InP wafers after annealing.

Sample No.	Phosphorus vapor pressure (atm)	Fe concentration (cm^{-3})	Resistivity ($\Omega \cdot cm$)	Carrier concentration (cm^{-3})	Mobility (cm^2/V·s)
1	as-grown	$\leq 1.0 \times 10^{14}$	3.2×10^{-1}	4.5×10^{15}	4340
2	25	1.2×10^{15}	4.7×10^{6}	3.2×10^{8}	4220
3	25	6.6×10^{14}	2.0×10^{7}	9.3×10^{7}	4460
4	25	3.6×10^{14}	1.1×10^{6}	1.4×10^{8}	4300
5	25	8.2×10^{14}	1.3×10^{7}	1.2×10^{8}	4020
6	25	2.1×10^{15}	5.0×10^{6}	3.1×10^{8}	4000
7	25	2.0×10^{15}	9.5×10^{6}	1.4×10^{8}	4560
8	25	3.6×10^{14}	1.5×10^{2}	9.3×10^{12}	4510
9	25	$\leq 1.0 \times 10^{14}$	1.1×10^{2}	1.3×10^{13}	4470

and are the best data ever reported for InP bulk material. These photoluminescence experiments therefore proved that the material purity was not deteriorated during high pressure annealing.

Another proof for the reduction of shallow donors arises from Fig. 9 and Fig. 10 [19], showing that the net shallow level concentration derived from CV measurements at 300K is affected by the change of the net carrier concentration due to annealing. Furthermore this change of shallow levels can be attributed to a change of shallow donors by an evaluation of the mobility and the

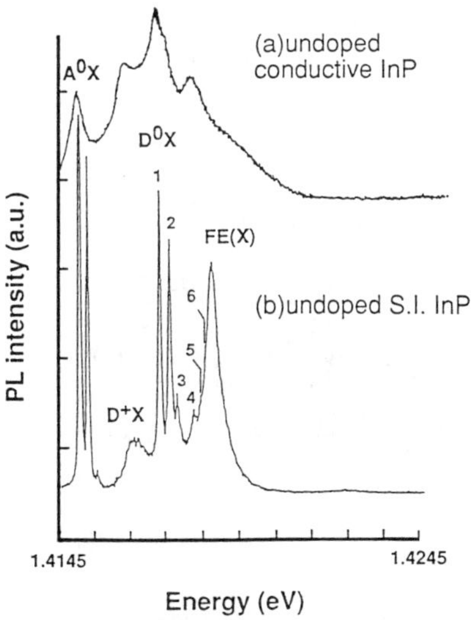

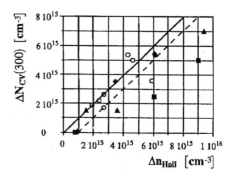

Fig. 8 Phosoluminescence (2K) of (a) undoped conductive InP and (b) undoped semi-insulating InP.

Fig. 9 Change of the net-concentration of shallow levels, detected by CV- measurements at 300K, versus the change of the carrier concentration Δn_{Hall}.

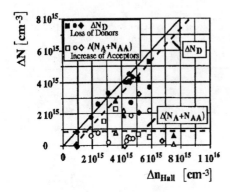

Fig 10. Decrease of the donor (ΔN_D) and increase of the total acceptor ($\Delta N_A + \Delta N_{AA}$) concentration versus the change of the carrier concentration Δn_{Hall} according to an evaluation of mobility.

compensation ratio respectively. As the main effect of high-pressure annealing is a reduction of the residual donor concentration it follows clearly that this is very effective way for preparing high-purity bulk material with extremely low carrier concentration.

Contribution of Deep Levels

It is pointed out that deep levels with the activation energy of about 0.4eV are formed during annealing. Hirano et al reported that the deep level of the activation energy of 0.44eV is formed by annealing in slightly Zn-doped InP wafers at 650°C for 3 hrs under phosphorus vapor pressure of 0.2 atm [6]. Bardeleben et al also reported, that an electron trap with an activation energy of 0.4eV was formed by wafer annealing at 720°C [14]. Hirt et al have investigated the deep levels and the compensation mechanism of annealed InP by DLTS measurements in detail [19]. It could be shown that two deep levels with activation energies of 0.4eV and 0.6eV are incorporated or created in undoped annealed InP (Fig. 11) The deeper level at 0.6eV could be identified as Fe, acting as the dominant compensating level in the annealed semi-insulating InP. The origin of the deep level at 0.4eV is not known yet. It is speculated as being related with native defects, since it was found in material annealed under a variety of conditions. Though it can be identified as a deep acceptor it does not contribute to the compensation mechanism of semi-insulating InP as it is not occupied due to its location well above the Fermi level. Without sufficient Fe it can either partially compensate the residual net carrier concentration or it can fully compensate the remaining carriers giving activation energies of about 0.4eV for the material, as observed by Hirano et al.

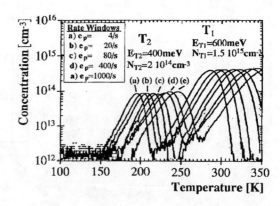

Fig 11 DLTS-spectra of undoped InP, annealed for 80h at 900 °C under phsophorus-pressure (5atm).

We can conclude that there is an intrinsic defect in annealed InP as in the case of GaAs but the compensation in undoped semi-insulating InP is caused by incorporated Fe.

Diffusion and Bulk Effects

The origin of the changes in the balance of electronic levels was studied by a detailed study on short time annealed InP samples by Hirt and Müller [24]. From DLTS measurements in the surface region of samples annealed under a slight phosphorus pressure (0.05atm) for 2.5hrs, the behavior of deep and shallow levels can be clearly distinguished (Fig. 12). Within a depth of about 150 μm, an out-diffusion of a part of the shallow donors ($\Delta N_D = 5.1 \times 10^{15} cm^{-3}$) can be detected. Approximately within the same scale an in-diffusion of Fe can be found. Though there is a drop of the Fe-concentration near the surface not understood by now, it can be seen quite clearly that the Fe concentration is decreasing about two orders of magnitude. From the nearly constant distribution of the deep level with 0.4eV

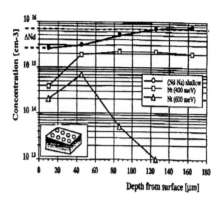

Fig. 12 Profile of the shallow and deep levels near the surface of a sample, annealed for 2.5h under phosphorus atmosphere. The bulk concentration of shallow levles was $7 \times 10^{15} cm^{-3}$.

activation energy, it seems very likely that this level is created within the volume and is not related to any diffusing species.

Taking into account, that semi-insulating behavior occurs, when the Fe concentration exceeds the concentration of shallow donors, the simultaneous out-diffusion of donors and the in-diffusion of Fe is equivalent to an in-diffusion of a region which shows semi-insulating behavior or at least high resistivity (Fig. 13).

From the analysis of the diffusion profiles it can be concluded, that both the mechanisms for the out-diffusion of donors and for the in-diffusion of Fe show diffusion coefficients in the range of $0.5-1 \times 10^{-7} cm^2/s$. Although these values are significantly higher than any previously reported values for diffusion of Fe or P, rapid diffusion mechanisms are possible for low concentrations of the involved species and especially in the presence of intrinsic defects [25, 26].

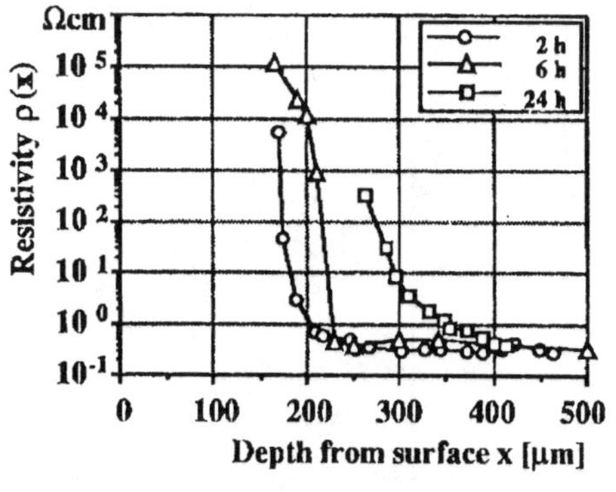

Fig. 13 Resisitivity distribution ρ(x) near the surface of short-time annealed samples measured by differential Hall-effect measurements. The samples were annealed for different time (2h, 6h, 24h) at 900 °C under phosphorus atmosphere.

Consideration by Shockley Diagrams

For consideration of the compensation mechanism, the Shockley diagram [27,28] is convenient. In the case of very pure undoped InP, we have to consider three levels, shallow donors due to Si, shallow acceptors due to Zn, deep acceptors due to Fe contamination. It is distinctive that the main shallow acceptor is Zn. It is known that the D-A pair peak due to Zn is clearly seen while D-A pair peaks due to other acceptor impurities such as Mg and Ca which can be observed in InP polycrystals are not seen [29,30]. The fact that the carrier concentration of the pure InP is $5 \times 10^{14} \mathrm{cm}^{-3}$ and the compensation ratio is about 0.2 suggests that the concentration of Zn (Na) is less than $1 \times 10^{14} \mathrm{cm}^{-3}$.

Fig. 14(a) shows a typical case of conventional Fe-doped material. The compensation takes place between shallow donors and doped Fe with higher concentration ranges. In the case of undoped semi-insulating InP by high pressure annealing, the Shockley diagram may become as shown in Fig. 14(b). There are three facts that 1) the Fe concentration must be higher than $5 \times 10^{14} \mathrm{cm}^{-3}$ for semi-insulation, 2) the residual shallow acceptor concentration is less than $1 \times 10^{14} \mathrm{cm}^{-3}$, 3) when it does not become semi-insulating with low Fe concentration, the carrier concentration is $1 \times 10^{13} \mathrm{cm}^{-3}$. The fact that the room temperature carrier concentration is $1 \times 10^{13} \mathrm{cm}^{-3}$ with a very low Fe concentration can be explained by the pinning of the Fermi-level by the deep acceptor of 0.4eV under the condition that the concentration of shallow donors is higher than that of shallow acceptors as shown by dotted lines. Here, the concentration of the shallow donors (Nd) which will be due to phosphorus vacancies is supposed to be 5×10^{14} cm^{-3} and that of shallow acceptors (Na) is $1 \times 10^{14} \mathrm{cm}^{-3}$. The concentration of 0.4eV deep acceptor (Naa') must be higher than $4 \times 10^{14} \mathrm{cm}^{-3}$ to obtain the experimental net carrier concentration of $1 \times 10^{13} \mathrm{cm}^{-3}$.

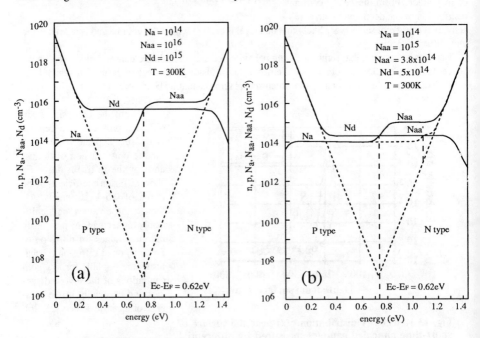

Fig 14 Schockley diagrams. (a) Conventional Fe doped InP. (b) Undoped semi-insulating InP.

Under this basis, if we have a slight amount of residual Fe exceeding the net concentration of shallow levels, semi-insulating behavior will occur for Fe concentrations above $5 \times 10^{14} \text{cm}^{-3}$ as shown in Fig. 14(b).

POSSIBLE MECHANISMS OF SEMI-INSULATION

From above mentioned experimental results, it is first concluded that high-pressure annealing is effective to reduce the net concentration of residual defects by reducing the shallow donor concentration which is probably due to phosphorous vacancies.

It was also found that the semi-insulating state was only achieved if the Fe concentration is higher than $5 \times 10^{14} \text{cm}^{-3}$, a concentration which is much lower than that for conventional semi-insulating Fe-doped InP. The semi-insulating behavior of annealed InP is thus concluded to be due to slight Fe contaminations and a large reduction of the concentration of shallow donors during high-pressure annealing.

Without sufficient Fe concentration, semi-insulating state cannot be achieved. However, due to compensation with the deep acceptor of an energetic level of 0.4eV below the conduction band, low residual carrier concentrations can occur.

The reproducibility of undoped semi-insulating InP is affected by the scattering of Fe contamination from run to run. The appropriate control of Fe contamination with the level of more than $5 \times 10^{14} \text{cm}^{-3}$ is therefore believed to be important for the reproducible preparation of "undoped" semi-insulating InP. Since the Fe contamination level for semi-insulation is extremely low, these semi-insulating materials can be called "undoped semi-insulating InP", as in the case of "undoped semi-insulating GaAs" for which it is known that the precise control of carbon contamination is important for stable production of semi-insulating crystals.

SUMMARY

Undoped semi-insulating InP was found to be obtained by high pressure annealing of high purity undoped InP. Residual Fe atoms which act as deep acceptors are originated by contamination during annealing. The minimum Fe concentration for realizing the semi-insulation of undoped conductive InP was found to be $5 \times 10^{14} \text{cm}^{-3}$. By detailed investigation of the balance of electronic levels, it was found that the donor concentration is reduced by annealing. Furthermore, the undoped semi-insulating InP shows a highly resolved photoluminescence spectrum, comparable to the best data reported for MOCVD material. From these results, it is concluded that the occurence of undoped conductive InP to semi-insulating state by high-pressure annealing is due to the compensation of shallow donors, the concentration of which is reduced by annealing significantly and the slight Fe contamination during high-pressure annealing.

ACKNOWLEDGEMENTS
The authors are grateful to Dr. K. Aiki and Dr. D. Hofmann for helpful discussions. Special thanks is also expressed to Drs. Y. Makita, A. Yamada and H. Yoshinaga for photoluminescence measurement and for useful discussions.

REFERENCES
[1] K. Kainosho, H. Shimakura, T. Kanazawa, T. Inoue and O. Oda, Proc. 16th International Symp. on GaAs and Related Compounds, (Inst. Phys. Conf. Ser. 106, 1990), pp25.
[2] D. E. Holmes, R. G. Wilson and P. W. Yu, J. Appl. Phys. 52, 3396 (1981).
[3] C. Miner, D. G. Knight, J. M. Zorzi, R. W. Streater, N. Puetz and M. Ikisawa, Proceedings for DRIP VI, 1993 Santander, to be published.

[4] P. B. Klein, R. L. Henry, T. A. Kennedy and N. D. Wilsey, "Defects in Semiconductors", edited by H. J. Bardeleben, Mater. Sci. Forum 10-12, 1259(1986).

[5] D. Hofmann, G. Müller, and N. Streckfuß, Appl. Phys. A48, 315 (1989).

[6] R. Hirano and T. Kanazawa, J. Cryst. Growth 106, 531 (1990).

[7] K. Kainosho, H. Shimakura, H. Yamamoto and O. Oda, Appl. Phys. Lett. 59, 932 (1991).

[8] O. Oda, K. Katagiri, K. Shinohara, S. Katsura, Y. Takahashi, K. Kainosho, K. Kohiro and R. Hirano, "Semiconductor and Semimetals", edited by R. K. Willardson and A. C. Beer (Academic, New York, 1990), Vol. 31, pp93.

[9] T. Amano, S. Kondo and H. Nagai, Ext. Abstracts (The 34th Spring Meeting) The Jpn. Soc. of Appl. Phys. and Related Socs. pp126.

[10] T. Inoue, K. Kainosho, R. Hirano, H. Shimakura, T. Kanazawa and O. Oda, J. Appl. Phys. 67, 7165 (1990).

[11] M. Morioka, K. Tada and S. Akai, Ann. Rev. Mater. Sci. 17, 75 (1987).

[12] G. Jacob, R. Coquille and Y. Toudic, Proc. 6th Conf. on Semi-Insulating III-V Materials, Toronto, 149 (1990).

[13] T. Inoue, H. Shimakura, K. Kainosho, R. Hirano and O. Oda, J. Electrochem. Soc. 137, 1283 (1990).

[14] H. J. von Bardeleben, D. Stievenard, K. Kainosho and O. Oda, J. Appl. Phys. 70, 7392 (1991).

[15] G. Müller, D. Hofmann, P. Kipfer and F. Mosel, Proc. 2nd International Conf. on InP and Related Materials, Denver, 21 (1990).

[16] G. Hirt, D. Hofmann, F. Mosel, N. Schäfer and G. Müller, J. Electronic Materials 20, 1062 (1991).

[17] N. Schäfer, private communication.

[18] O. Oda, H. Yamamoto, M. Seiwa, G. Kano, M. Mori, H. Shimakura and M. Oyake, Semicond. Sci. Technol. 7, A215 (1992).

[19] G. Hirt, S. Bornhorst, J. Friedrich, N. Schäfer and G. Müller, Proc. 5th International Conf. on InP and Related Materials, Paris, 313 (1993).

[20] G. Hirt, D. Wolf and G. Müller, J. Appl. Phys. 74, 5538 (1993).

[21] J. R. van Wazer, "Phosphorus and its compounds", vol. 1, (Interscience Publishers, New York, 1966), pp165.

[21] Y. Makita, A. Yamada, H. Yoshinaga, T. Iida, S. Niki, S. Uekusa, T. Matsumori, K. Kainosho and O. Oda, Proc. 4th International Conf. on InP and Related Materials, Newport, 626 (1992).

[22] H. Yoshinaga, Y. Makita, A. Yamada, T. Iida, S. Niki, T. Matsumori, K. Kainosho and O. Oda, Proc. 5th International Conf. on InP and Related Materials, Paris, 317 (1993).

[23] M. Razeghi, Ph. Maurel, M. Defour, F. Omnes, G. Neu, and A. Kozacki, Appl. Phys. Lett. 52, 117 (1988).

[24] G. Hirt and G. Müller, to be published

[25] J. A. van Vechten, J. Appl. Phys. 53, 7082 (1982).

[26] B. Tuck and P. R. Jay, J. Phys. D 11, 1413 (1978).

[27] W. Schockley, " Electrons and Holes in Semiconductors ", D. van Nostrand Co., Inc., Princeton (1950).

[28] S. Martin and G. Jacob, Acta Electronica 25, 123 (1983).

[29] E. Kubota, Y. Ohmori and K. Sugii, J. Appl. Phys., 55, 3779(1984).

[30] E. Kubota and A. Katsui, Jpn. J. Appl. Phys. , 24, L344(1985).

RELIABLE IMPURITY IDENTIFICATION IN InP

M. L. Schnoes*, T. D. Harris*, S. J. Pearton*, M. A. Di Giuseppe*, R. Bhat** and
H. M. Cox**
*Room 1A-332, AT&T Bell Laboratories, Murray Hill, NJ 07974
**Bell Communications Research, Red Bank, NJ 07701

ABSTRACT

Low temperature, resonantly excited photoluminescence (PL) has proven to be the method of choice for impurity identification in GaAs. InP has suffered from insufficient impurity binding energy data to benefit similarly. We will report results of selectively-excited donor-acceptor pair spectroscopy for acceptor identification in InP. Ion implantation doping of high purity InP is used for generation of known impurity samples. Progress toward a complete database of acceptor binding energies in InP is reported. We will discuss the results of high magnetic field low temperature resonant photoluminescence spectroscopy for donor identification in InP. The success of donor ion implantation studies will be included. This data should provide direction for efforts in growing high purity InP by MOCVD and gas source MBE.

INTRODUCTION

Indium phosphide and related compounds are candidate materials for high speed electronics and optoelectronic devices in telecommunication networks. In order to achieve the high purity base material upon which devices depend, identification of residual impurities is important. Low temperature (2K) photoluminescence (PL) is a sensitive tool for use in determining the active impurities in similar material. Previous work using the observation of free-to-bound (FB) and donor-acceptor pair (DAP) transitions have led to assignment of binding energies for the chemical species of several acceptors[1-5]. The difficulty with these data is recombination of pairs with different separations contribute to the DAP PL signal, and the dopant level dependant FB transitions involving electrons with a distribution of kinetic energy. Consequently, substantial uncertainty exists in these binding energies, with attendant uncertainty in identification[6-8]. Selective Pair Luminescence (SPL) overcomes this ambiguity in three important ways. Only specific pair spacings are excited and detected, yielding narrow peaks. All excited and emitting states are bound, overcoming the energy uncertainty of free carriers. Finally, several (typically 2-4) excited states from each acceptor are observable, reducing the possibility of incorrect peak assignment. SPL has been demonstrated for identification of a few acceptors in InP[9-11]. These experimental advantages are enhanced by the availability of effective mass theory[12-13], which permits the determination of any excited state energy from any correctly assigned peak or measured binding energy. Thus interfering phonon Raman spectral features, or overlapping SPL peaks can be confidently assigned. Some difficulties remain, such as the possibility of lattice defects with energies similar to shallow impurities. This is illustrated by the recent report of phosphorus antisites defects which are resonant with the bottom of the conduction band[14-15].

Donor identification is more difficult due to the small binding energy of the donors (7.4 meV vs. 35-55 meV for acceptors), and the attendant larger wave functions. The expected range of binding energies for different chemical species, referred to as the chemical shift, is ~0.1 meV[16]. Donor identification has been successfully performed using far infrared

113

photoconductivity (FIRPC) in high (3.6 T and 6.3 T) magnetic field[17-18]. The donor 1s-2p⁻ transition is detected by direct absorption. As many as 12 distinctly different donor binding energies have been observed, but only five have been assigned to specific donor species (Si, S, Ge, Te, and Se/Sn: these last having nominally identical binding energies)[17-18]. Donor identification has also been successfully performed using resonantly excited donor bound exciton PL in high (9.7 T) magnetic field[19-20]. The donor ground state and excited

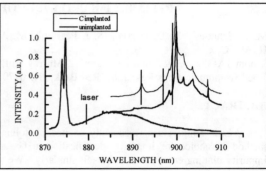

Figure 1 Non resonant and selective pair luminescence of as grown and C implanted InP.

states, known as two electron satellites (TES), are observed separately. Spectral separation between these two features is directly comparable to the FIRPC data. While the PL spectra are more complex, this method is more versatile. Impurity identification has only been based on doping during growth. We will explore the applicability of implantation doping to donor identification by resonantly excited magneto-photoluminescence.

EXPERIMENTAL SECTION

The samples used for this study were grown by hydride VPE, chloride VPE, and MOCVD, 1-3 um thick. Doping standards were prepared by implanting these samples in a non-channeling direction (7° tilt, 45° rotation) to a dose less than 10^{16} /cm^3, giving a peak level between 200-300 nm deep in the sample and the active range up to 500 nm. The samples were then rapid thermal annealed in a graphite susceptor for 10s at 750°C. The atmosphere was a phosphorous overpressure in a flowing argon ambient. Implants discussed here on various as grown material are Mg, C, Zn, Be, Ca, and Cd.

For the SPL experiments, the samples were mounted in a strain free manner within a Janis flowing helium cryostat, and submersed in superfluid helium (2K). A tunable titanium sapphire, (Ti:SA), laser was used for the excitation source. The light was collected and focussed into a Spex

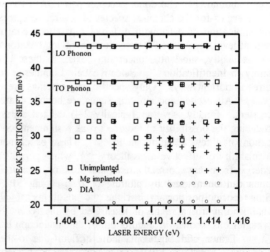

Figure 2 SPL peak position plotted vs. laser excitation energy for as grown and Mg implanted InP.

triple monochromater, dispersion 1.3 nm/mm, and detected using a LN_2 cooled Princeton Instruments CCD camera.

Magnetic photoluminescence (MPL) spectroscopy was used for donor spectroscopy. The samples were mounted strain free within a Janis magnetic flowing helium cryostat. The laser source was a Ti:SA laser tuned slightly above band gap ($0.002W/cm^2$) for donor ground state observation, or to the $(D^o,X)_{n=4}$ transition ($0.02W/cm^2$) for TES observation. Emission was collected and focussed into an 0.85 meter focal length, double monochromater, dispersion 0.4 nm/mm, and detected using a LN_2 cooled Photometrix CCD camera.

DISCUSSION

The recombination energy of a DAP is given by

$$h\nu_1 = E_g - (E_A + E_d) + e^2/(\varepsilon r) + f(r) \qquad \text{equation (1)}$$

while the energy of absorption into a DAP is given by

$$h\nu_2 = E_g - (E_A + E_d) + e^2/(\varepsilon r) + f(r)_{ex} + E_{ex} \qquad \text{equation (2)}$$

Where f(r) is the van der Waals forces between the donor and acceptor, and E_{ex} is the acceptor excited state shift. For pair separations with f(r)~0 the energy difference between equations (1) and (2) is then an accurate measure of the acceptor ground state to excited state energy. This condition holds for pair separations greater than the sum of the donor and acceptor Bohr orbits. A typical spectrum which illustrates this measurement is shown in Figure 1. For each excitation wavelength, there is a specific pair separation resonant for each excited state of each acceptor present in addition to Raman scattering from the optical phonons. Spectra of the sample as grown and after carbon implantation are shown. For reference a PL spectrum of the as grown sample excited with above gap energy is included in Figure 1. The enhanced visibility of acceptor transitions gained by resonantly exciting DAP transitions is obvious. At least three new transitions are detected in the implanted sample. That these new features are selectively excited pair luminescence, SPL, transitions can be assured by plotting their energy position against the laser energy for a range of laser energies, as shown in Figure 2, for as grown and Mg implanted samples. Substrate SPL signals can overwhelm the epilayer signal. We see no significant SPL

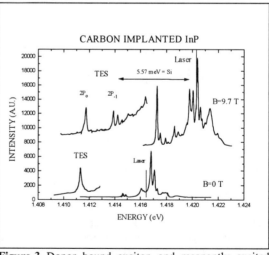

Figure 3 Donor bound exciton and resonantly excited TES spectra at both zero field and high field for C implanted InP.

from substrate alone and no change in SPL for as grown, annealed InP epilayers.

Mg was chosen specifically because it is among those few acceptors for which accurate excited state energies have been determined[11]. These are indicated by the extra peaks detected after implantation. In addition to those peaks expected from the known excited state positions of Mg, a well behaved set of acceptor excited states with much smaller binding energy than any previously is reported.

IMPURITY	BINDING		2S	$2P\frac{3}{2}$	$2P\frac{5}{2}\Gamma 7$
Mg	**Exp**		28.7	25	
	Calc	40.6	28.3	26.8	32.8
C	**Exp**		31	28	33.4
	Calc	44.6	31.1	29.5	36
Zn	**Exp**		32.4		38
	Calc	46.1	32.2	30.5	37.2
Be	**Exp**		28.7	24.8	30.4
	Calc	41.3	28.8	27.3	33.3
Ca	**Exp**		29	25.4	
	Calc	43	30.0	28.4	34.7
Cd	**Exp**			32.5	44.3
	Calc	53-56	37-39	35-37	43-45
DIA	**Exp**		23.2	20.5	
	Calc	33.2	23.2	21.9	26.8
EMA	**Calc**	23	16	15.2	18.6
	Calc	40	28	26.4	32.3

$\gamma_1 = 4.952 \quad \gamma_2 = 1.652 \quad \gamma_3 = 2.352$

This low binding energy acceptor was present in all implant doped samples, independent of the implanted chemical element. These data indicate that either the low binding acceptor arises from implant induced lattice damage or implant activation of an as grown inactive acceptor. Since the acceptor is identical in three different as grown and implanted samples, the former seems more likely and is thus labeled damage induced acceptor,(DIA).

Given that each acceptor present will exhibit two to four SPL features in addition to optical phonon Raman scattering, peak overlap is inevitable. Fortunately, a valid effective mass theory can predict the positions of all excited states for an acceptor from the measured energy of each excited state[12-13]. This redundancy of binding energy determination for SPL spectroscopy is among its most important advantages. Unfortunately no experimental verification of acceptor effective mass theory has been published for InP. In addition the range of published valence band parameters for InP is large, with no obvious method of choosing those most applicable to the problem at hand. Effective mass theory has been shown to be quantitatively accurate for acceptors in GaAs[21]. This model can not predict chemical shifts. Rather, effective mass theory predicts the position of each excited state as a fraction of the acceptor binding energy, in addition to the binding energy of the acceptor with zero chemical shift. There is no *a priori* reason to expect such an acceptor. For GaAs, C is a true effective mass acceptor. For InP, the effective mass binding energy calculations range from 20 to 50 meV.

In Table 1 on the preceding page is a summary of acceptor data taken by us to date. In addition we include calculated excited state energies using the most widely used band parameters. The most reliable acceptor binding energies published previously are for Mg and Zn. Our results are in excellent agreement for these elements. We show data here for Be, Ca and C which we believe to be of similar accuracy and certainty. For Cd, the as grown material was of marginal purity for implant doping. Since only the 2S excited state was detected there is still some uncertainty for the correct Cd binding energy. Work is in progress to complete data for Si, Ge, Sn, and Pb.

The high spectral resolution required for donor identification can be enhanced by using both low temperature and high magnetic field. The field induced donor wavefunction compression reduces line broadening by isolating the wavefunction from surrounding charged sites in the lattice while simultaneously increasing the binding energy chemical shift. The magnet used in this study is capable of 12T fields, 9.7T was used for comparison to existing magneto PL data. Spectra of donor ground and excited states at both zero field and 9.7 T are shown in fig 3. We rely on the prior studies for spectra assignments[19-20]. In figure 4 is an expanded view of the 2p⁻ TES for the as grown and C implanted samples. Each

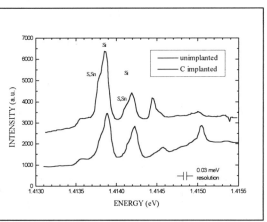

Figure 4 A comparison of 2p⁻ TES spectra for as grown and C implanted InP.

donor results in a doublet emission pattern. Peak positions agree well for the donors identified on the figure. The small differences in the two spectra can be attributed to small differences in the energy of excitation for the spectra. No new donor peaks can be detected for this sample. C doped InP is always n type as grown. Despite this unanimity of growth results, C acceptors are easily created by implant doping and we have failed to observe C donors. Several conclusions can be drawn and tested.

First, the implanted layer is less than 500 nm thick while the epitaxial layer for this sample is 3500 nm. Perhaps the C donor signal is too weak to detect. This possibility can be tested either by implanting thin epitaxial layers or by implants with a sufficient energy range to dope the majority of the thicker layers. Implant doping by our procedure may not

117

be ineffective for donor formation in InP. Since no C donor chemical shift data exists, the best alternative is to implant the only identified group IV donor not present in our as grown material, Ge. Implant doping studies with Ge will permit confirmation of the implant activation procedure.

CONCLUSIONS

We report the first precise binding energies for Be, C, and Ca, and confirm those previously reported for Zn, and Mg. The excited state energies of these acceptors with the exception of Zn do not conform to effective mass theory with existing band parameters. Finally, implant conditions which give strong acceptor signals have not created observable C donor levels. These conditions are best understood by Ge donor doping studies.

REFERENCES

1. P.J. Dean, D.J. Robbins, and S.G. Bishop, J. Phys. C 12, 5567 (1979).
2. E.W. Williams, W. Elder, M.G. Astles, M. Webb, J.B. Mullin, B. Straughan, and P.J. Tufton, J. .Electrochemical Society, 120, 1741 (1973).
3. Gernot S. Pomrenke, Y.S. Park, and Robert L. Hengehold, J. Appl. Phys. 52, 969 (1981).
4. A.M. White, P.J. Dean, K.M. Fairhurst, W. Bardsley, E.W. Williams and B. Day, Solid State Comm. 11, 1099 (1972).
5. J.D. Oberstar, B.G. Streetman, J. Appl. Phys. 53, 5154 (1982).
6. B.J. Skromme, G.E. Stillman, J.S. Oberstar, S.S. Chan, Appl. Phys. Lett. 44, 319 (1984).
7.Eishi Kubota, Yutaka Ohmore, and Kiyomasa Sugii, J. Appl. Phys. 55, 3779 (1984).
8. V. Swaminathan, V.M. Donnelly, and J. Long, J. Appl. Phys. 58, 4565 (1985).
9. P.J. Dean, D.J. Robbins, and S.G. Bishop, Solid State Comm. 32, 379 (1979).
10. D. Barthruff, and H. Haspeklo, J. of Lumin. 24/25, 181 (1981).
11. A.C. Beye, A. Yamada, T. Kamijoh, H. Tanoue, K.M. Mayer, N. Ohnishi, H. Shibata, and Y. Makita, Appl. Phys. Lett. 56, 349 (1990).
12. A. Baldereschi and Nunzio O. Lipari, Phys. Rev. B, 8, 2697 (1973).
13. A. Balderesch, and Nunzio O. Lipari, Phys. Rev. B, 9, 1525 (1974).
14. P.W. Yu, B.W. Liang and C.W. Tu, Appl. Phys. Lett , 61, 2443 (1992).
15. P. Dreszer, W.M. Chen, K. Seendripu et al, Phys. Rev. B, 47, 4111 (1993).
16. M.S. Skolnick and P.J. Dean, J. Phys C, 15, 5863 (1982).
17. C.J. Armistead, A.M. Davidson, P. Knowles etal, 1983 Proc. Int. Conf. on Appl. of High Magnetic Fields in Semiconductor Physics, Lecture Notes in Physics (Springer, Berlin, 1983), 177, 289.
18. C.J. Armistead, P. Knowles, S.P. Najda, and R.A. Stradling, J. Phys. C, 17, 6415 (1984).
19. P.J. Dean, and M.S. Skolnick, and L.L. Taylor, J. Appl. Phys, 55, 957 (1984)
20. M.S. Skolnick, P.J. Dean, L.L. Taylor et al, Appl. Phys. Lett, 44, 881 (1984).
21. T.D. Harris, J.K. Trautman, and J.I. Colonell, Materials Science Forum, 65-66, 21 (1990).

CORRELATION BETWEEN ELECTRICAL PROPERTIES AND RESIDUAL DEFECTS IN Se$^+$-IMPLANTED InP AFTER RAPID THERMAL ANNEALING

P. MÜLLER, T. BACHMANN, E. WENDLER, W. WESCH AND U. RICHTER*
Friedrich-Schiller-Universität Jena, Institut für Festkörperphysik, Max-Wien-Platz 1, D - 07743 Jena, Germany
* Labor für Mikrodiagnostik, Weinbergweg 2, D - 06120 Halle, Germany

ABSTRACT

$\langle 100 \rangle$ -semiinsulating InP was implanted with 600 keV Se-ions at temperatures between 300K and 425K with an ion dose of 1×10^{14} cm^{-2}. After capping the samples with about 120 nm siliconoxynitride annealing was performed at 700°C up to 975°C using a graphite strip heater system. The annealed samples were analyzed with Rutherford backscattering, electron microscopy and conventional Hall measurements. The results show, that a strong correlation exists between defects remaining after annealing (for instance dislocations, loops, microtwins) and the measured electrical properties. An implantation temperature $\geq$ 395K and annealing at least at 800°C for 50 s is necessary to obtain high performance electrically active layers. The activation of selenium in InP can be well described using a simple thermodynamical model. The model yields an activation energy of $E_A = (1.0 \pm 0.1)$ eV which can be understood as the energy necessary to split-up selenium-vacancy-complexes and a diffusion energy of $E_d = (2.0 \pm 0.2)$ eV representing material transport of the semiconductor material.

INTRODUCTION

Besides the well investigated GaAs, InP becomes more and more attractive for fabrication of microwave devices and semiconductor lasers[1]. As for the other III-V-compounds also, the most promising dopant technology seems to be ion implantation in combination with subsequent thermal processing to eliminate implantation damage and to create high performance electrically active layers. The implantation and annealing processing of InP, however, should differ somewhat of that of GaAs, because of the relative stability of defects produced during ion implantation, its lower melting point (lower annealing temperatures should be applied) and the redistribution of Fe in semiinsulating InP during annealing. Because of this, a careful adjusting of the desired damage introduced by ion implantation, which is primarily dependent on the dose and implantation temperature, to the structural change in InP and therefore for the annealing conditions is necessary.

Different elements have been implanted into InP to get n-type conducting layers, like Si[2-4], Se[2,3,5] and Ge[6,7]. A comparison of the results obtained by different authors shows that better electrical properties of the annealed InP-layers can be achieved if ion implantation is performed at elevated temperatures and amorphization is prevented. But a consequent investigation of the structural and electrical properties of InP-layers, implanted at temperatures between room temperature (300K) and 425K (implantation in the so called "transition region"), and the resulting correlation of these properties is missing up to now.

This paper deals with ion implantation of Se into semiinsulating InP-crystals at implantation temperatures mentioned above. The resulting implantation damage and the residual damage after rapid thermal annealing (RTA) is the subject of investigation. Furthermore, the electrical properties of the annealed samples are presented and compared with their structural properties. Attempts were made to apply an activation model developed for Se-implanted GaAs[8,9] for InP.

Mat. Res. Soc. Symp. Proc. Vol. 325. ©1994 Materials Research Society

EXPERIMENTAL CONDITIONS

$\langle 100 \rangle$ -semiinsulating, Fe-doped InP-wafers were implanted in a nonchanneling direction with 600 keV Se-ions at temperatures between 300K and 425K. The implanted dose was kept constant at 1×10^{14} Se cm^{-2}. After implantation part of the samples was capped with a silicon-oxynitride caplayer (thickness $\approx$ 120 nm) produced by dc-sputtering of silicon in an atmosphere containing a mixture of Ar and N_2. The following annealing was performed by means of a graphite strip heater system[10] at temperatures from 700°C up to 975°C for different times (20 s to 180 s). The given temperatures are the heater temperatures because these can be exactly measured. The typical sample temperature is estimated to be 10% below that of the heater. This was taken into account for the calculation of activation energies from the time and temperature dependent annealing.

After annealing the caplayer was removed using buffered HF-acid. Part of the samples was prepared for electrical measurements (van der Pauw technique[11]). For this purpose a $\approx$ 1 μm thick mesa structure was etched at the surface of the samples and ohmic contacts were alloyed at 400°C for 10 min using an eutectic mixture of Au-Ge.

The investigation of the as-implanted as well as of the annealed samples was performed with channeling Rutherford Backscattering Spectrometry (RBS) by means of 1.4 MeV He$^+$ at a backscattering angle of 170°. Further information about the kind of defects was obtained from Transmission Electron Microscopy (TEM) at 1 MeV accelerating voltage.

RESULTS AND DISCUSSION

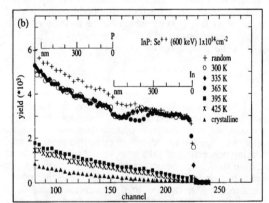

Fig. 1 RBS - spectra obtained on Se-implanted InP for different implantation temperatures.

Figure 1 shows the backscattering spectra obtained on the as-implanted InP. The implantation at 300K generates an amorphous surface layer of $\approx$ 480 nm thickness for the given dose of 1×10^{14} Se cm^{-2}. An increase of the implantation temperature up to 365K leads already to a thinning of the amorphized layer down to a thickness of $\approx$ 350 nm. A further increase of the temperature during ion implantation up to 395K results in a drastic decrease of the damage, i.e. amorphization is prevented and weakly damaged layers are created. Such layers contain mainly a mixture of point defects and point defect clusters, but we assume that also a certain concentration of larger defects (dislocations/dislocation loops) is present because this has been proved by means of energy dependent RBS-measurements for Si-implanted into InP performed at elevated temperatures[12].

Figure 2 shows the RBS - spectra obtained on annealed room temperature implanted InP-layers. An epitaxial recrystallization starting from the buried amorphous/crystalline (a/c-) interface is observed, however, only an approximately 160 nm thick layer recrystallizes perfectly at an annealing temperature of 750°C for 50 s. In the near surface region the back-scattering signal is nearly random-like indicating the presence of highly disorded structures within this layer. The TEM bright field micrograph (see fig. 3) shows an irregular structure which is identified as a microtwin structure of high density, indicated by extra cross-like spots in the diffraction pattern (inset of fig. 3). These microtwin clusters grow along the $\langle 111 \rangle$ -direction, seeding from defects originated during implantation on the a/c-interface[13].

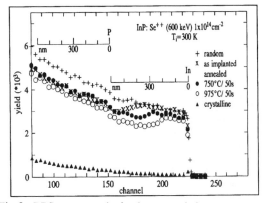

Fig.2 RBS - spectra obtained on annealed room temperature implanted InP samples.

A further increase of the annealing temperature up to 975°C (see fig. 2) enhances only the thickness of the perfectly recrystallized layer: the high backscattering yield in the near surface region remains. A cross-section micrograph made on this sample (see fig. 4) shows a high density of defects, consisting of a network of dislocations and microtwins extending from the surface up to a depth of ≈ 300 nm. A second defect band (dislocation loops) can be observed in a depth region of 420 to 500 nm, i.e. in the region of the former a/c-interface.

A somewhat different annealing behavior is seen in fig. 5 for the implantation at 365K. Epitaxial recrystallization occurs, but now from the depth and the surface, indicating that the near surface region was not fully amorphized. In this case a buried remaining damage peak occurs in the middle of the former amorphous layer. An increase of the annealing temperature up to 975°C does not lead to a perfectly restored crystalline layer.

Fig. 3 TEM - bright field image for an annealed (750°C/50s) room temperature implanted sample. The inset shows the diffraction pattern.

The annealing of the samples implanted at higher temperature was also investigated. Due to the absence of amorphous layers for implantation temperatures ≥ 395K the crystallization is going on without microtwin formation. This results in perfectly recrystallized layers with a backscattering yield nearly equal to that of the virgin material.

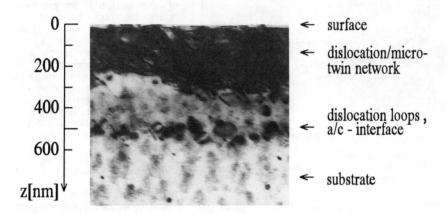

Fig. 4 Cross-section TEM - bright field image for an annealed (975°C/50 s) room temperature sample.

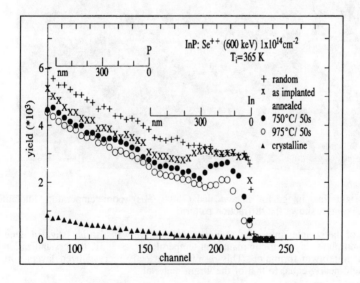

Fig. 5 RBS - spectra obtained on annealed InP samples implanted at 335K.

Electrical properties of the annealed layers

Figure 6 represents the electrical properties of the Se-implanted samples as a function of annealing and implantation temperature for a fixed time of 50 s. Additionally, time dependent annealing experiments at different annealing temperatures were also performed, but the results are not shown here.

Comparing the results for samples implanted at different temperatures it becomes obvious that the higher the implantation temperature, i.e. the lower the defect density after ion implantation, the better are the electrical properties of the annealed InP-layers. When increasing the implantation temperature from 300K to 425K the sheet carrier concentration (left part of fig. 6)

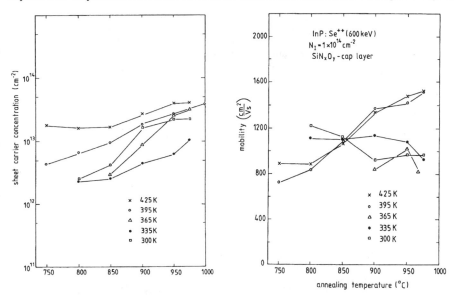

Fig. 6 Electrical properties (left: sheet carrier concentration, right: sheet mobility) of the annealed InP samples as a function of implantation and annealing temperature.

is enhanced by a factor of 4 from 1×10^{13} cm^{-2} to 4×10^{13} cm^{-2}, respectively, if the annealing is performed at 975°c for 50 s. A more drastic result was obtained for the sheet mobility (right part of fig. 6). For the samples containing an amorphous layer after implantation the mobility remains nearly constant at $\approx$ 1200 cm^2/Vs or decreases slightly with increasing annealing temperature. In the case of weakly damaged layers which were created for implantation temperatures $\geq$ 395K there is a strong increase of the mobility up to $\approx$ 1500 cm^2/Vs, indicating, that the nonperfect annealing (compare figs. 2 to 5) should be responsible for low mobilities resulting from additional scattering at defects.

An attempt was made to describe the annealing and activation behavior of Se in InP. For this purpose a simple thermodynamical model developed by Sealy et al.[8,9] for the activation of dopands in GaAs was applied. In this model the activation behavior is considered to be time and temperature dependent. The increase of activation with time at a given annealing temperature results in a saturation value of the activation at long times. From the latter results of the time independent region an activation energy E_A can be deduced, suggesting that this energy is required to split-up Se-vacancy-complexes (in GaAs: Se - V_{Ga}) which are responsible for the inactivity and to place the Se-atoms on nearby group V-vacancies (in GaAs: As, in InP: P), where they can act as donors. A careful analysis of the time dependent region of activation produces a kind of diffusion energy E_d of the donor and/or of the semiconductor constituents. The application of this model to our results for Se-implanted InP yields values for $E_d = (2.0 \pm 0.2)$ eV and $E_A = (1.0 \pm 0.1)$ eV. The same value for E_d has been obtained for Si-implanted InP[14] indicating that the diffusion energy should be connected with material transport of semiconductor constituents and not primarily with the diffusion of the donor. We propose that the value of $E_A = 1$ eV is connected (similar like in the case of GaAs) with the split-up of donor-vacancy complexes.

SUMMARY

For the creation of high performance electrically active InP-layers ion implantation should be carried out in such a way, that amorphization and strong damaging of the crystals is prevented. Perfect annealing can only take place if ion implantation leads to weakly damaged layers. The application of the activation model results in a very good description of the time and temperature dependent activation behavior of implanted Se in InP.

ACKNOWLEDGEMENTS

The authors would like to thank the technical staff of our group for their assistance, especially G. Lenk for carrying out ion implantation and K. Garlipp for photolithographical sample preparation.

REFERENCES

1. B.L. Sharma, Solid State Technology **11** (1989) 113.
2. J.P. Donnelly and C.E. Hurwitz, Appl. Phys. Lett. **31** (1977) 418.
3. J.P. Donnelly and C.E. Hurwitz, Solid State Electron. **23** (1980) 943.
4. H. Kräutle, J. Appl. Phys. **63** (1980) 4418.
5. J.D. Woodhouse, J.P. Donnelly, P.H. Nitishin, E.B. Owens and J.L.Ryan, Solid State Electron. **27** (1984) 677.
6. M.V. Ras, R.K. Nadella and O.W. Holland, J. Appl. Phys. **71** (1992) 126.
7. R.K. Nadella, M.V. Rao, P.S. Simons, P.H. Chi and H.B. Dietrich, J. Appl. Phys. **70** (1991) 7188.
8. B.J.Sealy, N.J. Barrett and R. Bensalem, J. Phys. D, Appl. Phys. **19** (1986) 2147.
9. N. Morris and B.J. Sealy, Inst. Phys. Conf. Ser. No. 91 (1987) chapter 2, 145.
10. T. Bachmann and H. Bartsch, Nucl. Instr. and Methods **B43** (1989) 529.
11. L.J. van der Pauw, Philipps Technische Rundschau **20** (1958/59) 230.
12. P. Müller, T. Bachmann, E. Wendler and W. Wesch, J. Appl. Phys., to be published.
13. F. Xiong, C.W. Nieh, D.N. Jamieson, T. Vreeland, Jr. and T.A. Tombrello, Mat. Res. Soc. Proc. vol. 100 (1988) 105.
14. P. Müller, PhD thesis, University of Jena, 1993.

DEEP LEVEL CHARACTERIZATION AND PASSIVATION
IN HETEROEPITAXIAL InP

*B. CHATTERJEE, * S. A. RINGEL, * R. SIEG, **I. WEINBERG and ** R. HOFFMAN
*Department of Electrical Engineering, Ohio State University, Columbus, OH 43202, USA
**NASA Lewis Research Center, Cleveland, Ohio 44135, USA

ABSTRACT

Deep levels in MOCVD grown p-InP on GaAs substrates have been investigated by Deep Level Transient Spectroscopy (DLTS). The effect of hydrogenation on the electrical activity of these levels has been studied through a combination of DLTS and Photoluminescence (PL) measurements. DLTS measurements indicate a drop of trap density from $\sim 5 \times 10^{14}$ cm^{-3} to $\sim 1 \times 10^{12}$ cm^{-3} after hydrogenation. Annealing at 400°C reactivated only the dopants, while temperatures above 600°C were necessary for deep-level reactivation. This combined with a logarithmic dependence on fill pulse time, indicate that at least one broad DLTS peak is associated with dislocations. The PL the DLTS results show that the dislocation related traps are passivated by hydrogen, preferentially over the dopants and that a wide annealing window exists for dopant reactivation.

INTRODUCTION

InP is a material of considerable interest for optoelectronic and photovoltaic applications. InP devices grown on Si, Ge or GaAs substrates are promising approaches to reduce weight, increase cell area and mechanical strength of InP based solar cells for space. However large lattice mismatch between these materials leads to the formation of misfit dislocations at the interface and subsequent threading into the growing InP layer. This leads to increased carrier recombination and lowers the cell efficiency. Dislocation densities less than 10^4 cm^{-2} are necessary to achieve solar cell efficiencies of 20%, in order to compete with their homoepitaxial counterparts [1]. Although a number of approaches have been proposed to reduce their concentration, dislocation densities are still far from the 10^4 cm^{-2} level. An alternative approach is to utilize hydrogen passivation, which is well known to passivate the electrical activities of both point defects and dislocations in many semiconductors. This paper describes the identification of deep levels associated with dislocations in InP and preliminary work on their passivation by plasma hydrogenation.

EXPERIMENTAL DETAILS

Figure 1 shows the structure of the heteroepitaxial p-InP grown on GaAs substrate. P-type InP about 1.9 μm thick, and doped to $1 \times 10^{17} cm^{-3}$ was grown in a low pressure MOCVD chamber on a (100) oriented n^+GaAs substrate. TMIN and TBP were used for the In and P sources, respectively and DEZn and SiH_4 were used for Zn (p) and Si (n) dopants. A layer of n-InP doped to $5 \times 10^{17} cm^{-3}$ was grown prior to the growth of p-InP to form a junction for DLTS measurement. Ohmic contacts to the p-InP and n^+GaAs were obtained by evaporating and alloying Au/Zn/Au and Ni/Ge/Au dots. The top ohmic contacts were mesa etched in a HCl:H_3PO_4:DI solution (2:2:1), to reduce the capacitance and facilitate DLTS measurements. Etch depths were confirmed using a DEKTAK profiler. Hydrogenation was performed in a plasma reactor chamber at 30 KHz, using a power density of 0.08 W/cm^2, temperature of 250°C, flow rate of 530 mTorr for periods between 45 minutes to 2 hours. All the samples were capped with

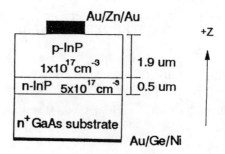

Figure 1: Device structure of heteroepitaxial p-InP.

100 - 200 $A°$ thick SiN_x layer, prior to H-exposure to prevent surface degradation and preferential loss of phosphorous. The nitride layer was removed in 10 % HF solution prior to metallization.

RESULTS

Figure 2 shows the DLTS plot obtained for a 10/s rate window, measured on a Biorad DL4602 spectrometer for the heteroepitaxial sample. Three peaks T1, T2 and T3 are clearly distinguishable. It was not possible to isolate the smaller peak situated on the shoulder of T1 due to its proximity to the much stronger T1 peak. Arrhenius' plots gave activation energies of ~ 0.80, 0.35 and 0.25 eV above the valence band , for T1, T2 and T3 respectively, noting that the broad T1 peak may in fact, be comprised of many smaller subpeaks. Depth resolved DLTS was performed by sequential variation of bias conditions. It was found that the concentration of T1 and T2 decreased sharply as we probe away from the buried interface towards the top surface in the +z direction [2]. In contrast, T3 remained unchanged with respect to the bias variations. These results suggest that T1 and T2 may be related to dislocations threading to the surface [3].

To gain further understanding about the nature of the traps, we performed DLTS under different fill pulse times and plotted the DLTS signal strength against the fill pulse time. Figure 3 shows that while T3 saturates quickly within 10 μs, T2 saturates after 0.1 ms and T1 does not saturate, even after 100 ms. Furthermore T1 has a much broader DLTS peak, than T2 or T3. A number of reports have been published on the capture dynamics of the dislocations as opposed to simple point defects for plastically deformed materials [4, 5]. Isolated point defects saturate quickly at very low fill pulse values, whereas traps located within the dislocation core or strain field give rise to broad DLTS peaks, whose amplitude has a logarithmic variation with the fill pulse time [4]. The capture rate for such traps are limited by a barrier height which increases with the number of captured carriers, making it difficult to saturate. Thus while T1 can be classified as a dislocation related trap by the 'barrier model', the observed saturation of T2 needs further investigation. The logarithmic dependence on t_{fill} however, suggests that T2 also may be related to dislocations. DLTS data taken at very short fill pulse times and TEM studies are underway currently to obtain further insight into the nature and origin of these traps.

Next we hydrogenated these samples and studied the effect of different hydrogen plasma exposure times on the passivation of these traps. The hydrogenated samples were annealed at progressively higher temperatures, to track the dopant and defect reactivation.

Figure 4 shows the effect of hydrogen exposure time on the electrical activity of the traps detected earlier through DLTS. A 2 hour hydrogenation effectively passivated all the traps resulting in a drop of trap concentration from 10^{14} to 10^{12} cm^{-3} range. The carrier concentration was reduced from ~ 1×10^{17} to 8×10^{15} cm^{-3}. It may be noted here

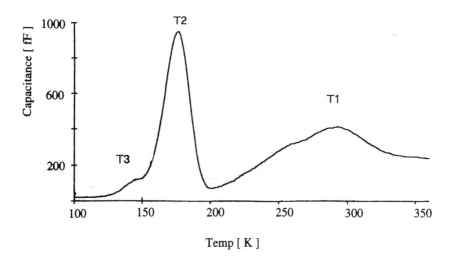

Figure 2: DLTS spectra for 10/sec rate window for buried interface junction for heteroepitaxial p-InP.

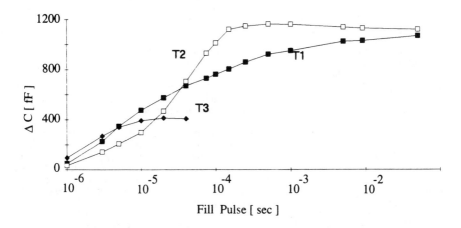

Figure 3: Variation of DLTS capacitance with fill pulse time for heteroepitaxial p-InP measured for 1000/s rate window

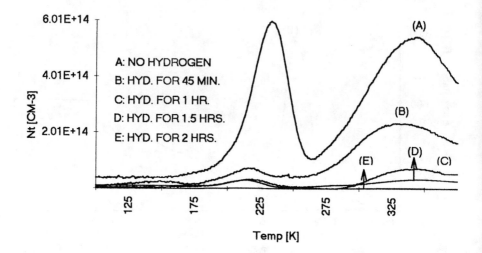

Figure 4: DLTS spectra for 1000/s rate window as a function of hydrogenation time. Note that the curves D and E are almost coincident with the temperature axis.

that the depletion region extends from the buried junction upwards in the +z direction.

Annealing at 380°C for ∼ 7 minutes was found to reactivate the carrier concentration to its original value. However the concentration of all the detected traps showed negligible increase as summarized in Table I. This is very noteworthy as the unwanted deep levels remain passivated while the desired doping concentration is recovered.

From Table I, a number of conclusions are apparent. First, the traps appear to be much more easily passivated than the dopants, as a function of exposure time. This is consistent with work of Hseigh et al [6], which reported enhanced hydrogen diffusion along dislocations in heteroepitaxial GaAs on Si. As can be seen from the Table, T1 and T2 drop to ∼ 5% of their initial values after 90 minute exposure, whereas the carrier concentration drop to ∼ 40% of its original value. In addition, the reactivation anneal temperature of 380°C is consistent with reports for Zn reactivation in homoepitaxial p-InP [7].

Secondly, it is seen from Table I that a 650°C anneal is necessary to achieve significant trap reactivation (anneals between 400 and 600°C were attempted, but did not show appreciable reactivation). A SiN_x cap layer was grown on the samples prior to the 650°C anneal to prevent phosphorous loss. This relatively high temperature is indicative of hydrogen - extended defect complexes and has previously been attributed to dislocation-hydrogen complex behavior in GaAs-Si and other systems [8]. This allows an extremely wide annealing window for subsequent device processing.

PL measurements were taken to support the results based on DLTS data. Figure 5 shows the PL spectra for a heteroepitaxial sample after various passivation and annealing treatments. The sample was first hydrogenated for 2 hours and then subsequently annealed at 400 and then at 650°C.

The peak C at 1.38 eV is due to the bound Zn acceptor to conduction band transition [9]. The broad band (marked A) around 1.3 eV has previously attributed to closely packed levels in the band-gap introduced as a result of dislocations [10]. The peak B is a replica of the Zn peak shifted by a LO phonon. The peak D situated higher than C

Table I: Effect of hydrogenation and subsequent annealing on the trap concentration. The post-hydrogen annealing was done only for samples hydrogenated for 2 hours. The dashes (-) in the table represents data too small to be read accurately.

Hydrogen. Time (min.)	Anneal Temp. (C)	Trap Concentration (x 1E12 cm-3)			Doping Conc. (x 1E15 cm-3)
		T1	T2	T3	
None	-	600	540	50	100
45	-	236	75	10	91
60	-	75	33	-	86
90	-	35	31	-	40
120	-	3.3	2.3	-	8
120	300	3.1	2.8	-	40
120	380	3.8	3.2	-	100
120	640	510	260	20	100

at 1.42 eV is due to the transition across the band gap.

Although it is evident from the figure that the acceptor- conduction band peak C increases very sharply after hydrogenation, in order to get an accurate quantitative comparison between the various measurements, it is necessary to measure the ratios of the peaks C and A. This method takes care of any differences introduced due to sample mounting and recalibration between the various measurements. It is seen from the figure that the ratio C/A increases from ~ 2 to ~ 92 after hydrogenation, which is an increase of one order of magnitude. The 400°C anneal reduced the ratio slightly to ~ 75, but the 650°C anneal reduced the ratio to ~ 13. These results clearly indicate that hydrogenation leads to very effective passivation of the dislocation related levels, decreasing the overall non-radiative recombination. Furthermore the PL results support our earlier claim that very high temperature anneals (> 600°C) are necessary to reactivate these levels. Such high temperature dissociation is further proof that the PL peak A is related to dislocations.

CONCLUSIONS

In conclusion it has been shown through PL and DLTS measurements, that the dislocations are effectively passivated by hydrogen and a wide annealing window exists within which it is possible to reactivate the dopants without reactivating the dislocations. Fill pulse measurements identified T1 as a dislocation, and T3 as point defect related peaks , corroborating the depth profile results. The PL characteristics showed a sharp increase in the band-acceptor transition and very sharp decrease in the dislocation assisted transition with hydrogenation. The dislocations reappeared only after anneals at temperatures above 600°C.

Further studies need to be conducted to pinpoint the exact nature of these traps and the complexes formed with these defects by the diffusing hydrogen. These results will be reported elsewhere [11].

The work for this project was supported by NASA Lewis Research Center through contract No. NAG3-1461.

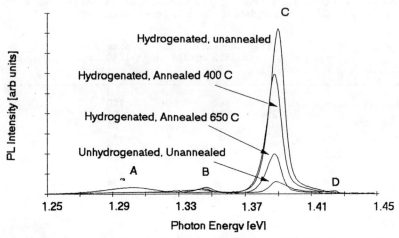

Figure 5: PL spectra of heteroepitaxial p-InP, hydrogenated for 2 hours and subsequently annealed under different temperature conditions. Measurement taken at 13°K and with an incident (argon) laser power of 10mW.

References

1) R. K. Jain and D. J. Flood, IEEE Trans. Electron Devices **40** (11), 1928 (1993).
2) B. Chatterjee, R. Sieg, S. Ringel, R. Hoffman and I. Weinberg, presented at the 1993 Electrochem. Soc. Conf. New Orleans.
3) S. J. Pearton, K. T. Short, A. T. Macrander, C. R. Abernathy, V. P. Mazzi, N. M. Haegel, M. M. Al-Jassim, S. M. Vernon and V. E. Haven, J. Appl. Phy. **65** (3), 1083 (1989).
4) T. Wosinski, J. Appl. Phys. **65** (4), 1566 (1989).
5) P. Omling, E. R. Weber, L. Montelius, H. Alexander and J. Michel, Phys. Rev. B **32** (10), 6571 (1985).
6) K. C. Hseigh, M. S. Feng, G. E. Stillman and N. Holonyak, Appl. Phys. Lett. **54** (4), 341 (1989).
7) G. R. Antell, A. T. R. Briggs, B. R. Butler, S. A. Kitching, J. P. Stagg , A. Chew and D. E. Sykes, Appl. Phys. Lett. **53** (9), 758 (1986).
8) M. J. Matragrano, G. P. Watson, D. G. Ast, T. J. Anderson and B. Pathangey, Appl. Phys. Lett. **62** (12), 1417(1993).
9) E. W. Williams, W. Elder, M. G. Astles, M. Web, J. B. Mullin, B. Straughan and P. J. Tufton, J. Electrochem. Soc. **120**, 1741 (1973).
10) M. Gal, A. Tavendale, M. J. Johnson and B. F. Usher, J. Appl. Phys. **66** (2), 968 (1989).
11) B. Chatterjee and S. A. Ringel, in preparation.

Photoluminescence Fatigue In Laterally-Ordered (GaP)$_2$/(InP)$_2$ Short-Period-Superlattices (SPS) Grown By Molecular Beam Epitaxy

X. C. Liu*, S. Q. Gu, E. E. Reuter, S. G. Bishop, A. C. Chen, A. M. Moy, K. Y. Cheng, and K. C. Hsieh

Center for Compound Semiconductor Microelectronics,
Materials Research Laboratory, and Beckman Institute
University of Illinois at Urbana-Champaign, Urbana, IL 61801

Abstract

Spontaneously laterally ordered (GaP)$_2$/(InP)$_2$ short period superlattices (SPS) grown by Molecular Beam Epitaxy (MBE) on nominal (100) GaAs substrates have been studied by photoluminescence (PL) spectroscopy. The samples studied included SPS comprising 110 pairs of (GaP)$_2$/(InP)$_2$ (total thickness ~90 nm) and multiquantum well structures in which quantum wells comprising 12 pairs of (GaP)$_2$/(InP)$_2$ SPS layers (thickness ~10 nm) are alternated with lattice-matched GaInP random alloy barrier layers. The 5K PL spectra include a ~1760 meV near-band edge band, and a much broader, lower energy (~1670 meV) luminescence band that exhibits an unusual fatiguing behavior; its intensity diminishes monotonically during continuous illumination by the exciting light. This fatigued PL state is metastable at low temperatures. In the quantum well structure, although the relative intensity of the lower energy band is significantly weaker in comparison to the higher one, the fatiguing behavior still exists. However the fatiguing rate is slower in quantum well structures than that observed in the thick SPS film.

Recently, it has been demonstrated[1] that long range, laterally ordered structures can be formed spontaneously during growth by a strain-induced lateral-ordering (SILO) process. For example, in the vertical short-period superlattices (SPS) of (GaP)$_2$/(InP)$_2$ grown by molecular beam epitaxy (MBE) on nominally (100) on-axis GaAs substrates, dark field cross sectional transmission electron microscopy (TEM)[2] reveals that the vertical SPS structure exhibits periodic dark and light fringes along the [110] direction with periodicities of the order of ~ 20 nm. The fringes are due to laterally periodic modulation of the averaged GaInP alloy composition as examined by energy dispersive x-ray (EDX) microanalysis. Lateral quantum wells perpendicular to the growth axis — effectively quantum wires along the [-110] direction — are therefore spontaneously formed in such laterally ordered vertical SPS structures. The deviation of the vertical SPS periodicity from 2a$_0$ caused by the large misfit strain between GaP and InP is believed[1] to be the major driving force of the lateral alloy composition modulation. Photoluminescence (PL) from the SILO structures exhibits strong linear polarization in the direction parallel to the long axis of the wires. This SILO[1] phenomenon has also been applied in fabricating strained quantum wire (QWR) GaInP devices, including QWR light-emitting diodes[3] (LED) and multiple QWR lasers[4]. The laterally ordered quantum wires in these devices manifested themselves in the strong linear polarization of the spontaneous emission from the LEDs and strongly anisotropic laser threshold current density for contact stripes oriented in the [110] and [-110] directions.

In this paper we present a study of PL spectroscopy at different temperatures and excitation intensities in a (GaP)$_2$/(InP)$_2$ SILO vertical SPS structure with total thickness of ~90 nm and a multiple quantum well laser structure composed of SILO materials. The 5 K PL spectra of the SPS sample exhibit a band at higher energy and two broad bands at lower energies. While all bands showed very similar pump-power dependent energy shifts to higher energies, the lower energy bands display an unusual fatiguing behavior; their PL intensities diminish monotonically during continuous illumination by the exciting light. The fatigued PL band is metastable at low temperatures, but the PL efficiency is recovered gradually as the temperature is elevated. For the

131

quantum well laser structure, qualitatively similar PL results are obtained with the PL energies blue-shifted due to quantum confinement in the vertical (growth) direction. The relative intensity of the fatiguing band is much weaker in comparison with the higher energy band edge luminescence. The fatiguing rate as a function of exposure time from the quantum well laser structure is also much slower than that from the SPS film. The observed pump-power dependent energy shift and the remarkably broad, deep PL bands that exhibit the highly unusual fatiguing behavior in the SPS film are interpreted in terms of the inevitable disorder associated with the vertical SPS and the lateral alloy composition modulations. The quantum confinement effects are invoked in explaining the quantitative differences between the PL spectra observed in the quantum well laser sample and those from the SPS film.

The samples used in this study were grown by MBE on (001) on-axis GaAs substrates. The SPS SILO structure (Sample #983) includes 110 pairs of $(GaP)_2/(InP)_2$, corresponding to a total thickness of ~90 nm. The multiple quantum well laser structure (Sample #900) consists of five wells sandwiched by 18 nm bulk $Ga_{0.5}In_{0.5}P$ alloy barriers. Each well includes 12 pairs of $(GaP)_2/(InP)_2$ SPS layers, making the thickness of the quantum well ~ 10 nm. Surrounding the five quantum wells are the two heavily doped 1μm-thick AlGaInP alloy and 180-nm-thick graded cladding layers commonly found in a laser structure. Similar structures and details of the growth procedure have been reported previously[2,4]. Cross-sectional TEM confirmed the formation of QWRs within the SILO materials oriented along the [-110] direction with lateral periodicity of the order of ~ 20 nm along [110]. In addition, strong linear polarization was obtained in low temperature PL spectra with PL polarized primarily parallel to the long-axis of the QWRs.

The excitation source used in the PL experiments reported here was provided by either the 514.5 nm line of an Ar-ion laser or monochromatic light created by a 150 W xenon arc lamp and a 0.22 m double grating monochromator. The luminescence spectrum was dispersed by a 1.0 m single-grating spectrometer and detected by a GaAs photocathode photomultiplier tube (PMT) in the photon-counting mode. The samples were mounted in a variable temperature optical cryostat and experiments were performed at liquid helium temperatures in a vapor-cooled environment.

Figure 1 shows the PL spectra obtained from the SILO film sample #983 at 5K excited with a low intensity (~ 10 μW) 515 nm monochromatic light before and after the sample was exposed to a 1.2 W unfocused 514.5 nm laser line for five minutes. The use of an unfocused laser beam rather than a focused one was to ensure uniform exposure coverage of the sample area from which the PL was excited. The spectrum exhibits a high energy band (band I) at ~1760 meV (linewidth ~35 meV) and a broad (~100 meV) band centered at ~1670 meV (band-II) with low energy tail extending to ~1370 meV (band-III). These three bands, each of which exhibits strong linear polarizations along the wire directions, all arise from the QWR layer in the SILO structure. The relatively sharp luminescence band at ~1490 meV is attributable to transitions associated with shallow acceptors in the GaAs substrate. While this substrate feature and band-I from the SPS film do not show any characteristic change before and after the strong laser exposure, the PL intensities of bands II and III are remarkably reduced (fatigued) after the sample is exposed to the intense light. In the quantum well laser sample (#900), the PL spectrum (Fig. 1) is dominated by a strong luminescence band with an energy blue-shifted by ~ 35 meV from band-I in the SPS film sample. The relative PL intensity of the lower energy tail in the quantum well sample is much weaker, apparently an effect due to the quantum confinement. Its intensity also exhibits fatiguing behavior, but to a somewhat lesser degree. Before examining further this novel PL fatiguing phenomenon[5] which is believed to be associated with defects and dislocations in the materials, we first discuss the non-fatiguing 1750 meV PL band-I.

The behavior of band-I in the SPS film (#983) and its highest energy position in the spectrum suggest that it is related to the energy gap of the laterally ordered SPS film. Its energy position, 1750meV, is about 250 meV below a typical "random" GaInP alloy lattice-matched to the GaAs substrate[6] with identical alloy composition. It should be noted that the magnitude of this energy difference (250 meV) is much greater than the band gap narrowing effects (<

150meV) that have been observed in the so-called spontaneously ordered GaInP alloy (CuPt structure)[7] with similar composition in which ordered monolayer SPS domains are embedded in the alloy. In addition to the further band gap narrowing expected for a strained SPS as explained by Zunger and co-workers,[8] one has to invoke the lateral ordering to explain the large (~ 100meV) extra band gap reduction. Interestingly, the energy position of band-I is very close to the band edge energy expected for a GaInP alloy with composition corresponding to the In-rich (~ 56%) regions of the lateral periodic modulation of composition caused by the SILO process,[2] as determined by EDX microscopy. This suggests that the red-shifted band edge luminescence band in these structures involves photoexcited carriers which thermalize to the lowest band gap, indium-rich (~56% In) volumes of the SPS (quantum wires with lateral confinement) before recombining radiatively. The relative blue-shift of the corresponding PL band in the quantum well laser sample (#900) is presumably attributable to additional quantum confinement by the GaInP alloy barriers in the vertical dimension.

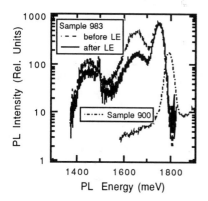

Fig. 1 PL spectra from the thin film sample #983 taken with ~ 10 μW of 515 nm light at 5 K. Data are acquired before and after the sample is exposed to a 1.2 W unfocused 514.5 nm laser line for five minutes. The PL spectrum from the multiple quantum well laser structure sample #900 is also shown.

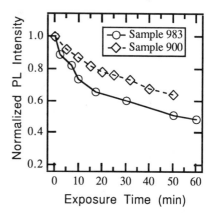

Fig. 2 Normalized peak intensity plots of the fatiguing luminescence bands as a function of exposure time for the thin film sample #983 and the multiple quantum well laser structure sample #900 at 5 K. The extent of fatiguing is slower in the quantum well sample than that in the thin film, although a stronger power is used for the former.

In marked contrast to band-I, the luminescence bands II and III are fatigued (diminished in intensity) dramatically under continuous excitation by intense, above-gap light as is evident in Fig. 1. Although the overall luminescence intensities of the bands are reduced, their lineshapes remain unaltered. Thus the PL peak intensities are good measures of the fatiguing phenomenon. We have determined[5] that the amount of intensity reduction depends on the "dose" of the exposure light (i.e., the photon flux defined as Power x time). Therefore, the observed fatiguing rate is qualitatively proportional to the intensity of the above-gap exciting light. The fatiguing was barely detectable at 5K in these samples for exposures of several hours to illuminating powers below ~ 20μW. In Fig. 2 we show the normalized peak intensity of band-II from sample #983 and the corresponding 1680 meV band from sample #900 as a function of continuous exposure to focused laser light with excitation powers of 100mW and 300mW, respectively. In both cases, the PL peak intensity exhibits a steep initial drop followed by a more gradual decrease during an extended continual exposure. However, the rate of the intensity diminution is much slower in the case of the multiple quantum well laser sample than that observed in the SILO SPS film, in spite of the fact that three times more excitation power was used in the former case. This implies that in the quantum well case there are fewer defect centers contributing to the fatiguing process. Apparently the disorder in spatial variation of composition and periodicity of lateral ordering which accompany the SILO process is somewhat suppressed in the 110 nm thick multi-quantum well SPSs in comparison to the disorder in the 900 nm thick continuous SPS.

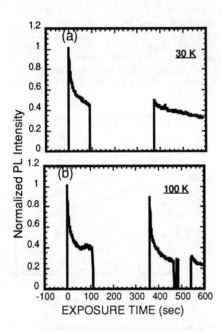

Fig. 3 Fatiguing and thermal recovery of the PL band in the thin film sample #983 at (a) 30 K and, (b) 100 K. Off and on indicate the times at which the laser light is interrupted and resumed, respectively.

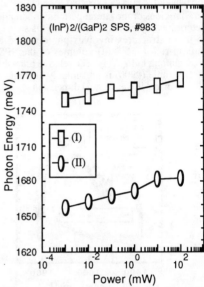

Fig. 4 PL energy shifts of bands I and II from the thin film sample #983 as a function of excitation power at 5 K. The solid lines are guides to the eye.

It has been shown that the fatigued state of the PL is metastable at low temperatures as shown in Fig. 1. However, the PL efficiency is recoverable at elevated temperatures after the illumination is terminated, as demonstrated in Fig. 3. We display in Fig. 3 the time dependence of the fatigue induced by 100mW of the 514.5 nm laser line at two different temperatures with the exposure source interrupted and then resumed during the data acquisition. It should be noted that the fatiguing experiments were performed on a sample that was freshly cooled down in a dark environment. Alignments and other preparations were done with minimal exposure to excitation light. Experiments at different temperatures were done with the excitation beam moved to different spots on the sample surface to eliminate possible effects from the previous exposures. At relatively low temperature (30K, Fig. 3a), the luminescence showed a typical sharp drop at the beginning of the exposure. We then interrupted the laser illumination and resumed it about three minutes later. The luminescence intensity appeared to have recovered only slightly relative to its fatigued level just before the interruption of excitation. This shows that there is no further fatiguing in the absence of excitation, and only minimal thermal recovery of the PL efficiency at low temperatures. When the same experiment is performed at 100K (Fig. 3b), a rather different picture is revealed. The luminescence intensity, after the resumption of the excitation, has recovered to a level which greatly exceeds the fatigued one just before the laser was blocked about three minutes earlier, and is nearly equal to its initial value.

Similar luminescence fatiguing phenomena[9-11] with low temperature metastable characteristics have been observed in other semiconducting materials and a variety of mechanisms involving metastable deep levels have been proposed[9-12]. These mechanisms invoke the presence of defects or disorder in the material. Although the mechanism for the PL fatigue observed in our samples cannot be specified definitively, it is apparent that the SILO structures contain disordered regions which could incorporate metastable deep levels. For example, the lateral ordering is caused by the deviation of the vertical SPS periodicity from n=2. EDX spectroscopy has shown that the interfaces of the compositionally modulated lateral alloy layers are not abrupt but graded, and involve some disorder. In addition, the lateral period of the ordering exhibits irregularities. One can envision deep traps or localized mid-gap states associated with the lateral interfaces in the SPS that could give rise to the metastable PL fatigue in these samples.

Fig. 4 shows the energy shifts of bands I and II of sample #983 as a function of laser pump power at 5K. Because band II has the property of fatiguing under strong laser light, a special procedure was adopted in acquiring the data. The procedure takes advantage of the fact that with high excitation power, the fatiguing rate was much faster at the beginning of the illumination and gradually slowed down (saturated) after the sample had been exposed for some time. The sample was first given a prolonged exposure to the highest laser power we were to use in order to saturate the fatiguing effect; the spectral shift as a function of excitation intensity data were acquired *after* the fatiguing effect had "saturated". This step was to preclude the possibility of a fatigue-related lineshape distortion during data acquisition which would affect the accuracy of peak position determination for band II. Subsequent to the saturation of the PL fatigue, the metastability of the fatigued PL state allowed us to acquire spectral data with sequentially decreased excitation powers. As shown in Fig. 4, both bands behave similarly with an energy shift of ~ 4meV for each decade of power increase. This shift, although much smaller than the "moving band"[7] observed in the spontaneously ordered CuPt GaInP structure, is explicable in terms of the disordered lateral variation in the average composition and band gap of the SPS. At low excitation intensities, photexcited carriers thermalize to the lowest band gap volumes of the laterally ordered SPS before recombining radiatively. With increasing pump power, some of the photoexcited carriers undergo "hot carrier" recombination in higher band gap volumes of the SPS giving rise to a blue shift in the PL spectrum. It is conceivable that the relatively broad PL linewidths in our samples (Fig. 1), compared with random alloys,[6] are also caused by the irregular nature of the SILO SPS layers.

In conclusion, we have observed a low-temperature metastable PL fatiguing phenomenon in structures containing laterally ordered vertical $(GaP)_2/(InP)_2$ SPS epilayers. The fatiguing phenomenon is believed to relate to deep metastable states associated with defects and disorder present in our samples. In samples comprising 110 nm thick SPS quantum wells separated by GaInP alloy barrier layers, the intensity of the fatiguing PL bands and their fatiguing rates are significantly lower than those observed in a continuous 900 nm thick SPS layer. This implies that the disorder in spatial variation of composition and periodicity of lateral ordering which accompany the SILO process is somewhat suppressed in the quantum well case and, consequently, there are fewer defect centers contributing to the fatiguing process. Although the fatigued state is metastable at low temperatures, its PL efficiency is recoverable as the temperature is increased. The existence of defects and compositional disorder is also evident from the spectral energies and linewidths of the PL spectra and the studies of energy shifts as a function of pump power. The relatively broad PL linewidths and the pump-power dependent energy shifts are explained in terms of variations in the lateral periodicity of the quantum well layers, and in the compositional disorder within the layers.

This work is supported by the National Science Foundation (ECD 89 43166 and DMR 89-20538), and the Office of Naval Research University Research Initiative Program (N00014-92-J-1519).

References:

* Presently at the Francis Bitter National Magnet Laboratory, Massachusetts Institute of Technology, Cambridge, MA 02139

[1] K. Y. Cheng, K. C. Hsieh, and J. N. Baillargeon, Appl. Phys. Lett. **60**, 2892 (1992).
[2] K. C. Hsieh, J. N. Baillargeon, and K. Y. Cheng, Appl. Phys. Lett. **57**, 2244 (1990).
[3] P. J. Pearah, E. M. Stellini, A. C. Chen, A. M. Moy, K. C. Hsieh, and K. Y. Cheng, Appl. Phys. Lett. **62**, 729 (1993).
[4] E. M. Stellini, K. Y. Cheng, P. J. Pearah, A.C. Chen, A. M. Moy, and K. C. Hsieh, Appl. Phys. Lett. **62**, 458 (1993).
[5] X. C. Liu, S. Q. Gu, E. E. Reuter, S. G. Bishop, A. C. Chen, A. M. Moy, K. Y. Cheng, and K. C. Hsieh, Phys. Rev. **B**, submitted.
[6] M. C. DeLong, D. J. Mowbray, R. A. Hogg, M. S. Skolnick, M. Hopkinson, J. P. R. David, P. C. Taylor, Sarah R. Kurtz, and J. M. Olson, J. Appl. Phys. **73**, 5163 (1993).
[7] M. C. DeLong, P. C. Taylor, and J. M. Olson, Appl. Phys. Lett. **57**, 620 (1990).
[8] S.-H. Wei, Alex Zunger, Appl. Phys. Lett. **56**, 662 (1990).
[9] S. G. Bishop, U. Strom, and P. C. Taylor, Phys. Rev. Lett. **34**, 1346 (1975).
[10] D. C. Chen, J. M. Viner, P. C. Taylor and J. Kanicki, in "Amorphous Silicon Technology-1992", Eds. M. J. Thompson, Y. Hamakawa, P. G. LeComber, A. Madan, and E. A. Schiff. (Materials Research Society, Pittsburgh, 1992), Vol. 258, p.661.
[11] D. K. Biegelsen and R. A. Street, Phys. Rev. Lett. **44**, 803 (1980).
[12] R. A. Street and N. F. Mott, Phys. Rev. Lett. **35**, 1293 (1975).

PHOSPHORUS-VACANCY-RELATED DEEP LEVELS IN GaInP LAYERS GROWN BY MOLECULAR BEAM EPITAXY

Z.C.Huang , C.R.Wie, J.A.Varriano*, M.W.Koch*, and G.W. Wicks*
Department of Electrical and Computer Engineering and Center for Electronic and Electro-optic Materials, State University of New York at Buffalo, Bonner Hall, Buffalo, NY 14260
*The Institute of Optics, University of Rochester, Rochester, NY 14627

ABSTRACT

Deep levels in lattice matched $Ga_{0.51}In_{0.49}P/GaAs$ heterostructure have been investigated by thermal-electric effect spectroscopy(TEES) and temperature dependent conductivity measurements. Four samples were grown by molecular beam epitaxy with various phosphorus (P_2) beam equivalent pressure(BEP) of 0.125, 0.5, 2, and 4×10^{-4} Torr. We report for the first time, to our knowledge, an electrical observation of phosphorus vacancy point defects in the GaInP/GaAs material system. The phosphorus vacancies , V_P, behave as an electron trap which is located at E_C-0.28$\pm$0.02 eV. We have found that this trap dominates the conduction band conduction when T> 220K, and is responsible for the variable-range hopping conduction when T < 220K. Its concentration decreases with the increasing phosphrous BEP. Successive rapid thermal annealing showed that its concentration increases with the increasing annealing temperature. Another electron trap at E_C-0.51eV was also observed only in samples with P_2 BEP less than 2×10^{-4} Torr. Its capture cross section is 4.5×10^{-15} cm^2 as obtained from the illumination time dependent TEES spectra.

INTRODUCTION

The lattice-matched $Ga_{0.51}In_{0.49}P/GaAs$ heterojunction system has received extensive attention as an important alternative to the AlGaAs/GaAs system in many applications[1-3]. Currently much attention is focused on the growth conditions, structural analysis and device envaluation. Although a few empirical studies have been done on defects in GaInP[4,5], almost no identification of point defects was done by these previous studies. In view of its potential technological applications, it is therefore worthwhile to perform a study of its basic properties in relation to defects. Such fundamental questions as the location of native defect levels in the gap, the identity of major native defects, and the effect of impurity incorporation on the crystal under various stoichiometric conditions have not been answered.

In this paper, we report some systemetic studies on deep levels in molecular beam epitaxy (MBE) grown $Ga_{0.51}In_{0.49}P/GaAs$ heterojunctions by varying the phosphorus (P_2) beam equivalent pressure (BEP) during the epitaxial growth. The thermo-electric effect spectroscopy (TEES) and the temperature dependent conductivity measurement were employed for the characterization. The variable-range hopping conduction was observed at T< 220K in this material system. Two electron traps were observed with the activation energies at 0.28$\pm$0.02eV and at 0.51eV. Their behaviors during various heat treatments and the origins will be discussed.

137

SAMPLE PREPARATION

Four $Ga_{0.51}In_{0.49}P$/GaAs heterojunctions, G41, G42, G43, and G44, were grown by MBE with varied P_2 BEP of 0.125, 0.5, 2.0 and 4.0×10^{-4} Torr, respectively. We used a valved, solid phosphorus cracker source for the phosphide growth[6]. A 0.5μm thick undoped $Ga_{0.51}In_{0.49}P$ layer was grown on an n-type GaAs(100) substrate, misoriented 4° towards (111)A. The carrier concentration for the GaAs substrate is about $9 \times 10^{17} cm^{-3}$. The growth temperature was 475^oC for the GaInP layer. Samples were grown $at \sim 1 \mu$m/hr with a P_2/III BEP ratio of approximately 150 to 250. The samples we studied in this paper have relatively high resistivity ranging from 10^3 to 10^5 Ω.cm at room temperature. Ohmic contact on the GaAs back surface was made by evaporating AuGe/Au and annealing at 420^oC for 15s. The Schottky diodes (area = 6.4×10^{-3} cm^2) were fabricated on the front GaInP surface by evaporating Au under the base pressure of 1×10^{-6} Torr.

Temperature-dependent conductivity measurement was performed using a standard system from MMR Technologies, Inc. Our setup for the TEES measurement was described in Ref.7. Data were taken under zero bias condition and under sufficient illumination time (5min.) when the samples were at the lowest temperature (60K). The heating rate was kept at 0.35K/s. The illumination was performed by He-Ne laser(1.96mW). The rapid thermal annealing (RTA) was employed using a commercial rapid thermal furnace (Heatpulse Mod.210) in a N_2 atmosphere by the proximity capping method for 30s.

RESULTS AND DISCUSSION

The temperature dependent conductivity data , $\sigma(T)$, for four GaInP/GaAs samples are shown in Fig.1. The $\sigma(T)$ curves clearly shows two different types of conduction. For 1000/T>4.5, the defect band hopping conduction dominates, the conductivity due to the hopping conduction obeys the relationship[8] of $\sigma = \sigma_0 exp(- b/T^{1/4})$, where b is a constant. Therefore, the conduction at 1000/T > 4.5 is due to variable-range hopping[9]. For 1000/T<4.5, the conduction band(CB) conduction becomes dominant. At room temperature, the Hall Effect measurement showed that the carrier mobility was about 1650-3300cm^2/Vs in our samples, indicating that the electrons are the conduction carriers in all samples. Thus the CB conduction in the 220-350K (or 2.8<1000/T < 4.5) region is due to the electrons excited from the deep donor levels whose energy level lies below the conduction band. The activation energies E_a, obtained from $\sigma(T) = \sigma_0 exp(-E_a/kT)$, are 0.26, 0.26, 0.27 and 0.29eV for samples G41, G42, G43 and G44, respectively. Although the conductivity difference between samples varies by a factor of 100 at the same temperature, their activation energies are very close to each other, suggesting that their CB conduction is due to the same donor level with an activation energy at E_C-0.28$\pm$0.02eV.

Because of the relatively high resistivity, a TEES technique under the zero bias condition was employed for defects characterization in these four samples. In TEES the driving force is the temperature gradient rather than the applied bias. The detailed method of TEES data analysis was described in Ref.7. Here the positive peak represents electron trap. The TEES spectra for four samples in the temperature range of 60-300K are shown in Fig.2. No traps were observed above 300K(up to 400K). Two main electron traps, trap A(128-145K) and trap B(255K), were observed. The trap A was observed in four samples, with slightly different peak position in each sample. Trap B was observed in samples G41 and G42 , but not observed in samples G43 and G44. Their activation energies can be obtained approximately from $E_a = kT_m ln(T_m^4/\beta)$[7], where T_m is the peak temperature, β is heating rate, and k is Boltzman constant. The obtained activation energies for trap A are 0.25, 0.26, 0.27 and 0.29eV for sample G41, G42, G43 and G44, respectively. These values are very close to the activation energy of the donor level obtained from the conductivity measurement in the temperature range of 1000/T< 4.5. We believe that it is trap A that

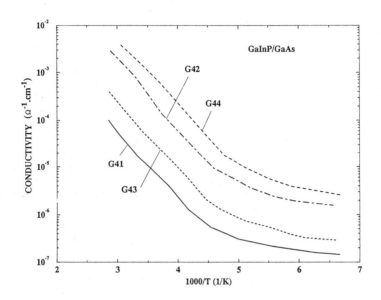

Figure 1. Conductivity vs temperature for four $Ga_{0.51}In_{0.49}P/GaAs$ samples.

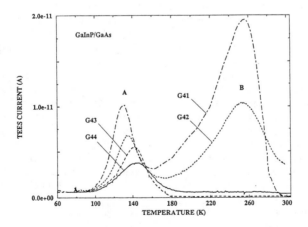

Figure 2. The TEES spectra for sample G41, G42, G43 and G44.

dominates the CB conduction in the range of $1000/T < 4.5$ in all samples. Trap B has an activation energy of 0.51eV. The absence of trap B in sample G43 and G44, and the different peak positions of trap A in each sample imply that these two traps are not originated from the GaAs substrate.

The trap concentration may be estimated from $Q = qN_tAL^*$, where Q is the total charge released from the trap which is equal to the area under a peak in TEES spectra ($= \int I/\beta \, dT$), N_t is the trap concentration, q is electronic charge, A is the area collecting the charge, and L^* is the effective depth which depends on the experimental condition[10]. In the TEES measurement, we assume that only the traps located within the light absorption length contribute to the released charge Q. We take $L^* = 0.25\mu m$ for the estimation of the trap concentration. The obtained trap concentrations from the TEES spectra in each sample are listed in Table I. The TEES signal was normalized by substracting the dark current. Table I shows that the concentration of trap A is in a range of $0.9\text{-}2.4 \times 10^{15} cm^{-3}$, and the concentration of trap B is about $12.3 \times 10^{15} cm^{-3}$ and $8 \times 10^{15} cm^{-3}$ in samples G41 and G42, respectively. It is well known that the concentration of native defects depends on the composition of MBE layer. The concentration of trap A decreases with increasing the P_2 BEP (as can be seen in Table I), suggesting that this trap may originate from the phosphorus vacancies, V_P, the antisites In_P or Ga_P, the interstitials Ga_i or In_i, or their complexes because the increase of P_2 BEP during the sample growth will decrease the concentrations of these defect species. However, the interstitials and the antisites are less probable causes for trap A as can be seen from the following considerations: i) It was reported that the interstitials, Ga_i or In_i, in the MBE grown samples are only present in Ga-rich or In-rich samples and found most in low temperature grown samples[11]. During our sample growth, the BEPs for Ga and for In were kept the same and the growth temperature was 475°C. ii) The antisites, In_P or Ga_P are acceptors, and the theoretically pridicted energy level is less than $0.12eV$[12] which is much smaller than the activation energy of trap A. In InP, the emission band around 1.10-1.20eV is attributed to the phosphorus vacancies (V_P) or V_P paired with an impurity atom which is a donor-like defect[13,14]. Therefore, we suggest that the phosphorus vacancies are responsible for the observed electron trap A.

In order to test the above idea further, a successive rapid thermal annealing was performed on each sample at various temperatures. One could expect an increased concentration of V_P after RTA because of the likely out-diffusion of phosphorus and an increased concentration of trap A if it is indeed due to V_P. As Fig. 3(a) shows, the TEES peak height of trap A increased with increasing the annealing temperature for sample G41, and the peak height of trap B was almost constant after RTA at 600°C, but the peak position shifted to a higher temperature. Fig.3(b) shows the concentration of both trap A and trap B vs the annealing temperature. It shows that the concentration of trap A increases with increasing the annealing temperature in all samples, and there is more increase in trap concentration in the sample with a higher P_2 BEP (during growth). This is consistent with our argument that V_P is involved in trap A. It seems that trap B is also related to phosphorus vacancies or V_P-related complexes in view of the facts that its concentration decreases with the increasing P_2 BEP in samples G41 and G42 as can be seen in Fig.1 and that it disappeared when P_2 BEP $> 2 \times 10^{-4}$ Torr (in samples G43 and G44). However the annealing did not affect its concentration, but only changed its energy position. This behavior is similar to the annealing behavior of EL6 group in GaAs[15]. Its configration may change due to the annealing. It might be a complex of several native defects rather than an isolated point defect. The identification of this trap needs further study.

It is noted that although trap B has a larger concentratoin than trap A in samples G41 and G42, the CB conduction at $1000/T < 4.5$ (or T>220K) is still dominated by trap A, as can be seen from the activation energy in Fig.1. In order to investigate the responsible trap for the hopping conduction for $1000/T > 4.5$ (or T < 220K) , we examined the TEES spectra and the conductivity under different illumination time. The illumination time dependent TEES spectra in sample

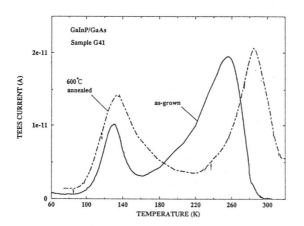

Figure 3(a). Comparison of TEES spectra between as-grown and 600°C annealed sample G41.

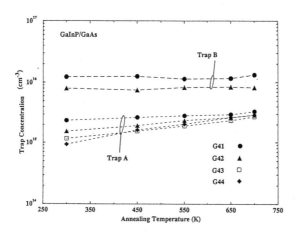

Figure 3(b). The estimated concentration of trap A and trap B vs annealing temperature in four samples.

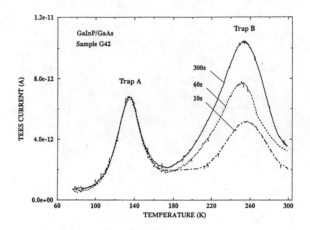

Figure 4. The illumination time dependent TEES spectra for sample G42.

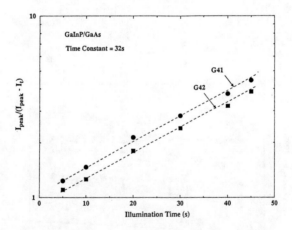

Figure 5. The plot of $I_{peak}/(I_{peak}-I_t)$] vs Δt of trap B for sample G41 (circle) and G42 (square).

G42 are shown in Fig.4. It shows that the trap filling rates, i.e., their free-carrier capture cross sections are different for these two traps. The concentration of trap A is almost independent of the illumination time. Trap B, however, is very sensitive to the illumination time, it needs considerably longer illumination to be completely filled. Similar results were observed in the other three samples. We did not, however, observe any change of conductivity under different illumination time. These results suggest that trap B has no contribution to the hopping conduction for T<220K. This can also be seen from Fig.1, in which the hopping conductivity has a similar temperature dependence in all samples, it does not matter whether trap B is present (in sample G41 and G42) or absent (in sample G43 and G44). Therefore we conclude that the trap A is responsible for the hopping conduction. The absence of any change for trap A under different illumination time can not be explained by the carrier trapping process [16,17]. Analogous phenomina were also observed by Fang et al[18] and Desnica et al[19] in the semi-insulating GaAs (T_3 in ref.18 and T_5 in ref.19).

The capture cross section σ_n of trap B can be obtained by the method suggested by Tomozane et al[16]. This method is based on the dependence of peak height on illumination period. The σ_n is given by

$$\sigma_n = eA/(I_{ph} \tau_t) \quad\dotfill(1)$$

where I_{ph} is the maximum photocurrent during the illumination through the sample, A is the area of the electrode, and τ_t is the time constant of the trap which can be obtained from

$$I_t = I_{peak} [1 - \exp(-\Delta t/\tau_t)] \quad\dotfill(2)$$

where I_{peak} is the maximum photocurrent of the trap under sufficient illumination time, I_t is the photocurrent level of the trap under illumination period Δt. As shown in Fig. 5, a good linear relation was obtained from the plot of $\ln[I_{peak}/(I_{peak}-I_t)]$ vs Δt for trap B in sample G41 and G42, from which a time constant of 32s was obtained. The capture cross section of trap B was then calculated to be 4.5×10^{-15} cm^2. In equations (1) and (2), we have ignored the effect of trap A due to its independence on illumination time. For the same reason, we could not get the capture cross section value for trap A from this method.

CONCLUSIONS

we observed two electron traps at 0.28 and 0.51eV in MBE-grown $Ga_{0.51}In_{0.49}P$/GaAs heterostructures by TEES measurement. We attribute the 0.28 eV trap to phosphorus vancancies, V_P, which are responsible for both CB conduction at T>220K and for the variable hopping conduction at T<220K. Its concentration decreases with the increasing phosphorus BEP and increases with the increasing annealing temperature. The illumination time dependent TEES showed that this trap at 0.28eV has a very fast filling rate. The 0.51eV trap was found only in the samples with P_2 BEP less than 2×10^{-4} Torr. This could be related to a V_P-related complex rather than a single native defect. The capture cross section of 0.51eV trap was 4.5×10^{-15}cm^2.

Acknowledgment

This work was supported in part by National Science Foundation through the Presidential Young Investigator Program under the grant number DMR 88-57403, and by the New York State Center for Advanced Optical Technology.

References

1. J.M.Olson, S.Kurtz, A.E.Kibbler and P.Faine, Appl. Phys. Lett., 56, 623 (1990)
2. H.Kawai, T.Kobayashi, F. Nakamura, and K.Taira, Electron. Lett., 25, 609 (1989)
3. M.I.Shikawa, Y.Ohba, H.Sugawara, M.Yamamoto, and T.Nakanishi, Appl. Phys. Lett., 48, 20 (1986)
4. S.L.Feng, J.C.Bourgion, F.Omnes and M.Razeghi, Appl. Phys. Lett., 59, 941 (1991)
5. E.C.Paloura, A.Ginoudi, G.Kiriakidis, N.Frangis, F.Scholz, M.Moser and A.Christon, Appl., Phys. Lett., 60, 2749 (1992)
6. G.W.Wicks, M.W.Koch, J.A.Varriano, F.G.Johnson, C.R.Wie, H.M.Kim and P.Colombo, Appl. Phys. Lett., 59, 342 (1991)
7. Z.C.Huang, K.Xie and C.R.Wie, Rev. Sci. Instrum., 62(8) 1951 (1991)
8. N.F.Mott, J. Non-cryst. Solods 1, 1 (1968)
9. D.C.Look, Z-Q. Fang, Phys. Rev. B 47, 1441 (1993)
10. K.Xie, Z.C.Huang and C.R.Wie, J. Electron. Mater., 20, 553 (1991)
11. T.A.Kennedy and M.G.Spencer, Phys. Rev. Lett., 57, 2690 (1986)
12. J.P.Buisson, R.E.Allen and J.D.Dow, Solid State Comm., 43, 833 (1982)
13. T.D. Thompson, J. Barbara, M.C. Ridgway, J. Appl. Phys., 71 6073 (1992)
14. J. Frandon, F. Fabre, G. Bacquet, J. Bandet and F. Reynaud, J. Appl. Phys., 59 1627 (1986)
15. S.K.Min, E.K.Kim, H.Y.Cho, J. Appl. Phys., 63, 4422 (1988)
16. M.Tomozane, Y.Nannichi, I.Onodera, T.Fukase and F.Hasegawa, Jpn. J. Appl. Phys., 27 260 (1988)
17. U.V.Desnica and B.Santic, Appl. Phys. Lett., 54, 810 (1989)
18. Z.Q.Fang, L. Shan, T.E.Schlesinger and A.G.Milnes, Mater. Sci. Engi., B5, 397 (1990)
19. U.V.Desnica, D.I.Desnica and B.Santic, Appl. Phys. Lett., 58, 278 (1991)

PART III

Defects and Impurities in SiGe/Si Heterostructures and Related Compounds

DEEP LEVEL DEFECTS, LUMINESCENCE, AND THE ELECTRO-OPTIC PROPERTIES OF SiGe/Si HETEROSTRUCTURES

PALLAB BHATTACHARYA*, SHIN-HWA LI*, JINJU LEE*, AND STEVE SMITH**
*Department of Electrical Engineering and Computer Science, University of Michigan, Ann Arbor, MI 48109-2122
**University of Dayton Research Institute, Dayton, OH 45469-0178

ABSTRACT

Deep levels and luminescence in SiGe/Si heterostructures and quantum wells have been investigated. We have studied the effects of Be- and B-doping on the luminescent properties of $Si_{1-x}Ge_x$/Si single and multiquantum wells. No new levels, or enhancement of luminescence, from that in undoped samples, is detected in samples which are selectively doped in the well-regions, implying that the observed luminescence in the undoped quantum wells is a result of alloy disordering. Slight enhancement of luminescence is observed in disordered wells and in quantum wires made by electron beam lithography and dry etching. Deep levels have been identified and characterized in undoped $Si_{1-x}Ge_x$ alloys. Hole traps in the p-type layers have activation energies ranging from 0.029-0.45 eV and capture cross sections (σ_∞) ranging from 10^{-9} to 10^{-20} cm^2. Possible origins of these centers are discussed. Some possibilities of obtaining enhanced electro-optic coefficients in SiGe/Si heterostructures are discussed.

INTRODUCTION

There is a lot of current interest in the development of $Si_{1-x}Ge_x$ alloys for use in electronic and optoelectronic devices and circuits[1]. It is of particular interest to integrate optoelectronic devices, in particular light sources, with Si-based digital or analog integrated circuits. The $Si_{1-x}Ge_x$ alloys are promising materials in this context. In order to realize electroluminescent devices with these alloys, it is important to grow high-quality materials and to understand and characterize their luminescent properties. To this end, some recent reports have been made on the luminescence from $Si_{1-x}Ge_x$ and $Si_{1-x}Ge_x$/Si quantum wells[2-14] grown by a variety of techniques.

SiGe crystals, like other semiconductors, will have intentional and unintentional impurity species at substitutional or interstitial sites, native defects, or combinations of both, which may give rise to deep levels in the forbidden energy bandgap. Deep levels can act as carrier trapping and recombination centers and are known to degrade the electronic properties and radiative efficiency of semiconductors, which will ultimately affect device performance. To our knowledge, there has been no explicit report on the properties of deep levels in undoped SiGe alloys.

In addition to sources and detectors, it is important to have electro-optic devices with SiGe. The electro-optic properties of SiGe are also largely unknown. We report here results from our recent studies of some of these fundamental material properties of the SiGe alloys grown by gas-source MBE (GSMBE) using disilane and solid Ge.

SiGe GROWTH BY GAS SOURCE MBE

A two-chamber RIBER 32 MBE with a vacuum load lock is used for our experiments. The growth chamber is provided with an ion pump which maintains a background vacuum of 10^{-10} Torr. The cryoshroud temperature is fixed at 77K and additional pumping is

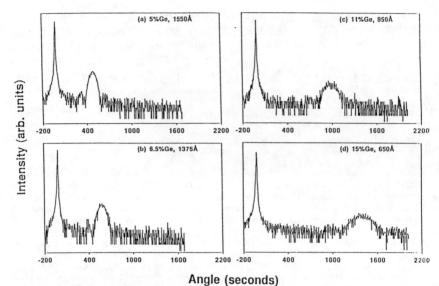

Angle (seconds)

Figure 1: Double crystal x-ray rocking curves for pseudomorphic $Si_{1-x}Ge_x$ layers with various Ge compositions and thicknesses.

provided during growth by a turbomolecular pump. The Si_2H_6 flow rate is controlled by a precision mass flow controller. Elemental germanium is effused from a resistively heated cell with a PBN crucible. During growth the substrate temperature is monitored by a pyrometer.

$Si_{1-x}Ge_x$ epitaxial layers were grown on (100)-oriented, boron-doped, p-type or phosphorus-doped, n-type silicon wafer with a resistivity between $2-5x10^3$ ohm-cm. The substrates were sequentially cleaned in (1) $1NH_4OH:1H_2O_2:5H_2O$, (2) $1HCl:1H_2O_2:3H_2O$, and (3) $1HF:50H_2O$ solutions for 10 min, 10 min, and 30 s, respectively. They were rinsed in deionized H_2O between each solution. Prior to growth, the surface oxide was removed by heating to 840°C for 10 min. At this point a clear (2 x 1) reflection high energy electron diffraction (RHEED) pattern is observed.

$Cuk_{\alpha 1}(004)$ x-ray rocking curve measurements were performed with a double-crystal x-ray diffractometer (XRD). Pendelosung oscillations were observed in the data. From these oscillations, both the perpendicular lattice constant and the layer thickness can be simultaneously determined[15]. Representative data obtained from single pseudomorphic layers are shown in Figure 1.

Typical surface morphology of the $Si_{1-x}Ge_x$ layers grown by Si_2H_6/Ge MBE is illustrated in Figure 2 for the thin pseudomorphic layers. It is nearly featureless. Only a few germanium clusters are observed. The density of these clusters increases with increasing layer thickness. We believe that they originate from the germanium cell. Comparable clusters, originating from the gallium cell, having been observed during the growth of GaAs.

DEEP LEVELS IN UNDOPED $Si_{1-x}Ge_x$

1-μm thick undoped $Si_{1-x}Ge_x$ layers were grown on (100)- oriented, B-doped p-type Si substrates for deep level measurements. Deep level transient spectroscopy (DLTS) measurements were made on Schottky diodes made on the different samples by

40 μm

Figure 2: Surface morphology for a thin (650Å), pseudomorphic $Si_{0.85}Ge_{0.15}$/Si layer. Germanium clusters are sometimes observed, as shown in this photomicrograph.

photolithographic techniques. The diodes were formed by an electron beam evaporated bilayer of Ni(100Å)/Au(5000Å). The nickel layer increased the adhesion sufficiently to allow the application of a gold bonding layer to the surface. Aluminum ohmic contacts were formed on both the surface of the epitaxial layer and the substrate backside. The diodes exhibited rectifying characteristics, with a typical reverse breakdown voltage of 15V.

The DLTS signal as a function of temperature for a fixed rate window is shown in Figure 3, for a sample with x = 0.14. The reverse bias applied to the diode is sufficient

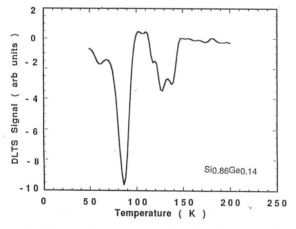

Figure 3: Deep level transient spectroscopy data showing peaks due to hole traps in p-type $Si_{0.86}Ge_{0.14}$.

to ensure that the deep levels that have been identified are bulk traps. The data indicate the presence of at least four majority-carrier (hole) traps. Arrhenius plots of these traps in different samples, from which the trap characteristics are derived, are shown in Figure 4. The characteristics of the traps are listed in Table I. It is apparent that most of the centers have either very small, or very large values of the capture cross section, σ_{∞}. Large capture cross sections are normally associated with recombination centers,

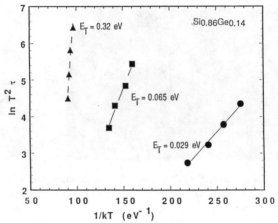

Figure 4: Arrhenius plots of three clearly identified hole traps in $Si_{0.86}Ge_{0.14}$.

while very small capture cross sections are usually attributed to traps that have a large lattice-relaxation, such as the D-X center in $Al_xGa_{1-x}As^{16}$.

The measurements reported above have been made on nominally undoped material. Capacitance-voltage (C-V) measurements have shown this material to be p-type. Since germanium is a Group IVA element, substitutional atoms should be isoelectronic with silicon. When the alloy is grown in the silicon-rich condition, germanium atoms occupy sites in the silicon diamond lattice structure. No deep levels are expected to arise from this incorporation of germanium atoms into the silicon lattice.

Since the $Si_{1-x}Ge_x$ alloys were not intentionally doped, one would first attribute the existence of deep levels to native defects. However, the number of candidates are extremely small. Self-interstitials anneal at temperatures well below 300K, as do most vacancies. Vacancy complexes can exist at higher temperatures and are known to give rise to deep levels in silicon[17]. An aluminum-vacancy complex and an aluminum-interstitial-aluminum substitutional pair were identified as having energies of $E_v+0.45eV$ and $E_v+0.25eV$, respectively, in electron irradiated, p-type silicon[18]. Dislocations are not considered to be a potential origin, since they produce a continuous band of levels, with high densities, rather than discrete states.

It is possible that unintentional dopants may be complexing with germanium atoms to produce these deep levels in $Si_{1-x}Ge_x$. Even though these films are relaxed, there is a small residual strain due to the difference in the size of the germanium atom. This is evident by the use of carbon to bring the lattice back to the silicon parameters. The extremely small extrapolated capture cross sections of some of the traps are similar to the vanishingly small cross sections reported for the DX centers in AlGaAs[16] due to a large lattice relaxation. It may be noted in Table I that some of the traps have very large values of α_∞, usually associated with recombination centers.

STEADY STATE AND TIME-RESOLVED PHOTOLUMINESCENCE FROM PERIODIC AND DISORDERED QUANTUM WELLS

Doping by isoelectronic impurities was proven to be a successful technique for enhancing the luminescence in indirect bandgap compound semiconductors such as GaP and GaAsP[19]. In the case of these compounds nitrogen (N) was found to be very suitable. Doping with N produced a series of deep levels in the energy bandgap and the k-selection rules for transitions involving these levels are relaxed. There are no suitable isoelectronic

Table I. Measured Characteristics of Majority-Carrier
Traps in Undoped p-Type $Si_{1-x}Ge_x$

Alloy Comp. (x)	Measured Activation Energy, E_T (ev)	Trap Concentration $(cm^{-3})(x10^{13})$	Capture Cross Section $\sigma_\infty(cm^2)$
0.06	No traps detected		
0.14	0.029	1.4	$1.96x10^{-20}$
	0.065	3.67	$8.5x10^{-20}$
	0.314	1.3	$1.3x10^{-11}$
0.19	0.041	8.0	$1.6x10^{-20}$
	A deeper trap	not well resolved	
0.26	<0.2	not well resolved	
	0.261	1.2	$8.55x10^{-15}$
	0.45	1.3	$5.0x10^{-9}$

dopants in Group IV of the Periodic Table for the SiGe alloys. We therefore chose to investigate the effects of doping with Be(Gr II) and B(Gr III). The latter is a well-behaved acceptor dopant for these materials. Similar relaxation of the k-selection rules can be obtained in disordered superlattices. We report here the luminescence measured in high-quality undoped and Be- and B-doped $Si_{1-x}Ge_x$/Si quantum wells and disordered superlattices.

The QW structures typically consist of a 100 nm undoped Si buffer layer, $Si_{1-x}Ge_x$/Si single quantum well (SQW) or multi-quantum well (MQW) and top undoped Si layer, typically 500Åthick. Two sets of samples were grown and studied. In the first, the entire SQW or MQW structure was undoped. In the second, the well regions were selectively doped with Be or B to a level of 1 x 10^{17} cm^{-3}. Well and barrier thicknesses vary from 20–100Å and 50–200Å, respectively.

High-resolution (~2Å) photoluminescence (PL) spectra of these samples were measured using the 488 nm line of an argon-ion laser of variable intensity. Luminescence signals were processed in a standard configuration using a 1-m Jarell-Ash spectrometer and a lock-in detection system with a liquid N_2-cooled Ge detector. Samples were cooled down to 18K with a closed-loop variable temperature He cryostat.

We will first present and discuss the PL spectra observed for the undoped QW samples. For clarity of presentation, the data will be presented in two regimes of photon energy. These correspond, respectively, to the spectral regions of the Si and SiGe bandedges. Figure 5(a) shows luminescence from the Si buffer and QW barrier regions (and possibly with some contribution from the Si substrate) in a MQW sample with 10 periods. The spectra are characterized by a series of sharp excitonic lines with linewidths ranging from 6–8 meV. These lines originate from free- and bound-exciton transitions with phonon participation (longitudinal and transverse acoustic and optical phonons). They are labeled by analogy with data published in the literature[2]. The broad transition labeled EHD is due to the electron-hole droplet and is not seen in all the samples. The transition at 0.998 eV is seen in all our samples and we believe it originates from a defect or impurity in the Si layers. All these transitions were identified in the PL spectra of an undoped Si wafer. Figure 5(b) shows typical luminescence observed from a 40Å $Si_{0.75}Ge_{0.25}$/75Å Si MQW with 10 periods. Note that the samples

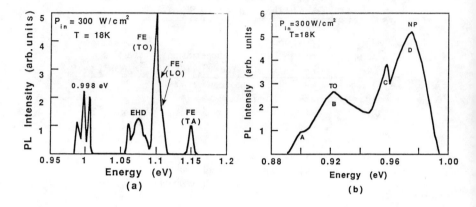

Figure 5: Measured low-temperature photoluminescence from (a) Si barrier and buffer layers, and (b) 40Å $Si_{0.75}Ge_{0.25}$ well regions of 10-period SiGe/Si undoped multi-quantum wells.

in Figures 5(a) and (b) are not the same. The peaks labeled A, B, C, and D correspond to the peaks a, b, c and d, observed and discussed by Vescan, et al.[12] Lines C and D are due to no-phonon (NP) transitions of bound excitons due to relaxation of k-selection rules by alloy scattering. From analysis of high-field transport data in $Si_{1-x}Ge_x$ alloys we have calculated the alloy scattering potential U_o in the valence band to be 0.6 eV[20]. The transitions C and D are believed to be a split doublet ($\sim$20 meV) caused by 2-3 monolayers intra-layer and interlayer well thickness variations. Transitions labeled A and B are due to TO_{Si-Si} phonon replicas (D-B and C-A are $\sim$59 meV) of the NP transitions[2]. No dislocation or defect related transitions[12] were observed at lower energies, indicating that all our sample had pseudomorphic quantum wells. The bandgap energies of the $Si_{1-x}Ge_x$ alloys can be estimated from the NP transition energies, taking into account the exciton binding energy. We have measured the shift in energy of the NP transitions as the well thickness is changed. A blue shift is observed with reduction of well width, confirming quantum confinement effects.

The photoluminescence spectra observed from the Be-doped sample are discussed next. Measurements were made with 100Å $Si_{0.9}Ge_{0.1}$/Si SQW and MQW samples with different Si barrier thicknesses. The spectra are essentially the same and are shown in Figure 6. The NP transition is observed at 1.041 eV and may primarily be band-to-band in nature due to the doping. The transition at 1.022 eV is possibly a doublet of the NP transition or a TA phonon replica. The peak at 0.998 eV is seen again. The TO_{Si-Si} phonon replica is observed at 0.977 eV. The overall intensity of the luminescence is approximately three times smaller than that of the undoped samples. Unlike N-doping in GaAsP, the luminescence intensity is not enhanced, or new transitions are not produced. These results suggest that the effect of alloy disordering to relax the selection rules is far more dominant in these alloys than the effect of doping, in as far as luminescence is concerned. In the B-doped samples ($\sim$5x10^{16} – 1x10^{17} cm^{-3}) the luminescence was quenched more severely and is not being shown here.

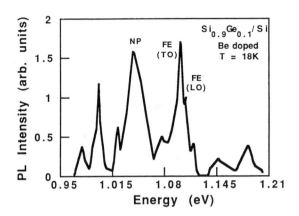

Figure 6: Low-temperature PL spectra of Be-doped 10-period $Si_{0.9}Ge_{0.1}/Si$ MQW.

To summarize, we have made a detailed study of high-resolution PL spectra measured in undoped and doped $Si_{1-x}Ge_x/Si$ MQW and SQW samples. The results suggest that doping in these alloys will not enhance PL efficiencies. Improvement of material quality or other techniques such as disordered quantum wells and quantum wires need to be explored.

Disordered superlattices with multiple periods were grown for PL measurements. The schematic of a typical structure is shown in Figure 7(a). The sequence number n is randomly varied between 1 and 3. The probability of each number is 1/3. The average quantum well structure has 85Å SiGe wells and 260Å Si barriers. Luminescence from such a sample is shown in Figure 7(b). No significant enhancement is observed in the low temperature PL which agrees with previous data [21-23]. However, some enhancement is observed at 77K and higher temperatures.

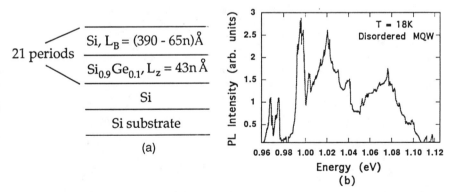

Figure 7: (a) Structure of undoped disordered superlattice, where n is a random integral number between 1 and 3. The probability of each number is 1/3; (b) Low temperature PL observed from disordered superlattice.

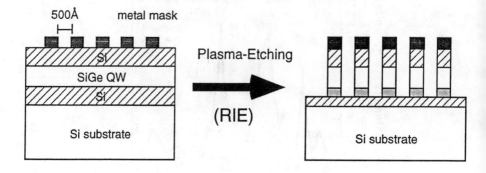

500Å   metal mask

Si

SiGe QW

Si

Si substrate

Plasma-Etching

(RIE)

Si substrate

Figure 8: Schematics for fabrication of SiGe/Si quantum wires.

It is equally important to note that the radiative recombination times have to be short enough (~1 ns) for such radiative transitions to be useful for a light source. We have therefore measured the steady state and transient luminescence properties of SiGe/Si quantum wells. The excitation source was a mW pulsed GaAs/AlGaAs laser with 100 ps pulsewidth and 50 KHz repetition rate. The temporal variation of the photoluminescence was measured with a New Focus InGaAs detector with large gain and response time of 1 ns. No significant decay signal could be observed, since the photoluminescence is very weak. The recombination lifetime in Si is ~1μs. Unless there is a significant reduction in the value of this parameter in SiGe, these materials will not be useful for practical luminescent devices.

LUMINESCENCE FROM SiGe/Si QUANTUM WIRES:

A possible technique for enhancing the oscillator strength of optical transitions is to use disordered quantum wires. Recently reported work on porous Si suggests that optical processes can be enhanced in quasi-one dimensional (quasi-1D) systems[24]. It has been shown[25] that electronic states can be localized in quasi-1D systems even with a small disorder. The disorder is expected to strongly localize electronic states near the band edges and this will enhance the optical transition rates. Multiquantum well samples, consisting of 85 Å wells and 260 Å barriers were first grown. The quantum wires are defined by reactive ion etching through nickel masks patterned by electron beam lithography (Figure 8). The wires so defined in a sample are shown in the photomicrograph of Figure 9(a). The width of such quantum wires made in our laboratory vary from 400–1000Å.

The steady-state, low-temperature photoluminescence spectrum observed from a typical quantum wire sample is shown in Figure 9(b). The transition observed at 1.1 eV is believed to be due to no-phonon (NP) transitions in the wire. For comparison, the NP transitions from the quantum wells of this particular wafer is shown by the dashed profile. A blue shift of 35 meV is clearly observed, indicating the existence of quasi-1D effects in a 900 nm wire.

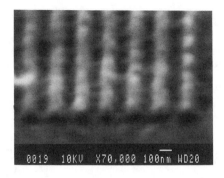

(a)

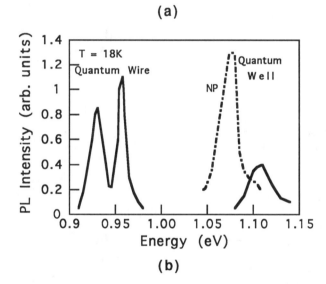

(b)

Figure 9: (a) SEM micrograph of $Si_{0.9}Ge_{0.1}$ quantum wires; (b) measured low-temperature photoluminescence spectra.

ELECTRO-OPTIC EFFECTS IN SiGe

It is well known that because of the presence of inversion symmetry, the electro-optic effect in bulk Si (and Ge) is negligible. However, in quantum wells made from the SiGe/Si system, we expect a larger electro-optic effect. It is important to note that the dielectric constant of Si is $11.9\varepsilon_o$ while that of Ge is $16.2\varepsilon_o$. This is an extremely large difference and if proper quantum wells were designed a large electro-optic effect can be expected. By using asymmetric quantum well structures, as shown in Figure 10, the linear electro-optic effect can be very strong. Experiments are in progress to measure these effects.

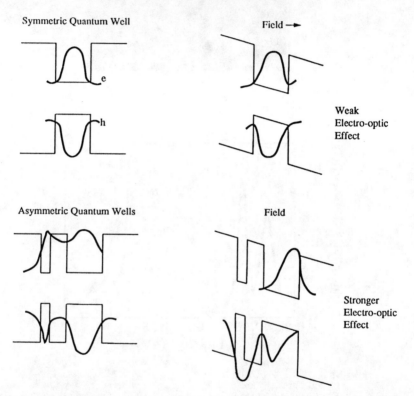

Figure 10: Schematic illustration of symmetric and asymmetric SiGe/Si quantum wells and the corresponding electro-optic effects.

CONCLUSIONS

Luminescence in quantum wells and quantum wires made with SiGe/Si heterostructures and deep levels in single SiGe layers have been characterized. It is clear that although luminescence is observed from the wells and wires, some mechanism to enhance the oscillator strength of the transitions is required. Deep levels have been identified and characterized in undoped $Si_{1-x}Ge_x$ alloys grown on silicon substrates. Hole traps in the p-type layers have activation energies ranging from 0.029–0.45 eV and capture cross sections (σ_∞) ranging from 10^{-9} to 10^{-20} cm². Possible origins of these centers are discussed.

ACKNOWLEDGEMENTS

The work at the University of Michigan is supported by the U.S. Air Force Office of Scientific Research and the Materials Research Laboratory, Wright-Patterson Air Force Base, under Grant No. AFOSR-91-0349. One of us (SRS) is supported by Wright Laboratory Materials Directorate, contract No. F33615-91-C-5603. We would like to thank G. Landis for sample preparation.

REFERENCES

1. J. C. Bean, J. Cryst. Growth **81**, 411 (1987).

2. J. Weber and M. I. Alonso, Phys. Rev. B, **40**, 5683 (1989).

3. K. Terashima, M. Tajima and T. Tatsumi, Appl. Phys. Lett., **57**, 1925 (1990).

4. J.-P. Noel, N. L. Rowell, D. C. Houghton and D. D. Perovic, Appl. Phys. Lett., **57**, 1037 (1990).

5. K. Terashima, M. Tajima, N. Ikarashi, T. Niino and T. Tatsumi, Jpn. J. Appl. Phys., **30**, 3601 (1991).

6. J. C. Sturm, H. Manoharan, L. C. Lenchyshyn, M. L. W. Thewalt, N. L. Rowell, J.-P. Noel and D. C. Houghton, Phys. Rev. Lett., **66**, 1362 (1991).

7. G. A. Northrop, D. J. Wolford and S. S. Iyer, Appl. Phys. Lett., **60**, 865 (1992).

8. D. J. Robbins, L. T. Canham, S. J. Barnett, A. D. Pitt and P. Calcott, J. Appl. Phys., **71**, 1407 (1992).

9. X. Xiao, C. W. Liu, J. C. Sturm, L. C.Lenchyshyn and M. L. W. Thewalt, Appl. Phys. Lett., **60**, 1720 (1992).

10. J. Spitzer, K. Thonke, R. Sauer, H. Kibbel, H.-J. Herzog and E. Kasper, Appl. Phys. Lett., **60**, 1729 (1992).

11. X. Xiao, C. W. Liu, J. C. Sturm, L. C. Lenchyshyn, M. L. W. Thewalt, R. B. Gregory and P. Fejes, Appl. Phys. Lett., **60**, 2135 (1992).

12. L. Vescan, A. Hartmann, K. Schmidt, C. Dieker, H. Lüth and W. Jäger, Appl. Phys. Lett., **60**, 2183 (1992).

13. J.-P. Noel, N. L. Rowell, D. C. Houghton, A. Wang and D. D. Perovic, Appl. Phys. Lett., **61**, 690 (1992).

14. S. Fukatsu, H. Yoshida, A. Fujiwara, Y. Takahashi, Y. Shiraki and R. Ito, Appl. Phys. Lett., **61**, 804 (1992).

15. Y. C. Chen, Ph.D. thesis, University of Michigan, 1992.

16. D. V. Lang, R. A. Logan and M. Jaros, Phys. Rev. **B19**, 1015 (1979).

17. O. F. Sankey and J. D. Dow, Phys. Rev., **B26**, 3243 (1982).

18. B. N. Mukashev, L. G. Kolodin, K. H. Nussupov, A. V. Spitsyn and V. S. Vavilov, Radia. Eff., **46**, 770 (1980).

19. See, for example, N. Holonyak, R. J. Nelson, J. J. Coleman, P. D. Wright, D. Finn, W. O. Groves and D. L. Keune, J. Appl. Phys., **48**, 1963 (1977), and publications therein.

20. S. H. Li, J. Hinckley, J. Singh and P. K. Bhattacharya, Appl. Phys. Lett., **63**, 1393 (1993).

21. A. Sasaki, J. Cryst. Growth, **115**, 490 (1991).

22. T. Yamamota, M. Kasu, S. Noda and A. Sasaki, J. Appl. Phys., **68**, 5318 (1990).

23. M. Kasu, T. Yamamoto, S. Noda and A. Sasaki, Jpn. J. Appl. Phys., **29**, 828 (1990).

24. I. Sagnes, *et al.*, Appl. Phys. Lett., **62**, 1155 (1993).

25. J. Singh, Appl. Phys. Lett., **59**, 3142 (1991).

ELECTRONIC CHARACTERIZATION OF DISLOCATIONS IN RTCVD GERMANIUM-SILICON/SILICON GROWN BY GRADED LAYER EPITAXY

P.N. GRILLOT,* S.A. RINGEL,* G.P. WATSON,** E.A. FITZGERALD** and Y.H. XIE**
*Dept. of Electrical Engineering, The Ohio State University, Columbus, OH 43210
**AT&T Bell Laboratories, Murray Hill, NJ 07974

ABSTRACT

Carrier trapping and recombination activity have been studied with DLTS and EBIC in RTCVD grown compositionally graded $Ge_{0.3}Si_{0.7}/Si$ heterostructures. DLTS peak height is found to vary with applied bias, and the bias conditions used indicate that at least one peak is present in the homoepitaxial Si buffer layer and perhaps the substrate as well. Variations in EBIC contrast as a function of reverse bias, and DLTS fill pulse experiments both indicate that the DLTS peaks observed are dislocation related. Moreover, the bias dependent decrease in DLTS peak height is observed to occur at different rates for different peaks, indicating a possible connection between certain DLTS peaks and dislocation orientation or type. Activation energies of one electron trapping center and one hole trapping center add up to roughly the expected bandgap in a relaxed Ge_xSi_{1-x} alloy with $x \leq 0.3$, indicating that the electron and hole trapping centers observed with DLTS may, in fact, be associated with the R-G center observed by EBIC.

INTRODUCTION AND EXPERIMENT

$Ge_{0.3}Si_{0.7}$ alloys were grown by RTCVD on p-type Si (100) using the graded layer epitaxy (GLE) technique. Specifically, a 1 μm Si buffer layer was grown, followed by a step-graded structure consisting of 10 discrete steps with a 3% change in Ge mole fraction at each interface. Finally, a 1 μm $Ge_{0.3}Si_{0.7}$ cap layer was grown. An average grading rate of 20% Ge/μm was maintained throughout the step graded region. GeSi epitaxy occurred at 800 °C, and all alloy thicknesses exceeded the equilibrium theory critical thickness. Subsequent material characterization indicates that the GeSi alloy layers are completely relaxed.[1]

All epitaxial films were nominally undoped, but C-V dopant profiling revealed a p-type conductivity at a concentration $\sim$ 2 x 10^{14} cm^{-3}. P-n junctions were formed in these samples by ion implantation with a dose of 5 x 10^{13} As$^+$/cm^2 at 50 kV, resulting in an As concentration $\sim$ 10^{19} cm^{-3} at a projected range of 300 Å. The As dopant was activated by a 60 sec. RTA at 700 °C. TEM and EBIC data clearly indicate the presence of a three-dimensional dislocation network in the compositionally graded region, as well as threading dislocations in the cap region and dislocation half loops (which have originated from misfit segments) in the Si buffer layer.[1] A more complete description of sample growth and processing conditions is reported elsewhere.[1]

Ohmic contacts were formed on these samples by evaporation of 1500 Å of Al on the p-type Si back surface and the $Ge_{0.3}Si_{0.7}$ n$^+$ implanted cap layer. Following metallization, I-V and C-V characterization were performed to ensure device integrity prior to performing DLTS and EBIC measurements. On those samples which exhibited satisfactory I-V and C-V characteristics, DLTS was performed with a Bio-Rad model 4602 Deep Level Transient Spectrometer, using the double boxcar method. EBIC data was obtained with a JEOL 6400 SEM which has been modified for EBIC measurements. SEM beam current was $\sim$ 1 nA, with an accelerating voltage

Mat. Res. Soc. Symp. Proc. Vol. 325. ©1994 Materials Research Society

of 20 kV. Bias values used for EBIC range from +0.5 volts to -4 volts, so the excitation volume extended beyond the depletion region edge in all EBIC measurements reported here.

RESULTS

In figure 1 a, we illustrate the deep level spectrum of the GeSi/Si GLE sample, which indicates the presence of at least two hole traps, D1 and D2. The bias conditions used to obtain this data were V_R = -1.1 volts, and V_F = -0.1 volts, where V_R is the quiescent reverse bias, and V_F is the diode bias during the fill pulse portion of the DLTS cycle. It should be noted that the Debye length in these low doped samples is ~ 0.4 μm, and the zero bias depletion width is ~ 2 μm.

Figure 1 b illustrates the DLTS spectrum of the same sample with an applied bias of V_R/V_F = -1.50 volts/-0.50 volts. Clearly, the DLTS peak height has decreased in figure 1 b, relative to that in figure 1 a. Moreover, the relative intensity of peaks D1 and D2 has inverted. That is, peak D1 is dominant in figure 1 a, while peak D2 is dominant in figure 1 b.

In figure 1 c, taken at bias conditions of V_R/V_F = -6.00 volts/-5.00 volts, peak D2 has disappeared completely, and peak D1 has decreased further in intensity. This trend is observed to continue with increasing reverse bias, as illustrated in figure 2.

As mentioned above, the zero bias depletion width in these samples is ~ 2 μm, while, for an applied bias of -0.5 volts, the depletion layer width is ~ 2.5 μm. This depletion width data indicates that the near-zero bias DLTS data in figure 1 a, where peak D1 is dominant, emphasizes the compositionally graded region, which contains a three-dimensional network of dislocations. However, for the DLTS data indicated in figure 1 b, where peak D2 dominates,

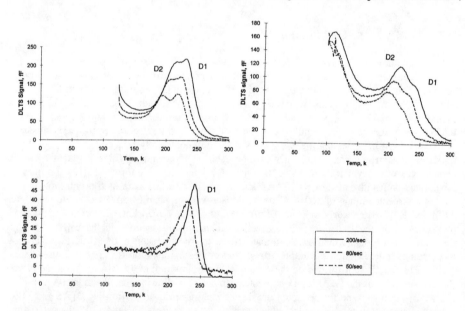

Fig. 1. DLTS spectrum under bias conditions of a.) V_R/V_F = -1.10 v/-0.10 v (depletion width, W $\approx$ 2 μm) b.) V_R/V_F = -1.50 v/-0.50 v (W $\approx$ 2.5 μm = alloy film thickness), and c.) V_R/V_F = -1.10 v/-0.10 v (W $\gg$ alloy film thickness).

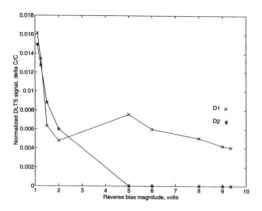

Fig. 2. DLTS peak height as a function of reverse bias magnitude.

the depletion region width indicates that DLTS measurements will tend to emphasize trapping near the edge of the compositionally graded region, as well as the Si buffer layer and may therefore be more sensitive to misfit dislocations lying in the (100) growth plane. For much larger reverse bias, i.e., $|V_R| > 4$ volts, all mismatched interfaces are deep within the depletion region, and the trapping centers associated with misfit dislocations in the compositionally graded region are not filled by the majority carrier injection pulse. Hence, for such a large reverse bias, as in figure 1 c, DLTS is sensitive to dislocation half loops in the Si buffer layer and substrate (which have been observed in these samples with TEM)[1], and peak D1 is again dominant. The overall bias dependence of the DLTS peak height is summarized in figure 2. While the ion implantation discussed above is expected to cause damage in the top 1000 Å of the Ge$_{.3}$Si$_{.7}$ cap region, these DLTS measurements probe the sample at a depth ≥ 2 μm. The DLTS results presented above are therefore not believed to be related to implant damage.

To further corroborate this DLTS depth dependence, we have investigated EBIC contrast as a function of bias. In figures 3 a and b, we display EBIC images obtained at zero bias, and at -4 volts, respectively. Note that the contrast was optimized individually for each EBIC image, hence figures 3 a and b should be interpreted as showing only qualitative trends in EBIC contrast. A more quantitative comparison is shown in figure 4, where the relative EBIC contrast has been corrected for gain and brightness adjustments at each applied bias. Clearly, the relative EBIC contrast decreases with applied bias, in agreement with the DLTS results presented above.

EBIC is a well established technique for the identification of dislocations in electronic materials. Hence the correlation between EBIC and DLTS bias dependence indicates that DLTS is most likely probing dislocation related trapping centers in the GeSi alloy layers and the Si buffer layer and substrate.

To better identify the physical sources of the observed DLTS peaks, we have examined the dependence of DLTS peak height on fill pulse time, t_p. For dislocation related trapping centers, it is well established that the DLTS peak height varies as $\log(t_p)$.[2-7] As illustrated in figure 5, we have indeed observed this same logarithmic dependence for the hole trapping centers D1 and D2, as well as an electron trapping center, D4. The observed saturation of DLTS peak height in figure 5 for $t_p > 1$ ms is in agreement with the results of Omling et al., on plastically deformed Si,[4] and this result will be discussed in more detail elsewhere.[8]

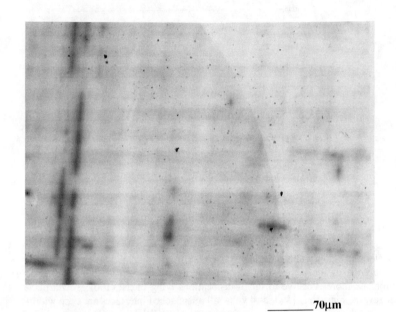

_____70μm

_____70μm

Fig 3. EBIC images at a.) zero volts and b.) -4.0 volts applied bias.

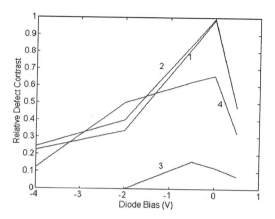

Fig. 4. EBIC relative defect contrast as a function of applied bias.

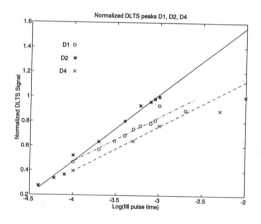

Figure 5. DLTS peak height vs. log(t_p). The linear relationship indicates that the observed DLTS peaks are dislocation related.

Arrhenius plots of DLTS peaks D1 (hole trapping center) and D4 (electron trapping center) indicate activation energies of $E_{D1} \approx 0.43$ eV and $E_{D4} \approx 0.64$ eV. The charge trapping characteristics of DLTS allow us to observe the electron trapping center under forward bias injection and the hole trapping center under reverse bias injection. However, additional DLTS data not reported here leads us to believe that the minority carrier electron trap is still present and active under reverse bias.[8] If trapping centers D1 and D4 are due to the same physical defect, then $E_{D1} + E_{D4} \approx E_g(GeSi)$, indicating that the trapping centers detected with DLTS may, in fact, be related to the R-G center which we detect with EBIC.

163

CONCLUSIONS

In conclusion, we have used DLTS to observe several trapping centers in compositionally graded RTCVD grown $Ge_{0.3}Si_{0.7}/Si$. Both the individual DLTS peak heights and the ratio of peak heights is observed to vary with applied bias, indicating that the observed peaks are associated with the spatially varying dislocation structure in these samples.

EBIC contrast, which is a well established technique for the detection of dislocations in semiconductors, indicates a bias dependence which is in qualitative agreement with that observed through DLTS. Moreover, DLTS peak height is observed to vary as $\log(t_p)$, and the activation energies of one electron trapping center and one hole trapping center add up to approximately the bandgap of a relaxed Ge_xSi_{1-x} alloy with $x \leq 0.3$.

The above facts indicate that the trapping centers observed by DLTS are related to dislocations, and may be related to the R-G centers observed by EBIC. We are currently extending this research to include other alloy compositions and grading schemes, and to present a more detailed treatment of the results given here. These results will be reported elsewhere.[8]

ACKNOWLEDGEMENTS

The authors are grateful to J.P. Pelz of The Ohio State University and M.L. Timmons of the Research Triangle Institute for helpful communications. This work supported in part by a grant from the OSU Center for Materials Research.

REFERENCES

1. G.P. Watson, E.A. Fitzgerald, Y. Xie, and D.P. Monroe, submitted to J. Appl. Phys.

2. G.P. Watson, D.G. Ast, T.J. Anderson, B. Pathangey, and Y. Hayakawa, J. Appl. Phys., **71**, 3399 (1992).

3. T. Wosinski, J. Appl. Phys., **65**, 1566 (1989).

4. P. Omling, E.R. Weber, L. Montelius, H. Alexander, and J. Michel, Phys. Rev. B, **32**, 6571 (1985).

5. L.C. Kimerling and J.R. Patel, in *VLSI Electronics*, vol. 12, edited by N.G. Einspruch, (Academic Press, Orlando, 1985) p. 223.

6. V.V. Kveder, Y.A. Osipyan, W. Schroter, and G. Zoth, Phys. Stat. Sol. A, **72**, 701 (1982).

7. T. Figielski, Solid State Electron., 21, 1403 (1978).

8. P.N. Grillot, S.A. Ringel, G.P. Watson, and E.A. Fitzgerald, manuscript in preparation.

SPACE DISTRIBUTION OF DEEP LEVELS IN SiGe/Si HETEROSTRUCTURE

Zhang Rong, Yang Kai, Gu Shulin, Shi Yi, Huang Hongbin, Wang Ronghua, Han Ping, Hu Liqun, Zheng Youdou and Li Qi
Department of Physics, Nanjing University, Nanjing 210008, CHINA

ABSTRACT

The small-pulse DLTS had been developed to measure distribution of deep levels in CVD grown SiGe/Si heterostructure before and after thermal processing at 800°C. Changes of defect states was found and after processing the original single deep level 0.62eV under the condition band split into two separated traps. A new weak deeper trap signal was found only in the just relaxed region. It could be Ge-related defect complex with misfit dislocations.

I. INTRODUCTION

SiGe/Si heterostructures have absorbed much attention of semiconductor scientists and engineers. Because of their novel properties, many new devices such as heterojunction bipolar transistor(HBT)[1], modulation doped field effect transistor (MODFET)[2], infrared detector[3-4], waveguide[5] and light emitting diode[6] have been fabricated. In these new device structures, the energy band of SiGe alloy layer can be controlled not only by changing composition, but also by altering thickness and strain in the layer. They are bases for Si "band-gap engineering". With the great progress in growth and band-gap engineering techniques further interesting devices are expected to be developed.

Deep traps in semiconductor have great influences on device properties. In semiconductor heterostructures, especially in the strained layer heterostructures, because of existence of misfit dislocations and different deep traps in different sublayer, both deep levels and their distribution are very important for understanding the behavior of heterostructure devices. A versatile and widely-used method to determine the deep levels in bulk semiconductor materials is deep level transient spectroscopy (DLTS). But it can not give the space distribution of heterostructures, especially in the case of interfaces and ultrathin multilayers. In this paper a new method--small pulse DLTS is developed to characterize deep levels and their space distribution in semiconductor heterostructures. Using the method, we report here for the first time the experimental results of the SiGe/Si heterostructure.

The DLTS spectra showed a majority-carrier trap in the SiGe/Si ML. In order to study the origin of this deep level, the sample was thermally processed at 800°C. After processing, the original trap split into two separated deep levels. A new weak deeper level signal was found only in the SiGe region, while the strong shallower level signal was found in whole sample. When reverse bias varied from 0.3V to 3.6V, and the magnitude of filling pulse maintain a constant of 100meV, it was found that the active energy of the deeper level varied between 0.54eV to 0.63eV and vanished when $Vr < 0.4eV$ and $Vr > 1.5eV$. A model was proposed to interpret the experimental results.

Mat. Res. Soc. Symp. Proc. Vol. 325. ©1994 Materials Research Society

II. EXPERIMENTS

The small-pulse DLTS method was put forward by Wendell D. Eadeas and Richard M.Swanson in 1984[7] in order to determine the distribution of interface state density in the forbidden gap. By choosing appropriate magnitude, the authors distinguished contribution from interface states with different energy in the forbidden gap. Detailed they set a small constant difference of Vr-Vp, varied Vr (and Vp) to measure the interface states with different energy. The small-pulse DLTS was a powerful tool to study the distribution of interface states with high sensitivity, high resolution and high signal to noise ratio.

The semiconductor heterostructure was different from homogenous bulk material, and its deep traps behave differently in different sublayers and at interfaces, so traditional DLTS technique can not distinguish distributions from different sublayers. By using small-pulse DLTS, we can control proper filling pulse to modulate "sampling space window" and measure the profile of deep levels in the depletion region. If the amplitude of reverse pulse was low enough, we may theoretically get a very thin "sampling space window" and distinguish different contribution of sublayers in the heterostructures from DLTS spectra. The calculated results showed that for Si under the 10^{16}/cm^3 doping condition, a 60mV smallgwulse corresponded to a 6nm "sampling space window", it is enough to detect special signal of deep levels in each sublayer in the semiconductor heterostructure multilayers.

The structure of measured sample was shown in Fig.1. The unintentional doped Si caplayer and the SiGe/Si multilayers were grown on 3-5 Ωcm n-type (100)Si substrate by RTP/VLP-CVD at 600°C[8]. SiH$_4$ and GeH$_4$ were used as reactant source gases, and the deposition chamber was kept with 2-8mTorr during growth. For distinguishing signal contributions between the metal/semiconductor interface and the ML region, a thick Si caplayer had been grown on the top of the SiGe/Si heterostructures. The X-ray diffraction spectrum was shown in Fig.2.

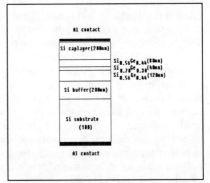

Figure 1 Schematic structure of the sample

III. RESULTS AND DISCUSSIONS

The typical DLTS spectrum of SiGe/Si heterostructure as prepared was given in Fig.3a. We can see a strong signal at T_p= 250K. The deep level parameters had been summarized in Table I. In order to study the origin of the deep trap, the sample was thermally processed in vacuum at 800°C for 15min. The DLTS spectrum of the sample after annealing was figured out in Fig.3b. It is obviously that the original peak had split into two separated peak named A and B respectively. The small-pulse DLTS spectrum with pulse height of 200mV and different reverse was bias shown in Fig.4. Peak B appeared only in the region between 0.9V and 1.5eV. The finer measurements with pulse height of 100mV gave active energy, capture cross-section and intensity in Fig.5, Fig.6, Fig.7 and Fig.8.

From Fig.5 and Fig.6, we could see that both energy level and capture cross section varied in different sublayers, and the interfaces between sublayers could be clearly seen. The traps in sublayer a and c were located at 0.42eV under the conduction band edge, and the trap in sublayer b was located at 0.53eV under the conduction band edge, while in Si the trap was 0.62eV under the CB. Similarly, the capture cross section of the trap behaved differently in different sublayers and showedsame trends. A shallower trap had smaller capture cross section than a deeper one indicated that they had different defect microstructure. We took attention to Fig.7 and Fig.8, and found that both trap A and B concentrated in

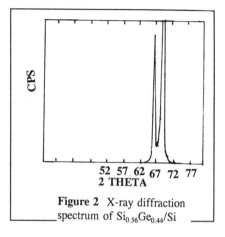

Figure 2 X-ray diffraction spectrum of $Si_{0.56}Ge_{0.44}/Si$

sublayer b. This fact indicated that the sublayer was relaxed during annealing and produced defect complex related to misfit dislocations.

In the heterostructure as prepared, the thickness of sublayer c was much larger than the critical value about 12nm, so sublayer c was fully relaxed and sublayer b were strained. After thermally processing, the sublayer b became relaxing, and the defects in the sublayer b combined with misfit dislocation to form defect complex, while the deep level moved from 0.62eV to 0.60eV and 0.56eV respectively. The deep trap in sublayer a and c changed from 0.62eV to 0.42eV, the stable defect state in SiGe alloy[9].

Table II listed the intensity of deep level before annealing and sum of intensity of trap A and trap B after annealing. The near number of deep traps before and after annealing indicated that both peak A and B were originated from

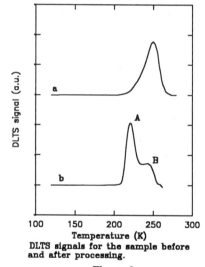

DLTS signals for the sample before and after processing.

Figure 3

the deep trap existing in the sample as prepared, and the misfit dislocation didn't introduced new deep traps and only combined with original traps to form defect complex. Fig.9 gave the relative magnitude of peak A and B as a function of depth. It is constant of 0.37 confirmed that trap A and trap B were Si and Ge related respectively.

Nagesh[9] et al. found that the mid-gap level was pinned to the valence band, so the composition of alloy could be deduced from the energy level data of DLTS. Using this method, we calculated the composition of SiGe alloy as x=0.42 for both sublayer a and c, and x= 0.22 for sublayer b. The data of sublayer b was lower than the designed x value maybe because that the defect in the sublayer b was the defect complex and differed from those in sublayer a and c.

IV. CONCLUSIONS

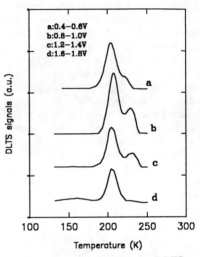

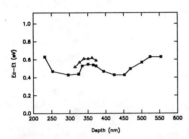

Figure 5 Enery level of trap A and B

Figure 4 Small-pulse DLTS spectrum with different bias.

It had been demonstrated that the small-pulse DLTS method were a useful experimental tools for studying the deep traps and their distribution in the SiGe/Si heterostructures. These properties allowed a comprehensive analysis of the defect in different space region. Using the method we had measured the trap profile in the SiGe/Si multilayer, resulting in distribution of energy level and capture cross section as well as trap densities. The results of pre- and post annealing suggested that thermal processing could change the state of deep traps and form defect complex. By using the small pulse DLTS we also obtained the composition in different sublayers.

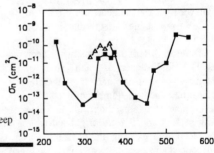

Table I Comparisen of total intensities of deep traps pre- and post annealing

Vr-Vp(V)	N_t/N_d	$(N_t^A+N_t^B)/N_d$
0.4-0.8	6.4%	6.1%
0.6-1.2	15.6%	15.1%
1.2-1.9	7.6%	5.5%

Figure 6 Capture cross section of trap A and B

Table II Data of deep trap before annealing

	Vp-Vr(V)	Ec-Et(eV)	N_t/N_d	$\sigma_n(cm_2)$
1	0.4-0.8	0.60	6.4%	1.3e-12
2	0.6-1.2	0.62	15.6%	7.5e-12
3	1.2-1.9	0.58	7.6%	1.2e-12

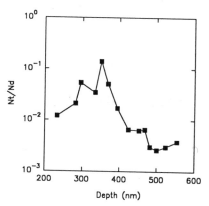

Figure 7 Intensity of trap A

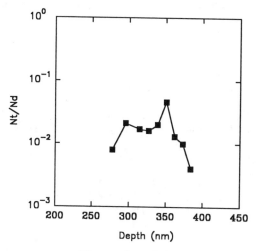

Figure 8 Intensity of trap B

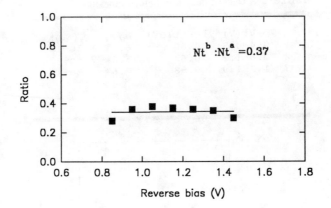

Figure 9 Ratio of intensity of A to B

REFERENCES:

1. G.L.Patton,J.H.Comfort,B.S.Meyerson,E.F.Crabbe,G.J.Scilla, E.D.Fresart,J.M.C.Stork,J.Y.C. Shen,D.L.Harame,J.N.Burchartz, IEEE Electron Device Lett.,11,171(1990)
2. U.Konig etal.,Electronics Lett.,27(16),1405(1991),ibid,28(2), 160(1992)
3. T.L.Lin and J.Maserjian,Appl.Phys.Lett.,57,1423(1990)
4. B.Y. Tsaur,C.K.Chen and S.A.Marino,IEEE Electron Device Lett., 12,293(1991)
5. R.A.Soref,F.Namavar and J.Lorenzo,Optics Lett.,15,270(1990)
6. Q.Mi,X.Xiao,J.C.Sturm,L.C.Lenchyshyn and M.L.W.Thewalt, Appl.Phys.Lett.,60,3177(1992)
7. Wendwll D.Eadeas and Richard M Swanson, J.Appl.Phys.,56, 1774(1984)
8. R.Zhang etal,Appl.Surf.Sci.,48/49,356(1991)
9. V.Nagesh,H.G.Grimmeiss,E.L.Heliqvist,K.L.Ljutovich and A.S.Ljutovich,Semicond.Sci.Technol.,5,556(1990)

PHOTOLUMINESCENCE OF SIGE ALLOYS IMPLANTED WITH ERBIUM

H.B.Erzgräber,* P.Gaworzewski,* K.Schmalz,* D.Krüger,* T.Morgenstern,* and A.Osinsky,** M.Vatnik**
* Institute of Semiconductor Physics, 15230 Frankfurt (Oder), Germany
**Ioffe Physico-Technical Institute, 194021 Petersburg, Russia

ABSTRACT

The 1.54µm emission of erbium implanted SiGe alloys was investigated as a function of oxygen and germanium concentrations as well as defect densities. Samples of SiGe layers grown by atmospheric pressure chemical vapor deposition technique which contain high concentrations of grown-in oxygen in the 10^{19} - 10^{20} cm^{-3} range and 2 - 24 % Ge were chosen for the experiments. By using rapid thermal annealing in nitrogen atmosphere, a high optical activation of Er was found by photoluminescence. In samples with low defect densities nearly one third of the implanted Er ion dose could be detected in form of electrically active donors by spreading resistance measurements. The behavior of different SiGe : Er layers, based on sample property and annealing condition, suggests that more than only one type of luminescent Er-complex is present.

INTRODUCTION

Erbium is proposed to be a very promising candidate for the development of optical elements like waveguides, light emitting diodes or lasers . The emission of Er^{3+} is the result of an internal 4f transition, it is centred at about 1.54 µm near the absorption minimum of optical fibres, and is therefore of great interest for optical communication technologies. A lot of investigations was performed to study the optical and electrical properties of Er-implanted Si and to enhance the efficiency of Er related luminescence.[1-10] In the results it was shown that the intensity of the Er-related 1.54 µm emission depends significantly on the presence of additional impurities such as C, N, F or especially O. Er - impurity complexes are formed and the electrical properties as well as the concentration of optically active centres are remarkably influenced[6,7,11-13].

Up to now only little information[14] is available about the properties of Er doped SiGe alloys although it should have the potential as optoelectronic material. The formation of Si/SiGe:Er/Si quantum wells (QW) should increase the efficiency of Er pumping due to the higher concentrations of excitons bound to the Er atoms in the QW. However, at first some more knowledge of Er-luminescence excitation and quenching mechanisms in SiGe is needed.

In this study we investigate the PL spectra of Er implanted atmospheric pressure chemical vapor deposition (APCVD)-SiGe layers, containing high grown-in oxygen content in the 3×10^{19} - 5×10^{20} cm^{-3} range. Changes of the PL spectra and the intensities of the main Er lines as well as the Er-related donor concentrations are studied in dependence on the SiGe layer properties and the rapid thermal annealing (RTA) conditions.

EXPERIMENTAL

The main characteristics of the SiGe/Si samples, used in our experiments, are given in Table 1. The epitaxial SiGe layers were grown in an APCVD process at 1000 K on four inch (100) p-type silicon wafers [15]. Erbium ions were implanted at room temperature with 70 KeV using the $^{166 - 168}$ Er isotopes. The Er dose was 1.3×10^{13} cm^{-2}, resulting in a peak Er volume density of about 5×10^{18} cm^{-3} in the SiGe layer in a depth

Table 1. Characteristics of SiGe-samples used for Er implantation

Sample	x_{Ge} at %	d_{SiGe} nm	grown-in oxygen/cm^3	defect density/cm^2 (before Er impl.)
A	2	60 nm	$2 - 3 \times 10^{20}$	$< 10^4$
B	12.4	89 nm	$2 - 5 \times 10^{20}$	ca. 10^6
C	14	69 nm	3×10^{19}	$< 10^4$
D	24	58 nm	2×10^{20}	6.6×10^4
E	11.8	84nm	$2 - 3 \times 10^{20}$	

of about 40 - 45 nm. RTA was performed in dry nitrogen or argon-hydrogen atmosphere at temperatures of 850 - 950 °C for 10 - 600 seconds.

The Er-related photoluminescence (PL) spectra were investigated using the 488 nm line of a 6W CW argon ion laser. The samples were immersed in liquid helium. The emitted light was dispersed in a 0.65 m single grating monochromator and detected by a liquid-nitrogen-cooled germanium detector. The Er profiles as well as the grown-in oxygen content were controlled by secondary ion mass spectroscopy (SIMS). The Er related carrier concentrations were determined by means of spreading resistance (SR) and Hall measurements. Au Schottky diodes were prepared for deep level transient spectroscopy (DLTS).

RESULTS AND DISCUSSION

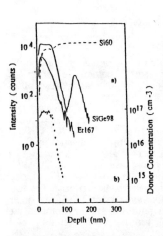

Fig. 1. SIMS profile of Er distribution in the SiGe layer of sample C (a) and SR profile of Er-related donors measured after RTA at 900 ° C for 50 sec (b).

Figure 1a shows the typical SIMS profile of Er distribution, measured for the as-implanted state of sample C. The profile remains nearly unchanged after RTA under the conditions used in our experiments. The distribution of Er-related donors, which are attributed to the formation of electrically and optically active Er-O complexes, is also demonstrated (Figure 1b). It is seen that both profiles are located inside the SiGe layer.

The PL spectra of an Er-implanted SiGe layer with low Ge concentration (sample A) which was annealed in N_2 at 900 °C for 50, 100 or 400 s, are shown in Figure 2. More than 8 emission lines are seen, leading to the assumption that the Er atoms are located in a lower than tetrahedral (T_d) symmetry [16] and form different types of luminescent Er complexes. The arrows give the energy positions which correspond to the first four lines measured by Tang et al. [17] for CZ-Si with Er incorporated in T_d symmetry. Besides the most intense emission line at 1536 nm the other lines at 1555, 1574, and 1598 nm are detected with much lower intensity than in Si, often visible in the shoulder of a neighboured line only. Some essential lines which are typical for our Er-implanted SiGe layers are found at wavelengths of 1544, 1552, 1566, and 1572 nm. Comparing the spectra in Figure 2, we observe changes in the luminescence intensities after longer annealing. A time dependent decrease of the lines at 1544 or 1552 nm, which follow the main Er line at 1536 nm next in intensity, is found. The intensity reduction follows the square root of annealing time t_a; the slope depends on the temperature (Figure 3). Obviously, the intensity reduction is stronger in samples with higher Ge and O concentrations. The dependence on the Ge content is shown in Figure 4 for the samples A,B and D, having oxygen in the same order of magnitude.

Samples with high grown-in oxygen content had been used in our experiments to be sure that a large amount of Er atoms is able to link with O, forming high concentrations optically active Er-O complexes and the corresponding electrically active donor

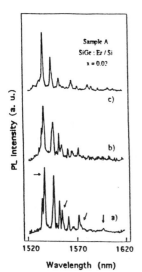

Sample A
SiGe : Er / Si
x = 0.02

c)

b)

a)

1520   1570   1620

Wavelength (nm)

PL Intensity (a. u.)

Fig. 2. Er - related PL spectra of sample A , after RTA in N$_2$ at 900 °C for 50 s (a), 100 s (b), and 400 s (c).

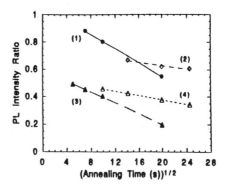

Fig. 3. PL intensity of emissions at 1544 nm (1), (2) and 1552 nm (3), (4), normalized on the most intense Er line at 1536 nm as a function of annealing time and temperature. (900 °C : (1), (3), 850 °C : (2), (4))

states. The correlation between the Er-related donors formed during annealing in N$_2$ at 850 or 900 °C for 25 - 600 s, and the PL intensity of the 1536 nm Er line is shown in Figure 5 (curve 1) for sample A. A nearly linear dependence is found, supporting the assumption that the formation of high donor concentrations is a key for increased Er luminescence. The distribution of the donors in the SiGe layer of sample A is shown in Figure 6 for one of the best activation results obtained after RTA at 850 °C for 200 s. It results in an electrically active Er dose of 4.5×10^{12} cm^{-2}, about one third of the implanted Er ion dose of 1.3×10^{13} cm^{-2}. The maximal Er-related donor concentrations of the different samples are listed in Table 2. Large differences of some orders of magnitude and even concentrations below the detection limit were found for the same annealing con-

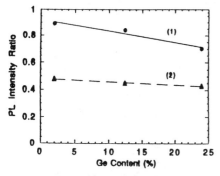

Fig. 4. PL intensity of 1544 nm (1) and 1552 nm lines (2) normalized on the most intense Er line at 1536 nm in dependence on the Ge concentration of the SiGe:Er samples, measured after RTA at 900 °C for 50 s.

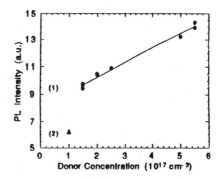

Fig. 5. PL intensity vs. maximal Er donor concentration, as extracted from SR profiles of sample A after RTA in N$_2$: (1) at 850 °C and 900 °C for 25 - 600 s, (2) at 950 °C for 10 s.

ditions. If it is correct to correlate electrical with optical activations, one has to expect, that the concentrations of optically active Er (and therefore the luminescence intensity) must be quite low in samples with low (or even without) Er-related donor concentrations. But, comparing the luminescence in dependence on the pump power, less differing intensities are observed (Figure 7). The lumi -

Table 2. Maximal Er-related donor concentrations

Sample	Donor Concentration (cm^{-3})
A	$1 - 8 \times 10^{17}$
B	$\leq 2.5 \times 10^{15}$
C	$0.4 - 1 \times 10^{17}$
D	$4 - 5 \times 10^{15}$

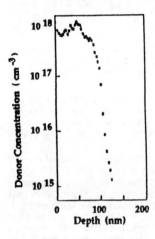

Fig. 6. Distribution of Er-related donors in the SiGe layer of sample A, obtained after RTA in N_2 at 850 °C for 200 s.

luminescence. Additional broad band luminescence (BBL) is also not sufficient to suppress the Er luminescence, as shown in Figure 8.
Probably, another fact additionally enhances the differences in the pumping efficiency of our SiGe samples. I. Yassievich and L.C. Kimerling[19] have shown by calculations that the excitation mechanism of rare-earth atoms in semiconductors occurs most likely through a resonant Auger process in which an e-h pair recombines and, as a result, the excitation of a 4f shell electron in the Er atom is achieved. They concluded, if Er has a charge or exists in a charged complex, free carriers acquire a greater probability to occur in the vicinity of the charged Er atoms and therefore also the probability of Er excitation must be increased.Taking into account these consid-

nescence intensity increases continuously with the pump power in all samples. Powers higher than 500 mW were avoided to be on the safe side that the sample temperature is not increased by the laser light. By this means, only a fraction of the existing optically active Er could be excited; the saturation levels which might give some more informations on the differences of optically active sites in the samples, could not be determined. The results suggest that the pumping efficiency is a very deciding factor for high luminescence intensity at a fixed power below the saturation level. Crystallographic defects and oxygen atoms in too high concentrations which are not linked to Er atoms provide additional centres for nonradiative e-h recombination, reducing the pumping efficiency of Er luminescence. This should also explain the different behavior of samples B and D (having higher defect densities and very high oxygen content) in comparison to sample C with low defect density and an order of magnitude lower O concentration.
One should expect that the formation of dislocations (detected by the defect related D-lines in the PL spectra) due to the relaxation of SiGe during RTA is a further reason for a strongly reduced Er pumping efficiency. But, as recently reported[18] the appearance of D-lines is not a guarantee for the disappearance of Er

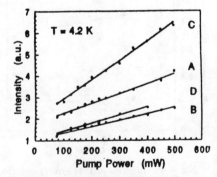

Fig. 7. PL intensity of the 1536 nm line of samples A - D, measured at 4.2 K in dependence on the pump power (after RTA in N_2 at 850 °C for 200 s).

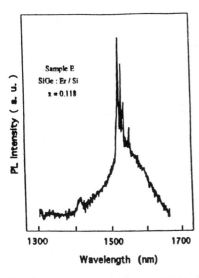

PL intensity (a. u.)

1300   1500   1700

Wavelength (nm)

Fig. 8. PL spectra of sample E after RTA in N₂ at 900 °C for 400 s. The Er- related luminescence is coexistent with BBL and D-line luminescence.

the donor formation after annealing. Figure 9 shows the DLTS spectra of sample C measured at reverse bias V_R = -1V for different pulse biases, indicating that the deep level concentration N_T is maximal near the surface. Assuming a shallow donor concentration, induced by Er implantation and annealing, of about 10^{17} cm^{-3} as revealed by the SR-measurements for this sample (see Figure 1b), a level concentration N_T of about 10^{13} cm^{-3} is roughly estimated. The thermal activation energy of the electron emission rate obtained from the Arrhenius-plot of e_n /T^2 is about 0.34 eV. In difference to our results the Er-implanted Si samples investigated by J.L.Benton et al. [9] exhibit nine levels in relative high concen-

erations we have to expect that in samples A and C the pumping efficiency should be increased also due to the much higher Er-related donor concentrations compared to samples B and D.

But, the fact that Er luminescence is also found in samples without Er-related donors suggests that these donors are not imperative for luminescence. The results let us assume the existence of more than one type of luminescent Er. This is in agreement with experiments where Er was introduced in p- or n-type Si by diffusion from Er-containing surface films. [20] Instead of donors, acceptor states were formed, but nevertheless Er luminescence could be measured.

We have no hint from our experiments that Er-related acceptors are possibly formed during longer annealing, compensating the donor states. Therefore we assume that uncharged luminescent Er might be at least coexistent in our SiGe layers.

Additionally, deep level transient spectroscopy was performed to study the electrically active defects which are related to Er implantation in SiGe. Au-Schottky diodes were used for the DLTS -investigations of the Er-implanted SiGe layer of n-type conductivity due to

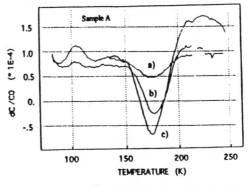

Fig. 9. DLTS spectra of sample C, measured at Au-Schottky diodes for reverse bias V_R = - 1 V and pulse biases V_1 = - 0.5 (a), V_1 = 0 V (b), V_1 = 0.5 V (c).

trations. Maybe that in our case the RTA procedure leads to a more complete annealing of Er-implantation induced defects (which might be less severe due to lower implantation energy) compared to the furnace heat treatment used by Benton.

SUMMARY

Er implanted APCVD SiGe layers show similar PL spectra as they are found in CZ-Si co-implanted with Er and O. The similarities concern the energetic positions of the spectra, not the intensity ratios. Less resolution and line broadening are seen with increased Ge content. As expected, the high grown-in oxygen content of the SiGe layers enhances the formation of optically active Er-O complexes. In samples with low crystallographic defect densities a nearly linear correlation was found between the PL intensity of the main Er line at 1536 nm and the Er-related shallow donor concentration. Using appropriate RTA conditions to annihilate implantation damage and to activate Er, maximal donor concentrations as high as 8×10^{17} cm^{-3} are obtained. About one third of the implanted Er ion dose of 1.3×10^{13} cm^{-2} is electrically activated in this case. The results obtained for our different SiGe:Er samples suggest the existence of more than one luminescent Er-complex and show that the formation of Er-related donor states is not imperative for Er luminescence. We assume that uncharged luminescent Er-complexes are at least coexistent in the SiGe :Er layers.

ACKNOWLEDGMENT

We wish to thank G.Weidner and Mrs. S. Hoppenrath of the Institut of Semiconductor Physics for performing rapid thermal annealing.

1. H. Ennen, J. Schneider, G. Pomrenke, and A. Axmann, Appl. Phys. Lett. **43**, 943 (1983)
2. H. Ennen, G.S. Pomrenke, A. Axmann, K. Eisele, W. Haydl, and J. Schneider, Appl. Phys. Lett. **46**, 361, (1985)
3. D. Moutonnet, H. L'Haridon, P.N. Favennec, M. Salvi, M. Gauneau, F.A. D'avitaya, and J. Chroboczek, Mater. Sci. and Engineer. **B4**, 75 (1989)
4. T. Oestereich, C. Swiatkowski, and I. Broser, Appl. Phys. Lett. **56**, 446 (1990)
5. Y.H. Xie, E.A. Fitzgerald, and Y.J. Mii, J. Appl. Phys. **70**, 3223 (1991)
6. J. Michel, J.L. Benton, R.F. Ferrante, D.C. Jacobson, D.J. Eaglesham, E.A. Fitzgerald, Y.-H. Xie, J.M. Poate, and L.C. Kimerling, J. Appl. Phys. **70**, 2672 (1991)
7. J. Michel, L.C. Kimerling, J.L. Benton, D.J. Eaglesham, E.A. Fitzgerald, D.C. Jacobson, J.M. Poate, Y.-H. Xie, and R.F. Ferrante, Mater. Sci. Forum **83-87**, 653 (1992)
8. S. Coffa, F. Priolo, G. Franzo, V. Bellani, A. Carrera, and C. Spinella, to be publ. in Phys. Rev. B
9. J.L. Benton, J. Michel, L.C. Kimerling, D.C. Jacobson, Y.-H. Xie, D.J. Eaglesham, E.A. Fitzgerald, and J.M. Poate, J. Appl. Phys. **70**, 2667 (1991)
10. D.J. Eaglesham, J. Michel, E.A. Fitzgerald, D.C. Jacobson, J.M. Poate, J.L. Benton, A. Polman, Y.-H. Xie, and L.C. Kimerling, Appl. Phys. Lett. **58**, 2797 (1991)
11. P.F. Favennec, H. L'Haridon, D. Moutonnet, M. Salvi, and G. Gauneau, Jap. J. Appl. Phys. **29**, L524 (1990)
12. F. Priolo, S. Coffa, G. Franzo, C. Spinella, A. Carrera, and V. Bellani, J. Appl. Phys. **74**, 1 (19993)
13. D.L. Adler, D.C. Jacobson, D.J. Eaglesham, M.A. Marcus, J.L. Benton, J.M. Poate, and P.H. Citrin, Appl. Phys. Lett. **61**, 2181 (1992)
14. J.-P. Noel, N.L. Rowell, T.E. Jackman, D.D. Perovic, and D.C. Houghton, Mater. Res. Soc. Ext. Abstr. **EA-21**, 11 (1990)
15. T. Morgenstern, I. Babanskaja, G. Morgenstern, K. Schmalz, P Gaworzewski, P. Zaumseil, D. Krüger, K.Tittelbach-Helmrich, and H.B. Erzgräber, Phys. Concepts and Mater. for Novel Optoel. Dev. Appl., Trieste (1993)
16. Y.S. Tang, Z. Jingping, K.C. Heasman, and B.J. Sealy, Sol. State Commun. **72**, 991 (1989)
17. Y.S. Tang, K.C. Heasman, W.P. Gillin, and B.J. Sealy, Appl. Phys. Lett. **55**, 432 (1989)
18. H.B.Erzgräber, G.Kissinger, D. Krüger, T. Morgenstern, K. Schmalz, J. Schilz, M. Kurten, A. Osinsky, Proc. 17th ICDS, Gmunden Austria , 1993, in press
19. I.N. Yassievich, L.C. Kimerling, Semicond. Sci. Technol. **7**, 1 (1993)
20. N.A. Sobolev, O.V. Alexandrov, B.N. Gresserov, G.M. Gusinskii, V.O. Naidenov, E.I. Sheck, V.I. Stepanov, Yu.V. Vyzhigin, L.F. Chepnik, and E.P. Troshina, Solid State Phe - nomena **32 - 33**, 83 (1993)

STRONG VISIBLE PHOTOLUMINESCENCE IN SILICON NITRIDE THIN FILMS DEPOSITED AT HIGH RATES

Sadanand V. Deshpande and Erdogan Gulari
Department of Chemical Engineering
and
Steven W. Brown and S.C. Rand
Department of Applied Physics
University of Michigan, Ann Arbor, MI 48109.

ABSTRACT

Amorphous silicon nitrogen alloy (a-Si:N_x) thin films have been deposited using a novel hot filament chemical vapor deposition (HFCVD) technique. In this method, a hot tungsten filament is used to decompose ammonia to obtain highly reactive nitrogen precursor species which further react with disilane to form silicon nitride thin films. This allows for very high deposition rates ranging from 600 Å/min to 2500 Å/min at low substrate temperatures. These films deposited at high rates show strong photoluminescence (PL) at room temperature in the visible region when excited with the 457 nm line of Ar^+ ion laser. Intrinsic defects introduced into the amorphous silicon nitride matrix due to the rapid deposition rates seem to give rise to the visible PL. The PL intensity is at least 8-10 times stronger than silicon nitride films deposited by conventional plasma enhanced CVD. PL peak position of this broad luminescence was varied in the visible region by changing the film stoichiometry (Si/N ratio). The PL peak energy also scales predictably with the refractive index and optical band gap of the films. These samples showed reversible PL fatigue and also have band edge tail states characteristic of amorphous materials.

INTRODUCTION

Amorphous silicon nitride (a-SiN_x:H) is widely used in the microelectronics industry for various applications including: oxidation mask, dopant diffusion barrier, gate dielectric in field effect and thin film transistors, encapsulant for III-V semiconductors, interlevel dielectric, charge storage layer in MNOS non-volatile memories and final passivation layer for device packaging. Low temperature (25-400°C) deposition schemes have been particularly important due to the increasing complexity of semiconductor processing and a need for thermal budgeting. Several different low temperature deposition methods have been employed in the past, which include plasma enhanced CVD, remote plasma CVD, ECR-plasma and sputtering. Another novel deposition technique has also been reported that employs a hot metal filament to activate the reactant gases. The hot filament activation of gases allows low temperature deposition (200-400°C) of a variety of films, including amorphous silicon, boron nitride, gallium nitride, silicon nitride and silicon carbide. We have deposited good quality silicon, aluminum and titanium nitride films in the past with the hot filament method [1-3].

The properties of silicon nitride films obtained are strongly dependent on the method of deposition. The deposition parameters such as deposition temperature, reactor pressure, film deposition rate, and plasma conditions (rf power, substrate bias) significantly affect the film properties. Various characterization techniques have been utilized to study the relevant properties of these films. Electron spin resonance (ESR) has been a useful technique in measuring the density of dangling bond centers (both silicon and nitrogen) in a-SiN_x:H. The electron and hole trapping centers associated with dangling bonds in these films have also been studied [4,5]. Some studies have reported visible photoluminescence (PL) from silicon nitride films deposited by the plasma enhanced method [6,7]. The bonding configuration (and

177

hydrogen content) in the films has been measured with infrared absorption. The hydrogen content has also been determined from elastic recoil detection measurements [8]. The hydrogen content in the films is known to affect their etch rates in standard etchants such as buffered hydrofluoric acid (BHF).

In the present study, various properties of silicon nitride films deposited by Hot Filament assisted CVD (HFCVD) method are being reported. The films have been characterized by FTIR, elastic recoil detection, BHF etch rates, photoluminescence, and optical band gap measurements. The properties of our HFCVD films have been compared with those of films deposited by a conventional plasma enhanced CVD technique.

DEPOSITION AND CHARACTERIZATION OF FILMS

The experimental system used in this study was the same as reported earlier [1,2]. Briefly, the reactor consists of a cold-wall, six-inch, six-way stainless steel chamber capable of processing a single two-inch wafer. The reactor pressure was measured with a capacitance manometer and controlled by an exhaust throttle valve. Reactant gases were fed into the reactor via two inlets: one to feed ammonia which flowed over the filament and the other was a gas dispersal ring that bypassed the hot filament to deliver disilane along with the carrier gas. The substrates were clamped to a stainless steel susceptor heated by two cartridge heaters. The substrate temperature was monitored using a type C thermocouple clamped to the susceptor surface close to the substrate. The filament was a resistively heated tungsten wire (dia~0.25 mm) placed 4 cm away from the substrate. The filament temperature was measured with an optical pyrometer through a quartz viewport on the reactor. The silicon nitride films were deposited onto 2-inch p-type silicon (100) wafers. The substrates were cleaned with solvents and stripped of the native oxide with 10% HF just prior to deposition. The deposition parameters used in this study are summarized in Table I.

Table I. Deposition parameters for a-SiN$_x$:H Thin Films

Parameter	Set point(s)
Reactor pressure	0.5 torr
Substrate temperature	245 - 375°C
Filament temperature	1500-1700°C
Ammonia Flow rate	80 sccm
Carrier gas flow rate:	
Hydrogen	230 sccm
Disilane flow rate	1.1 - 3.2 sccm

The hydrogen content in the films was measured by infrared absorption of the NH- and SiH-stretching bands and also with elastic recoil detection (ERD) from a 2.0 MeV He^{2+} beam line. The film etch rates in buffered HF solution were determined by monitoring thickness change with an ellipsometer. Photoluminescence experiments were performed using an ISA THR 1000 spectrometer with 5 Å resolution. The PL was excited with the 457 nm line of an Ar$^+$ laser at 100 mW power. The PL intensity was measured in a back-scattering geometry. The optical band-gap of the films was determined by depositing films on UV-grade quartz and measuring absorption with a Cary UV-Visible spectrophotometer.

RESULTS AND DISCUSSIONS

Table II summarizes the properties of films deposited at three different substrate temperatures. The hydrogen content in the films was determined from the infrared absorption crossections reported elsewhere [9]. These measurements were compared with the hydrogen concentrations measured with elastic recoil scattering (using 2 MeV He^{2+} ions). As shown in Table II, the hydrogen content measured with the two techniques are in close agreement. The film etch rates in buffered HF correlate well with the hydrogen content (and substrate temperatures) in the films. The low etch rates of these films indicate that the films are of higher density than the PECVD films.

Table II. Effect of Substrate Temperature on Film Properties

Substrate Temperature	Deposition Rate (Å/min)	Refractive Index	FTIR Hydrogen Content (%)	ERD Hydrogen Content (%)	BHF Etch Rates (Å/min)
375°C	630.0	2.08	6.0	9.5	25.2
300°C	585.0	2.04	7.5	13.4	51.7
245°C	540.0	2.04	14.6	17.7	79.7
350°C (PECVD 1)	56.0	1.88	12.0	---	518.0

The a-SiN$_x$:H films deposited by HFCVD has shown strong visible photoluminescence at room temperature. The PL peak energy can be shifted to the blue by increasing the optical band gap. This was accomplished by decreasing the disilane flow rate, and therefore, decreasing the silicon content (or Si/N ratio) in the films. The changes in film properties with disilane flow rate are included in Table III. As shown in Figure 1, the PL peak shifts to the blue with increasing band gap of the films. Also the relative PL intensity decreases with the increasing optical band gap. This may be due to two possible reasons: 1) the excitation at 457 nm is not as efficient for wider band gap films, or 2) there are larger number of non-radiative decay paths in these films. Annealing studies at a temperature (800°C) much higher than deposition temperature have revealed that these defects are very stable even after 4.25 hrs. of annealing in both argon and hydrogen atmosphere (see Figure 2). It should be noted that the PL intensity of the HFCVD films is at least 8-10 times stronger than the PECVD films.

Further studies are needed to understand the exact origin of these optically active defects. We have also observed weak electroluminescence (red color) from these films using a simple p-i-n diode structure. We could not fabricate bright luminescent structures due to the lack of appropriate doped wide band gap layers for efficiently injecting carriers into the intrinsic silicon nitride layer. One study has recently reported the fabrication of LED's with silicon nitride as the active layer [10]. These optically active defects could be potentially useful for use in large area electroluminescent displays.

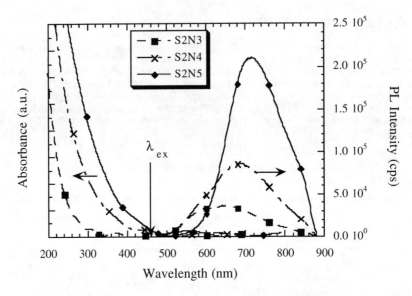

Figure 1. Optical band gap and room temperature PL intensity of SiNx films.

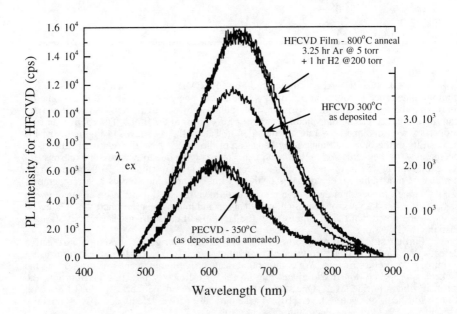

Figure 2. Effect of high temperature annealing on PL from HFCVD and PECVD films.

Table III. Effect of Silicon source gas flow rate on room temperature
Photoluminescence in Silicon Nitride Films

Film #	Disilane Flow Rate (sccm)	Optical Band Gap (eV)	PL Peak Position(s) (eV)	Relative intensity
S2N1	1.1	4.74	2.2 & 1.92	0.07
S2N2	1.6	4.39	2.08	0.20
S2N3	2.1	4.17	2.05	0.56
S2N4	2.7	3.03	2.01	0.64
S2N5	3.2	2.43	1.76	1.0
PECVD 1	(Low silane)	5.33	2.23	0.41
PECVD 2	(High silane)	4.40	1.98	1.0

SUMMARY

In this paper, we have shown that silicon nitride thin films deposited with a novel HFCVD process have low hydrogen content compared to films obtained by conventional low temperature plasma deposition and, therefore, have low etch rates in BHF. The optical band gap can be varied by changing the film stoichiometry. These films show strong visible photoluminescence. The optically active defects are stable even after prolonged high temperature annealing. The nature of the luminescence centers in these films is presently being studied in further detail and will be presented in the near future.

ACKNOWLEDGMENTS

The authors wish to acknowledge the NSF Center for Ultrafast Optical Science (STC PHY 8920108) for partial support of this study. We appreciate the help from Dr. Victor Rotberg of Michigan Ion Beam Laboratory with Elastic Recoil Detection and RBS analysis.

REFERENCES

1. S.V. Deshpande, J.L. Dupuie, and E. Gulari, *Appl. Phys. Lett.* **61**, 1420 (1992).
2. J.L. Dupuie and E. Gulari, *J. Vac. Sci. Technol.* **A10**(1), 18, (1992).
3. S.V. Deshpande and E. Gulari, *Materials Research Society*, Fall Meeting, Symposium N, Boston, MA (1993).
4. Y. Kamigaki, S. Minami, and H. Kato, *J. Appl. Phys.* **68**(5), 2211 (1990).
5. W.L. Warren, J. Kanicki, F.C. Rong, and E.H. Poindexter, *J. Electrochem. Soc.* **139**(3) 880 (1992).
6. N.R.J. Poolton and Y. Cros, *J. Phys. I France* **1**, 1335 (1991).

7. I.G. Austin, W.A. Jackson, T.M. Searle, P.K. Bhat, and R.A. Gibson, *Phil. Mag. B* **52**(3), 271 (1985).
8. A. Markwitz, M. Bachmann, H. Baumann, K. Bethge, E. Krimmel, and P. Misaelides, *Nucl. Instr. and Meth. B* **68**, 218 (1992).
9. W.A. Lanford and M.J. Rand, *J. Appl. Phys.* **49**, 2473 (1978).
10. W. Boonkosum, D. Kruangam and S. Panyakeow, Symposium A, presented at the 1993 Spring Meeting, San Francisco, CA (1993).

OCCURRENCE OF GROUND AND EXCITED-STATE IMPURITY BANDS IN SILICON INVERSION LAYERS: ELECTRIC FIELD EFFECTS

O. HIPÓLITO*, SALVIANO A. LEÃO* AND A. FERREIRA DA SILVA**
* Departamento de Física e Ciência dos Materiais, Instituto de Física e Química de São Carlos, Universidade de São Paulo, 13560-250 São Carlos, SP, Brazil
** Laboratório Associado de Sensores e Materiais, Instituto Nacional de Pesquisas Espaciais, Caixa Postal 515, 12225-000 São José dos Campos, SP, Brazil

ABSTRACT

We investigate the effects of charged impurities in n-type Silicon inversion layers. We show the occurrence of ground and excited-state impurity bands as a function of electric field and concentration. Also the effects of disorder and for given impurity concentrations, the lowest excited band play an essential role in the optical and transport measurements. For high electric fields the impurity bands go to the ideally 2D separated bands.

Impurities in Silicon inversion layers have been the subject of considerable experimental and theoretical interest [1,2]. The main purpose of this work is to investigate the ground and excited-state impurity bands as a function of electric field and impurity concentration, subject to disorder in comparison to the unperturbed first conduction subband (UFCS) edge. In the wake of a recente work on ground charged impurities n-type Si-inversion layers [2], we assume the Hamiltonian in a random, one-body tight-binding approximation, with monovalent impurities as

$$H = E_\lambda^B \sum_i |i><i| + \sum_{i\neq j} V_{ij}|i><j|, \qquad (1)$$

where E_λ^B is the binding energy, which is taken as our energy origin below the edge of the UFCS, λ are the states (ground and lowest excite states) to be considered, and V_{ij} is the random energy integral, i.e., hopping integral.

The impurity bandwidths and the density of states are calculated from the Green's functions

$$G_{ij}^{(\pm)}(\omega) = <0|a_i \frac{1}{\omega - H \pm i\epsilon} a_j|0> \qquad (2)$$

183

with configuration averaging over the random distribution of impurities [2]. The density of states is given by

$$\xi^{(\pm)}(\omega) = Z_\pm < G_{ii}^{(\pm)}(\omega) >, \qquad (3)$$

where $< ... >$ means configuration averaging and $Z_\pm = \omega \pm i\epsilon$, and

$$\eta(\omega) = \frac{N_{ox}\xi(\omega)}{4\pi^2\omega^2} \int \frac{V^2(\vec{K})d^2K}{1 - N_{ox}\xi(\omega)/\omega V(\vec{K})}, \qquad (4)$$

where N_{ox} is the inversion layer impurity concentration and $V(\vec{K})$ is the Fourier transform of V_{ij}.

Defining

$$\frac{\xi(\omega)}{\omega} = \left\{ Na^{*2}\left[u(\omega) + is(\omega)\right] \right\}^{-1} \qquad (5)$$

where a^* is the effective Bohr radius, given by $a^* = 22\text{Å}$, we easily found the bandwidth values and the density states, now in the form

$$D(\omega) = \frac{-1}{\pi}Im\left[\frac{\xi(\omega)}{\omega}\right]. \qquad (6)$$

For the Si-SiO$_2$ system we adopt $R_y^* = 42meV$. We get the binding energy from variational solution for the isolated impurity as [3]

$$E_\lambda^B = < \psi_\lambda | -\nabla_{x,y}^2 - \nu\nabla_z^2 + \frac{\delta}{2} + \varepsilon z - 2\Phi(\vec{r})|\psi_\lambda > \qquad (7)$$

where ε is the external electric field, δ e ν are related the dielectric constant and the effective mass respectively and $\Phi(\vec{r})$ is the screened Coulomb potential [3].

Then we get E_λ^B from the relation

$$E_\lambda^B = E_0 - E_\lambda \ , \qquad \lambda = 2p_o \ , \ 3d_{\pm 1}, \qquad (8)$$

where E_0 is the energy value without impurity. The wave functions are described by

$$\psi_\lambda(\vec{r}) = \Phi_\lambda(x,y)\varphi(z), \qquad (9)$$

184

where

$$\Phi_{2p_o}(R) = \left(\frac{a_o^2}{2\pi}\right)^{1/2} \exp(-a_o R/2) \tag{10}$$

represents the ground state, and

$$\Phi_{3d_{\pm 1}}(R) = \left(\frac{a_1^4}{12\pi}\right)^{1/2} (x \pm iy)\exp(-a_1 R) \tag{11}$$

represents the lowest excited state and $\vec{R} = \vec{R}(x, y)$. The function

$$\varphi(z) = \left(\frac{b^3}{2}\right)^{1/2} \exp(-bz), \tag{12}$$

is the lowest electric subband. The parameters a_0, a_1 and b are determined variationally by Eq. (7).

The Fourier transforms of the hopping terms V_{ij} are obtained for these states respectively as

$$V_{2p_o}(\vec{K}) = -\frac{\left(E_{2p_o}^B + a^{*2}K^2\right)\pi}{2a_o^2\left(\frac{1}{4} + \frac{K^2 a^{*2}}{a_o^2}\right)^3}\left[a^{*2}R_y^*\right] \tag{13}$$

and

$$V_{3d_{\pm 1}}(\vec{K}) = -\left(E_{3d_{\pm 1}}^B + a^{*2}K^2\right)\pi\left\{\frac{1}{3a_1^2 F(\vec{K})^3}\left[\frac{9}{16F(\vec{K})^2} - \frac{6}{4F(\vec{K})} + 1\right]\right\} \tag{14}$$

where

$$F(\vec{K}) = \left(\frac{1}{4} + \frac{K^2 a^{*2}}{a_1^2}\right). \tag{15}$$

In Fig. 1 we show the density of states for $2p_o$ and $3d_{\pm 1}$ bands for different impurity concentrations (N_{ox}) with an electric field of $\varepsilon = 10^3$ esu and depletion concentration $N_d = 1.9\times 10^{12}$ cm^{-2} presented in the Si-SiO$_2$ inversion layer. Increasing N_{ox} the bandwith will enlarge and eventually cross the UFCS. The $3d_{\pm 1}$ band will overlap more with the $2p_o$ band, increasing faster its peak and crossing the UFCS first than the $2p_o$ band. These behaviour will play an essential role in the optical and transport measurements [1-4,5]. For high electric fields the impurity bands go to the ideally 2D separated bands [6,7] with no screening as shown

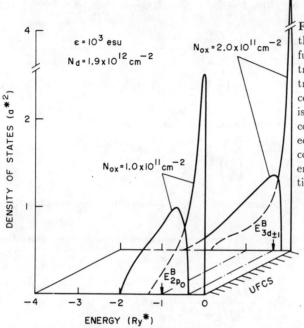

Figure 1. Density states for the $2p_o$ and $3d_{\pm1}$ states as a function of impurity concentration (N_{ox}) with fixed electric field (ε) and depletion concentration (N_d). The origin is set at the unperturbed first conduction subband (UFCS) edge. The dot-dashed lines correspond to the binding energies $E_{2p_o}^B$ and $E_{3d_{\pm1}}^B$ respectively.

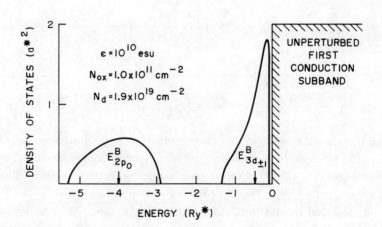

Figure 2. Density of states for high electric field $\varepsilon = 10^{10}$ esu with $N_{ox} = 10 \times 10^{11}$ cm^{-2}.

in Fig. 2. We found for $\varepsilon = 10^{10}$ esu, $E_{2D}^B \simeq 4R_y^*$ and $E_{3d_{\pm1}}^B = \frac{4}{9}R_y^*$ as in the previous works [6,7].

References

[1] T. Ando, A. B. Fowler and F. Stern, Rev. Mod. Phys. **54**, 437 (1982).

[2] O. Hipólito and A. Ferreira da Silva, Phys. Rev. B**47**, 10918 (1993)

[3] O. Hipólito and V. B. Campos, Phys. Rev. B**19**, 3083 (1979).

[4] A. Hartstein and A. B. Fowler, Phys. Rev. Lett. **34**, 143 (1975).

[5] A. B. Fowler and A. Hartstein, Phylos. Mag. B**42**, 949 (1980) and references there in.

[6] A. Ghazali, A. Gold, J. Serre, Phys. Rev. B**39**, 3400 (1989).

[7] I. C. da Cunha Lima and A. Ferreira da Silva, Mat. Sci. Forum. **65**, 79 (1990).

EFFECT OF PRESSURE ON ARSENIC DIFFUSION IN GERMANIUM

S. MITHA*, S. D. THEISS*, M. J. AZIZ*, D. SCHIFERL** AND D. B. POKER†.
*Division of Applied Sciences, Harvard University, Cambridge, MA 02138
**Los Alamos National Laboratory, Los Alamos, NM 87545
†Oak Ridge National Laboratory, Oak Ridge, TN 37831

ABSTRACT

We report preliminary results of a study of the activation volume for diffusion of arsenic in germanium. High-temperature high-pressure anneals were performed in a liquid argon pressure medium in a diamond anvil cell capable of reaching 5 GPa and 750° C, which is externally heated for uniform and repeatable temperature profiles. The broadening of an ion-implanted arsenic profile was measured by Secondary Ion Mass Spectrometry. Hydrostatic pressure retards the diffusivity at 575° C, characterized by an activation volume that is +15% of the atomic volume of Ge. Implications for diffusion mechanisms are discussed.

INTRODUCTION

Bulk diffusion in crystalline solids is mediated by point defects, such as vacancies in metals. The diffusivity can be perturbed by changing the equilibrium concentration of point defects with hydrostatic pressure. High pressure typically reduces the diffusion constant in metals, suggesting a vacancy mechanism for diffusion [1-3]. High pressure techniques can also be used to study the defects mediating diffusion in semiconductors. The slow rate of diffusion in elemental semiconductors however, requires very high temperatures to achieve measurable diffusivities. High temperature, high pressure devices suitable for these experiments have proved difficult to make. Werner et al [4] used a compressed gas cell to measure the activation volume of self diffusion of germanium in intrinsic and doped germanium. However they were limited to a maximum pressure of 0.6 GPa. In this paper we report preliminary results of a technique capable of attaining much higher pressures with less risk of contamination, applied to the diffusion of arsenic in germanium.

The diffusivity of a random-walking species in a cubic system is:

$$D = \tfrac{1}{6}\lambda^2\,\Gamma \tag{1}$$

where λ is the jump distance and Γ is the jump rate. In the transition state theory [5] the jump rate is:

$$\Gamma = P\nu\exp\left(-\Delta G_m/kT\right) \tag{2}$$

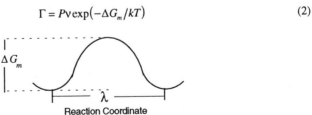

Reaction Coordinate

Fig. 1. Standard free energy vs. system configuration as atom jumps across barrier to new position. ΔG_m is the barrier to migration.

where ΔG_m is the free energy of migration ν is the attempt frequency, kT has the usual meaning and P is the probability that there is a defect in a position for the jump to take place. Assuming random site occupation, the probability that the defect exists next to the jump site is:

$$P = \exp(-\Delta G_f^0 / kT) \qquad (3)$$

where ΔG_f^0 is the standard free energy of formation the defect in the process depicted in Fig. 2.

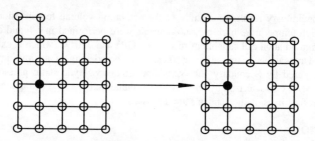

Fig 2. Vacancy Formation

The final form the diffusivity is:

$$D = a\nu\lambda^2 \exp(-\Delta G^* / kT) \qquad (4)$$

where $\Delta G^* = \Delta G_f^0 + \Delta G_m$ and a is a geometrical factor that takes into account the crystal structure and the diffusion mechanism [6].

The familiar activation enthalpy comes from the dependence of ΔG^* on reciprocal temperature. The activation volume comes from the dependence of ΔG^* on pressure using the thermodynamic identity

$$V = \left.\frac{\partial G}{\partial P}\right|_T, \qquad (5)$$

which, when applied to Eq. (4), yields

$$\Delta V^* = -kT\left.\frac{\partial \ln D}{\partial P}\right|_T + kT\left.\frac{\partial \ln(a\nu\lambda^2)}{\partial P}\right|_T \qquad (6)$$

where ΔV^* is the activation volume. The second term is has been shown to be small [4]. Hence the activation volume is determined experimentally by a measurement of the pressure-dependence of the diffusivity:

$$\Delta V^* \approx -kT\left.\frac{\partial \ln D}{\partial P}\right|_T \qquad (7)$$

and

$$\Delta V^* = \Delta V_f^0 + \Delta V_m \qquad (8)$$

where ΔV_f^0, the formation volume, is the volume change in the system upon formation of one defect in its standard state. ΔV_m, the migration volume, is the additional volume change when the defect reaches the saddle point in its migration path.

For the above analysis we have assumed that the defects are in thermodynamic equilibrium at the temperatures and pressures under which the experiment is conducted. If the time scale of the experiment is too short for the defects to equilibrate, then the volume of migration term will become disproportionately important. For example, in the extreme case where the concentration of defects is constant, the measured activation volume will be equal to the volume of migration.

EXPERIMENTAL METHOD

The experimental procedure is to anneal germanium samples with known initial arsenic depth profiles at various pressures for a fixed temperature. Diffusion is measured *ex situ* by sputter sectioning using Secondary Ion Mass Spectrometry (SIMS).

Samples are prepared by ion implantation of As^+ at 500 keV with a dose of 2×10^{14} ions/cm^2 at 77 K into Ge (100) wafers 50 µm thick. The resultant depth profile of As is shown by SIMS to be approximately gaussian in shape with a peak concentration of 7×10^{18} atoms/cm^3 at 215 nm depth, with a FWHM of 67 nm. The wafers are subsequently implanted at 77 K with $^{71}Ge^+$ at 250 keV to a dose of 5×10^{14} ions/cm^2 and at 500 keV to a dose of 8×10^{14} ions/cm^2, in order to completely amorphize the implanted layer which is necessary for the subsequent restoration of defect-free crystal by Solid Phase Expitaxial Growth (SPEG). SPEG occurs at temperatures too low for measurable diffusion to occur; we found that approximately 15 minute anneals at 450° C restores crystallinity in our samples [7]. The samples are then cleaved into small pieces approximately 150 µm by 150 µm to fit into the Diamond Anvil Cell (DAC). Diffusion anneals were for various pressures at 575 degrees C for 30 minutes.

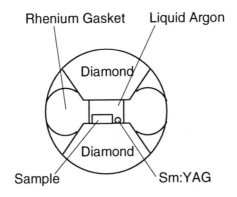

Fig 3. Pressure Chamber in Diamond Anvil Cell

The high pressure device is a modified Merrill-Bassett [8] DAC. The cell body is constructed from Haynes 230, a nickel based superalloy. The pressure is generated by forcing two diamonds together on a metal 'gasket' with a hole in it, which serves as a pressure chamber (Fig 3). The gasket deforms plastically, simultaneously forming a seal and the chamber walls. We use rhenium as a gasket material due to its high strength and ductility at both ambient and high temperature. The pressure chamber is loaded cryogenically with liquid argon as a pressure transmitting medium. Liquid argon is both chemically inert and hydrostatic because it does not solidify until 7 GPa at 575° C, the temperature of our experiment. The cell is externally heated in an inert atmosphere. The furnace has an optical window for *in situ* pressure measurement.

Pressure is measured using fluorescence peak shifts of samarium doped yttrium aluminum garnet (Sm:YAG). Typically pressure is measured in a DAC by using ruby fluorescence peak shifts, but these peaks are not visible beyond about 250° C. Hess and Schiferl [9] have shown that Sm:YAG may be used to measure pressure at least up to 800° C. The Sm:YAG is excited by the 488 nm line of Ar ion laser. The spectrum is acquired using a spectrometer and a diode array. Pressure was calculated by simultaneously fitting the 617 nm, 616 nm and 610 nm lines. The fit was based on Hess and Schiferl's calibration of the 617 nm and 616 nm lines. The 610 nm line was calibrated against the other two at room temperature and ambient pressure. By assuming that the temperature and pressure shifts add linearly we developed a protocol for fitting all three lines. This procedure was necessary for robust and repeatable fits of the Sm:YAG spectrum.

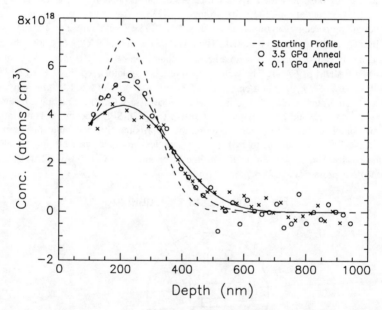

Fig. 4. Arsenic depth profiles for two samples annealed at different pressures for identical temperatures and durations.

Diffusion is measured by Secondary Ion Mass Spectroscopy (SIMS). We used a VG Ionex 1170X magnetic sector SIMS. The primary beam was 16 keV Cs^+, rastered to produce a flat crater bottom. To remove crater wall effects, the secondary ions were collected from an area that was gated electronically to cover only the central 5% of the area of the crater. The $AsGe^-$ ion, which has a stronger signal than the As^- ion, was tracked for the arsenic concentration profile. The as implanted depth profiles are gaussian within our experimental resolution. Typical profiles are shown in Fig. 4. The top 100 nm of the profiles are lost due to a surface transient artifact of

SIMS. The amount of diffusion was determined by fitting gaussians to the broadened depth profiles, using a no-flux boundary condition at the surface. The solid curve in Fig. 4 is a fit to the 3.5 GPa anneal and it gives a diffusivity of 1.5×10^{-14} cm^2/sec. The dashed curve is a fit to the 0.1 GPa anneal and gives a diffusivity of 3.7×10^{-14} cm^2/sec. Pressure clearly reduces the diffusivity of As in Ge.

RESULTS

We summarize our preliminary results in Fig. 5. The error bars on the point at 2.0 GPa are unusually large because the SIMS primary beam spontaneously changed its intensity in mid-profile, reducing the depth calibration accuracy. A least-squares fit of Eq. (7) to the data yields ΔV^* to be $+2.0 \pm 1.0$ cm^3/mole, which is $15\% \pm 7\%$ of Ω_{Ge}, the atomic volume of germanium.

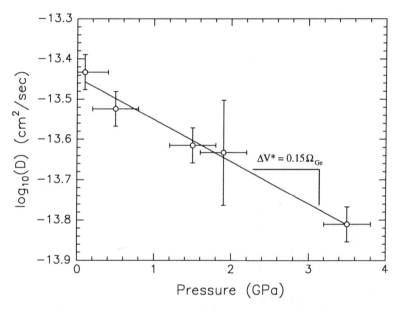

Fig. 5. Diffusion of As in Ge vs. pressure for 30 minute anneals at 575 degrees C

DISCUSSION

We have found $\Delta V^* = +0.15\,\Omega_{Ge}$ for As diffusion at 575° C. Dopant diffusion in Ge is generally believed to occur by a vacancy mechanism [6]. The vacancy formation volume has been calculated in Si, and various calculations result in formation volumes (ΔV_f^0) between $+0.75\,\Omega_{Si}$ and $+1.1\,\Omega_{Si}$ [10-14]. One might then expect a similar sized formation volume for vacancies in germanium, in which case Eq. (8) would imply a migration volume of about $-0.6\,\Omega_{Ge}$ to $-0.95\,\Omega_{Ge}$ for the Ge vacancy. Negative migration volumes for point defects in covalent networks are consistent with the effects on pressure on SPEG of Si and Ge [15,16], and the interdiffusion of amorphous Si-Ge multilayers [17]. Our result would then be consistent with a vacancy mechanism if the vacancy mobility is enhanced by pressure, but not as strongly as the vacancy concentration is reduced by pressure.

Werner et al [4] found ΔV^* for self diffusion in intrinsic germanium to vary from 0.24 Ω_{Ge} at 603° C to 0.41 Ω_{Ge} at 803° C. Combined with the effects of doping on self diffusion they concluded that self diffusion in Ge is mediated by two charge states of vacancies, and inferred activation volumes of +0.28 Ω_{Ge} for the negatively charged vacancy and +0.56 Ω_{Ge} for the neutral vacancy. They do not separate the migration and formation components of the activation volume. Note that as a donor the arsenic atom in our study will be positively charged and will tend to associate with the negatively charged vacancy, which has the smaller activation volume. Work is in progress to determine whether the defect concentrations are in equilibrium during these experiments, by repeating the experiments at the same temperatures for significantly longer times. If the defect concentrations are in equilibrium then there should be no change in the apparent diffusivity with time.

ACKNOWLEDGMENTS

We thank John Martin of MIT Materials Characterization Facilities for his assistance in SIMS depth profiling. This research was supported by NSF-DMR-89-13268. Work at ORNL was supported by Division of Materials Sciences, DOE under contract DE-AC05-84-OR21400 with Martin Marietta Energy Systems, Inc.

REFERENCES

1.   H. Mehrer and A. Seeger, Crys Latt Def, **3,** 1 (1972)
2.   R.H. Dickerson, R.C. Lowell and C.T. Tomizuka, Phys. Rev. **137,** A613 (1965)
3.   J.N. Mundy, Phys. Rev. B, **3,** 2431 (1971)
4.   M. Werner, H. Mehrer and H.D. Hochhiemer, Phys. Rev. B, **32,** 3930 (1985)
5.   G.H. Vinyard, Phys. Chem. Solids, **3,** 121 (1957)
6.   W. Frank , U. Gösele, H. Mehrer and A. Seeger, in Diffusion in Crystalline Solids, edited by G.E. Murch and A.S. Norwick (Academic Press, Orlando, FL, 1984), p. 63
7.   G.L. Olsen and J.A. Roth, Mat. Sci. Reports, **3,** 1 (1988)
8.   L. Merrill and W.A. Bassett, Rev. Sci. Instrum. **45,** 290 (1974)
9.   N. Hess and D. Schiferl, J. Appl. Phys. **71,** 2082 (1992)
10.   F.P. Larkins and A.M. Stoneham, J. Phys. C, **4,** 143 (1971); **4,** 154 (1971)
11.   G.A. Baraff, E.O. Kane and M. Schluter, Phys. Rev. B, **21,** 5662 (1980)
12.   U. Lindefelt, Phys. Rev. B, **28,** 4510 (1983).
13.   M. Scheffler, J.P. Vigneron and G.B. Bachelet, Phys. Rev. B, **31,** 6541 (1985).
14.   A. Antonelli and Bernholc, Phys. Rev. B, **40,** 10643 (1989)
15.   F. Spaepen and D. Turnbull, AIP Conf. Proc. **50,** 73 (1979)
16.   G.Q. Lu, E. Nygren and M.J. Aziz, J. Appl. Phys. **70,** 5323 (1991)
17.   S.D. Thiess, S.Mitha, F. Spaepen and M.J. Aziz, in Crystallization and Related Phenomena in Amorphous Materials, edited by M. Libera, P. Cebe, T. Haynes, J. Dickenson. (Mater. Res. Soc. Proc. **321,** in press. 1993)

PART IV

Doping and Defects in III-V Semiconductor Materials for Device Applications

CARBON DOPING OF InGaAs FOR DEVICE APPLICATIONS

G.E. STILLMAN, S.A. STOCKMAN,* C.M. COLOMB, A.W. HANSON,** AND M.T. FRESINA
University of Illinois at Urbana-Champaign, Department of Electrical and Computer Engineering, Microelectronics Laboratory, Urbana, IL
*current address: Hewlett-Packard Company, Optoelectronics Division, San Jose, CA
** current address: M/A-COM, Lowell, MA

ABSTRACT

Carbon has gained wide acceptance as a p-type dopant for GaAs-based device structures due to its low atomic diffusivity. Carbon doping of InGaAs, however, is complicated by the amphoteric nature of C and difficulty in incorporating C efficiently during epitaxial growth. We have achieved hole concentrations as high as 7×10^{19} cm^{-3} in CCl$_4$-doped InGaAs grown at low temperature by MOCVD. Growth-related issues include the effect of CCl$_4$ on the alloy composition due to etching during growth, and the incorporation of hydrogen, which passivates the C acceptor and reduces the hole concentration during growth and during the post-growth cool-down. The effect of H passivation on minority carrier transport has been characterized by the zero-field time-of-flight technique. High frequency InP/InGaAs HBTs with a C-doped base have been demonstrated with $f_t = 62$ GHz and $f_{max} = 42$ GHz, which is comparable to the best performance reported for MOCVD-grown InP/InGaAs HBTs.

1. INTRODUCTION

Heavily-doped p-type GaAs and In$_{0.53}$Ga$_{0.47}$As are important for use in GaAs/AlGaAs and InGaAs/InP n-p-n heterojunction bipolar transistors (HBTs), where a thin, low-resistance base region is required for high-speed operation. Recently, metalorganic molecular beam epitaxy (MOMBE)[1] and metalorganic chemical vapor deposition (MOCVD)[2] have been shown to be capable of producing heavily carbon-doped GaAs. Carbon has been shown to have a significantly lower atomic diffusion coefficient in GaAs than has been measured for other p-type dopants.[3,4] Consequently, carbon p-type doping in GaAs-based HBTs has been adopted as a solution to base dopant redistribution and degradation problems encountered with Be and Zn.

While the use of a p-type C-doped In$_{0.53}$Ga$_{0.47}$As base in InP/InGaAs HBTs is of great interest, prior to this work it was not known if high doping levels could be obtained, or if carbon has a low diffusivity in InGaAs as in GaAs. Initial attempts at C-doping of In$_{0.53}$Ga$_{0.47}$As by MOMBE resulted in highly compensated n-type material, which was attributed to the amphoteric nature of carbon.[5] Subsequent studies by Ito and Ishibashi[6] employing a heated graphite strip as the carbon source for MBE-grown In$_x$Ga$_{1-x}$As ($0 \le x \le 1$) resulted in p-type material for $x < 0.6$, and n-type conduction for $x > 0.6$. The hole concentration near $x = 0.53$ was limited to about 10^{17} cm^{-3}. Abernathy et al.[7] found that n-type or p-type conduction could be obtained in MOMBE-grown InGaAs using TMGa as the source of Ga and C. The conduction type was dependent on whether TMIn or elemental In was used as the In source. More recent investigation of carbon doping by MOMBE[8] using TMGa and solid In found a type conversion from p to n occurring at $x = 0.8$. A hole concentration of 1.2×10^{18} cm^{-3} was reported for $x = 0.54$, and was in approximate agreement with the carbon concentration deduced using SIMS, indicating that the hole concentration may be limited by carbon incorporation during growth rather than self-compensation. Chin et al.[9] have achieved hole concentrations as high as 3×10^{19} cm^{-3} for x = 0.5 using CCl$_4$ during growth by gas source molecular beam epitaxy (GSMBE). They found that low growth temperatures produced the most heavily doped layers, and short post-growth anneals lead to an increase in the hole concentration. They speculated that reversible carbon acceptor passivation by hydrogen during growth was responsible for the observed annealing behavior.[9]

2. GROWTH OF CARBON-DOPED InGaAs BY MOCVD

In the work described here, carbon doped In$_x$Ga$_{1-x}$As ($0 < x \leq 0.53$) has been grown by LP-MOCVD using CCl$_4$ as the carbon source.[10] Trimethylindium (TMIn), trimethylgallium (TMGa) or triethylgallium (TEGa), and 100% AsH$_3$ were used as the growth precursors, and a mixture of 2000 ppm CCl$_4$ in H$_2$ was used as the carbon source. Carbon-doped layers were grown at $425°C \leq T_g \leq 575°C$, and V/III ratios as low as five were employed. All growths were carried out on semi-insulating (100) GaAs or InP substrates misoriented 2° toward (110). The In composition of the layers was determined by double crystal x-ray diffraction (DCXD). The hole concentration and mobility were determined by van der Pauw - Hall effect measurements.

The etch rates for GaAs and InAs due to CCl$_4$ in the MOCVD reactor have been measured under conditions similar to those used for growth. InAs and GaAs substrates were capped with SiO$_2$ using a low-temperature plasma-enhanced chemical vapor deposition (PECVD) process. Standard photolithographic procedures were then used to define SiO$_2$ dots, which were used as a mask for the CCl$_4$ etching of the substrates. Samples were etched by injecting CCl$_4$ under typical growth conditions (AsH$_3$ flow and substrate temperature), except for the absence of column III sources. After etching with CCl$_4$, the SiO$_2$ was removed using a buffered HF solution, and the etched depth was measured using a surface profilometer.

The etch rates for InAs and GaAs are plotted as a function of reciprocal temperature in Figure 1 for CCl$_4$ and AsH$_3$ flow rates of 4.5×10^{-6} mol/min and 1.1×10^{-3} mol/min, respectively. The etch rate for InAs is significantly higher than for GaAs, and varies exponentially with substrate temperature. The etch rate can be described by

$$R_{etch} \propto \frac{[CCl_4]}{\sqrt{[AsH_3]}} e^{-(1.7eV)/kT}$$

and appears to be similar to the etching of GaAs[11] and InP[12] which occurs in the presence of CCl$_4$. We have found no evidence of significant gas-phase reaction between CCl$_4$ and the organometallic column III sources. Etching during growth of CCl$_4$-doped InGaAs results in reduced incorporation efficiency for In.

The incorporation of carbon in InGaAs is approximately proportional to the CCl$_4$ partial pressure ([CCl$_4$]) and inversely proportional to the AsH$_3$ partial pressure ([AsH$_3$]). This trend is similar to that observed in GaAs, with the exception that much lower growth temperatures are required to achieve significant C concentrations in InGaAs. However, a high [CCl$_4$] and low V/III ratio (low [AsH$_3$]) also contribute to rapid etching of In from the surface during growth, as shown in Equation (1), making compositional control difficult. Thus, a trade-off between C incorporation and etch rate must be considered when choosing the CCl$_4$ flow rate and V/III ratio. The incorporation of C in InGaAs is also extremely temperature-dependent. Low substrate temperatures result in increased carbon doping levels. The etch rate also

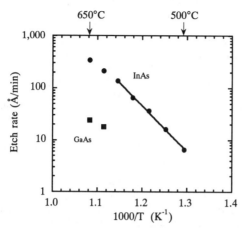

Figure 1. Etch rate for (100) GaAs and InAs substrates in the presence of CCl_4 as a function of reciprocal temperature. The AsH_3 and CCl_4 partial pressures were typical of those used for growth of heavily C-doped GaAs. The InAs etch rate is significantly higher than that for GaAs.

decreases with decreasing temperature, so high C incorporation and low etch rates can be achieved simultaneously by growing CCl_4-doped InGaAs at low substrate temperatures ($T_g < 550°C$).

A number of factors influence the alloy composition, uniformity, and doping level of CCl_4-doped InGaAs grown by MOCVD. An Emcore GS3100 growth chamber was employed for growth on 2-inch InP substrates, and the axis of rotation (1000 rpm) for the wafer was coincident with the center of the wafer. The substrate temperature was estimated using a thermocouple, located beneath the rotating Mo susceptor, which was calibrated using the Al/Si eutectic (577°C). The susceptor was heated from below by a graphite resistance heater. A slight radial nonuniformity in temperature (about 5°C ± 2°C from the center to the wafer edge) was verified by observation of the formation of the Al/Si eutectic on 2 inch silicon substrates.

The MOCVD growth of GaAs using TMGa and AsH_3 is mass-transport limited for temperatures greater than 575°C. For $T_g < 575°C$, however, the growth rate decreases with decreasing temperature and the growth of GaAs becomes kinetically-limited.[13] For the case of CCl_4-doped InGaAs, a substrate temperature below 550°C is necessary to achieve efficient incorporation of C and In, as described above. This makes it necessary to grow the ternary alloy InGaAs in the kinetically-limited growth regime. The lower thermal stability of TEGa relative to TMGa enables mass-transport-limited growth of GaAs at lower temperatures.[13] Thus, the incorporation of Ga from TEGa is more efficient and less sensitive to temperature than the incorporation of Ga from TMGa.

The MOCVD growth of InAs using TMIn ($In(CH_3)_3$) is mass-transport-limited at lower temperatures than for GaAs using TMGa ($Ga(CH_3)_3$) due to the fact that the In-CH_3 bond is weaker than the Ga-CH_3 bond.[14] In the case of heavily CCl_4-doped InGaAs, however, the CCl_4 partial pressure is comparable to the TMIn partial pressure, and the In incorporation is reduced by the etching reaction described above. This results in a decrease in the In incorporation efficiency with increasing substrate temperature. The alloy composition of

InGaAs grown under these conditions is highly temperature-dependent due to the fact that In incorporation becomes less efficient, while Ga incorporation becomes more efficient, with increasing temperature. For growth of heavily CCl_4-doped $In_xGa_{1-x}As$ ($x \sim 0.53$, $p_{annealed} \sim 1 \times 10^{19}$ cm^{-3}) at $T_g \sim 520°C$, $R_g \sim 1.5$ $\mu m/hr$, and V/III ~ 5, using TMGa as the Ga source, the lattice mismatch ($\Delta a/a$) changes by approximately -6×10^{-4} for an increase in substrate temperature of 1°C. This is equivalent to a change in composition with respect to growth temperature (T_g) of $\Delta x/\Delta T_g = -0.009$ $(°C)^{-1}$. Substitution of TEGa for TMGa under these conditions results in a reduction of the sensitivity of alloy composition to temperature. This is due to the fact that Ga incorporation is less temperature dependent, and $\Delta x/\Delta T_g$ is reduced to -0.005 $(°C)^{-1}$.

The fact that the alloy composition (x) of CCl_4-doped $In_xGa_{1-x}As$ is dependent on the growth conditions (T_g, CCl_4 flow, AsH_3 flow) means that each time one of these parameters is changed to achieve a different doping level, the flow rate of TMIn or the Ga source must also be adjusted to obtain $x \sim 0.53$. For the conditions used here, the dependence of x on T_g places severe restrictions on the degree of control over substrate temperature stability, reproducibility and uniformity which are required in order to repeatably achieve lattice match to InP over large areas. Fluctuation in T_g of less than 1°C during growth of a thick (0.5-1.0μm) CCl_4-doped InGaAs layer results in significant broadening of the epitaxial layer peak when the sample is analyzed using double crystal x-ray diffraction (DCXD). In addition, very tight control over T_g is required to achieve run-to-run reproducibility in terms of lattice match to InP.

Figure 2 shows the effect of a small variation in substrate temperature across a 2 inch (50 mm) diameter InP wafer on the compositional uniformity for three cases. The lattice mismatch is plotted as $-\Delta a/a_c$, where a_c is the lattice constant of the epitaxial layer in the center of the wafer (r = 0 mm), rather than $-\Delta a/a_o$, where a_o is the lattice constant of the InP substrate, so that changes in lattice constant across the wafer can be directly compared. The data were compiled from DCXD rocking curves taken at several locations on a series of HBT structures grown on 2 inch InP substrates. The lattice mismatch of the subcollector/collector and the base region was then estimated by comparison with simulated curves obtained using dynamical x-ray diffraction theory.

For the case of undoped InGaAs grown at $\sim 625°C$ the composition is extremely uniform across the wafer, with a variation in $\Delta a/a_c$ of less than 1×10^{-4}. For the case of CCl_4-doped InGaAs grown using TMGa as the Ga source ($T_g \sim 520°C$, V/III ~ 5, [C] $\sim 1 \times 10^{19}$ cm^{-3}), a severe compositional nonuniformity is apparent. This is primarily the result of the slight temperature nonuniformity described above. The substrate temperature is 5°C ($\pm$2°C) higher near the edge (r ~ 22 mm) than at the center of the wafer (r = 0 mm). Thus, the CCl_4-doped layer has a lower InAs mole fraction (x) near the edge than at the center. When TEGa is substituted for TMGa for growth of the CCl_4-doped InGaAs base region, the compositional uniformity is improved. This is the result of the decreased sensitivity of composition to temperature, as described above. The base sheet resistance has also been measured across 2 inch wafers, and varies by less than 10% from r = 0 mm to r = 22 mm, suggesting that the doping level and thickness of the base are relatively constant.

3. CARBON AND HYDROGEN INCORPORATION DURING MOCVD GROWTH

Figure 3 shows the hole concentration as a function of growth temperature for CCl_4-doped $In_xGa_{1-x}As$ ($x \sim 0.53$) grown using TMGa or TEGa. The CCl_4 and TMIn molar flow rates were kept constant, and the flow of the Ga source was adjusted for each growth temper-

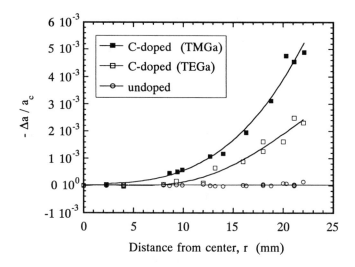

Distance from center, r (mm)

Figure 2. Lattice mismatch with respect to the center of a 2" diameter wafer for InGaAs grown on InP. For the case of undoped InGaAs grown at $\sim 625°C$, the composition is extremely uniform, with a variation in $\Delta a/a_c$ of less than $1x10^{-4}$. For the case of CCl_4-doped InGaAs grown using TMGa as the Ga source ($T_g \sim 520°C$, [C] $\sim 1x10^{19}$ cm^{-3}), a severe compositional nonuniformity is apparent. When TEGa is used as the Ga source ($T_g \sim 520°C$, [C] $\sim 1x10^{19}$ cm^{-3}), the uniformity is improved because the incorporation of Ga is less sensitive to temperature than when TMGa is employed as the Ga source.

ature to achieve lattice match to InP. When using TMGa under these conditions, lattice-matched epitaxial layers could be grown in the temperature range of $520°C \leq T_g \leq 560°C$. A maximum doping level of p $\sim 1x10^{19}$ cm^{-3} was achieved for $T_g = 520°C$. The use of TEGa allowed growth at temperatures as low as $T_g = 450°C$, where a maximum doping level of p $\sim 7x10^{19}$ cm^{-3} was achieved.

It is clear from the data of Figure 3 that the incorporation of C in InGaAs is highly dependent on the growth temperature. This trend is similar to that observed for GaAs and AlGaAs, for C incorporation from TMAs[4] and CCl_4[2,15,16] except that much lower substrate temperatures are required to achieve high C incorporation during MOCVD growth of InGaAs. Buchan et al[15] have proposed that the incorporation of C in GaAs from halomethane sources is controlled by competition between decomposition and desorption of adsorbed carbon-containing species, where both processes are temperature dependent. We propose that in the case of CCl_4, the decomposition of adsorbed CCl_y ($1 \leq y \leq 4$) is not a rate-limiting factor for temperatures as low as 450°C. As shown in Fig. 3, the carbon incorporation is decreased for increasing substrate temperatures due to the increased desorption rate of the carbon-containing species. The low substrate temperatures required to obtain high concentrations of C acceptors in InGaAs relative to GaAs and AlGaAs can be explained by considering the relative strengths of the In-C, Ga-C, and Al-C bonds. The average bond strength for In-CH$_3$ (~ 47 kcal/mol) is significantly lower than for Ga-CH$_3$ (~ 59 kcal/mol) or Al-CH$_3$ (~ 66 kcal/mol) in the case of the metal alkyls TMIn, TMGa, and TMAl.[14,17] Desorption of CCl_y ($1 \leq y \leq 3$) is more efficient from an InGaAs surface than from GaAs, since the In-CCl_y bond is weaker than Ga-CCl_y, leading to reduced incorporation of C acceptors (C on As sites). It has been reported that

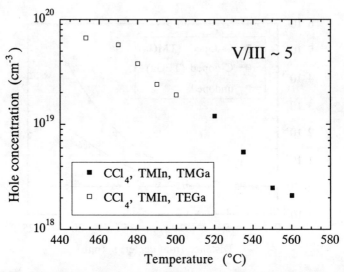

Figure 3. Maxmimum hole concentration achieved for CCl₄-doped InGaAs grown on InP as a function of growth temperature. The Hall effect measurements were performed after a post-growth anneal at 400°C in N_2 for 5 min.

the addition of only a few percent In to GaAs to form $In_xGa_{1-x}As$ ($0 < x \leq 0.03$) can drastically reduce the incorporation efficiency of C during MOCVD growth at $T \geq 600°C$.[18] This behavior may be explained by the enhanced surface mobility of CCl_y ($1 \leq y \leq 3$) at higher substrate temperatures and the efficient desorption of the carbon-containing species from a surface site where it is bonded to In, rather than Ga. These arguments may be extended to account for the efficiency of carbon incorporation which we have observed in the As-based materials[2,10,16] grown by MOCVD using CCl₄: $[C_{As}]_{AlAs} > [C_{As}]_{AlGaAs} > [C_{As}]_{GaAs} > [C_{As}]_{InGaAs}$.

Unintentional hydrogen passivation of C acceptors in InGaAs has been found to occur during both growth and the post-growth cool-down phase.[19] The major difference between growth of C-doped InGaAs and GaAs is that lower growth temperatures are required to achieve high carbon incorporation in InGaAs. The fraction of C acceptors passivated is shown as a function of the hole concentration after annealing ($p_{annealed}$) in Figure 4. The passivation was reversed by annealing at 400°C for 5 min in N_2, and the fraction of acceptors passivated was estimated by comparing the hole concentration before and after annealing (H leaves the crystal through the surfce during the anneal).

The effect of re-cooling these C-doped InGaAs layers in the same ambient used during the original post-growth cool-down was also studied. Samples were first annealed at 400°C in N_2 to reverse the original passivation. They were then heated to 500°C, held at that temperature for 2.5 minutes, cooled to 250°C in AsH_3/H_2 ($[AsH_3] = 3x10^{-4}$), and then cooled the rest of the way to room temperature in H_2. Ohmic contacts were then made, and Hall measurements were performed. As in the case of C-doped GaAs, we found that the fraction of C acceptors passivated returned to nearly the same value as that measured immediately after growth, as shown in Figure 4.

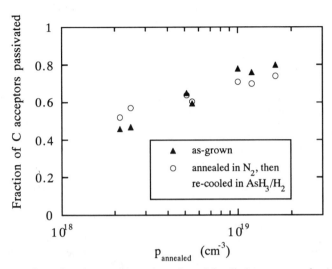

Figure 4. Fraction of carbon acceptors passivated by hydrogen as a function of hole concentration after annealing for several C-doped $In_{0.53}Ga_{0.47}As$ samples. The layers are all between 5000 and 8000 Å thick. The triangles represent the fraction of acceptors passivated immediately after growth, while the circles represent the fraction of acceptors passivated after annealing the sample in N_2 to reverse the passivation, then heating to 500°C and re-cooling in an AsH_3/H_2 mixture ($[AsH_3]/[H_2] = 3 \times 10^{-4}$).

The trend of increasing degree of passivation with increasing doping level shown in Figure 4 is similar to the trend observed for GaAs.[19] However, the C-doped InGaAs layers are much more highly passivated than GaAs layers for comparable doping level, thickness and cooling ambient. In fact, for $In_xGa_{1-x}As$ where x~0.7, n-type conduction is observed after growth, but the layers become strongly p-type ($p > 5 \times 10^{17}$ cm^{-3}) after annealing in N_2. Acceptor passivation has also been observed to occur as a result of cooling in PH_3/H_2 and 100% H_2 ambients, although PH_3 is a less efficient source of atomic H than AsH_3, and H_2 is less efficient than both hydride sources.

Hydrogen incorporation and acceptor passivation in InP/InGaAs HBTs with a carbon-doped base have also been investigated. The as-grown hole concentration for an 8000Å-thick base doping calibration layer was 3×10^{18} cm^{-3}. After annealing at 400°C in N_2, the hole concentration increased to 1.2×10^{19} cm^{-3}. The base sheet resistance, R_S, in an HBT structure with a 1000Å-thick base was measured for three cases. In the first case, the emitter was removed and no anneal was performed, and a sheet resistance of 1350 Ω/sq was measured. This corresponds to an estimated hole concentration of ~ 7×10^{18} cm^{-3}. In the second case the HBT was annealed after etching off the emitter and $R_S = 800$ Ω/sq ($p \sim 1.2 \times 10^{18}$ cm^{-3}) was measured, indicating nearly complete reversal of the passivation. In the last case, the HBT was annealed before etching off the emitter, and R_S, was 1620 Ω/sq. Thus, the passivation was not reversed in the case where the n-type emitter structure was left intact during the anneal.

The acceptor passivation in the base of these HBT structures is a result of incorporation of H during growth of the base region, and is unaffected by the cooling ambient due to the presence of the n-type InP emitter. It also appears that H is unable to escape the base during an anneal at 400°C if the emitter is in place. This is likely due to trapping of positively ionized hydrogen (H⁺) in the base region by the built-in fields at the base-emitter and base-collector p-n junctions.

4. CHARACTERIZATION OF CARBON-DOPED InGaAs

Majority carrier (hole) transport in heavily C-doped InGaAs lattice-matched to InP was characterized using van der Pauw-Hall effect measurements, while minority carrier transport was characterized using the zero-field time-of-flight technique. The most striking results of this study are related to the effects of unintentional H passivation on hole and electron transport and electron lifetime. The ultimate test of material quality is the actual performance of HBTs with a C-doped InGaAs base, and this topic is discussed in detail below.

The hole mobility for carbon-doped InGaAs layers is comparable to that for Be-doped layers over the entire doping range studied. Thus, even for the most heavily C-doped layers, where $p_{annealed} \sim 7 \times 10^{19}$ cm^{-3} and $[C] \sim 1.7 \times 10^{20}$ cm^{-3}, the high hole mobility suggests low C self-compensation. Similarly, the high activation after annealing in N_2 (shown in Figure 5) suggests a maximum possible self-compensation ratio of $N_D/N_A \sim 0.37$ at $[C] \sim 10^{20}$ cm^{-3}, and $N_D/N_A \sim 0.22$ at $[C] \sim 10^{19}$ cm^{-3}. In terms of majority carrier transport, C appears to be a well-behaved acceptor in $In_{0.53}Ga_{0.47}As$, as it is in GaAs, in spite of the fact that it is potentially an amphoteric dopant in $In_{0.53}Ga_{0.47}As$.

Preliminary characterization of minority carrier transport in C-doped InGaAs has been performed using the zero-field time-of-flight (ZFTOF) technique. ZFTOF, which consists of measuring the transient photovoltage generated in a p⁺-n diode illuminated by a picosecond

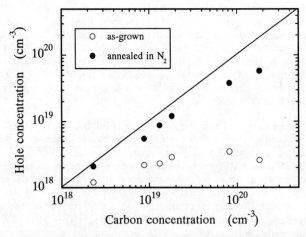

Figure 5. Variation in hole concentration (determined by Hall effect) with total carbon concentration (determined by SIMS) for InGaAs grown on InP. The line represents the case of 100% activation of C atoms as acceptors. The as-grown hole concentration is limited by hydrogen passivation, which can be reversed by annealing at 440°C in N_2 for 5 min.

light pulse at the surface of a thick p^+ layer, measures the electron transport under conditions similar to those present in an HBT, without the presence of high electric fields. This technique has been used to study electron transport in heavily Be-doped GaAs by Lovejoy et al.[20] and in C-doped GaAs by Colomb et al.[21]

The p^+-n diodes used for these measurements were annealed in N_2 to reverse the H passivation in the 8000 Å-thick p^+-InGaAs layer before fabrication of mesa diodes. Fixed parameters which are included in the model to calculate a theoretical curve for the transient photovoltage include the surface recombination velocity, p^+ layer thickness, and device capacitance. Parameters adjusted to produce a fit to the transient voltage are the electron lifetime (τ_n) and diffusion coefficient (D_n). We have measured $\tau_n \sim 0.15$ ns and $D_n \sim 23$ cm^2/s, which correspond to a diffusion length of $L_n \sim 0.6$ μm, for a sample grown using TMGa as the Ga source, with $p_{annealed} \sim 1.2 \times 10^{19}$ cm^{-3}. Samples grown with TEGa have also been characterized, and exhibit nearly identical behavior. Further ZFTOF studies have shown that the electron transport in C- and Be-doped InGaAs is comparable for p-type doping levels as high as 3×10^{19} cm^{-3}. These preliminary results appear promising for application of C-doped InGaAs to minority carrier devices such as HBTs.

The room temperature hole mobility (μ) is plotted as a function of the hole concentration for C-doped InGaAs samples as-grown and after a post-growth anneal in N_2 in Figure 6. All samples are 5000-7000Å thick, and were cooled in the same AsH$_3$/H$_2$ ambient following growth. The mobility decreases monotonically with increasing hole concentration among samples which have been annealed in N_2 to reverse the H passivation. This behavior is typical of heavily doped semiconductors, and is a result of increased ionized impurity scattering with increasing acceptor concentration. In the as-grown case, the hole concentration saturates at $\sim 3 \times 10^{18}$ cm^{-3} as the C-doping level is increased, as described above. If the H passivation

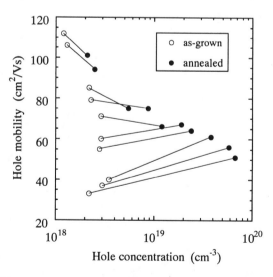

Figure 6 300K hole mobility as a function of hole concentration for C-doped InGaAs as-grown, and after annealing in N_2 to reverse the hydrogen passivation. The lines indicate pairs of data points which correspond to the same epitaxial layer.

mechanism were purely a neutralization effect (formation of a neutral H-C pair), then the as-grown mobility should be a function of only $p_{as-grown}$, and should be independent of the concentration of neutral H-C pairs. However, the as-grown mobility continues to decrease with increasing C-doping level, even though $p_{as-grown}$ remains constant at $\sim 3 \times 10^{18}$ cm^{-3}. These results suggest that the H-C complex which passivates the C acceptors acts as a scattering center. Another possibility is that some degree of compensation of C acceptors (C$^-$) by H donors (H$^+$) occurs, thereby increasing the total ionized impurity concentration and reducing the mobility.

The effects of partial hydrogen passivation on minority carrier mobility in C-doped InGaAs are illustrated in Figure 7(a). The data were obtained using the ZFTOF technique on diodes which were annealed at 440°C in N$_2$ before processing and on diodes which were processed without annealing (as-grown). The electron mobility of C-doped InGaAs after annealing is comparable to that of Be-doped InGaAs grown by GSMBE over the doping range studied ($p_{annealed} \sim 2 \times 10^{18}$ to 3.8×10^{19} cm^{-3}). However, the as-grown C-doped samples ($p_{as-grown} \sim 2 \times 10^{18}$ to 3×10^{18} cm^{-3}) have an electron mobility of only ~ 300 cm^2/Vs. For comparison, Fig. 7(a) also shows that an annealed sample with $p_{annealed} \sim 2 \times 10^{18}$ cm^{-3} exhibits an electron mobility of more than 900 cm^2/Vs. These results indicate that the presence of hydrogen passivation in the most heavily doped InGaAs layers degrades the minority carrier mobility through introduction of additional scattering centers, similar to the results for hole transport described above. These results are somewhat surprising, since partial H passivation has been reported to cause no degradation in majority carrier transport in p-type GaAs[22] and no reports on the effect of H on minority carrier transport are found in the literature. A direct comparison of minority carrier transport before and after annealing has not been carried out for the case of C-doped GaAs, but the data described here for C-doped InGaAs suggest that the low electron mobilities previously reported for heavily C-doped GaAs (p > 3×10^{19} cm^{-3})[21] may be related to the high degree of passivation present in the diode structures.

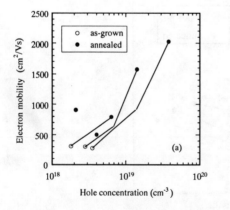 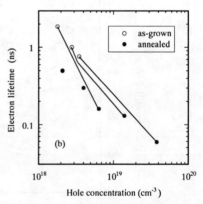

Figures 7(a) Minority carrier mobility as a function of hole concentration of C-doped InGaAs as-grown, and after annealing in N$_2$ to reverse the hydrogen passivation. (b) Minority carrier lifetime as a function of hole concentration as-grown, and after annealing. The lines indicate pairs of data points which correspond to the same epitaxial layer. Data were obtained using the ZFTOF technique by Colomb et al.[25]

Several authors have noted that intentional hydrogenation of GaAs and AlGaAs/GaAs quantum well structures causes a significant increase in photoluminescence intensity.[23,24] They have attributed this behavior to hydrogen passivation of defects which act as nonradiative recombination centers. We have compared the electron lifetime, τ_n, for as-grown and annealed C-doped InGaAs measured using the ZFTOF technique[25], and these results are shown in Figure 7(b). The lifetime decreases dramatically upon reversal of the passivation by annealing in N_2. Most of this decrease can be explained by an increase in the radiative recombination rate due to the increase in the hole concentration. However, it is also observed that τ_n for an annealed sample with $p_{annealed} \sim 2x10^{18}$ cm^{-3} is about 3 times lower than for as-grown samples with $p_{as\text{-}grown} \sim 2x10^{18}$ cm^{-3}. Thus, the partial hydrogen passivation of C acceptors appears to result in an increase in τ_n, likely as a result of passivation of point defects which would otherwise contribute to non-radiative recombination. It is also plausible that these hydrogen-defect complexes could contribute additional scattering, thus causing the degradation in μ_h and μ_e described above.

5. InGaAs/InP HBTs WITH A CARBON-DOPED BASE

We have reported on the dc operation of InP/InGaAs heterojunction bipolar transistors (HBTs)[26] grown by MOCVD with a C-doped InGaAs base and have recently fabricated high-frequency devices. The HBT structure used is shown in Table I. No undoped spacer layer was employed at the base-emitter junction of this device and it was fabricated with an all wet chemical etch, triple-mesa process. The base metallization was self-aligned to the emitter sidewall with a 0.2 μm spacing. All contacts were non-alloyed Ti/Pt/Au. A maximum common-emitter current gain of β_{max} = 250 was typical for the devices tested. An f_t and f_{max} of 62 and 42 GHz respectively were measured for a 2x5 μm^2 emitter device (Figure 8). This cutoff frequency is comparable to the best reported for MOCVD-grown InP/InGaAs HBTs.[27] These results indicate good electron transport through the carbon-doped InGaAs base region, and that carbon is a suitable p-type dopant for high-frequency InP/InGaAs HBTs.

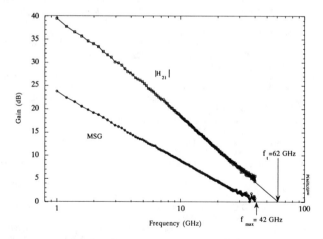

Figure 8. Frequency dependence of current gain H$_{21}$ and maximum stable gain (MSG) for an InP/InGaAs C-doped HBT grown by LP-MOCVD with a 2 x 5 μm^2 emitter. For I_C = 12.93 mA and V_{CE} = 1.2 V.

Table I. Epitaxial Device Structure and Growth Parameters for C-doped InP/InGaAs HBT

Layer	Material	Thickness (Å)	Dopant	Type	Carrier Conc. (cm-3)
Emitter Cap	InGaAs	1500	Si	n^+	2×10^{19}
Emitter	InP	1000	Si	N	5×10^{17}
Base	InGaAs	1000	C	p^+	5×10^{18}
Collector	InGaAs	2200	Si	n^-	$\sim 10^{15}$
Etch stop	InP	200	Si	n^+	5×10^{18}
Subcollector	InGaAs	3000	Si	n^+	1×10^{19}
Substrate	InP	—	Fe	SI	—

This work was supported by the National Science Foundation under contracts ECD 89-43166 and DMR 89-20538 and by SDIO/IST contract DAAL 03-92-G-0272 administered by the Army Research Office. Supplementary support was provided through the AASERT program under contract DAAH 04-93-G-0172.

REFERENCES

1. M. Weyers, N. Putz, H. Heinecke, M. Heyen, H. Luth, and P. Balk, J. Electron. Mat. 15, 57 (1986).
2. B.T. Cunningham, M.A. Haase, M.J. McCollum, J.E. Baker, and G.E. Stillman, Appl. Phys. Lett. 54, 1905 (1989).
3. N. Kobayashi, T. Makimoto, and Y. Horikoshi, Appl. Phys. Lett. 50, 1435 (1987).
4. T.F. Kuech, M.A. Tischler, P.J. Wang, G. Scilla, R. Potemski, and F. Cardone, Appl. Phys. Lett. 53, 1317 (1988).
5. M. Kamp, R. Contini, K. Werner, H. Heinecke, M. Weyers, H. Luth, and P. Balk, J. Cryst. Growth 95, 154 (1989).
6. H. Ito and T. Ishibashi, Jpn. J. Appl. Phys. 30, L944 (1991).
7. C.R. Abernathy, S.J. Pearton, F. Ren, W.S. Hobson, T.R. Fullowan, A. Katz, A.S. Jordan, and J. Kovalchick, J. Cryst. Growth 105, 375 (1990).
8. E. Tokumitsu, J. Shirakashi, M. Qi, T. Yamada, S. Nozaki, M. Konagai, and K. Takahashi, Third International Conference on Chemical Beam Epitaxy, Oxford, UK, Sept. 1991, paper G2.
9. T.P. Chin, P.D. Kirchner, J.M. Woodall, and C.W. Tu, Appl. Phys Lett. 59, 2865 (1991).
10. S.A. Stockman, A.W. Hanson, and G.E. Stillman, Appl. Phys. Lett. 60, 2903 (1992).
11. M.C. Hanna, Z.H. Lu, and A. Majerfeld, Appl. Phys. Lett., 58, 164 (1991).
12. A.E. Kibbler, S.R. Kurtz, and J.M. Olson, J. Cryst. Growth 109, 258 (1991).
13. C. Plass, H. Heinecke, O. Kayser, H. Lüth, and P. Balk, J. Cryst. Growth 88, 455 (1988).
14. S.J.W. Price, In Comprehensive Chemical Kinetics, edited by C.H. Bamford and C.F.H. Tipper (Elsevier, New York, 1972), vol. 4, pp. 197-259.
15. N.I. Buchan, T.F. Kuech, G. Scilla, and F. Cardone, J. Cryst. Growth 110, 405 (1991).
16 B.T. Cunningham, J.E. Baker, and G.E. Stillman, Appl. Phys. Lett. 56, 836 (1990).
17. B.T. Cunningham, J.E. Baker, S.A. Stockman, and G.E. Stillman, Appl. Phys. Lett. 56, 1760 (1990).
18. P. Enquist, J. Appl. Phys. 71, 704 (1992).
19. S.A. Stockman, A.W. Hanson, S.M. Lichtenthal, M.T. Fresina, G.E. Höfler, K.C. Hsieh, and G.E. Stillman, J. Electron. Mater. 21, 1111 (1992).
20. M.L. Lovejoy, B.M. Keyes, M.E. Klausmeier-Brown, M.R. Melloch, R.K. Ahrenkiel, and M.S. Lundstrom, Jpn. J. Appl. Phys, 30, L135 (1991).
21. C.M. Colomb, S.A. Stockman, N.F. Gardner, A.P. Curtis, G.E. Stillman, T.S. Low, D.E. Mars, and D.B. Davito, J. Appl. Phys. 73, 7471 (1993).
22. R. Rahbi, B. Pajot, J. Chevallier, A. Marbeuf, R.C. Logan, and M. Gavand, J. Appl. Phys. 73, 1723 (1993).
23. W.C. Dautremont-Smith, J.C. Nabity, V. Swaminathan, M. Stavola, J. Chevallier, C.W. Tu, and S.J. Pearton, Appl. Phys. Lett. 49, 1098 (1986).
24. J.M. Zavada, R. Voillot, N. Lauret, R.G. Wilson, and B. Theys, J. Appl. Phys. 73, 8489 (1993).
25. C.M. Colomb, S.A. Stockman, S.L. Jackson, N.F. Gardner, S. Varadarajan, A.W. Hanson, M.T. Fresina, A.P. Curtis, and G.E. Stillman (to be published).
26. A.W. Hanson, S.A. Stockman, and G.E. Stillman, IEEE Electron Device Lett., 13, 10 (1992).
27. R.N. Nottenburg, Y.K. Chen, T. Tanbun-Ek, R.A. Logan, and D.A. Humphrey Appl. Phys. Lett., 55, 171 (1989).

CONTACT-RELATED DEEP STATES IN Al-GaInP/GaAs INTERFACE

Z.C.Huang and C.R.Wie
Department of Electrical and Computer Engineering and Center for Opto-Electronics of Materials, State University of New York at Buffalo, Buffalo, NY 14260

ABSTRACT

Deep levels have been measured in molecular beam epitaxy grown $Ga_{0.51}In_{0.49}P$/GaAs heterostructure by double correlation deep level transient spectroscopy. Gold(Au) and Aluminum (Al) metals were used for Schottky contact. A contact-related hole trap with an activation energy of 0.50-0.75eV was observed at the Al/GaInP interface, but not at the Au/GaInP interface. To our knowledge, this contact-related trap has not been reported before. We attribute this trap to the oxygen contamination, or a vacancy-related defect, V_{In} or V_{Ga}. A new electron trap at 0.28eV was also observed in both Au- and Al-Schottky diodes. Its depth profile showed that it is a bulk trap in GaInP epilayer. The temperature dependent current-voltage characteristics (I-V-T) show a large interface recombination current at the GaInP surface due to the Al-contact. Concentration of the interface trap and the magnitude of recombination current are both reduced by a rapid thermal annealing at/or above 450°C after the aluminum deposition.

I. Introducttion

The lattice-matched $Ga_{0.51}In_{0.49}P$/GaAs system has attracted much interest recently because of many optoelectronic applications. It is widely used in laser diodes[1], solar cells[2], and as an important alternative to the AlGaAs/GaAs system for the modulation-doped field-effect transistors and heterojunction bipolar transistors[3,4]. Most of these studies have been concentrated on the effect of growth conditions, the structural characteristics and device performances[5,6,7,8]. Very little work has been done on the electrical defects[9,10]. Defects in this material system appear to be dependent on the growth technique and growth conditions. The deep levels in this system are still far from being fully understood.

We used a double-correlation deep level transient spectroscopy (DDLTS) technique to investigate the deep levels in molecular beam epitaxy (MBE) grown $Ga_{0.51}In_{0.49}P$/GaAs material by using gold(Au) or aluminum(Al) as the Schottky contact. We report a contact-related hole trap at the Al/GaInP interface and an electron trap at E_c-0.28eV in the GaInP epilayer. The current-voltage (I-V) characteristics show that this interface trap acts as a recombination center in the as-deposited diodes. Its concentration could be reduced through a rapid thermal annealing (RTA).

II. Experiment

The $Ga_{0.51}In_{0.49}P$/GaAs heterojunction was grown by MBE. Details of the MBE growth method for this material have been reported elsewhere[11]. Briefly, a 0.5μm thick undoped $Ga_{0.51}In_{0.49}P$ layer was grown on an undoped n-type GaAs(100) substrate, misoriented 4° towards (111)A. The carrier concentration is about $2x10^{16}$ cm^{-3} for the $Ga_{0.51}In_{0.49}P$ epilayer, and is about $9x10^{15}$ cm^{-3} for the GaAs substrate based on the capacitance-voltage (C-V) measurement. Ohmic contact on the GaAs back surface was made by evaporating AuGe/Au, and annealing at 420°C for 15s. The Schottky diodes (area = $6.4x10^{-3}$ cm^2) were fabricated on front GaInP surface by evaporating Au or Al under the base pressure of 1×10^{-6} torr. In order to avoid any contamination of impurities from the source metal, a shutter was placed for a few seconds during the initial evaporation before opening it for deposition.

Our DDLTS system was described in detail elsewhere[12]. The depth profiles were obtained by using a voltage pulse pair with a slightly different amplitude to define the spatial observation window[13]. The current-voltage (I-V) measurement was performed by using a HB 4145B Semiconductor Parameter Analyzer. The annealing was employed using a commercial rapid thermal furnace (Heatpulse Mod. 210) in N_2 atmosphere by the proximity capping method for 30s.

III. Results and Discussions

The DDLTS spectra for both gold and aluminum diodes are shown in Fig.1 (by spectrum(a) and spectrum(b), respective-ly). Both data were taken under a reverse bias of 1.5V, and a voltage pulse pair of 0.5V and 1.0V. One electron trap (A at 190K) and one hole trap (B at 320-365K) were observ- ed in the Al-contact diodes (curve (b)). The trap A is also observed in the Au contact diodes (curve (a)) at the same temperature position. The trap B, however, is not present in the Au contact diodes.

Absence of trap B in the Au-contact diodes indicates that this trap is neither from the bulk nor from the GaInP/GaAs interface. This trap must originate from the Al/GaInP interface. To veryfy this, we have changed the voltage pulse pair during the DDLTS measurement. In Fig.1, the spectrum (c) was taken from the Al-contact diode, but under a pulse pair of 0.5V and 1.5V and the same reverse bias (1.5V) as in spectrum(b). A peak shift was clearly observed for trap B, suggesting that this trap is distributed in energy. No peak shift was observed for trap A as can be seen in Fig.1. From the variation of the emission rate with the inverse of temperature for trap A , an activation energy of 0.28eV and a capture cross section of $4.0 \times 10^{-16} cm^2$ are obtained. Fig.2 shows the depth profile of trap A obtained by DDLTS in GaInP epilayer as well as in GaAs substrate. It shows that its concentration is almost constant in the GaInP epilayer, and decreases rapidly near the GaInP/GaAs interface, indicating that trap A is a bulk trap in GaInP layer.

Trap B has an activation energy of 0.50-0.75eV. Feng et al[9] have observed two electron traps at 0.075eV and at 0.9eV in their MOVPE grown GaInP layers. Paloura et al[10] reported only one electron trap at 0.7-0.9eV in their MOVPE grown $Ga_{0.51}In_{0.49}P$/GaAs heterojunctions. It seems that the defects in this material system are dependent on the growth conditions. To our knowledge, the 0.28eV trap is reported for the first time in MBE-grown GaInP/GaAs samples. A systemetic study showed that this trap is a native defect in GaInP, and is attributed to phosphorus vacancies, V_P It dominates the electron concentration in the conduction band in semi-insulating materials. The concentration BEP of trap A decreased with increasing the P_2 BEP (during the growth) and increased with increasing the annealing temperature (as can be seen in Fig.6). A more detailed study will be described elsewhere[14].

Early X-ray photoemission spectroscopy mesurements showed that an ideally clean surface of III-V semiconductors exhibits no intrinsic surface states in the energy gap. The presence of the hole trap B in our Al-contact diodes may be due to defects formed near the interface during deposition of the Al metal. Spicer and co-workers[15] reported that less than a monolayer of metal or oxygen was sufficient to perturb the semiconductor, producing lattice defects at or near the surface, and these defects in turn produced surface states which pin the Fermi-level in the resulting Schottky diodes. In our case, the formation of the surface states depends on the metal. Although a shutter was used to eliminate the possible contamination of impurities during the initial evapora-

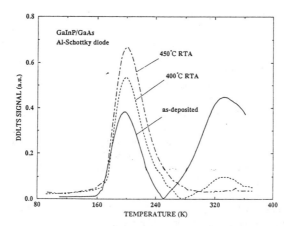

Figure 1. The DDLTS spectra for Au- and Al-contact diodes. (a) Au-contact diode, V_R=-1.5V; pulse pair: 0.5V, 1.0V. (b) Al-contact diode, V_R=-1.5V; pulse pair: 0.5V, 1.0V. (c) Al-contact diode, V_R=-1.5V; pulse pair: 0.5V, 1.5V.

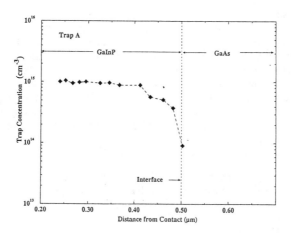

Figure 2. The depth profile for trap A obtained from DDLTS.

tion of metal, this interface trap still existed. This trap may have the same origin as the 0.7-0.9eV trap in MOVPE grown samples observed by others[9,10], although in their case it was an electron trap. However, since the trap level is near the midgap, it could act as electron trap or hole trap depending on its capture process of carriers. We believe that trap B is extrinsic and discuss a few possible causes in the following.

First, the oxygen contamination. It has been reported in GaAs[16] that Al-Schottky may introduce oxigen contamination at the interface, and the measured DLTS signal from bulk defects maybe distorted by using Al-Schottky contact[17]. This oxigen contamination is easier to form at the Al-Schottky interface than at the Au-Schottky interface and can be greatly reduced by a thermal treatment which is indeed true for trap B in our samples, as we will describe later. Second, a reaction between the metal and GaInP. The hole trap at the Al/GaInP interface indicates that these surface states are donor-like (positively charged when unoccupied). Since the cation vacancies are donor-like states, the vacancies V_{In} or V_{Ga} may be a possible origin for trap B. Brillson et al[18] have reported that In diffuses into an overlaying Al or Ni metal from InP. If this is true in our case, one could expect some V_{In} or V_{Ga} at the GaInP surface causing the surface states. Third, the limitation of the measurement system. If the interface states are too fast, they can not be detected by the DLTS technique. In our case the initial delay is greater than 0.1ms. A simulation of DLTS based on our material system showed that no DLTS signal can be observed if the time constant of the states is less than 1×10^{-5}s in the measured temperature range (300K-400K). We have assumed that the DLTS signal can be detected only when $C(t_1)-C(t_2) > 10^{-3} C(t)|_{t=0}$, where t_1 and t_2 are the time windows, C(t) is transient capacitance at time t. It can not be ruled out that trap B exists in both Al and Au-Schottky diodes, and for Au-Schottky, however, it cannot be measured becase it is too fast, while in Al-Schottky it is relatively slow because a thin oxide layer (due to the oxygen contamination) may cause a much slower capture process[19].

These deep states can affect the current transport mechanism in Schottky diodes. Fig.3 shows the forward I-V characteristics for both Au and Al-contact diodes at 300K. The ideality factor for Au-contact diodes was 1.03 in the most linear region (0-0.4V) at 300K, indicating that the thermionic emission is the dominant mechanism. Whereas for the as-deposited Al-contact diodes , the thermionic emision is valid only in the lower voltage region (V<0.3V). In a higher voltage range (V>0.3V), current is dominated by another mechanism. By fitting the I-V-T curves in the higher voltage region, we found that a thermionic emission of electrons followed by interface recombination explains the data well for the Al-contact diodes. The fitting was based on an interface recombination model suggested by Rothwarf[20]. For our Schottky diodes, the interface recombination current is given by

$$J_{IR}(V) = q\sigma_n v_{th} N_{is} N_d exp(-V_{dn}/kT) \{ exp \left(\frac{V}{nkT}\right) - 1 \} \quad ....(1)$$

where σ_n, v_{th} are the electron capture cross section and its thermal velocity, N_{is} and N_d are interface state density and the carrier concentration in GaInP epilayer, respectively, V_{dn} is the built-in voltage, T is temperature, and k is the Boltzman constant. With the interface states considered, the ideality factor n is given by:

$$n = 1/[1-N_{is}(qV_{dn}/2\varepsilon_n N_d)^{1/2}] \quad ....(2)$$

where ε_n is the dielectric contant. A detailed procedure for the I-V-T fitting using eq.(1) was described elsewhere[21]. The best fitting results are shown in Fig.4 for an Al-contact diode. Only one N_{is} value can fit all the curves obtained at different temperatures. The interface state density was 7.5×10^{11} cm^{-2} and the capture cross section was 2.6×10^{-16} cm^2. It should be pointed out that the equations (1) and (2) are only approximate, in which the recombination currents in bulk and at GaInP/GaAs interface are ignored. Here we have assumed that the interface recombination current is the only current transport process in Al-contact diodes in the larger voltage region(V>0.3V).

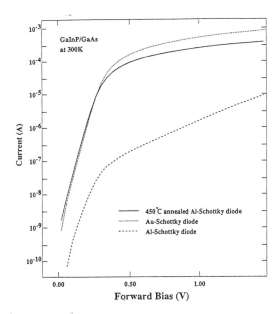

Figure 3. The I-V characteristics at 300K for Au-Schottky diodes (dotted line), Al-Schottky diodes (dashed line) and for 450°C annealed Al-Schottky diodes (solid line).

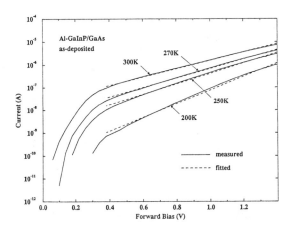

Figure 4. The I-V-T fitting for the Al-Schottky diodes (as-deposited) by using the interface recombination model.

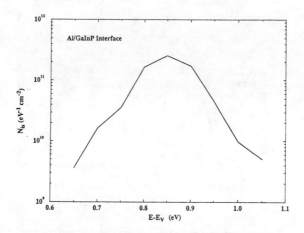

Figure 5. Energy profile of interface state density obtained from DDLTS measurement.

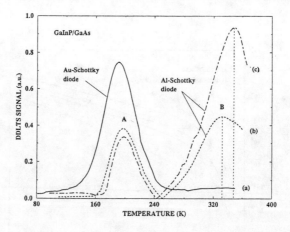

Figure 6. The DDLTS spectra for the annealed Al-contact diodes, showing the reduction of trap B.

The energy distribution of the interface state density N_{is} at Al/GaInP interface can be obtained approximately by assuming that all DDLTS signal in high temperature range (300K-400K) is due to the interface traps. The energy position was determined by the voltage pulse pair and $N_{is} = N_{it}$ d where, N_{it} is the trap concentration obtained from the DDLTS signal and d is the effective depletion thickness related to the pulse pair. The obtained energy profile is shown in Fig.5, with a maximum of 2.6×10^{11} eV^{-1}.cm^{-2} located at E_V+0.85eV. For comparison, the average interface state density obtained from Fig.5 is about 1.1×10^{11}cm^{-2} which is smaller than the value obtained from I-V-T fitting (by a factor of 6). The difference is probably due to the fact that fast states do not contribute to the DLTS signal. In the DDLTS measurement, only the states with emission time longer than 1×10^{-5}s are detected, while all interface states contribute to the recombination current.

Rapid thermal annealing was performed on the Al-Schottky diodes. The DDLTS spectra are shown in Fig. 6. After a 400°C RTA the concentration of the interface trap B was greatly reduced. The trap signal completely disappears after a 450°C annealing. For the 450°C annealed Al-contact diodes the I-V characteristics (see Fig.3) also showed that the diodes had an ideality factor of 1.04 in the most linear voltage region (0-0.4V) at 300K, showing that the thermionic emission becomes the dominant mechanism after a 450°C RTA. This is consistent with the DDLTS results shown in Fig.6. These results confirmed that the interface trap B acts as recombination centers under a forward bias condition. The current level also increased about 2-3 orders after a 450°C annealing in Al/GaInP diodes (see Fig.3), which is probably due to the elimination of the thin oxide layer by annealing. Watanabe et al [22] and we also have reported recently the reduction of vacancy or oxygen related defects by thermal annealing in the GaAs/AlGaAs[23] and in the GaInAs/GaAs[21] heterojuctions, respectively.

The Schottky barrier height was increased slightly after the thermal annealing. The Schottky barrier height, measured using the I-V-T method[24], was 0.93 and 0.92eV for as-deposited Au and Al -GaInP/GaAs diodes , respectively. For the 450°C RTA annealed Al-contact diodes, the barrier height was 0.95V. Sinha et al[25] and Chino[26] reported earlier about the barrier height increase by thermal annealing in Ti/GaAs and Al/Si diodes. Inter-diffusion or dissipation of impurities by thermal diffusion away from the Al/GaInP interface could be a reason for the observed increase in the barrier height.

IV. Conclusions

In summary, the following was observed: (1) One electron trap at 0.28eV was found in GaInP epilayer. (2) One contact related hole trap at 0.50-0.75eV was observed in Al-contact diode, but not in Au-contact diodes. It acts as recombination centers when the diode is forward biased. The interface state density at Al/GaInP interface is about 7.5×10^{11} cm^{-2}, obtained from the I-V-T analysis by simulation fitting. (3) Reduction of this interface trap and of the recombination current was observed after the rapid thermal annealing. (4) The Schottky barrier height of Al-GaInP/GaAs diodes was increased by about 0.03V after the 450°C RTA.

Acknowledgement

This work was supported in part by the National Science Foundation under the grant number DMR-8857403 through the Presidential Young Investigator Program. Authors also acknowledge Prof. G. Wicks for providing the MBE samples used in this study.

215

References

1. M.Ishikawa, Y.Ohba, H.Sugawara, M. Yamamoto, and T. Nakanishi, Appl. Phys. Lett., **48**, 20 (1986)
2. J.M. Olson, S. Kurtz, A.E. Kibbler and P. Faine, Appl. Phys. Lett., **56**, 623 (1990)
3. Y.J. Chan, D. Pavlidis, M.Razeghi, M. Jaffe, and J. Singh, in <u>Proceedings of the Fifteenth International Symposium On GaAs and Related Compounds</u>, Atlanta, edited by J.S.Harris, p 459 (1989)
4. H. Kawai, T. Kobayashi, F. Nakamura, and K. Taira, Electron. Lett., **25**, 609 (1989)
5. R.H. Horng, D.S. Wuu, and M.K. Lee, Appl. Phys. Lett., **53**, 2614 (1988)
6. J.R. Shealy, C.F. Schaus, and L. Eastman, Appl. Phys. Lett., **48**, 242 (1986)
7. O. Ueda, M. Takeshi and J. Komeno, Appl. Phys. Lett., **54**, 2312 (1989)
8. S.J. Hsieh, E.A. Patten, and C.M. Wolfe, Appl. Phys. Lett., **45**, 1125 (1984)
9. S.L. Feng, J.C. Bourgoin, F. Omnes and M. Razeghi, Appl. Phys. Lett., **59**, 941 (1991)
10. E.C. Paloura, A. Ginoudi, G. Kiriakidis, N. Frangis, F. Scholz, M. Moser and A. Christou, Appl. Phys. Lett., **60**, 2749 (1992)
11. G.W. Wicks, M.W. Koch, J.A. Varriano, F.G. Johnson, C.R. Wie, H.M. Kim and P. Colombo, Appl. Phys. Lett., **59**, 342 (1991)
12. Z.C. Huang, C.R. Wie, D.K. Johnstone, C.E. Stutz and K.R. Evans, J. Appl. Phys., **73**, 4362 (1993)
13. H. Lefevre and M. Schulz, Appl. Phys., **12**, 45 (1977)
14. Z.C.Huang , C.R. Wie and G.W. Wicks, submitted.
15. W.E. Spicer, I. Lindau, P. Skeath, C.Y. Su, and Patrick Chye, Phys. Rev. Lett., **44**, 420 (1980)
16. J.C. Bourgion, D. Stievenard, D. Deresmes and J.M. Arroyo, J. Appl. Phys., **69**, 284 (1991)
17. F. Hasegawa, M. Onomura, C. Mogi and Y. Nannichi, Solid State Electron., **31**, 223 (1988)
18. L.J. Brillson, Phys. Rev. Lett., **40**, 260 (1978)
19. S. Wang, <u>Fundamentals of Semiconductor Theory and Device Physics</u>, Prentice Hall, Inc., 1989 p.302
20. A. Rothwarf, IEEE Trans. Electron. Devices, **29** 1513 (1982)
21. Z.C.Huang and C.R.Wie, Solid State Electron., **36**, 767 (1993)
22. K. Watanabe, H.Yamazaki, and Kohji Yamada, Appl. Phys. Lett., **58**, 934 (1991)
23. K. Xie, C.R. Wie, D. Johnson, G.W. Wicks, J. Electron. Mater., accepted for publication.
24. Sze in " <u>Physics of Semiconductor Devices</u>" , second edition, John Wiley & Sons, 1981 p.285
25. A.K. Sinha, T.E. Smith, M.H. Read, and J.M Poate, Solid State Electron., **19**, 489 (1976)
26. K. Chino, Solid State Electron., **16**, 119 (1973)

STRUCTURAL DEFECTS IN PARTIALLY RELAXED InGaAs LAYERS

P.MAIGNE*, J.-M.BARIBEAU**, A.P.ROTH** AND C.L.BOON*.
*Communications Research Centre, P.O Box 11490, Station H, Ottawa, Ontario,K2H 8S2, CANADA.
**Institute for Microstructural Sciences, National Research council, Ottawa, Ontario, K1A 0R6, CANADA.

ABSTRACT

The development of novel optoelectronic devices using thick, defect free, strain free buffer layers mismatched to the substrate requires an understanding of the mechanisms responsible for the relaxation of the elastic strain. Using X-ray diffraction, we have studied the structural properties of partially relaxed InGaAs layers as a function of thickness and substrate misorientation and measured the residual strain. This study shows that strain free layers are difficult to achieve even for thicknesses well above the critical layer thickness. Our measurements also show that the relaxation of the strain induces geometrical effects such as a triclinic distortion of the epilayer unit cell and a tilt with respect to the substrate. These deviations from the ideal structure may seriously degrade the quality of optoelectronic active layers grown on relaxed InGaAs buffer layers.

INTRODUCTION

The design of optoelectronic devices would gain another degree of flexibility if the growth of low defect density, strain-free buffer layers, lattice mismatched with respect to the substrate could be routinely achieved. This would also open new possibilities in device structures, since the thickness of active layers would not be limited. This improvement would be particularly beneficial for the highly strained InGaAs/GaAs system because of its potential applications for HBT[1] and MODFET[2] structures. However, due to a poor understanding of the elastic strain relief mechanisms, strain free buffers are difficult to optimize. In particular, several studies[3,4,5] have shown that some residual strain is present even for layer thickness well above the critical layer thickness and at the present time, it is difficult to predict the final residual strain from lattice mismatch, layer thickness and growth conditions. It is therefore necessary to assess the quality of partially relaxed layers in order to estimate whether relaxed buffer layers would be appropriate for the subsequent growth of active layers. In addition, another difficulty arises from the fact that the structural defects induced by the relaxation process, in particular the final symmetry of the epilayer unit cell, have not been fully investigated. For this purpose, we have carried out a systematic study of the geometry of InGaAs epilayers with different thickness and indium composition using high resolution X-ray diffraction. Our data show that the crystallographic symmetry of the epilayer is no longer tetragonal but triclinic. The magnitude of triclinic distortion has been measured for all layers as a function of substrate misorientation and layer thickness.

EXPERIMENTAL PROCEDURE

The samples used in this study are $In_xGa_{1-x}As$ layers grown by OMVPE in a horizontal reactor at low pressure. The growth conditions were set up to obtain indium composition in the range x= 0.05-0.23. The GaAs substrates were either (100) nominally flat or (100) offcut by 2°. The nominal thickness of the layers varies from 20 to 3000 nm.

The structural properties of the samples were measured using a Rigaku double crystal X-ray diffractometer and a CuK_α radiation (a set of vertical slits is used to eliminate the $K_{\alpha 2}$ wavelength). [400] rocking curves were recorded as a function of the azimuthal angle ω with a 45° step (in this study, when the incident X-ray beam is aligned along the [010] direction, the azimuthal angle is arbitrarily set to 0). The resulting set of 8 different peak spacings was then least squares fitted in order to obtain a value of the lattice parameter in the growth direction, c_2, as well as the amplitude, α, and the direction, ξ, of the epitaxial tilt (the direction of the tilt is measured by the angle between its projection on the sample surface and the [010] direction). Residual strain measurements have been performed by recording [422] or [511] rocking curves, in the low geometry, along the four different <011> directions. Averaging these 4 values leads to the in-plane lattice parameter, $<a_2>$. The indium composition for each sample is then deduced from c_2 and $<a_2>$, as well as a relaxation coefficient defined by $R = (<a_2>-a_0)/(a_r-a_0)$, where a_0 represents the substrate lattice parameter and a_r, the bulk lattice parameter of the epilayer.

RESULTS AND DISCUSSION

Residual strain measurements.

The residual strain in each sample has been measured using symmetrical and asymmetrical reflections. In order to compare layers with different indium composition, we have plotted in Fig.1 the variation of the relaxation coefficient as a function of normalized thickness $\tau=(h-h_c)/h_c$, where h represents the thickness of the layer and h_c represents the critical layer thickness[6] for the particular In composition. In qualitative agreement with the Dodson and Tsao model for strain relaxation,[7] we note two different trends in the range of thickness and In composition investigated. First a quasi-exponential variation of R, for small values of τ, followed by a region where R reaches a plateau, measured at 86%. It is worth noting that even for a layer thickness about 250 times larger that the critical layer thickness a residual elastic strain of about 15% is found, showing that in that range of thickness a strain-free, thick buffer layer is difficult to achieve.

Comparison with other similar works can be instructive. Results from the literature have been transposed into R versus τ data. First a study by Dunstan[8] et al. shows an excellent agreement with our data, in particular the value of the relaxation coefficient reached in the plateau is within few percent of our experimental value. On the other hand, a study by Krisnamoorthy et al.[4] indicates for comparable value of τ a higher residual strain or lower value of the relaxation coefficient measured at 70%. It is interesting to note first that in the three studies, the samples under investigation cover about the same range of thickness and In composition and second that neither the growth temperature nor the growth technique are likely explanations for the observed differences. This comparison emphasizes again that for thick layers, the residual strain is difficult to predict and does not depend solely upon layer thickness and layer composition.

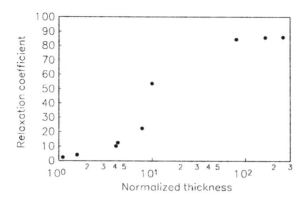

Fig.1. Relaxation coefficient versus $\tau=(h-h_c)/h_c$

Geometry of the epilayer unit cell.

1- Epitaxial tilt

It has been shown for various systems[9] that (100) planes of epitaxial layers, lattice mismatched with the substrate, are tilted with respect to the (100) planes of the substrate. The magnitude and direction of the tilt in the samples investigated are reported in Table I.

Two models have been developed to account for this tilt. Using geometrical considerations, Nagaï[10] first proposed, in the case of fully strained layers grown on misoriented substrates, a mechanism where the (100) epilayer planes have to be tilted in order to avoid a discontinuity of the atomic planes at each interface between steps. This model can be applied to strained mismatched layers grown on misoriented substrate. This is the case only for sample 2 where Nagaï's model predicts a tilt of 205 arcsec toward the [0-10] direction, assuming a substrate offcut angle of 2° and an In composition of 22.5%. These predictions are in excellent agreement with our results.

Second, Olsen and Smith[11] developed a model where the epitaxial tilt is induced by strain relaxation. In this model, the epitaxial tilt is the result of an imbalance of the orientation of the Burgers vectors associated with the formation of 60° type dislocations. Preferential orientation of Burgers vectors is due to substrate misorientation, where the inclined interface results in larger strain relieving component of Burgers vectors along particular directions. The magnitude of the tilt induced by the relaxation of the strain is given by:

$$\alpha = \frac{[(fR)\ \sin(\pi/4-\rho)]}{[n\ \cos(\rho) + \cos(\pi/4-\rho)\ \cos(\pi/4)]} \tag{1}$$

where (fR) is the portion of the lattice mismatch accommodated by interface dislocations, n the ratio of the density of edge type to 60° type dislocations and ρ the substrate offcut angle. It is worth pointing out that both models require a substrate misorientation in order to explain an epitaxial tilt.

Table I: Thickness t, nominal substrate misorientation ρ, tilt magnitude α and tilt direction ξ measured with respect to the [010] direction. β, γ, δ are the parameters corresponding to the triclinic distortion described in Fig.2 .The uncertainties have been estimated to ±10", ±2°, ±0.05°, ±35" and ±10° for α, ξ, β, γ and δ respectively.

sample	t (nm)	ρ (degree)	α (arcsec)	ξ (degree)	β(degree)	γ(arcsec)	δ(degree)
1	20	< 0.5	152	47	-	83	35
2	20	2	218	180	-	190	0
3	40	< 0.5	-	-	-	53	200
4	40	2	153	159	-	419	19
5	80	< 0.5	496	26	-	86	127
6	80	2	1212	38	0.12	452	10
7	3000	2	133	18	0.06	54	293
8	3000	2	370	128	0.12	95	153
9	3000	2	628	341	0.06	205	319

Although an accurate comparison with this model requires, in addition to the X-ray diffraction data, an identification of the type and a measure of the dislocations densities, our results tend to support the existence of another tilting mechanism: first, we notice that sample 5 exhibits a tilt of about 500 arcsec, despite the fact that the offcut angle of the substrate has been measured to be less than 0.25°. In this case, the ratio between strain relieving components of the Burgers vectors on each side of the interface is less than 1% and it is unlikely that such a small difference would lead to a preferential orientation for all Burgers vectors. Therefore an almost complete cancellation of the tilt component of the Burgers vectors is expected. Second, Eq.(1) shows that the magnitude of the tilt is expected to increase with the amount of elastic strain relieved. This is clearly not the case in our samples where a thin layer (sample 6) displays a tilt which is about 10 times larger than that of a thick, more relaxed layer (sample 7). In addition, for samples 7 to 9 the density of misfit dislocations, which is derived from the value of R (assuming only 60° type dislocations) is equal to $1.8 \ 10^5$, $3.3 \ 10^5$ and $4 \ 10^5/cm^2$ respectively, as the tilt magnitude is increasing more rapidly. If the magnitude of the tilt is proportional only to the densities of 60° type misfit dislocations, α would be expected to level off since previous studies[12] have shown that edge type dislocations which do not contribute to the tilt, can outnumbered 60° type dislocations as the layer further relaxes. Consequently, these results suggest that another mechanism also contributes to the epitaxial tilt in our partially relaxed samples.

2- Triclinic distortion

In order to characterize the geometry of partially relaxed unit cell, we have derived the absolute positions of the substrate and epilayer peaks and hence the peak spacing as a function of substrate misorientation, epilayer tilt and crystallographic symmetry of the epilayer unit cell using the following equation:

$$\sin[\theta_b(hkl)] = \sin[\theta(hkl)] \cos[\Phi(hkl)] + \cos[\theta(hkl)] \sin[\Phi(hkl)] \cos[\chi(hkl)] \qquad (2)$$

where $\theta_b(hkl)$ is the Bragg angle, $\theta(hkl)$ the angle between the surface and the incident X-ray beam, Φ(hkl) the angle between the surface normal and the normal to the (hkl) planes and χ(hkl) the angle between the projections of the normal to the (hkl) planes and the incident beam on the surface. The lattice parameter in the growth direction c_2, the tilt magnitude α and the tilt direction ξ are first obtained from [400] symmetrical measurements. The epilayer unit cell is then characterized by four additional parameters: $<a_2>$ the average in-plane lattice parameter, β the angle by which the angle between in-plane vectors of the unit cell differs from 90°, γ and δ, the magnitude and direction of a triclinic distortion causing the [100] InGaAs direction to be no longer perpendicular to the (100) InGaAs planes. A diagram of the unit cell is represented in Fig.2. Applying Eq.(2) to the substrate and epilayer peak positions, the four parameters $<a_2>$, β, γ, δ can be adjusted to match the experimental [hkl] peak splittings. The results are summarized in Table I.

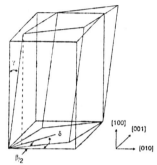

Fig.2. diagram of the triclinic distortion

The first deviation from a tetragonal symmetry is characterized by β and is related to an asymmetry in the formation of misfit dislocations along [011] and [0-11] directions.

The second deviation is characterized by γ and δ. Such a distortion is found in all our layers although it is small in most cases. Analysis of the data presented in Table I suggests, for thin layers, a strong correlation between γ and the misorientation of the substrate. A triclinic distortion has been observed before in the AlGaAs/GaAs strained system[13], with, however two major differences. The amplitude of the distortion was small compared to our results, (about 20 arcsec, roughly the same value as the epilayer tilt), and the direction of the distortion was 180° away from the direction of the epitaxial tilt. In this case, since the tilt is assumed to be equivalent to a rigid body rotation, the effect of the distortion is to keep the [100] epilayer direction parallel to the [100] substrate direction. Sample 2 is the only sample almost completely strained and grown on misoriented substrate and indeed the results from this sample are consistent with the study by Lieberich et al[13], since the amplitude of the distortion is almost equal to the amplitude of the tilt and since the direction of the distortion is exactly opposite to the tilt direction. However, except for sample 2, the triclinic distortion measured in our samples result probably from a different mechanism related to strain relaxation, because the magnitude of the distortion is different from that of the tilt magnitude and no correlation can be established between the direction of the distortion and the direction of the tilt.

CONCLUSIONS

In conclusion, we have characterized the structural properties of partially relaxed layers. Our results show a triclinic distortion of the epilayer unit cell which has been correlated to the substrate misorientation and is probably induced by strain relaxation. These deviations from the ideal structure may seriously degrade the quality of optoelectronic layers grown on thick InGaAs layers, since they are related to strain relaxation. However, it should be kept in mind that the major obstacle for device performances is the propagation of dislocations through the active layers of the devices.

REFERENCES

1- L.P.Ramberg, P.M.Enquist, Y.K.Chen, F.E.Najjar, L.F.Eastman, E.A.Fitzgerald, K.L.Kavanagh. J.Appl.Phys. **61**, 1234, (1987).
2- S.M.Liu, M.B.Das, C.K.Peng, J.Klem, T.S. Henderson, W.P.Kopp, H.Morkoc IEEE Trans. Electr. Dev. ED- 33(5), 576, (1986)
3- A.V.Drigo, Y.Aydiuli, A.Carnera, F.Genova, C.Rigo, C.Ferrari, P.Franzosi, G.Salviati. J.Appl.Phys. **66**, 1975 (1989).
4- V.Krisnamoorthy, Y.W.Lin, L.Calhoun, H.L.Liu, R.M.Park. Appl.Phys.Lett. **61**, 2680 (1992).
5- D.I.Westwood, D.A.Woolf. J.Appl.Phys. **73**, 1187 (1993)
6- J.W.Matthews, A.E.Blakeslee. J.Cryst.Growth, **27**, 118 (1974).
7- B.W.Dodson, J.Y.Tsao. Appl.Phys.Lett. **51**, 1710 (1987).
8- D.J.Dunstan, P.Kidd, L.K.Howard, R.H.Dixon. Appl.Phys.Lett. **59**, 3390 (1991)
9- A.Pesek, K.Hingerl, F.Riesz, K.Liscka. Semicond. Scie. Technol. **6**, 705 (1991).
10- H.Nagai. J.Appl.Phys. **45**, 3789 (1974).
11- G.H.Olsen, R.T.Smith. Phys.Stat.Sol.(a) **31**, 739(1975).
12- K.B.Breene. J.Electron.Mat. **21**, 409 (1992).
13- A.Lieberich, J.Levkoff. J.Cryst.Growth **100**, 330 (1990)

DISLOCATION DISTRIBUTION IN GRADED
COMPOSITION InGaAs LAYERS

S.I.Molina[a], G.Gutiérrez[b], A.Sacedón[b], E.Calleja[b] and R.García[a]
[a]Dep. de Ciencia de los Materiales, Ingeniería Metalúrgica y Química Inorgánica, Univ. de Cádiz, Apdo. 40   11510 Puerto Real (Cádiz), Spain.
[b]Dep. de Ingeniería Electrónica, Univ. Politécnica de Madrid, Ciudad Univ. 28040 Madrid, Spain.

The defect distribution of a graded composition InGaAs layer grown on GaAs by MBE has been characterized by TEM (XTEM, PVTEM, HREM). The observed configuration does not correspond completely with that theoretically predicted. Dislocation misfit segments are in a quantity much bigger than in constant composition layers. Dislocation density is quite uniform up to a certain layer thickness t_1. Few dislocations are observed between this t_1 thickness and a larger thickness t_2. Dislocation density is below the detection limit of XTEM for thicknesses bigger than t_2. Some dislocations are observed to penetrate in the GaAs substrate.

Several mechanisms (reactions between 60° dislocations, Hagen-Strunk and modified Frank-Read processes) are proposed to explain the interactions of dislocations in the epilayer and their penetration in the substrate.

Introduction

Though the linearly graded composition layers have constituied the subject of several studies for different materials ($GaAs_{1-x}P_x$[1], $InGa_{1-x}P$[2], $In_xGa_{1-x}As$[3], Si_xGe_{1-x}[4]), their use as buffer layers has been much less intense compared to the uniform composition layers and the superlattices. Nevertheless, the dislocation density measured in these layers (uniform composition and superlattices) near the top surface is higher that this density in linearly graded composition layers[5,6].

The continuosly graded composition layers possess several advantages to be used as buffer layers for the dislocation filtering. The dislocation pinning is reduced because an interface between two materials of very different composition has not to be present. Also, the force exerted on the propagated dislocations is larger than in uniform composition layers; this is due to that in the mentioned graded composition layers the residual strain is much larger in the nearnesses of the surface[7].

The dislocation distribution of linearly graded composition InGaAs layers grown on GaAs has been studied by TEM in this paper, and this distribution has been compared with the dislocation configuration theoretically predicted.

Experimental

Graded composition 1μm thick $In_xGa_{1-x}As$ layers have been grown on GaAs (001) substrates by Molecular Beam Epitaxy. The grading is linear and the composition changes between x=0, at the surface of the substrate, and x=0.30, at the top of the grown layers. The nominal composition profile has been verified by SIMS measurements[8]. The epilayer thickness measured by XTEM is 1031 ± 34 nm. Figure 1 shows the scheme of the studied heterostructures.

The specimens studied by cross-sectional TEM (XTEM) have been prepared by conventional Ar$^+$ ion-milling techniques. A solution of Br$_2$/CH$_3$OH has been used to prepare, by chemical etching, the planar view TEM (PVTEM) specimens. The TEM results have been obtained by using two transmission electron microscopes, a JEOL 2000 EX, for the HREM work, and a JEOL 1200 EX, for the conventional dislocation analysis by XTEM and PVTEM.

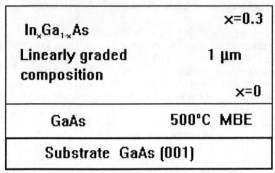

Fig. 1 Scheme of the graded composition layer.

Dislocation distribution

Dislocations are detected by XTEM up to a maximum epilayer thickness; this maximum thickness will be named t_m. Therefore, the dislocation density for regions of the epilayer with thickness t larger than t_m has to be smaller than 10^7 cm^{-2}. This parameter t_m is different for each cross section (a and b); t_{ma} and t_{mb} correspond to the cross sections a and b, respectively. $t_{ma} = 611 \pm 20$ nm and $t_{mb} = 927 \pm 28$ nm. Figure 2 shows a XTEM image taken from the cross section a. For the cross section b the maximum thickness t_{mb} only is reached for the dislocations in some lateral regions of the epilayer; one of these regions is that shown in the right part of the figure 2. The dislocation lines are near or in the (001) growth plane, along the two <110> directions contained in this plane.

Interactions between dislocations have been observed. Figure 3 shows an example of a XTEM image corresponding to an interaction of four 60° dislocations. This configuration is explained considering initially a disposition of a couple of dislocations for each <110> direction contained in the plane (001). A couple is constituied of dislocations **a** and **b**, and the other of **c** and **d**. Each couple is situated in two different planes {111}. Reactions have place between dislocations in the same plane {111} (reaction of **a** and **b**, and of **c** and **d**), and also between dislocations in different planes {111} (reaction of **a** and **d**, and of **b** and **c**). The analysis of this interaction has been carried out using stereoscopic procedures. The explained configuration can be considered as a multiple Hagen-Strunk interaction[9].

60° and edge perfect dislocations have been identified from the HREM images corresponding to cross section preparations, making Burgers' circuit analysis. Not all the dislocations can be characterized in this way because for completing this analysis is necessary that the dislocation lines are along the <110> direction parallel to the incident electron beam. Figure 4 presents an example of two edge dislocations identified by HREM analysis.

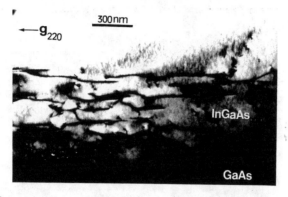

Fig. 2 XTEM image that shows the dislocation distribution in the graded composition layer.

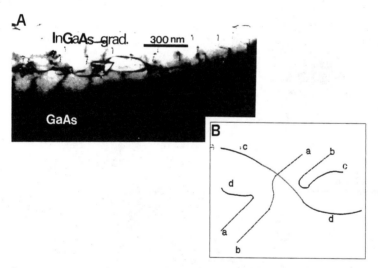

Fig. 3 The representation of B) explains the interaction between dislocations of the region arrowed in the XTEM image of A).

PVTEM images verify that dislocations are mostly situated along the <110> ortogonal directions. Figure 5 shows a PVTEM image. Typical configurations of interactions between two 60° dislocations are observed in these images; the image of figure 5, in the zone arrowed with the letter **c**, shows an example of this configuration.

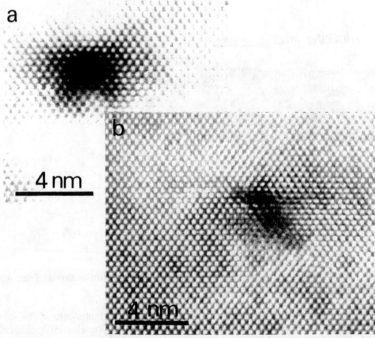

Fig. 4 XTEM image taken from a very thinned region of the graded composition layer. The HREM images (a and b) correspond to the two edge perfect dislocations arrowed in the XTEM image.

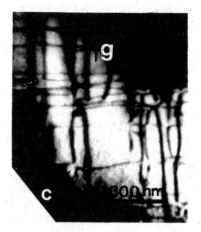

Fig. 5   BF (220) PVTEM image of the graded composition layer.

Dislocations penetrate in the substrate in some regions. Several configurations have been observed for these dislocations. The main configurations are: a) Two 60° dislocations have penetrated in the substrate and they have interactioned, resulting a triangular shape disposition. An example of this behavior, determined using the g·b and g·bxu invisibility criteria on TEM images, is described by the following reaction of dislocations:

$$\frac{a}{2} \cdot [011] + \frac{a}{2} \cdot [10\overline{1}] \dashrightarrow \frac{a}{2} \cdot [110] \quad (60° + 60° = edge)$$

b)F. K. LeGoues et al.[4] have proposed a "modified Frank-Read" mechanism to explain the penetration of dislocations in Si substrates from Si_xGe_{1-x} epilayers and they confirmed that this mechanism is extrapolable to the $GaAs\backslash In_xGa_{1-x}As$ system. The configuration observed in $Si\backslash Si_xGe_{1-x}$ XTEM images for these authors have been observed in the present material.

Comparation between theoretical and experimental behavior

J.Tersoff[7] has developed a model that predicts the dislocation distribution for a graded composition layer. This distribution consists of a completely relaxed epilayer thickness (from thickness t=0 up to t=t_{mt}) and a remaining epilayer (its upper part, from t=t_{mt} up to t=$t_{total\ epilayer\ thickness}$) dislocation free. This distribution is similar to the experimentally observed though the dislocation rich thickness is not completely relaxed. Table 1 shows the values of t_{mt} if only are considered 60° or edge perfect dislocations for the studied heterostructure. The accuracy of these values is mainly due to the unaccuracy of the dislocation cutoff distances; however, this accuracy is quite good (6.1 - 8.2 nm). These values of t_{mt} correspond to a situation in which the

Table 1   Values of t_{mt} and ∂_{cr} considering 60° and edge perfect dislocations

Dislocation	t_{mt}(nm)	δ_{cr}(cm⁻²)
60°	874.5 ± 8.2	$10.6 \cdot 10^9$
edge	915.5 ± 6.1	$5.3 \cdot 10^9$

dislocation density (∂) leads to a complete relaxation of the material for $t \leq t_{mt}$. The mean dislocation density measured for both cross sections is similar and it results ∂ = 5.0 ± 2.0 $\cdot$ 10^9 cm⁻². The densities corresponding to a complete relaxation (∂_{cr}) are indicated in table 1 if only a type of dislocation (60° or edge) is considered. This is, the measured density is equal or smaller than the corresponding to a complete relaxation. This means that the theoretically expected values of t_m must be equal or larger than the calculated values of t_{mt}. A comparison between the experimental values of t_m (t_{ma} = 611 ± 20 nm and t_{mb} = 927 ± 28 nm) and the theoretically predicted for a equilibrium situation (see values of t_{mt} in table 1), if the above explained considerations relative to the dislocations density are taken into account, shows that the system behaves as (or at least is near) a equilibrium system if only some regions of a cross section (b) is considered ($t_{mb} \approx t_{mt}$) while the behaviour for the other cross section is quite out of the equilibrium situation ($t_{ma} \ll t_{mt}$). Therefore, the graded composition layer behaves as a metastable system in some regions. Other facts that contribute to the differences between the theoretical and experimental values of t_m are the interaction between dislocations because these interactions have been observed but are not taken into account in the model of Tersoff.

Acknowledgements

This work has been supported by the "Comisión Interministerial de Ciencia y Tecnología", CICYT, Project MAT 1205/91, by the "Junta de Andalucía", under the group 6020, and by the CE, under the BLES ESPRIT Project. The work has been carried out at the Electron Microscopy Facilities of The University of Cádiz.

References

[1]M.S.Abrahams, L.R.Weisberg, C.J.Buiocchi and J.Blanc, J. Mat. Sci. **4**, 233 (1969).
[2]M.S.Abrahams, C.J.Buiocchi and G.H.Olsen, J. Appl. Phys. **46**(10), 4259 (1975).
[3]J.C.P.Chang, J.Chen, J.M.Fernandez, H.H.Wieder and K.L.Kavanagh, Appl. Phys. Lett. **60** (9), 129 (1992).
[4]F.K.LeGoues, B.S.Meyerson,J.F.Morar and P.D.Kirchner, J.Appl.Phys. **71**,4230(1992).
[5]F.K.LeGoues, B.S.Meyerson y J.F.Morar, Phys. Rev. Lett. **66**, 2903 (1991).
[6]E.A.Fitzgerald, Y.H.Xie, M.L.Green, D.Brasen, A.R.Kortan, J.Michel, Y.J.Mii y B.E.Weir, Appl. Phys. Lett. **59**, 811 (1991).
[7]J.Tersoff, Appl. Phys. Lett. **62**(7), 693 (1993).
[8]M.Maier (private communication).
[9]W.Hagen and H.Strunk, Appl. Phys. **17**, 85 (1978).

RAMAN SCATTERING BY LO PHONON–PLASMON COUPLED MODE IN HEAVILY CARBON DOPED P-TYPE InGaAs

MING QI[*,+], JINSHENG LUO[*],
JUNICHI SHIRAKASHI[**], EISUKE TOKUMITSU[**],
SHINJI NOZAKI[**], MAKOTO KONAGAI[**] AND KIYOSHI TAKAHASHI[**]
[*]Xi'an Jiaotong University, Dept. of Electronic Engineering, Xi'an 710049, China
[**]Tokyo Institute of Technology, Dept. of Physical Electronics, Tokyo 152, Japan

ABSTRACT

The Raman scattering from LO phonon–plasmon coupled (LOPC) mode in heavily carbon doped p-type $In_xGa_{1-x}As$ grown by metalorganic molecular beam epitaxy (MOMBE) was studied experimentally. Only one LOPC mode appears between the GaAs-like and InAs-like LO modes was observed. The peak position of the LOPC mode is near the GaAs-like TO mode frequency, and is not sensitive to the hole concentration. The intensity of the mode increases with increasing the carrier concentration while the two LO modes decrease and become unvisible under the higher doping level. The hole concentration dependence of the linewidth and intensity of the LOPC mode is very similar to that in p-type GaAs. It was shown that the plasmon damping effect plays a dominant role in the p-type doping case.

INTRODUCTION

In a polar semiconductor, the longitudinal–optical (LO) phonons will couple strongly with collective oscillations of the free–carrier system (plasmons) due to the macroscopic electric fields associated with both kinds of elementary excitations. Raman scattering by LO phonon–plasmon coupled (LOPC) modes in the binary semiconducters[1–4], especially in GaAs[5–7], has been studied extensively by many investigators, and has been proposed as a nondestructive technique for determining doping level of the material[8]. Up to now, however, only a few works have been done on the ternary and multinary compounds though they are of considerable interest for their applications. Because of the "two-mode" behavior in many ternary III–V compounds, the feature of the LOPC modes will be different from that in the binary materials as have been shown in $Al_xGa_{1-x}As$ alloys reported previously[9–11]. The $In_xGa_{1-x}As$ system is another important material for the fabrication of high–speed and optoelectric devices, therefore an understanding of the plasmon–phonon interactions in the material is of significant importance.

In this work, we experimentally studied the Raman scattering by coupled hole plasmon–LO phonon mode in heavily carbon doped p-type $In_xGa_{1-x}As$ with $x \approx 0.3$ and various hole concentrations. The frequency, intensity and linewidth of the LOPC mode in the first–order Raman spectroscopy measurements were analysed based on the experimental results.

EXPERIMENTAL

The carbon doped p-type $In_xGa_{1-x}As$ layers used in the present study were grown on (001) SI–GaAs substrates by metalorganic molecular beam epitaxy (MOMBE) using trimethylgallium (TMG), solid indium and solid arsenic as source materials. TMG was kept at $0\,°C$ during the growth and introduced by helium as a carrier gas. The growth temperature was varied from 450 to $530\,°C$ with the pressure–equivalent beam fluxes of 1.8×10^{-5}, 7×10^{-6} and 2×10^{-7} mbar for TMG, arsenic and indium, respectively. The hole concentration of the samples were controlled by growth temperature. Details concerning the

Mat. Res. Soc. Symp. Proc. Vol. 325. ©1994 Materials Research Society

growth characteristics have been discussed elsewhere[12].

Indium compositions were determined by X−ray diffraction (XRD) measurements. Because the thicknesses of all samples are $0.9 \sim 1.3 \mu m$, misfit strain induced by lattice mismatch between $In_xGa_{1-x}As$ epilayer and GaAs substrate is assumed to be completely relaxed. The hole concentrations of the carbon doped p−type $In_xGa_{1-x}As$ layers were determined by Hall measurements using the van der Pauw method at room temperature. The Raman spectra were excited with the 514.5nm line of an Ar−ion laser and recorded in the standard backscattering geometry.

RESULTS AND DISCUSSION

The $In_xGa_{1-x}As$ ternary alloy has two sets of optical phonons[13, 14], and the two longitudinal branches of the phonons will couple with the carrier plasmons. Figure 1 shows

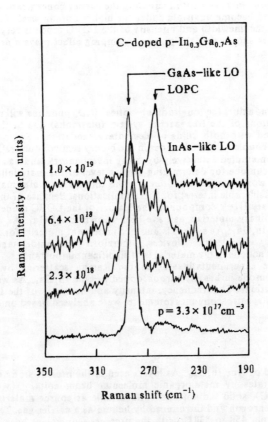

FIG.1. Raman spectra from the carbon doped p−type $In_{0.3}Ga_{0.7}As$ samples with various hole concentrations.

230

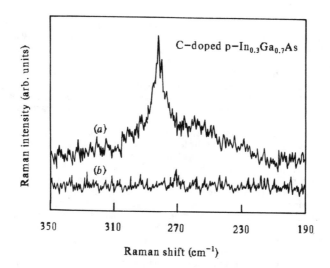

FIG.2. Raman spectra from a carbon doped p-type $In_{0.3}Ga_{0.7}As$ sample in two different scattering configurations:

(a) $\bar{z}(\frac{1}{\sqrt{2}}(x+y), \frac{1}{\sqrt{2}}(x+y))z$, (b) $\bar{z}(\frac{1}{\sqrt{2}}(x+y), \frac{1}{\sqrt{2}}(x-y))z$.

the Raman spectra obtained at room temperature from the carbon doped p-type $In_xGa_{1-x}As$ samples with the indium composition $x \approx 0.3$ and different hole concentrations from 10^{17} to $10^{19}cm^{-3}$. The two peaks observed at high frequency ($282cm^{-1}$) and low frequency ($232cm^{-1}$) are so called GaAs-like and InAs-like LO modes. Another peak appears at the intermediate frequency is assigned to the LOPC mode. It has been found by Yuasa et al.[10, 11] that in $Al_xGa_{1-x}As$, the coupled plasmon-LO phonon modes consist of three branches at high, low and intermediate frequencies. However, we did not observed so many modes in our samples. In addition to the two LO phonons, only one peak is clearly present near the GaAs-like transverse (TO) mode frequency ($262cm^{-1}$) in each of the spectra.

It is well known that in backscattering geometry, only LO-like scattering is allowed while TO scattering should be forbidden from a (100) oriented surface except for highly disordered materials because of the breakdown of the wave vector conservation rule. Such "forbidden" TO scattering has been reported by several authors in experiments dealing with (100) faces, particularly in heavily doped materials[15, 16]. In order to identify the mode appears near the GaAs-like TO mode frequency, we have recorded polarized spectra in different scattering configurations as shown in Figure 2. No violation of the selection rules is observed in our measurements, i.e. all peaks distinctly appear in the "allowed" configuration, whereas they perfectly join the background noise in the "forbidden" one. Thus the intermediate peak behaves as an LO mode in the sense of the selection rules, and corresponds to the scattering by LO phonon-plasmon coupled mode. Although a "forbidden" TO scattering has been measured by Olego and Cardona[17] in highly Zn-doped p-type GaAs, it is here evidenced that carbon doping does not result in a sufficient amount of disorder into the host crystal to cause activation of the TO mode.

The frequency of the LO phonon–plasmon coupled mode as a function of hole concentration in carbon doped p–type $In_{0.3}Ga_{0.7}As$ is shown in Figure 3. When the samples have the same indium composition, the peak position of the LOPC mode is not sensitive to hole concentration though a slight shift to the high frequency direction can be observed. The facts that the LOPC mode is present near the GaAs–like TO mode and the absence of distinguishable high and low frequency branches imply that plasmon damping effect plays a dominant role in the phonon–plasmon coupling because this effect can even destroy the collective carrier behavior as has been found in beryllium and carbon doped p–type GaAs[18–20].

The damping constant of the plasma oscillation can be evaluated by the carrier scattering rate $\tau^{-1} = e / \mu m^*$. The small hole mobility μ due to the large hole effective mass m^* introduces the large damping constant, so the coupling between polar lattice vibrations and carrier plasma oscillations in the p–type doping case will take place through the interaction between the LO phonon and a strongly damped plasmon mode. The large plasma–damping constant shifts the frequency and broadens the linewidth of the coupled mode. This damping effect is more serious in $In_xGa_{1-x}As$ than that in GaAs because there exist some additional carrier scattering mechanisms, such as alloy scattering and dislocations induced by lattice mismatch between the epilayer and substrate. We have indeed found the lower hole mobilities in p–type $In_xGa_{1-x}As$ as compared with those of p–type GaAs under the same doping level.

Figure 4 shows the hole concentration dependence of the linewidth and intensity of the

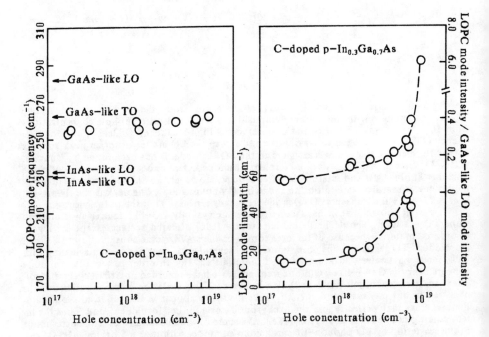

FIG.3. Hole concentration dependence of the LOPC mode frequency.

FIG.4. Changes of the LOPC mode linewidth and intensity ratio of the LOPC and GaAs–like LO modes versus hole concentration.

LOPC mode, which is very similar to that for p-type GaAs[19, 20]. With increasing the hole concentration, a large scattering rate, it means a large plasma damping, results in the large broadening of the coupled mode. However, the phonon conponent of the LOPC mode is enhanced for the highst doping level so that the Raman signal is no more a broadened structure but a well defined peak located at the frequency of the GaAs-like TO mode. From Figure 1, we can also see the notable change of the intensities of the LOPC and two LO modes versus the doping level. The intensity of the LOPC mode increases with increasing the hole concentration while the two LO modes decrease and become unvisible. The increase of the LOPC mode intensity suggests the reduced influence by plasmon as the intensity of the LOPC structure is dependent on the phonon strength of the coupled mode. The intensity of the two LO modes is due to the unscreened-LO modes from the depletion layer present in the vicinity of the surface, therefore they are reduced as the space-charge zone narrowing with increasing carrier concentration. In Figure 4 the intensity ratio of the LOPC and GaAs-like LO modes is also plotted, a rapid increase can be observed when the hole concentration is higher than $7 \times 10^{18} cm^{-3}$, accompanying a sharp reduction of the linewidth of the LOPC mode because the phonon content has become dominant in the coupled mode.

CONCLUSION

We have studied the Raman scattering from the LO phonon-plasmon coupled mode in different carbon doped p-type $In_xGa_{1-x}As$ grown by MOMBE. In addition to the GaAs-like and InAs-like LO modes, only one LOPC mode appears near the GaAs-like TO mode frequency was observed. The hole concentration dependence of the linewidth and intensity of the LOPC mode for p-type $In_xGa_{1-x}As$ is very similar to that for p-type GaAs. It has been shown that the plasmon damping effect plays a dominant role in the p-type doping case.

+Present address: State Key Laboratory of Functional Materials for Informatics, Shanghai Institute of Metallurgy, Chinese Academy of Sciences, Shanghai 200050, China.

REFERENCES

1. M. V. Klein, B. N. Ganguly, and P. J. Colwell, Phys. Rev. B6, 2380 (1972).
2. D. T. Hon and W. L. Faust, Appl. Phys. 1, 241 (1973).
3. G. Irmer, V. V. Toporov, B. H. Bairamov, and J. Monecke, Phys. Stat. Sol. (b) 119, 595 (1983).
4. M. Geihler and E. Jahne, Phys. Stat. Sol. (b) 73, 503 (1976).
5. G. Abstreiter, R. Trommer, M. Cardono, and A. Pinczuk, Solid State Commun. 30, 703 (1979).
6. H. Shen, F. H. Pollak, and R. N. Sacks, Appl. Phys. Lett. 47, 891 (1985).
7. M. Gargouri, B. Prevot, and C. Schwab, J. Appl. Phys. 62, 3902 (1987).
8. K. Wan, J. F. Young, R. L. S. Devine, W. T. Moore, A. J. S. Thorpe, C. J. Miner, and P. Mandeville, J. Appl. Phys. 63, 5598 (1988).
9. O. K. Kim and W. G. Spitzer, Phys. Rev. B20, 3258 (1979).
10. T. Yuasa, S. Naritsuka, M. Mannoh, K. Shinozaki, K. Yamanaka, Y. Nomura, M. Mihara, and M. Ishii, Phys. Rev. B33, 1222 (1986).
11. T. Yuasa and M. Ishii, Phys. Rev. B35, 3962 (1987).
12. J. Shirakashi, T. Yamada, M. Qi, S. Nozaki, K. Takahashi, E. Tokumitsu, and M. Konagai, Jpn. J. Appl. Phys. 30, L1609 (1991).
13. M. H. Brodsky and G. Lucovsky, Phys. Rev. Lett. 21, 990 (1968).
14. G. Lucovsky and M. F. Chen, Solid State Commun. 8, 1397 (1970).
15. K.Murase, S.Katayama, Y.Ando, and H.Kawamura, Phys. Rev. Lett. 33, 1481 (1974).

16. R. Dornhaus, R. Farrow, and R.Chang, Solid State Commun. **35**, 123 (1980).
17. D. Olego and M. Cardona, Phys. Rev. **B24**, 7217 (1981).
18. K. Wan and J. F. Young, Phys. Rev. **B41**, 10772 (1990).
19. A. Mlayah, R. Carles, G. Landa, E. Bedel, and A. M. Yague, J. Appl. Phys. **69**, 4064 (1991).
20. M. Qi, J. S. Luo, M. Konagai, and K. Takahashi, in Proc. of 21st Int. Conf. on Phys. of Semicon., Beijing, 1992; M. Qi, J. S. Luo, J. Shirakashi, E. Tokumitsu, S. Nozaki, M. Konagai, and K. Takahashi, Acta Physica Sinica (in chinese) **42**, 963 (1993).

DEEP LEVEL CHARACTERIZATION OF
LP-MOCVD GROWN Al$_{0.48}$In$_{0.52}$As

F. DUCROQUET*, G. GUILLOT*, K. HONG**, C.H. HONG**, D. PAVLIDIS** AND
M. GAUNEAU***
* Laboratoire de Physique de la Matière (URA CNRS 358), INSA-LYON, bât 502,
69621 Villeurbanne cedex, France
** Solid-State Electronics Laboratory, The University of Michigan, Beal avenue, Ann Arbor,
MI 48109-2122, USA
*** CNET - France-Telecom, 22301 Lannion cedex, France

ABSTRACT

Deep levels in unintentionally doped Al$_{0.48}$In$_{0.52}$As layers epitaxially grown on InP substrates by low-pressure MOCVD have been investigated as a function of growth temperature (T$_g$ ranging from 570 to 690°C). Two different origins for the residual carrier concentration are deduced depending on T$_g$: i) low growth temperatures favor the creation of a deep donor located at E$_c$-(0.13±0.04)eV; ii) At higher T$_g$, a preferential incorporation of a shallow donor occurs, which can be attributed to silicon by SIMS measurements. The oxygen contamination deduced by SIMS and the electrical characteristics of the AlInAs layers do not appear to be correlated.

INTRODUCTION

The large band gap material Al$_{0.48}$In$_{0.52}$As, lattice-matched both to InP and GaInAs has received in recent years increasing interest for optoelectronics and microwave devices.[1,2] Undoped AlInAs layers grown at low temperature by molecular beam epitaxy (MBE) have low background doping and have been widely used as buffer or barrier layers in the fabrication of heterostructure field effect transistors. However, it is reported that depending on the growth conditions, surface roughness and clustering can take place.[3] A large concentration of deep levels is also commonly observed.[4-5] More recently, very high performance HEMT structures have been achieved by metalorganic chemical vapor deposition (MOCVD),[6-8] despite the difficulty to grow Al-based III-V compounds by this technique which generally results in high background concentration. The latter appears to depend on growth temperature and V/III ratio and has values which differ depending on whether they are deduced by capacitance-voltage measurements (n$_{C-V}$) or Hall measurements (n$_{Hall}$).[9-10] This indicates the presence of deep levels in the layers the origin of which is not clearly understood up to now. Correlation with the oxygen concentrations measured by secondary ion mass spectroscopy (SIMS) has been reeported, leading to conclude that oxygen is involved in the creation of deep levels in AlInAs.[11-12] However, the effect of arsenic lattice defects is also suggested as origin for these defects.[13]

In this study, the evolution of deep levels in non-intentionally doped AlInAs grown by LP-MOCVD is reported for different growth temperatures and compared to oxygen and silicon SIMS profiles.

MOCVD GROWTH AND CHARACTERIZATION PROCEDURES

The epitaxial layers were grown on semi-insulating (Fe) or n$^+$(S)-doped InP substrates in a vertical reactor at reduced pressure (60 Torr). They consisted of a 300nm-thick InP buffer and an 1μm-thick lattice-matched AlInAs layer. Trimethylaluminium (TMA), trimethylindium (TMI), arsine (100%) and phosphine (100%) were used as source materials. The typical growth rate was 1μm/h. The investigated structures were grown at different substrate temperatures

235

ranging from 550 to 690°C. Full details of the growth approach will be reported elsewhere.[14] The electrical properties of these layers have been investigated from Schottky diodes which were made by evaporating Pt/Ti/Pt/Au for the Schottky contact. Ohmic contacts were formed by depositing Ge/Au/Ni/Ti/Au on the n^+-substrates, or on the top AlInAs layer in the case of S.I. substrates.

Deep levels have been characterized by admittance and deep level transient spectroscopy (DLTS). Admittance spectroscopy[15] was conducted for frequencies ranging from 1KHz to 100KHz by using a double-phase lock-in amplifier. DLTS experiments were carried out with an 1MHz Sula Technologies capacitance meter and capacitance transients were processed by a A-B correlation technique.[16]

CHARACTERIZATION RESULTS AND DISCUSSION

The residual carrier concentration at 300K and 77K deduced from C-V measurements at 1MHz is plotted in Figure 1 as a function of growth temperature. Two different trends are observed. For T_g lower than 630°C, a large discrepancy is observed between n_{C-V} (300K) and n_{C-V} (77K) indicating that at room temperature, the free carriers are provided by a high density of deep donors. On the contrary, no significant decrease of n_{C-V} is observed for higher T_g and the background carrier concentration is found in this case to remain as high as $10^{17} cm^{-3}$. This high value is maintained even when samples are cooled down to 10K. This suggests that in the layers grown at higher temperatures the free carriers are introduced by shallow donors.

The capacitance versus temperature characteristics (Figure 2) show only one step which corresponds to one main deep donor level E1. The concentration of this defect is deduced from the capacitance drop and decreases rapidly when T_g increases. The E1 level was determined from admittance spectroscopy measurements and was found to be located at $E_c-(0.13\pm0.04)eV$ (Figure 3). Its activation energy (E_a) was found to decrease slightly when high T_g's are used (see Table 1). This trend agrees with the results for the E1 level reported by Luo et al[13], and the characteristics of the C level which is located at $E_c-0.07eV$ and has been reported by Naritsuka et al.[12].

Deep defects present in the AlInAs layers were investigated by DLTS. A typical DLTS spectrum is shown in Figure 4 and suggests the presence of three traps E1, E2 and E3; E1 could only be observed in samples for which a high enough fraction of the free carriers is provided by shallow donors. The activation energies and capture cross-sections of all three traps are summarized in Table I.

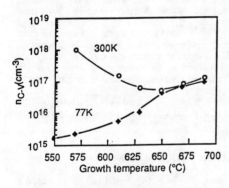

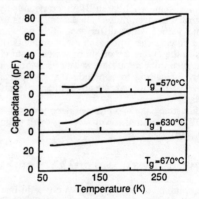

Figure 1: Evolution of the residual carrier concentration deduced from C-V at 300K (o) and 77K(♦) as a function of the growth temperature.

Figure 2: Evolution of the capacitance vs temperature for different growth temperatures.

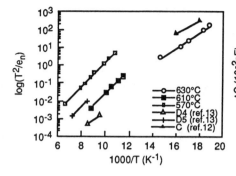

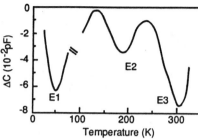

Figure 3: Arrhenius plot of E1 level.

Figure 4: Typical DLTS spectrum on undoped MOCVD grown AlInAs layers.

The Arrhenius plots of E2 and E3 are given in Figure 5 and demonstrate a large dispersion of signatures. The dispersion appears to be more pronounced for E2 in spite of the fact that it is E3 that shows a large variation in activation energies (0.50 to 0.67eV) and capture cross-sections (10^{-12} to 8.10^{-15} cm^2). These two "families" of defects are commonly observed in MOCVD and MBE layers and correspond to the E2 and E3 levels reported by Luo et al.[13] or the ES1 and EA3 levels reported on MBE layers by Hong et al.[4] Their concentrations ranged from 10^{15} to 2.10^{16} cm^{-3}.

The evolution of the defect concentration with T_g is given in Figure 6 and shows a minimum for a growth temperature around 650°C. As can be seen, this temperature corresponds also to a minimum for the background carrier concentration. The choice of T_g also seems to impact the crystalline quality as suggested by the resolution of the DLTS peaks. These could be better resolved in high T_g samples while layers grown below 600°C, show a DLTS spectrum with large and indefinite response. The latter indicates a high defect concentration. The more pronounced and continuous increase of the capacitance with temperature at low T_g's (figure 2) is also a consequence of the high defect density. Overall, the results confirm that low temperature growth deteriorates the crystalline quality of the material.

The concentrations of levels E2 and E3 are typically one order of magnitude lower than the free carrier concentration. This suggests that the excessively high n_{C-V} values can not only be explained by the presence of deep levels as already reported[12] but should be related to both deep and shallow donors.

Table I: Activation energy, capture cross-section and concentration of deep levels in AlInAs.

	T_g	a. 570°C	b. 610°C	c. 630°C	d. 650°C	e. 650°C	f. 670°C	g. 690°C
n_{c-v} at 300K(cm^{-3})		10^{18}	2.10^{17}	6.10^{16}	5.10^{16}		8.10^{16}	1.10^{17}
E1	E_a(eV)	0.14	0.12	0.10				
	σ_n(cm^2)	3.10^{-14}	4.10^{-13}	2.10^{-13}				
	N_t(cm^{-3})	7.10^{17}	9.10^{16}	2.10^{16}				
E2	E_a(eV)		0.35	0.32	0.40	0.40	0.40	
	σ_n(cm^2)		1.10^{-14}	5.10^{-15}	3.10^{-15}	2.10^{-13}	1.10^{-14}	
	N_t(cm^{-3})		9.10^{16}	2.10^{16}	1.50^{15}	2.10^{15}	3.10^{15}	
E3	E_a(eV)		0.60	0.52	0.50	0.67	0.65	0.57
	σ_n(cm^2)		5.10^{-14}	8.10^{-15}	1.10^{-15}	1.10^{-12}	1.10^{-14}	8.10^{-15}
	N_t(cm^{-3})		2.10^{16}	7.10^{15}	2.10^{15}	2.10^{15}	4.10^{15}	3.10^{15}

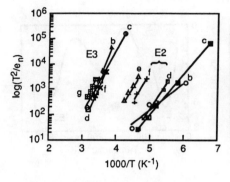

Figure 5: Arrhenius plots of E2 and E3.

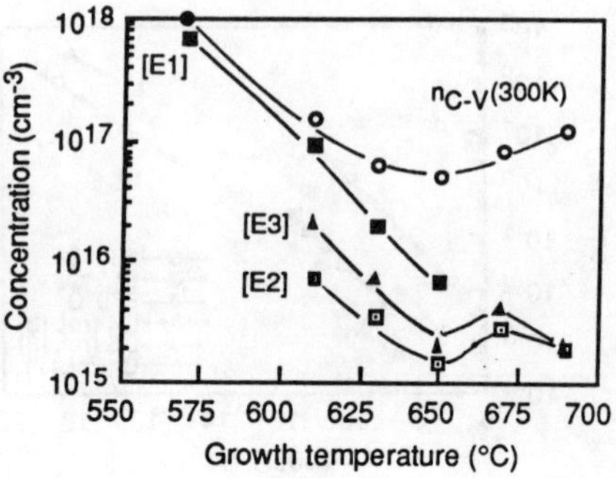

Figure 6: Evolution of the defects concentration as a function of T_g.

In order to determine the origin of shallow donors, SIMS measurements have been carried out to analyze the residual impurities in the AlInAs layers, such as carbon [C], oxygen [O] and silicon [Si]. [C] was always found below the detection limit ($\approx 10^{17} cm^{-3}$) while [O] and [Si] profiles are shown in Figure 7. Although a correlation has previously been reported between n_{C-V} and [O] leading to attribute the residual carrier concentration to an oxygen related defect[9,10], in our case, no evidence of such correlation could be found; indeed, no significant variation of the oxygen incorporation with T_g is observed. A possible reason for this is the relatively low V/III ratio (~45) used for the AlInAs growth. These results suggest that oxygen alone can not explain the increase of carrier concentration when T_g decreases. On the other hand, [Si] incorporation is favored when T_g increases. Since [Si] is known to act as a shallow donor, if we assume that all the Si atoms are ionized, then the free carrier origin can be explained by the sum of the deep donors E1, principally at low T_g and shallow donors related to Si which dominate at high Tg as shown in Figure 8.

The results reported here do not allow the determination of the origin of the E1, E2 and E3 levels. These defects have, however, already been analyzed in Si-doped or undoped MBE samples and Si-doped MOCVD samples, and seem to be related to the AlInAs material itself. Previous studies reported that annealing can reduce the concentration of these defects which could therefore indicate that they can be attributed to arsenic related defects.[17] The presence of excess arsenic at low growth temperature could then explain the strong increase of E1 when Tg decreases.

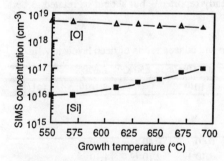

Figure 7: Evolution of [O] and [Si] SIMS concentrations with the growth temperature.

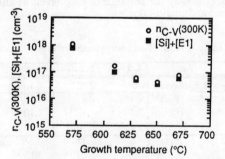

Figure 8: Comparison between the free carriers at 300K and carriers provided by the deep donor E1 and shallow donor related to Si.

CONCLUSION

Deep levels have been investigated as a function of the growth temperature in MOCVD grown AlInAs layers. The background carrier concentration at room temperature shows a minimum for T_g around 650°C. The origin of the free carriers is explained by two mechanisms: i)the creation of a deep donor favored at low growth temperatures, ii) the incorporation of silicon which acts as a shallow donor and can become the dominant cause of the high residual carriers when T_g increases.

The authors want to thank E. Bearzi for electrical measurements and M.Sacilotti for fruitful discussions. One of the authors (FD) is grateful to NATO for its financial support and to Prof. G. Perachon for his comprehension.

This work is supported by NASA (contract no. NAGW 1334), URI (contract no. DAAL 0.-92-G-0109) and NSF/CNRS (contract no. INT-9217513).

REFERENCES:

[1] D. Pavlidis, Materials Science on Engineering, **B20**, 1-8 (1993).
[2] Y. Kwon, D. Pavlidis, T. Brock, G.I. Ng, K.L. Tan, J.R. Velebir and D.C. Streit, Proc. of the 5th Int. Conf. on InP and Related Materials, 465 (1993).
[3] W.P. Hong, P.K. Bhattacharya, and J. Singh, Appl. Phys. Lett, **50**, 618 (1987).
[4] A.T. Macrander, S.J. Hsieh, F. Ren, and J.S. Patel, J. Crystal Growth, **92**, 83 (1988).
[5] H.Hoenow, H.G.Bach, J.Böttcher, F.Gueissaz, H.Künzel, F.Scheffer and C.Schramm, Proceedings of the 4th Int. Conf. on InP and Related Materials, Newport (RI), 136 (1992).
[6] L. Aina, M. Mattingly, M. Serio and E. Martin, J. Crystal Growth, **107**, 932 (1991).
[7] G.I. Ng, D. Pavlidis, Y. Kwon, T. Brock, J.I. Davies, G. Clarice and P.K. Rees, Proc. of the 4th Int. Conf. on InP and Related Materials, 18 (1992).
[8] Y.H. Jeong, D.H. Jeong, W.P. Hong, C. Caneau, R. Bhat, and J.R. Hayes, Jpn. J. Appl. Phys., **31**, L66 (1992).
[9] R. Bhat, M.A. Koza, K. Kash, S.J. Allen,W.P. Hong, S.A. Schwarz, G.K. Chang, and P. Lin, J. Crystal Growth, **108**, 441 (1991).
[10] M. Kamada, H. Ishikawa, S. Miwa and G.E.Stillman, J. Appl. Phys., **73**, n°8, 4004 (1993).
[11] I. Gyuro, W. Kuebart, J.K. Reemtsma, H. Großkopf, F. Grotjahn, D. Kaiser, F. Buchali, Proceedings of the MOVPE conference, Malmöe, (1993).
[12] S. Naritsuka, T. Noda, A. Wagai, S. Fujita and Y. Ashizawa, J. Crystal Growth, **131**, 186 (1993).
[13] J.K. Luo, H. Thomas, S.A. Clark and R.H. Williams, J. Appl. Phys., **74**, n°11, 6726 (1993).
[14] K. Hong, D. Pavlidis, Y. Kwon and C.H. Hong, Proc. of the 6th Int. Conf. on InP and Related Materials, (1994).
[15] D.L. Losee, J. Appl. Phys., **46**, 2204 (1975).
[16] G.L. Miller, D.V. Lang, L.C. Kimmerling, Ann. Rev. Mater. sci., 377 (1977).
[17] J.K. Luo and H.Thomas, Proceedings of the 5th Int. Conf. on InP and Related Materials, Paris, 175 (1993).

HYDROGEN PASSIVATED CARBON ACCEPTORS IN GaAs AND AlAs : NO EVIDENCE FOR CARBON DONORS.

B. R. DAVIDSON*, R. C. NEWMAN*, R. E. PRITCHARD*, T. J. BULLOUGH+, T. B. JOYCE+, R. JONES° AND S. ÖBERG#.
*IRC Semiconductor Materials, The Blackett Laboratory, Imperial College of Science, Technology and Medicine, London SW7 2BZ UK.
+Department of Materials Science and Engineering, PO Box 147, Liverpool University, Liverpool L69 BX, UK.
°Department of Physics, University of Exeter, Exeter EX44 QL, UK.
#Department of Mathematics, University of Luleå, Luleå, S95187, Sweden.

ABSTRACT

GaAs and AlAs layers grown by CBE and doped with either ^{12}C or ^{13}C have been passivated with hydrogen or deuterium. Infrared absorption lines due to hydrogen stretch modes, symmetric A_1 modes and "carbon-like" E modes of $H-C_{As}$ and $D-C_{As}$ pairs have been assigned for both isotopes in both hosts. Comparisons have been made with new *ab initio* local density functional theory and simple harmonic models. Anticrossing behaviour is found for the two types of coupled E modes in the two hosts. The dynamics of the $H-C_{As}$ centre are very similar in GaAs and AlAs.

INTRODUCTION

The passivation of shallow acceptors by atomic hydrogen in III-V compounds is now well known [1]. The hydrogen may be introduced by exposing heated samples to a hydrogen plasma, by heating samples to a higher temperature in an atmosphere containing hydrogen or hydrogen compounds, or the hydrogen may be incorporated during the growth of epitaxial layers. Most of the work relating to the structure of hydrogen acceptor pairs using optical methods, has involved infrared absorption (IR) measurements of the hydrogen (or deuterium) stretch modes in samples containing natural carbon. When pairing occurs with another impurity of low mass there is the possibility of obtaining further information by observing localized vibrational modes of the impurity in its passivated state.

In this paper we report comprehensive measurements for $H-C_{As}$ pairs in GaAs and AlAs. C is a group IV impurity that appears to occupy only As-lattice sites rather than group III lattice sites. The passivating hydrogen atom occupies a bond-centred position where it is bonded to the C_{As} impurity while the neighbouring Ga/Al atom becomes three fold co-ordinated (Fig. 1) [2]. Such a pair could clearly give rise to two A_1 modes and two doubly degenerate LVMs (E modes) that are all IR and Raman active. In the first IR investigations of $H-^{12}C_{As}$ pairs in GaAs [3,4] only the stretch (antisymmetric A_1) modes were discussed, but additional lines labelled X (at 452.7cm^{-1}) and Y (at 562.6 cm^{-1}) were then detected [5]. Later, the isotopic analogue of line X for $D-C_{As}$ pairs was observed (Table) and assignments were tentatively made to the symmetric longitudinal A_1 modes [6] by comparison with the results of *ab initio* local

241

density function (LDF) theory [2]. This assignment was confirmed by Raman scattering measurements [7], which yielded the mode symmetry directly. Further analysis of the Y lines led to severe problems of interpretation [8]. We have now resolved these problems and made assignments of all the modes in GaAs and extended the investigations to AlAs, making use of samples grown by chemical beam epitaxy (CBE) and doped with either ^{12}C or ^{13}C. We also present updated calculated frequencies for GaAs using the same larger basis set used for AlAs [9] (Table).

Table .   **LVM Frequencies of C_{As}, C_{Ga} and H-C_{As} in GaAs and AlAs**

Mode	Experiment (cm^{-1})		Theory (cm^{-1})	
	GaAs	AlAs	GaAs	AlAs
$^{12}C_{As}$ (T_d)	582.8	631.5	544	594
$^{13}C_{As}$ (T_d)	561.8	608.5	(a)	575
$^{12}C_{III}$ (T_d)	N.D.	N.D.	538	529
H-$^{12}C_{As}$ (str)	2635.2	2558.1	2950	2885
H-$^{13}C_{As}$ (str)	2628.5	2549.7	2942	2877
D-$^{12}C_{As}$ (str)	1968.6	1902.6	2154	2111
D-$^{13}C_{As}$ (str)	1958.3	1894.4	2144	2100
H-$^{12}C_{As}$ ($A_1,^{12}X^H$)	452.7	487.0	456	466
H-$^{13}C_{As}$ ($A_1,^{13}X^H$)	437.8	477.2	440	453
D-$^{12}C_{As}$ ($A_1,^{12}X^D$)	440.2	479.8	442	454
D-$^{13}C_{As}$ ($A_1,^{13}X^D$)	426.9	471.2	428	442
H-$^{12}C_{As}$ ($E^-,^{12}Y_2^H$)	N.D.	670.8	888	740
H-$^{13}C_{As}$ ($E^-,^{13}Y_2^H$)	N.D.	652.9	883	725
D-$^{12}C_{As}$ ($E^-,^{12}Y_2^D$)	637.2	656.6	707	684
D-$^{13}C_{As}$ ($E^-,^{13}Y_2^D$)	616.6	635.3	693	662
H-$^{12}C_{As}$ ($E^+,^{12}Y_1^H$)	562.6	ND	553	559
H-$^{13}C_{As}$ ($E^+,^{13}Y_1^H$)	547.6	ND	536	551
D-$^{12}C_{As}$ ($E^+,^{12}Y_1^D$)	466.2	ND	495	437
D-$^{13}C_{As}$ ($E^+,^{13}Y_1^D$)	463.8	ND	487	436

N.D., not detected : (a), not available : III=Ga for GaAs, III=Al for AlAs

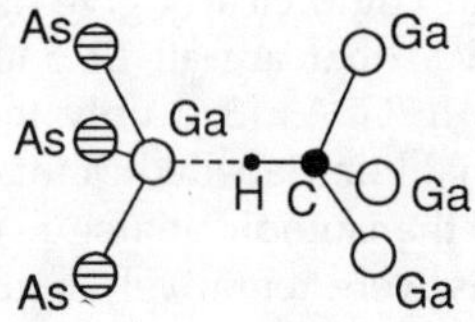

Fig. 1.   Model of the H-C_{As} pair in GaAs showing the hydrogen atom in the bond-centred position only weakly coupled to the neighbouring Ga atom.

EXPERIMENTAL DETAILS

The GaAs samples were grown at 540°C on semi-insulating GaAs (001) by CBE at Liverpool University using cracked AsH_3 and triethylgallium with $^{12}CBr_4$ or $^{13}CBr_4$ as a dopant gas [10,11]. The AlAs layers were grown at 550°C using trimethylaminealane or dimethylethylaminealane as a precursor, again with cracked AsH_3 and CBr_4 as a dopant gas [9].

Carbon doping levels and the doped layer thicknesses were $1.4 \times 10^{19} cm^{-3}$ and 3-4μm for GaAs:C and 1μm and $\sim 10^{20} cm^{-3}$ for AlAs:C. Further details of Hall and SIMS measurements have been given elsewhere [9,11]. Hydrogen was detected in the as-grown AlAs:C by the presence of absorption from the H-$^{12}C_{As}$ stretch mode at $2558 cm^{-1}$ (Table), but similar absorption was not found in the GaAs:C samples. FTIR spectra were recorded at $0.25 cm^{-1}$ or 0.1 cm^{-1} resolution with the samples at $\sim 10K$ using Bruker IFS113v and IFS120 interferometers. Spectra were recorded for as-grown samples and following treatments in an RF plasma (13.56MHz, 2mbar, 40W, 350°C, 3-6h) to introduce either hydrogen or deuterium.

RESULTS

The spectra of as-grown GaAs:C showed LVM absorption features due to $^{12}C_{As}$ or $^{13}C_{As}$ in the form of Fano derivative profiles due to a local spectral redistribution of the free-carrier background absorption (Fig. 2)[5] : no other resolved structure was detected in this spectral range. AlAs:C samples also showed LVM absorption due to the C_{As} impurities as Fano profiles, but there were in addition satellite absorption lines. These lines

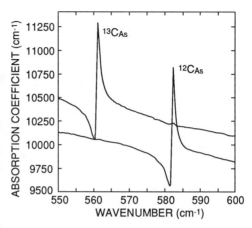

Fig. 2. The infrared absorption of $^{12}C_{As}$ and $^{13}C_{As}$ LVM's in GaAs epitaxial layers grown by CBE.

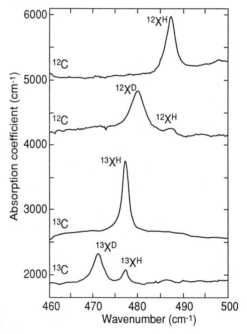

Fig. 3. Infrared absorption spectra of plasma-treated AlAs samples showing the symmetric $A_1(X)$ modes (see also Table).

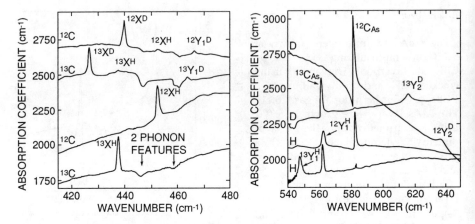

Fig. 4. Infrared absorption spectra of plasma-treated GaAs samples showing the symmetric $A_1(X)$ and $E^+(Y_1)$ modes of the various isotopic combinations of hydrogen-carbon pairs and the antisymmetric $E^-(Y_2)$mode of deuterium-carbon pairs. Other features are due to incomplete matching of the two-phonon absorption of the samples and the undoped reference GaAs.

might be due to C_{As}-C_{As} second neighbour pairs but other possibilities such as C_{As}-O_{As} cannot be ruled out : a full discussion has been given elsewhere [9,12].

After plasma treatments all samples showed strong stretch modes of H-C_{As} pairs (Table) which are the anti-symmetric A_1 modes where the two impurities vibrate out of phase. No satellites were detected which could be attributed to passivated C_{As} complexes in either GaAs or more particularly in AlAs where satellites were present near the LVM from isolated C_{As}. The symmetric A_1 modes (labelled X), where the two atoms vibrate in phase, have been detected for all four isotopic combinations for both host crystals (Table). Spectra showing these lines in AlAs : C are shown in Fig. 3. X-lines due to H-C_{As} pairs observed in the as-grown ^{12}C and ^{13}C doped AlAs samples were not removed by the deuterium passivation treatment. In these modes the displacements of the H and C atoms are very similar because the force constant of the connecting bond is strong (~360 N/m) so that its length remains almost unchanged. Consequently, the frequencies of H-$^{13}C_{As}$ pairs should lie close to those of D-$^{12}C_{As}$ pairs for both hosts, as was found (Fig. 3, Table).

In GaAs:C LVMs at 562.6 and 547.6 cm^{-1} are attributed to the transverse "carbon-like" modes of H-$^{12}C_{As}$ and H-$^{13}C_{As}$ respectively because the relatively large isotopic shift (15cm^{-1}) is comparable with that (21cm^{-1}) between the modes of isolated $^{12}C_{As}$ and $^{13}C_{As}$ acceptors (Table). Surprisingly corresponding modes of D-$^{12}C_{As}$ and D-$^{13}C_{As}$ were not immediately apparent. The enigma was resolved by realising that the paired H and D atoms should also give rise to doubly degenerate wag-modes with the same E-symmetry and that there would be a strong interaction between the two types of mode (anti-crossing) if their frequencies were comparable. Modes of D-$^{12}C_{As}$ and D-$^{13}C_{As}$ have now been detected. Two of these modes occur at 637.2 and 616.6 cm^{-1} (Table) and again have a large isotopic separation (21cm^{-1}) but their frequencies are *higher* than those of their isotopic hydrogen analogues. Another pair of weak lines at 466.2 and 463.8 cm^{-1} with a small splitting (2.4cm^{-1}) were attributed to deuterium wag-like modes (symmetric E$^+$ modes) of D-^{12}C$_{As}$ and D-^{13}C$_{As}$ respectively (Fig. 4). Fitting these observations to a simple model of two coupled oscillators (the coupled pendulum problem) led to the results shown in Fig. 5, where the mass of the hydrogen is used as an independent variable. The figure demonstrates that the two E$^-$ modes are strongly admixed with the H-wag (E$^-$ mode) lying above the E$^+$ mode of the transverse H-C$_{As}$ pairs, while the D-wag mode (E$^+$ mode) lies below the transverse (E$^-$ modes) of D-C$_{As}$ pairs. To date, the E$^-$ modes of H-C$_{As}$ pairs have not been detected but theory [2] indicates that their dipole moments should be small. The observation of both E$^+$ and E$^-$ modes for D-C$_{As}$ pairs results from the admixing of the two wavefunctions which allows the dipole moment of the "carbon-like" transverse mode to be shared with the "hydrogen-like" mode.

A similar analysis has been made for AlAs : C (Fig. 5) but is less definitive since only E$^-$ modes have been detected (Table). However, the isotopic splitting of the modes of H-^{12}C$_{As}$ and H-^{13}C$_{As}$ is smaller than that for D-^{12}C$_{As}$ and D-^{13}C$_{As}$, leading to the conclusion that the "hydrogen-like" modes lie below the "carbon-like" modes and that there is a significant interaction only between the modes of the hydrogen pairs.

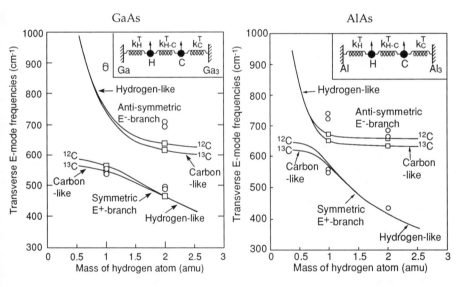

Fig. 5. Vibrational transverse E modes of the coupled masses shown in the insets, used as a simple model to represent the H-C$_{As}$ pair, with force constants for GaAs (AlAs) of k_H = 9.2 (11) N/m, $k_{H\text{-}C}$ = 20.5 (10) N/m and k_C = 243.4 (293) N/m. The squares are the experimental frequencies and the circles are the results of the *ab initio* calculations. The important feature is the strong anticrossing of the modes which changes their effective identities as the mass increases.

DISCUSSION AND CONCLUSIONS

The measured frequencies of the modes of $H\text{-}C_{As}$ pairs in GaAs and AlAs are compared with LDF-theory in the Table. The results of the updated theory with the larger basis set are in better agreement with the measured frequencies for $H\text{-}C_{As}$ in GaAs than those obtained previously [2]. Consequently it is now possible to make a more meaningful comparison of the frequencies for the two host crystals (Fig. 5). Thus, all the observed LVMs in the mid-IR spectral region found for GaAs:C have been assigned either to isolated C_{As} acceptors or the various modes (A_1 stretch, A_1 symmetric, E^+ and E^-) of the four possible isotopic combinations of $H\text{-}C_{As}$ pairs. The lower values of the measured stretch mode frequencies compared with theory are almost certainly due in part to anharmonicity. No mode was detected from C_{Ga} donors and so there is no evidence for the amphoteric behaviour of carbon. The same conclusions apply to AlAs but satellite lines due to unknown carbon complexes are found around the line from isolated carbon atoms, but none of these lines is likely to be due to C_{Al} donors as the observed isotopic shifts are too small [9]. In conclusion we find that the dynamics of the $H\text{-}C_{As}$ pair are very similar in GaAs and AlAs host crystals.

ACKNOWLEDGMENTS

The UK authors thank The Science and Engineering Research Council, UK for financial help and SÖ thanks the Swedish NSRC for financial support.

REFERENCES

1.   R. Rahbi, B. Pajot, J. Chevallier, A. Marbeuf, R. C. Logan and M. Gavand, J. Appl. Phys. **73**, 1723 (1993).
2.   R. Jones and S. Öberg, Phys. Rev. B, **44**, 3673 (1991).
3.   B. Clerjaud, F. Gendron, M. Krause and W. Ulrici, Phys. Rev. Lett. **65**, 1800 (1990).
4.   D. M. Kosuch, M. Stavola, S. J. Pearton, C. R. Abernathy and J. Lopata, Appl. Phys. Lett. **57**, 2561 (1990).
5.   K. Woodhouse, R. C. Newman, T. J. de Lyon, J. W. Woodall, G. J. Scilla and F. Cardone, Semicond. Sci. Technol. **6**, 300 (1991).
6.   K. Woodhouse, R. C. Newman, R. Nicklin, R. R. Bradley and M. J. L. Sangster, J. Cryst. Growth **120**, 323 (1992).
7.   J. Wagner, M. Maier, Th. Lauterbach, K. H. Bachem, A. Fischer, K. Ploog, G. Morsch and M. Kamp., Phys. Rev. B, **45**, 9120 (1992).
8.   M. J. Ashwin, B. R. Davidson, K. Woodhouse, R. C. Newman, T. J. Bullough, T. B. Joyce, R. Nicklin and R. R. Bradley, Semicond. Sci. Technol. **8**, 625 (1993).
9.   R. E. Pritchard, B. R. Davidson, R. C. Newman, T. J. Bullough, T. B. Joyce, R. Jones and S. Öberg 1993 (unpublished work).
10.   B. R. Davidson, R. C. Newman, T. J. Bullough and T. B. Joyce, Semicond. Sci. Technol. **8**, 1783 (1993).
11.   B. R. Davidson, R. C. Newman, T. J. Bullough and T. B. Joyce, Phys. Rev. **B** in press.
12.   B. R. Davidson, R. C. Newman, R. E. Pritchard, D. A. Robbie, M. J. L. Sangster, J. Wagner, A. Fisher and K. Ploog, Semicond. Sci. Technol **8**, 611 (1993).

Recent Developments in Doping Techniques for Compound Semiconductors

J. E. Cunningham
AT&T Bell Labs Holmdel, NJ
and
W. T. Tsang
AT&T Bell Labs, Murray Hill, NJ.

Abstract

We report new methods to dope compound semiconductors. First, we demonstrate the concept of doping engineering whereby it becomes possible to tailor the activation energy of the dopant in a host semiconductor for the first time. In this application, the band offset of a thin, sacrificial semiconductor is used to lower the activation energy of the dopant below its value in the host semiconductor. This allows the freedom to control dopant activity in ways not accessible to a uniformly placed dopant. We chose δ-Be-AlGaAs/GaAs as a model example and show the hole binding energy is reduced by a factor of five. Secondly, we demonstrate overcoming the p-type solubility limit in GaAs by use of monolayer δ-Be in a GaAs base of an HBT. Here, an effective hole concentration of $> 10^{21}$cm^{-3} is measured in real devices. We present a qualatative view of doping solubility limitations that are controlled by surface processes.

I. Introduction

The necessity to dope a semiconductor is an embodiment common to nearly all semiconductor devices. Unfortunately, few if any elements are found to efficiently dope some semiconductors. The circumstances are particularly severe for compound semiconductors where the choice of dopant is often limited to a single element or none. Some of the limitations arises from an incompatibility between dopant and semiconductor host. Issues such as a rapid impurity diffusion coefficient or a high vapor pressure clearly restrict the choice. When the above considerations are applied to GaAs only beryllium and carbon produce successful p-type behavior such that holes are bound with energies predicted by the "effective mass approximation". Other possible elements like Mn enter GaAs as a deep acceptor and hence have limited, if any, utility for device applications [1]. The situation is aggravated for the more polar semiconductors like ZnSe of the II-VI 's where only nitrogen has proven, to date, to have just marginal efficiency as a p-type dopant [2-3].

Even where the activity of a dopant in the semiconductor can be regarded as successful they often have limited solubility at high concentrations. This may exclude certain classes of devices where a high solubility of dopant is required to realize the full potential offered by the semiconductor. One example of this is Heterojunction Bipolar Transistors consisting of Be in a GaAs base where dopant solubility, when uniformly distributed in the base, is limited to $\sim 10^{20}$ cm^{-3} [4]. Often times dopant limitations result from the lack of a physical foundation to predict dopant activity in the host. Other than rules on atomic valence it has not yet been possible to quantitatively predict dopant/host behavior in advance. Instead, most dopant behavior has been either measured or determined empirically.

Mat. Res. Soc. Symp. Proc. Vol. 325. ©1994 Materials Research Society

We describe within two doping concepts that allow the doping bottlenecks given above to be surmounted. First, we demonstrate the concept of doping engineering whereby it becomes possible to tailor the activation energy of the dopant in a host semiconductor for the first time. In this application, the band offset of a thin, sacrificial semiconductor is used to lower the activation energy of the dopant below its value in the host semiconductor as was originally proposed by Tsang [5]. This allows the freedom to control dopant activity in ways not accessible to a uniformly placed dopant. We chose δ-Be-AlGaAs/GaAs as a model example and show the hole binding energy is reduce by a factor of five. Secondly, we demonstrate overcoming the p-type solubility limit in GaAs by use of monolayer δ-Be in the GaAs base of an HBT. Here, an effective hole concentration of $> 10^{21}$cm^{-3} has been measured in real devices [6]. We show how doping can be improved by controlling surface processes.

II. Doping engineering

When dopants are placed into a host semiconductor excess carriers are bound with a characteristic ionization energy also called an activation energy. This energy level falls within the forbidden band gap of the host semiconductor. Often times the characteristic activation energy of the dopant can be described by the "effective mass approximation" which for shallow hydrogenic like impurities ionizes according to an effective Rydberg constant that is corrected for the effective mass and dielectric constant of the semiconductor.

$$E_i = \frac{1}{2}[2\varepsilon]^{-2} \frac{m^* e^4}{h^2} \qquad (1)$$

Here, ε is the permitivity of the semiconductor and m* is the electron or hole effective mass. Because of the smaller dielectric constant for II-VI compared to III-V semiconductors the ionization energy of dopants in the former host are significantly higher than in the latter. Generally, eqn. (1) remains valid for many dopant/host combinations since the electronegativity differences between dopant/host enter as a second order correction to eqn. (1). This also implies that the ionization energy of a given dopant is fixed and cannot be changed when it is uniformly placed in a semiconductor.

In Fig. 1 we show band diagrams corresponding to an n-type dopant having a fixed E_A. The dopant has been placed uniformly in semiconductor host A. On the basis provided above, E_A of the dopant is given by eqn. (1). On the other hand Fig. 1b corresponds to a different case in which donor atoms are confined to a lattice plane placed inside a semiconductor host B that is only a few monolayers thick i.e., about 4 ML. The case of Fig. 1b corresponds to δ doping a quantum well, (qw), and E_A will not be given by eqn. (1). Instead, the net effect of quantum confinement on dopant binding changes in ways previously addressed both experimentally [7,8] and theoretically [9]. In this case it is known that dopant E_A increases but has a dependence on both qw width and position in the well. The general idea is that the carrier wavefunctions are quantum confined in the well and thus have smaller spatial extent about their parent ion than their Bohr radius in host A. Therefore, dopants depicted by case b, have increased binding of carriers. However, to conceptually simplify the DE description, this quantum mechanical effect and also the potential energy shifts of the energy level due to the δ doping potential have not been considered in the placement of the energy level of the dopant in Fig 1b. If on the other hand the donors are confined to semiconductor

248

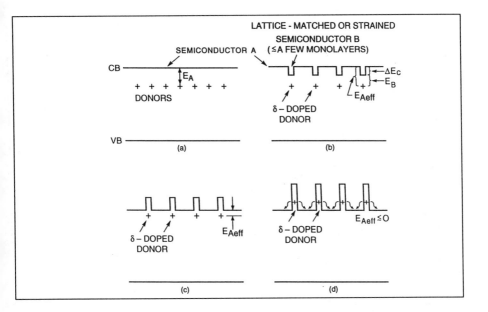

Fig.1 Normal n-type doping with the dopant having a fixed EA in a uniform semiconductor host A. (b), (c), and (d) illustrate the new concept of doping engineering under three different conditions.

host B they will have an activation energy E_B relative to the conduction band edge of semiconductor B. However, in comparison to semiconductor A, the donor effective activation energy is $E_{Aeff} \approx \Delta E_C + E_B$, where ΔE_C is the conduction band discontinuity between semiconductors A,B. Since semiconductor B is made very thin the composite structure behaves as if having a donor doping with an effective activation energy E_{Aeff} in the uniform semiconductor host A. Thus by varying ΔE_C through use of a different semiconductor B one can change the E_{Aeff}. As an example, semiconductor A can be InP and semiconductor B can be InGaAs. Because semiconductor B is thin, then strained layers can be used without incurring dislocations. The equivalent 3-dimensional (3-D) doping concentration is then determined from the 2-dimensional (2-D) density and the periodicity of semiconductor B layers.

One potential application of case b is the p-type doping problem in ZnSe. This could be accomplished via δ-P doping ultrathin qw composed of ZnTe embedded in a ZnSe matrix. This is because P effectively dopes ZnTe ($\approx 10^{18}$ cm^{-3}) [10] but ineffectively dopes ZnSe. Unfortunately, the binding energy of P in ZnTe is 70 meV and would increase by the valance band discontinuity of ZnTe / ZnSe. Nevertheless, it still might be possible to use ZnSeTe alloys in place of ZnTe, provided δ doping remains electrically active in the alloy, to reduce the magnitude of the discontinuity.

In the above case E_{Aeff}, invariably, will be larger than E_A. We consider next application of DE to reduce the E_{Aeff} of dopants. This is shown in Fig. 1c where semiconductor B has a wider bandgap than the host semiconductor A. The E_{Aeff} can be made smaller than E_A. Actually, in Fig. 1d E_{Aeff} becomes negative implying no energy is needed to ionize the donor atoms. In fact, this can be considered as the limiting case of modulation doping in which the thicknesses of both the impurity doping is confined to a lattice plane and the wide bandgap semiconductor layers (a few ML) are pushed to the very thinnest limit. An example of cases (c) and (d) can be realized in GaAs/δ -Al$_x$Ga$_{1-x}$As or InP/δ -Ga$_x$In$_{1-x}$P for lattice-matched and strained-layer systems, respectively. In addition, DE differs from modulation doping (MD) [11] as we illustrate in Fig. 2. In (MD) the carriers are spatially removed from the parent acceptors and are no longer bound by Coulomb attraction to its ion. In many MD structures the dopant is displaced from the

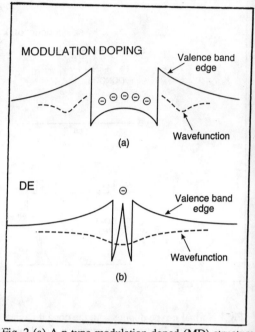

Fig. 2 (a) A p-type modulation doped (MD) structure. (b) A p-type doping engineering (DE) structure.

heterointerfaces by distances that are larger than an impurity Bohr radius. Here, the free carriers occupy energy eigenstates at the bulk hetero-interface as our example illustrates in Fig. 2(a). However, in DE we chemically change only the dopant location for distances smaller than the impurity Bohr radius. The quasi free carriers still remain partially bound to their parent acceptor but with a reduced activation energy as illustrated in Fig. 2b.

In essence, doping engineering permits the freedom to design the E_A of a dopant in a semiconductor. In fact, it may prove to be particularly useful for achieving dopings in semiconductors in which a normally employed doping process is not successful. In addition to the previous case of (ZnSe/δ -ZnTe$_x$Se$_{1-x}$), other possible applications of DE can be employed to p-type doping problem in ZnSe. Compare P doping ZnTe and ZnSe. Electronegativity, η, difference reveals that P(η=2.1) would prefer to bond to Zn(η=1.6) as opposed to Te(η=2.1). Hence, P would incorporate on the Te sublattice and bond to Zn to produce good p-type dopant activity. In ZnSe the situation differs in that P electronegativity also differs significantly with Se (η =2.4) and hence bonding on either sublattice may occur. Hence, simple incorporation energetics show that P in ZnSe has limited utility as a p-type dopant in agreement with experimental measurements. On the other hand, atomic nitrogen doping in ZnSe has been known to successfully produce p- type electrical activity. On electronegativeity grounds N (η=3.0) prefers bonds with Zn over bonds with Se by a factor five and therefore incorporates mainly on the Se

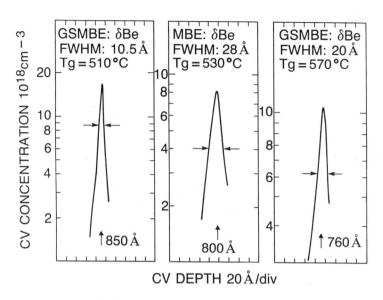

Fig. 3 C-V from δ - Be doped GaAs at 0.03 ML.

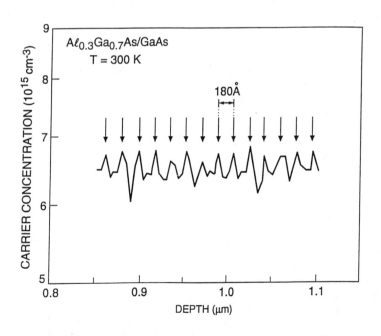

Fig. 4. C-V hole concentration as a function of depth for the DE sample.

sublattice to produce p- type electrical activity. To improve p-type doping efficiencies in ZnSe under nitrogen doping one may apply doping engineering to reduce E_{Aeff} by application of ZnS. N in ZnS can be expected to have lower E_{Aeff} by the valence band offset. In addition, N has improved anion site occupation in ZnS over the case in ZnSe.

To demonstrate the concept of doping engineering, we have grown a multilayer structure consisting of GaAs/δ-Be-Al$_{0.3}$Ga$_{0.7}$As. TEM shows the AlGaAs thickness was 15 Å and the GaAs layer was 210 Å. The center of the AlGaAs layer was δ-doped with Be (a 2-D concentration of 2×10^{10} cm^{-2}). This combination should result in an equivalent bulk hole concentration of 10^{16} cm^{-2} in GaAs. For comparison, the same periodically modulated structure was also grown but it was uniformly doped with Be at the 10^{16} cm^{-3} level.

The δ doping was accomplished using Group III interruptions during epitaxial growth. The layers were grown by Gas Source Molecular Beam Epitaxy at a relatively low growth temperature of 500°C where redistribution of Be through segregation and diffusion can be suppressed. Capacitance-Voltage (C-V) profiles of higher δ-Be (10^{13}cm^{-2}) layers under these growth temperatures are shown in Fig. 3. The Full Width Half Maximum (FWHM) of the profile is 11 Å and thereby illustrates that Be is highly confined to the δ plane [12]. The profile FWHM corresponds to the wavefunction of carriers in the δ-potential [13]. The measured profile width is among the narrowest for Be at this concentration and does not change as the growth temperature decreases below 500°C. The C-V hole concentration from the δ Be of the DE growth as a function of depth is shown in Fig. 4. Sharp spikes as a function of depth are observed at a periodicity of 180 Å i.e., close to the period length measured by TEM. The FWHM of each spike is large because Be concentration was low such that carrier wavefunction along the growth direction was only weakly confined by the δ-potential. This produces a very weak depth dependence to the holes and their wavefunction should overlap considerably so that the DE should exhibit vertical and longitudinal transport characteristics close to the homogeneous doping case.

As a practical demonstration of the DE concept, we perform variable temperature Hall effect on the structures specified above. Fig. 5 shows the carrier concentration versus reciprocal temperature as circles for the DE sample and triangles for the homogeneously doped specimen. In these investigations we have chosen the Be

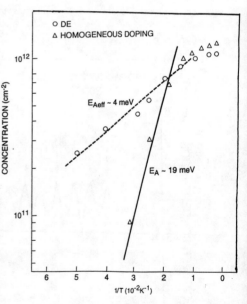

Fig. 5. A plot of hole concentration measured by the technique for both the uniformly doped and DE sam as a function of inverse temperature.

concentration to be an order of magnitude lower then the critical concentration corresponding to a Mott transition of holes in GaAs. By doing so, our variable temperature results should reflect the true ionization energy of the dopant instead of reduced energy activation caused by banding effects. Indeed, the thermal behavior of the homogeneously doped sample shows this to be the case. Fig. 5 illustrates that hole freeze-out commences at temperatures just below 100° K. By 30 °K the hole concentration has fallen to one-tenth of its room temperature value. Clearly, this behavior is the expected thermal characteristics of lightly Be doped GaAs. By contrast, the thermal dependence of the carrier concentration is observed to be much weaker for the DE sample. For instance at 30 °K the DE concentration has only fallen by a factor of two of its room temperature value. One further feature of the dependence of carrier concentration on temperature is that both structures exhibit Arrhenius behavior at very low temperature. We may analyze their dependences to quantify more precisely the change in impurity ionization induced by the DE concept. However, to simplify the analysis we assume the thermal dependence of the carrier concentration is associated with only carrier freeze out whereas other thermal dependences from interface/surface depletion and band transport factors are negligible in comparison. As Fig. 5 shows both the DE and homogeneous doping may be fit by an Arrhenious expression at very low temperature as evidenced by the solid and dashed lines. From their slopes we estimate the impurity ionization is 19.0 meV for homogeneous doping and 4 meV for DE. The 19 meV energy is in reasonable accord for the known binding energy of Be (25meV) in GaAs.

The reduction in Be ionization energy induced by the DE concept is remarkable since we have reduced the impurity binding energy by a factor five its value in conventional doping. For comparison, changes in acceptor binding energy in GaAs brought about by different chemical dopants are only about 30% of the acceptor binding energy. The 4 mev of residual binding energy in the DE case arises because the carriers remain quasi bound to the parent ions, an effect we have not considered in the analysis. The measured hole mobilities of the uniformly doped and DE samples are 380 cm^{-2} /Vsec and 300 cm^{-2}/Vsec at room temperature and 3800 and 4300 cm^{-2}/Vsec at 77 °K, respectively. Hence, our DE structure retains bulk-like transport characteristics and not the enhanced mobilities of MD structures. The DE contains, of course, the flexibility in design of the ionization energy.

III. Doping at high concentrations

A widegap emitter layer, when used in a heterojunction bipolar transistor [14], HBT, allows the use of extremely high base doping to improve high speed device performance. For several reasons δ doping can be advantageous in the above application. First, the carrier transit time in the base of an HBT may be reduced when the base consist only of an atomically thin δ - doping region. In particular, δ− Be HBTs have been successfully demonstrated in ref [15] at the 0.04 monolayer (ML) δ -Be level and by us at the 0.4 ML level [6]. Secondly, Be, when introduced by homogeneous doping techniques, is known to have limited solubility in the lattice of GaAs. Generally, the solubility limit for homogeneous doping occurs at Be concentrations above 10^{20} cm^{-3}. Above the solubility limit, Be is known to degrade the lattice. From results shown below we show how the process of lattice degradation could be minimized by the δ doping process. In addition, we demonstrate that an equivalent homogeneous hole density above 10^{21} cm^{-3} is achieved for δ−Be at the 0.4 ML level. Thirdly, the requirement of low sheet resistivity in the base becomes more achievable per dopant atom for the δ doping technique. This follows from the mobility enhancements of the δ–dopants over those inferred from equivalent concentrations for homogeneously doping [16]. As we show below sheet resistivities below 300 Ω-cm are obtained in a single δ–Be plane at the 0.4 ML level.

IV. Doping solubility limit

Continuously doping GaAs with Be at concentrations above mid 10^{19} cm^{-3} can produce haze in the epitaxial layer (morphological roughness) [17,18]. Therefore there is an effective limit in homogeneous doping whereby the base doping of an HBT may not increased further. Empirically we show an approximate doping limit in Fig. 6 whereby the survey shows homogeneously doped HBT gain has abruptly terminated above a Be concentration 10^{20} cm^{-3} [19]. At this concentration the HBT gain has fallen to about 3 and use of higher Be concentrations often lead to catastrophic loss in HBT gain. The doping solubility limit could arises from Be precipitates in the lattice that are caused by inclusion of dopants clusters into GaAs [18,20]. The physical process of how this happens during crystal growth is not precisely known but we outline, qualitatively, one distinct possibility below.

It is known by several investigations that out-diffusion of Be can occur during growth (dopant segregation) and its amplitude increases with Be concentration [21,22]. These results suggest that significant amounts of Be begin to accumulate on the growth surface when Be concentrations approach the homogeneous doping solubility limit. In addition, there is evidence to support the occurrence of Be dimerization (dopant pairing) on GaAs surfaces at high concentrations where it forms lateral domains locally enriched in Be surface content [23,24]. When GaAs deposition proceeds on surfaces in this state then a component showing a 3-D growth mode can be detected from faceting that is observable in RHEED. Growth appears to consist of a combination of growth modes with the majority component being a 2-D mode that occurs on surface domains largely denuded of Be and a weaker 3-D mode for growth on surface domains enriched in Be. In general, the 3-D growth mode increases with increasing out diffusion of the dopant. Now, when the rate of both Be outdiffusion and Be deposition to the surface are sufficiently high (as is the case for

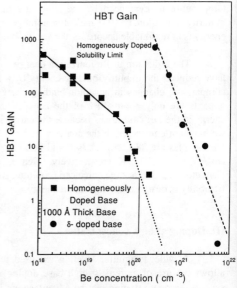

Fig. 6. Gain of an HBT versus Be concentration when either δ -doped or uniformly doped.

homogeneous doping above 10^{20} cm^{-3}) then this 3-D growth mode is amplified. Continued deposition under these conditions would lead to a rough morphology such that the defected areas (from 3D growth modes) contain locally enriched Be precipitates in an doped GaAs matrix.

Examples illustrating this behavior are shown in Fig. 7. They are SIMS profiles of δ-Be near the 0.5 ML level on structures grown at 500° C with either As₄ or AsH₃. The Be profile consist mainly of a central peak that corresponds to profiling though the δ -Be plane. Superimposed on this main Be peak is a shoulder with a forward edge which has a long, linear decay toward the surface of ≈700 Å/decade. Notice that it is not on the profiles of the δ -AlAs plane that were placed 700 Å beneath the δ -Be plane. This shoulder accompanying the Be profile occurs simultaneously with the faceting that was detected in RHEED (up to 700 Å of GaAs growth atop the δ -Be plane). We associate this profile component with 3-D growth mode that occurs for growth atop the δ - Be plane [25]. Our picture of a 3-D mode is also supported by RHEED damping and recovery dynamics of GaAs growth upon initially δ - Be doped surfaces that Iimura et. al. reported [26]. The measurements clearly reveal degradation in RHEED dynamics upon increasing δ - Be content in support of the above surface induced roughening. The profiles of Fig. 7 further reveal that the 3-D component is reduced during arsine growth in comparison to the case of As₄ growth. The improved confinement of δ -Be at the δ -plane as well as the reduction in the 3-D growth mode component for arsine growth appear to arise from a supersaturation of Ga vacancies in the subsurface region during arsine growth. This is because Ga vacancies are inversely proportional to Group V overpressure. Consequently, Ga vacancy formation is increased by use of lower molecular arsenic species derived from arsine [22]. High concentrations of Ga vacancies in the subsurface of the crystal provide more bulk sites for Be incorporation and reduce the surface content of Be. This then leads to a reduction in the 3-D

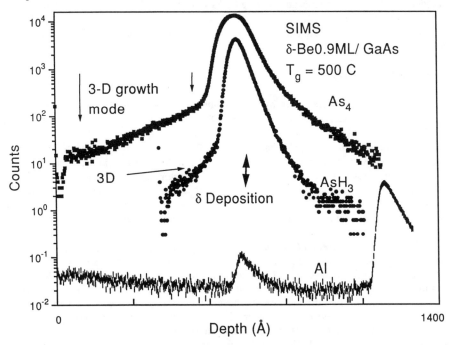

Fig. 7 SIMS from 0.5 ML δ -Be doped GaAs grown with As₄ or AsH₃.

growth component. Notice the 3-D component in the SIMS recovers faster for AsH$_3$ than As$_4$. Low growth temperatures can also reduce the 3-D component because of reduced Group V re-evaporation from the surface [27] and also because the rate of dopant segregation is observed to be smaller [21,22].

Dopants introduced by gaseous methods can also be advantageous to overcome doping problems. For example, InP has been difficult to dope p-type. When InP is doped with diethyl Zn during Chemical Beam Epitaxy (CBE), Tsang et. al [28] achieve higher p-type concentrations over conventional doping techniques. In general, CBE gas doping has improved p-type solubility over Metal Organic Chemical Vapor Deposition (MOCVD) techniques. This could be the result of the dopant, when on the surface, retaining radical bonds with organic groups during CBE [29]. The dopant is therefore unable to pair together to form dimers. This could eliminate the dopant clustering process on the surface and thereby suppress the 3-D growth mode. In MOCVD complete pyrolysis of the dopant occurs in the gas phase and the dopant then reaches the surface in atomic form. Then a dopant might then be subject to similar dimerization processes and therefore MOCVD may encounter a similar solubility problems.

On the basis of Fig. 7 an improved Be solubility can be expected in δ-doping over homogeneous doping. This is because the δ-doping process automatically leads to a very high supersaturation of Ga vacancies which in turn reduces the 3-D growth mode component. In addition, Be deposition is suspended during GaAs growth atop the δ plane and minimizes those conditions leading to amplification of the 3-D growth mode. In fact, results from RHEED dynamics also indicate less degradation occurs on δ - Be doped surfaces than for uniformly doped techniques [26].

V. Large Area Devices

HBT structures were grown by GSMBE methods in which δ– Be concentrations near 1 ML were placed in a GaAs base between 50 to 200 Å in thickness. The emitter layer consisted of a 1000 Å layer of graded AlGaAs x= 0.3 to 0.0 doped n-type to 5 x10^{17} cm sup $^{-3}$, then followed by Si δ-doped GaAs and epitaxial Al metal for non-alloyed ohmic contact. The collector was 5000 Å GaAs and Si doped to 5x10^{16} cm $^{-3}$. The base and emitter components of the structure were grown at 500 C atop a n⁻ -GaAs collector layer on an n- type substrate. A description of selective contacting, processing and device planarization has been reported elsewhere by Goossen and Kuo [30,31]. We show here, systematic dependence of this device structure on the concentration of δ Be doping in the base as it varies from 0.09 to 0.9 ML.

In Fig. 8a we present a Gummel plot of the emitter-base junction, eb, and corresponding collector response for the 0.4 ML device at a 200Å base layer width. From the I-V characteristics of the bc junction the diode ideality factor, n, is n=1.08 and is a general feature of the bc diode quality found in all the remaining structures listed above. The eb junction also appeared perfectly ideal (i.e. n=1.0 to 1.2) for driving currents above 100 microamps. Hence, nearly perfect diffusive transport from injector into the base occurred in our devices under typical HBT operation. At lower currents, the Gummel plots of the eb junction reveals components in the diode I-V having ideality in the range (1.7- 1.9). The contribution of the higher order ideality components to the I-V relation varies considerably with the δ-Be concentration. At the 0.08, 0.9 δ-Be level the n ≠ 1 components are visible below the 100 picoamps, microamp level respectively. This strong δ-Be dependence of the n≠1 components in the eb diode I-V, and also their low temperature characteristics, are indicative of tunneling assisted transport processes in the AlGaAs emitter much like those found in homogeneously doped heterojunction diodes reported previously by Chand [32]. Tunneling assisted processes at the low current level in the eb junction could arise from traps formed in the AlGaAs owing to the low growth temperature employed to confine Be to the base. Nevertheless, our devices work as excellent HBTs at temperatures as low 2 K and hence the number of traps is insufficient to affect the device performance under typical operation.

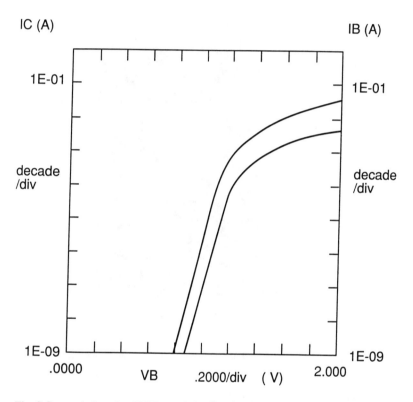

Fig. 8 Gummel plot of an HBT containing δ -Be at 0. 1 ML in the base of an HBT.

An example of the output performance of an HBT containing δ-Be at the 0.4 level is shown in Fig 9. The current gain is 10 but is uniform versus collector current and the device has low output conductance. The gain falls to 5 when the base width is 50 Å because of hole back injection presumably due to hole tunneling out of the δ-Be base. The dependence of the gain on δ-Be concentration is strong but surprisingly systematic. Our measurements give gain (250, 30, 15, 0.10) versus δ-Be concentration (0.08, 0.16, 0.4 and 0.9 ML), respectively. Gain versus δ-Be concentration is plotted in Fig. 6. The 2-D concentration in the base has been converted to a 3-D concentration by taking the ratio of the areal concentration of Be to the Full Width Half Maximum of δ–Be profiles that were measured with SIMS. For comparison, we also show in Fig. 6 the case of gain versus homogeneous doping that was obtained from ref [33]. One distinct feature of the gain versus doping relationship is that δ doping opens up HBT devices to effective doping concentrations well beyond the solubility limit of homogeneous doping. Fig. 6 shows that good gain is maintained in such devices. The gain drops below ten at an effective doping concentration in the upper $10^{21} \mathrm{cm}^{-3}$, nearly two orders of magnitude greater than the solubility limit. Even for gains near ten the sheet resistivity is below 300 Ω per sq. for δ-Be at the 0.4 ML level.

$$I_C = 10\text{mA/DIV}$$

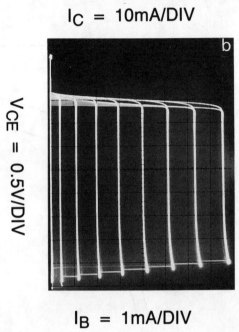

$$V_{CE} = 0.5\text{V/DIV}$$

$$I_B = 1\text{mA/DIV}$$

Fig. 9 Output characteristics of HBT containing an δ -Be at 0.4 ML in the base.

The strong gain reduction in the device at higher δ-Be concentration is indicative of limitations comprising higher order p-type dependences such as Auger processes or hole back injection [33,34]. At 77 K the 0.4 ML device has a current gain of 100 and sheet base resistance of 150 ohms/sq. Thus the δ-Be HBT device appears well suited for high performance applications.

VI. Conclusion

To conclude we have presented two methods to overcome doping problems in compound semiconductors. First, DE is a method to flexibly change the activation energy of a dopant in a semiconductor host. We demonstrated this concept in the AlGaAs/GaAs system and outlined possible applications to other problematic semiconductors. Use of δ–Be in GaAs at the monolayer level was demonstrated to overcome the solid state solubility limit for a uniform dopant. This method however requires careful attention to surface conditions during the δ–doping process.

References

[1]. M. Ilegiums, J. Appl. Phys., 48, 1278 (1977).

[2].T. Mitsuyu, K. Ohkawa, and O. Yamazaki, Appl. Phys. Lett., 49, 1348 (1986).

[3]. J. Qiu, H. Cheng, J. M. DePuydt and M. A. Haase, J. Crystal Growth 127, 286 (1993).

[4]. J. E. Cunningham, T. Y. Kuo, A. Ourmazd, K. Goossen, W. Jan, F. Storz, F. Ren, C. G. Fonstad, J. Crystal Growth 111 515 (1991).

[5]. W. T. Tsang, E. F. Schubert and J. E. Cunningham, Appl. Phys. Lett., 60, 115 (1992).

[6]. K. W. Goossen, J. E. Cunningham, T. Y. Kuo, W. Jan and C. G. Fonstad, Appl. Phys. Lett., 59, 682 (1991).

[7]. G. Bastard, Phys. Rev. B 24, 4714 (1981).

[8]. R. L. Green and K. K. Bajaji, Solid State Commun. 53, 1103 (1985).

[9]. W. T. Masselink, Y. C. Chang, H. Morkoc, D. C. Reynolds, C. W. Litton, K. K. Bajaji, and P. W. Yu solid State Electron. 29, 205 (1986).

[10]. Y. Hishida, H. Isshii, T. Toda, and T. Niina, J. Crystal Growth, 95 517 (1989).

[11]. R. Dingle, H. L. Stormer, A. C. Gossard, and W. Wiegmann, App. Phys. Lett. 37, 805, (1978).

[12]. J. E. Cunningham, T. H. Chiu, A. Ourmazd, W. Jan and T. Y. Kuo, J. Crystal Growth, 105, 111, (1990).

[13]. E. F. Schubert, J. B. Stark, B. Ullich and J. E. Cunningham, Appl. Phys. Lett., 52, 1508 (1988).

[14]. H. Kroemer, Proc. IEEE 70, 13 (1982).

[15]. R. J. Malik, L. M. Lunardi, J. F. Walker and R. W. Ryan, IEEE Electron Device Letters EDL-9, 7 (1988).

[16]. E. F. Schubert J. E. Cunningham, W. T. Tsang Sol. State. Comm. 63, 591 (1987).

[17]. N. Duhamel, P. Henoc, F. Alexandre and E.V.K. Rao, Appl. Phys. Lett. 39, 49 (1981).

[18]. M. Illegems, J. Appl. Phys. 48,1278 (1977).

[19]. J.L. Lievin, C. Dubon-Cheavallier, F. Alexandre, G. Le Roux, J. Dangla and Ankri, IEEE Electron Device Letters, EDL-7 129 (1986).

[20]. T. J. Drummond , W. G. Lyons, R. Fisher, R. E. Thorne, H. Morkoc, C. G. Hopkins and C. A. Evans, Jr., J. Vac. Sci. Technol. 21 957 (1982).

[21]. J. E. Cunningham, M. D. Williams, T. H. Chiu , W. Jan and F. Storz, J. Vac. Sci. And Technol. B. 10, 866 (1992).

[22]. E. F. Schubert, J. M. Kuo, R. Kroff, H. S. Luftmann, L. C. Hopkins, and N. J. Sauer, J. Appl. Phys. 67 1969 (1990).

[23]. J. E. Cunningham, K. W. Goossen, T. H. Chiu, M. D. Williams, W. Jan and F. Storz, Appl. Phys. Lett. 62, 1236 (1993).

[24]. A. Ourmazd, J. Cunningham, W. Jan and J. A. Rentschler, Appl. Phys. Lett. 56 854 (1990).

[25]. J. E. Cunningham, M. Williams, T. H. Chiu, W. Jan, F. Storz and E. Westerwick, J. Crystal Growth, 10, 306 (1992).

[26]. Y. Iimura and Mitsuo Kawabe, J. J. Appl. Phys. 25, L81 (1986).

[27]. R. A. Hamm, D. A. Humphery, R. N. Nottenburg, M. B. Pannish, Y. K. Chen, Appl. Phys. Lett., 54, 2586, (1989).

[28]. W. T. Tsang, F. S. Choa, N. T. Ha, J. Elect. Mat., 20, 541, (1991).

[29]. T. H. Chiu, J. E. Cunningham, A. Robertson and D. Malm, J. Crystal Growth, 105, 155 (1990).

[30]. T. Y. Kuo, K. W. Goossen, J. E. Cunningham, A. Ourmazd, C. G. Fonstad, F. Ren and W. Jan, Electron. Lett. 26, 1187, (1990).

[31]. K. W. Goossen, T. Y. Kuo, J. E. Cunningham, W. Y. Jan, F. Ren and C. G. Fonstad, IEEE-Electron Device Transactions, 38, 2423 (1991).

[32]. N. Chand, R. Fisher, J. Klem and H. Morkoc, J. Vacum Sci. and Technol. B4, 2 (1986).

[33]. S. Tiwari and S. Wright, Appl. Phys. Lett. 56, 563 (1990).

[34]. R. Jalaki, R. N. Notenberg, A. F. Levi, R. A. Hamm, M. B. Pannish, D. Sivco and A. Y. Cho, Appl. Phys. Lett. ,56 , 1460 (1990).

EFFECTS OF ION-IRRADIATION AND HYDROGENATION ON THE DOPING OF InGaAlN ALLOYS

S. J. PEARTON, C. R. ABERNATHY, W. S. HOBSON AND F. REN
AT&T Bell Laboratories, Murray Hill, New Jersey 07974

ABSTRACT

Carrier concentrations in doped InN, $In_{0.37}Ga_{0.63}N$ and $In_{0.75}Al_{0.25}N$ layers are reduced by both F^+ ion implantation to produce resistive material for device isolation, and by exposure to a hydrogen plasma. In the former case, post-implant annealing at $450-500°C$ produces sheet resistances $> 10^6 \Omega/\square$ in initially n^+ ($7 \times 10^{18} - 3 \times 10^{19}$ cm^{-3}) ternary layers and values of $\sim 5 \times 10^3$ $\Omega/\square$ in initially degenerately-doped (4×10^{20} cm^{-3}) InN. The evolution of sheet resistance with post-implant annealing temperature is consistent with the introduction of deep acceptor states by the ion bombardment, and the subsequent removal of these states at temperatures $\leq 500°C$ where the initial carrier concentrations are restored. Hydrogenation of the nitrides at $200°C$ reduces the n-type doping levels by 1-2 orders of magnitude and suggests that unintentional carrier passivation occurring during cool down after epitaxial growth may play a role in determining the apparent doping efficiency in these materials.

III-V nitride alloys are excellent candidates for photonic devices operating in the blue region of the spectrum.[1-5] Tremendous progress has been made in the preparation and characterization of binary (InN, GaN, AlN) and ternary (AlGaN, InGaN) III-V nitrides on a variety of substrates.[6-12] There is still much scope for work on realization of reproducible ohmic contacts to some of these nitrides, and on the growth of p^+ layers particularly in GaN. Mg-doped layers with hole concentrations as high as 3×10^{18} cm^{-3} after low energy electron-beam irradiation have been reported but the as-grown films had much lower (2×10^{15} cm^{-3}) doping levels.[5] Van Vechten et al.[13] have proposed this result can be explained by de-bonding of hydrogen from passivated Mg acceptors, increasing the p-type doping density. The unintentional passivation of dopants in other III-V materials due to residual hydrogen trapped during the growth process is a common phenomenon with gas-source epitaxial techniques.[14-18]

In this letter we report on reversible changes in doping of InN, InGaN and InAlN layers as a result of either ion implantation or hydrogen plasma exposure. Selective area implantation of inert ions is widely used in GaAs-technology for device isolation purposes, since the implanted region may be made highly resistive.[19] This technique will also prove useful in III-V nitride materials due to its ability to retain the planarity of the active element area. It is also necessary to understand the role of hydrogen in determining apparent doping efficiencies in the nitrides, since it is a major component of the metalorganic group III sources used for many of the epitaxial growth techniques.

Mat. Res. Soc. Symp. Proc. Vol. 325. ©**1994 Materials Research Society**

The nitride layers were grown at ~500°C on semi-insulating GaAs substrates using group III metalorganics (trimethylindium, triethylgallium and trimethylamine alane respectively for In, Ga or Al sources) and atomic nitrogen generated by a 200W, 2.45 GHz Electron Cyclotron Resonance (ECR) source (Wavemat MPDR) on an Intevac Gas Source Gen II Molecular Beam Epitaxy system.[20] The growth rates were 50-75Å · min^{-1} and the films were polycrystalline columnar in nature. The InN films were strongly n-type (3.9×10^{20} cm^{-3}, room temperature mobility $\mu = 77$ cm^{2}V^{-1}sec^{-1}) as-grown, and annealing under a N$_2$ ambient up to 450°C did not alter the conductivity level. Similarly, both of the ternary alloys were also n-type (In$_{0.57}$Ga$_{0.63}$N $n = 3.4 \times 10^{19}$ cm^{-3}, $\mu = 12$ cm^{2}V^{-1}sec^{-1}, and In$_{0.75}$Al$_{0.25}$N: $n = 7.3 \times 10^{18}$ cm^{-3}, $\mu = 20$ cm^{2}V^{-1}sec^{-1}) and these levels were unchanged by annealing at 450°C.

Fluorine ions were implanted at multiple energies (40, 100, 200 and 300 keV) and two sets of doses (8, 9, 10 and 20×10^{13} cm^{-2} respectively or 8, 9, 10 and 20×10^{14} cm^{-2} respectively) in order to produce a uniform nuclear stopping damage profile throughout the 0.4 µm thick nitride layers. The implant temperature was ~80°C and the samples were not amorphized by this treatment. These are typical of the isolation schemes used for highly doped GaAs films. Alloyed (400°C, 30 sec) HgIn eutectic contacts for Hall measurements were prepared on some of the samples prior to implantation so that subsequent low temperature (<400°C) anneals could be examined. Other non-contacted samples were post-implant annealed for 60 sec at temperatures up to 600°C in an N$_2$ ambient with the implanted faces covered by low temperature (75°C) plasma-enhanced chemically vapor deposited SiN$_x$. The conductivity of the implanted and annealed layers was obtained from Hall measurements.

Hydrogen plasma exposures were performed at 200°C for 1h using a 0.5 Torr, 13.56 GHz discharge. Some of the samples were coated with 200Å of the low temperature SiN$_x$ prior to plasma treatment in anticipation of possible surface degradation induced by the high hydrogen flux present in this environment. It is well-known that H$_2$ discharges preferentially remove group V species from III-V materials and in the case of InP for example, this can lead to the presence of In droplets on the surface.[21] In all cases the alloyed HgIn Hall contacts were already in place prior to the hydrogen plasma exposure.

Figure 1 shows the evolution with post-implant annealing temperature of the sheet resistance of InN layers implanted with multiple energy F$^+$ ions at the two different sets of doses. For the lower doses ($8 \times 10^{13} - 2 \times 10^{14}$ cm^{-2}) the resistance of the implanted material is increased by about an order of magnitude over its as-grown value, and post-implant annealing reveals a recovery stage centered around 350°C. This behaviour is consistent with creation of deep compensating acceptors by the energetic ions, which trap free electrons and increase the resistance of the InN. The subsequent annealing of these acceptors restores the initial conductivity of the material. At the higher dose condition ($8 \times 10^{14} - 2 \times 10^{15}$ cm^{-2}) the initial increase in sample resistance is greater since more of the free electrons are trapped, and subsequent annealing up to ~450°C actually produces a further increase in resistance. This is the behaviour typically observed in implant-isolated GaAs, and it is commonly ascribed to reduction of hopping probabilities for trapped carriers.[19] Above 450°C enough of the trap sites have been removed that electrons are returned to the conduction band of the InN and the conductivity returns towards its initial, unimplanted value. The maximum sheet resistance obtained is only ~5×10^3 Ω/□, far below the values required for electronic device isolation (>10^6 Ω/□). A wider range of experiments is needed to determine whether this is a function of our particular material or is an intrinsic effect related to the position in the bandgap of the defect states created in the InN by the implantation process. For example, in InP high dose implantation pins the Fermi level in the upper half of the bandgap, leading to limiting resistivities of $10^3 - 10^4$ Ωcm for initially n-type material.[22-24]

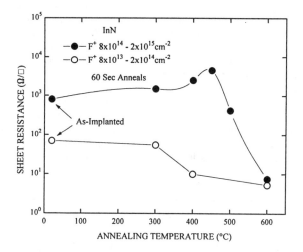

Fig. 1.   Sheet resistance of n^+ InN layers implanted with multiple energy (40, 100, 200 and 300 keV) F^+ ions at two different sets of doses (8, 9, 10 and 20 $\times 10^{13}$ cm^{-2} or 8, 9, 10 and 20 $\times 10^{14}$ cm^{-2}, respectively), as a function of post-implant annealing temperature (60 sec anneals).

Results from F^+ implanted InGaN and InAlN are shown in Fig. 2. In both of these materials the layer resistances after implantation are approximately two orders of magnitude greater than the as-grown values ($1-1.5 \times 10^3$ Ω/□), and they increase to practical values for device isolation after annealing at $450-500°C$. These results are particularly relevant to integration of photonic devices such as lasers or quantum well detectors with electronic devices such as heterojunction bipolar transistors based on III-V nitride heterostructures. It will be interesting to examine whether thermally stable implant isolation can be obtained in this material system by implanting species which create chemical, rather than damage-related, deep level states.

Turning to the effects of hydrogen plasma exposure we did indeed observe preferential loss of nitrogen from unprotected InN surfaces. Figure 3 shows optical micrographs of InN surfaces after 1h, 200°C H_2 plasma treatments. At top the unprotected surface shows a large density of In droplets, whereas hydrogenation through a thin SiN_x encapsulating layer (subsequently removed in a CF_4/O_2 barrel reactor) did not lead to surface deterioration (Fig. 3, bottom). The surfaces of InGaN and InAlN were much more resistant to plasma-induced degradation and did not show the presence of In droplets after the hydrogenation. Changes in the electrical properties of the III-V nitride materials as a result of the hydrogenation are shown in Table 1. The initial doping densities are reduced by 1-2 orders of magnitude after the plasma treatment due to hydrogen passivation of donor impurities. The electron mobilities also increased by factors of 40-80% after hydrogenation indicating true dopant passivation. Heating at 200°C for 1h in H_2 gas, or exposing the samples to a He discharge did not produce measurable changes in the carrier concentrations, directly implicating the atomic hydrogen. The source of the n-type conductivity in III-V nitrides is still controversial, although native defects such as nitrogen vacancies have been suggested.[1,2] Post-hydrogenation annealing at 450°C for 5 min under a N_2 ambient produced substantial reactivation of the passivated donors. True thermal dissociation energies for the hydrogen-door complexes can only be obtained under conditions where retrapping is minimized, such as in the reverse-biased depletion region of a Schottky diode structure.[25] This type of measurement is not possible in the present samples because of their high doping levels. The temperature at which reactivation of passivated donors occurs will be a function of the thermal history, sample geometry, annealing ambient and doping levels. It is likely that in some cases where hydrogen is a component of the growth chemistry, passivation of dopants may be significantly reducing the apparent carrier concentration in the material. Post-growth annealing under a N_2 ambient to remove this hydrogen would be necessary to produce the true doping density.

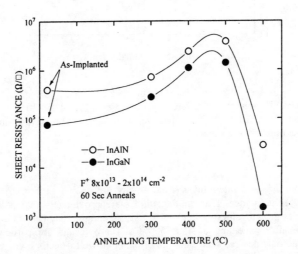

Fig. 2.   Sheet resistances of initially n-type InGaN or InAlN layers implanted with multiple energy (40, 100, 200 and 300 keV) F^+ ions at doses of 8, 9, 10 and 20 × 10^{13} cm^{-2} respectively, as a function of post-implant annealing temperature (60 sec anneals).

Table 1.   Electrical characteristics of InGaAlN alloy layers before and after hydrogenation for 1h at 200°C.

Material	Initial Doping Density (cm^{-3})	After H Plasma (cm^{-3})	After H Plasma and 450°C Anneal (cm^{-3})
InN	3.9×10^{20}	2.5×10^{19}	3.4×10^{20}
InAlN	7.3×10^{18}	6.8×10^{16}	5.4×10^{18}
InGaN	3.4×10^{19}	4.6×10^{18}	2.0×10^{19}

10 μm

Fig. 3. Optical micrographs (1000X magnification) of InN surfaces after direct exposure to a H₂ plasma for 1h at 200°C (top) or after hydrogenation through a 200Å SiNₓ cap which has subsequently been removed (bottom).

In conclusion we have produced reversible changes in the carrier concentration in n^+ InN, InGaN and InAlN thin films using either ion implantation or hydrogenation, followed by annealing in both cases. The implant isolation results show similar characteristics to GaAs in that post-growth annealing is required to produce the maximum sheet resistances in the III-V nitrides. In ternary compounds with initial doping levels $\leq 3 \times 10^{19}$ cm^{-3}, implantation with F^+ ions produces maximum sheet resistances $> 10^6$ $\Omega/\square$, which are practical values for electronic device isolation. We also found direct evidence for efficient hydrogen passivation of n-type conductivity in the nitride films, with reactivation of the passivated dopants being significant for 450°C anneals in a N₂ ambient.

REFERENCES

1. R. F. Davis, Proc. IEEE *79* 702 (1991).

2. S. Strite and H. Morkoc, J. Vac. Sci. Technol. B*10* 1237 (1992).

3. J. I. Pankove, Mat. Res. Soc. Symp. Proc. *162* 515 (1990).

4. I. Akasaki, H. Amano, M. Kito and K. Hiramastsu, J. Lumin. *48/49* 666 (1991).

5. S. Nakamura, M. Senoh and T. Mukai, Jap. J. Appl. Phys. *30* L1708 (1991).

6. R. C. Powell, G. A. Tamasch, Y.-W. Kim, J. A. Thornton and J. E. Greene, Mat. Res. Soc. Symp. Proc. *162* 525 (1990).

7. T. Y. Sheng, Z. Q. Lu and G. J. Collins, Appl. Phys. Lett. *52* 576 (1988).

8. A. Wakahara and A. Yoshida, Appl. Phys. Lett. *54* 709 (1989).

9. T. Lei, M. Faniculli, R. J. Molnar, T. D. Moustakas, R. J. Graham and J. Scanlon, Appl. Phys. Lett. *59* 944 (1991).

10. T. L. Tansley and C. P. Foley, J. Appl. Phys. *60* 2092 (1986).

11. K. Kubata, Y. Kobayashi and K. Fujimoto, J. Appl. Phys. *66* 2984 (1989).

12. M. A. Khan, J. M. Van Hove, J. N. Kuznia and D. T. Olsen, Appl. Phys. Lett. *58* 2408 (1991).

13. J. A. Van Vechten, J. D. Zook, R. D. Horning and B. Goldenberg, Jap. J. Appl. Phys. *31* 3662 (1992).

14. G. R. Antell, A. T. R. Briggs, B. R. Butler, S. A. Kitching, J. P. Stagg, A. Chew and D. E. Sykes, Appl. Phys. Lett. *53* 758 (1988).

15. S. Cole, J. S. Evans, M. J. Harlow, A. W. Nelson and S. Wang, Electronics Lett. *24* 813 (1988).

16. M. Glade, D. Grutzmacher, R. Meyer, E. G. Woelk and P. Balk, Appl. Phys. Lett. *54* 2411 (1989).

17. B. Clerjaud, Physica B*170* 383 (1991).

18. S. J. Pearton, W. S. Hobson and C. R. Abernathy, Appl. Phys. Lett. *61* 1588 (1992).

19. S. J. Pearton, Mat. Sci. Rep. *4* 313 (1990).

20. C. R. Abernathy, P. Wisk, S. J. Pearton and F. Ren, J. Vac. Sci. Technol B (March/April 1993).

21. W. C. Dautremont-Smith, J. Lopata, S. J. Pearton, L. A. Koszi, M. Stavola and V. Swaminathan, J. Appl. Phys. *66* 1993 (1989).

22. J. D. Woodhouse, J. P. Donnelly and G. W. Iseler, Solid State Electron. *31* 13 (1988).

23. M. W. Focht, A. T. Macrander, B. Schwartz and L. C. Feldman, J. Appl. Phys. *55* 3859 (1984).

24. M. V. Rao, R. S. Baba, H. B. Dietrich and P. E. Thompson, J. Appl. Phys. *64* 4755 (1988).

25. T. Zundel and J. Weber, Phys. Rev. B*39* 13549 (1989).

ANNEALING EFFECT ON PHOTOLUMINESCENCE PROPERTIES
OF Be-DOPED MBE GaAs

Hajime Shibata, Yunosuke Makita, and Akimasa Yamada
Electrotechnical Laboratory, 1-1-4 Umezono, Tsukuba 305, JAPAN

ABSTRACT

Low temperature (2° K) photoluminescence (PL) measurements were performed for the wavelength (λ) region from 800 to 1600 nm on Be-doped GaAs grown by molecular beam epitaxy (MBE) for wide range of net hole concentration at room temperature (p) up to 2×10^{20} cm^{-3} after annealing at 850 °C for 20 minutes. The annealing procedure formed several new PL emission bands associated with deep levels appeared between 0.4 and 1.4 eV. The most intensive new PL band appeared at around 1.2 eV with a band width of about 0.2 eV for all samples investigated. Its peak position and band width were observed to shift slightly toward lower energy side and increase, respectively with increasing p. An additional two sharp bands appeared overlapping the main band at around 1.31 and 1.35 eV for p less than 1×10^{18} cm^{-3}. New broad band formation was also observed at around 1.0 eV in the same samples. In addition, a new prominent PL band was found at about 1.35 eV in the as-grown samples with p = $5 \sim 6 \times 10^{17}$ cm^{-3}, which disappeared entirely after annealing. The formation and annihilation mechanism of these deep levels after annealing can be presumably attributed to Be redistribution from Ga sites to As sites and interstitial sites to Ga or As sites, driven by the formation of As vacancies (V$_{As}$) due to As evaporation from samples during annealing.

INTRODUCTION

Systematic study of optical properties such as PL of impurity doped GaAs for a wide range of impurity concentration is important for fabrication of electronic and opto-electronic devices. Particularly, investigation of annealing effects on relaxation of defects introduced during epitaxial growth of heavily doped samples is important to estimate the reliability of devices such as heterojunction bipolar transistors (HBTs) or laser diodes, as well as to know the defect formation mechanism during epitaxial growth.

Recently we found several new strong PL emissions in the near-band-edge emission region (λ = 800~850 nm) for heavily acceptor-doped GaAs [1-7]. The most significant emissions were labeled by [g-g], [g-g]α, [g-g]β and [g-g]γ. Their emission mechanisms have not been clearly understood up to now, although we proposed previously a new concept which was based on radiative recombination due to acceptor-acceptor pair [4,6,7].

Be has been used as an effective p-type dopant for heavily doped GaAs grown by MBE, though its high diffusivity has been pointed out to be the major drawback compared with carbon [8,9]. Optical and electrical properties of Be-doped GaAs have been reported by many authors up to Be concentration as high as 2×10^{20} cm^{-3} [10-16]. In this report, for the investigation of the influence of dopant diffusion and defects relaxation on heavily doped semiconductor optical properties, annealing effects on low temperature PL properties were studied in Be-doped GaAs grown by MBE for wide range of p up to 2×10^{20} cm^{-3}. The results obtained were analyzed based on the formation of V$_{As}$ in samples due to As evaporation during annealing, and resultant Be redistribution from Ga sites to As sites and interstitial sites to Ga or As sites.

Mat. Res. Soc. Symp. Proc. Vol. 325. ©1994 Materials Research Society

EXPERIMENTAL

The samples used in this study were grown by a commercial MBE system (RIBER 2300) on Cr-doped semi-insulating (100) GaAs wafers. For all samples, the substrate temperature and the V/III flux ratio of As₄ to Ga were kept at 550 ℃ and 2.0 respectively. The growth duration of Be-doped layer was 2 hours, followed by the growth of undoped buffer layer for 30 minutes. The total thickness of as-grown layer was about 5.4 μm, which was constant for all p values. The doping level was controlled by changing the Be effusion cell temperature from 417 ℃ to 870 ℃. We also fabricated undoped GaAs layers under the same growth conditions to check the concentration of background contaminants in the MBE chamber. Annealing was done at 850 ℃ for 20 minutes in a liquid phase epitaxy furnace under a purified hydrogen atmosphere at ambient pressure. The PL measurements were performed at 2° K using the 514.5 nm Ar⁺ laser line under conditions of extremely low excitation density (~ 1 W/cm²). The PL signal was analyzed by a 1-m focal length monochromator (SPEX 1704) and detected by a photomultiplier tube with a cooled GaAs photocathode.

RESULTS

Figures 1 and 2, respectively, show typical PL spectra as a function of p in the same samples both as-grown and after annealing with p around 10^{17} cm^{-3}. Strong emissions observed in near-band-edge emission region in both Figs. 1 and 2 were assigned as [g-g], (e-Be°), and their phonon replicas. Appearance of the [g-g] band strongly ensures us that background donor concentration in our samples is extremely low (of the order of $\sim 10^{14}$ cm^{-3}), since it has been demonstrated that the intensity of [g-g] band can be easily quenched by the incorporation of small amount of donor ("optical compensation effect") [2,6,7].

The main feature in Fig. 2 is the appearance of an intensive and wide new PL emission band whose peak position is located at about 1.2 eV with a band width of about 0.2 eV. Since this new band was observed for all samples investigated (up to p = 2×10^{20} cm^{-3}), the appearance of the 1.2 eV band after annealing is a common phenomena for a wide range of p. Additionally with increasing p, its peak position and band width were observed to shift slightly toward the lower energy side and increase in width, which strongly suggests that there is a relation between the formation of the 1.2 eV band and Be incorporation into GaAs.

Additionally in Fig. 2, two intensive and sharp bands were observed to overlap the 1.2 eV band at around 1.31 and 1.35 eV in samples with p less than 1×10^{18} cm^{-3}. Since these two bands were observed in pairs in almost all cases, there may be a significant relation between emission mechanisms of these two bands. A new broad band was observed about 1.0 eV in samples with p less than 1×10^{18} cm^{-3}. This new emission band had a very small intensity but a large band width of around 0.15 eV.

In addition, a new prominent PL band was found in the as-grown samples (in Fig. 1) at around 1.35 eV in samples with p around $5 \times 10^{17} \sim 1 \times 10^{18}$ cm^{-3}, which disappeared entirely after annealing (in Fig. 2). Since this new band was not observed in slightly Be doped samples including undoped one or p-type GaAs doped with any other acceptor species except Be, it is strongly suggested that there is a close relation between the formation of this 1.35 eV band and the incorporation of Be impurity into GaAs. The emission mechanism of the 1.35 eV band in Fig. 1 seems to be completely different from that of the 1.35 eV band in Fig. 2, because their spectral features such as band width differ so much each other. Since the band width of 1.35 eV band in Fig. 1 is much larger than that in Fig. 2, we refer them

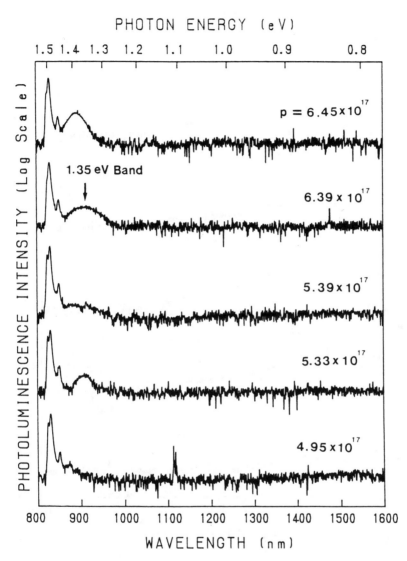

Figure 1. 2° K phptpluminescence spectra of Be-doped
as-grown GaAs at various hole concentrations.

"1.35 eV broad band" and "1.35 eV narrow band", respectively hereafter.

DISCUSSION

Since the LO phonon energy of GaAs is estimated to be around 36 meV, it
is reasonable to assign the 1.31 eV band in Fig. 2 to the LO phonon replica

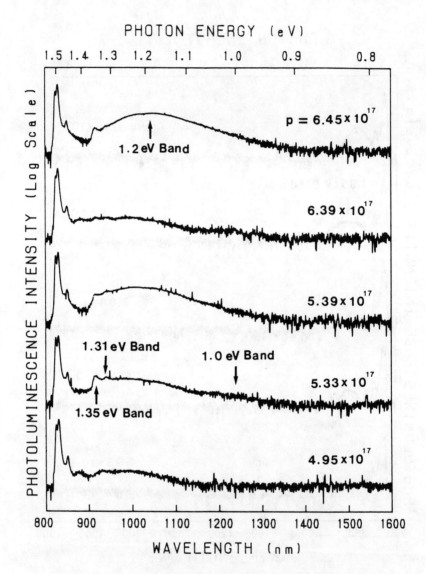

PHOTON ENERGY (eV)

Figure 2.   2° K photoluminescence spectra of the same
samples shown in Fig. 1 after annealing.

of the 1.35 eV narrow band in the same figure, taking into account the relation of emission intensity among them.   The appearance of new emission band
around 1.35~1.36 eV in GaAs after annealing has been previously reported by
several authors using cathodoluminescence [17] and PL (accompanied with a
~1.33   eV emission band) [18,19].   It is strongly suggested that the emission mechanism of the 1.35~1.36 eV band has a close relation with Ga

vacancies (V_{Ga}), in all of these reports. Therefore it seems to be reasonable to estimate here also in this report that the 1.35 eV narrow band in Fig. 2 has a close relation with V_{Ga}.

There have been only a few reports on the appearance of broad band emission around 1.2 eV in GaAs as-grown [20] or annealed [21] samples. The emission mechanism is suggested to have a strong relation with the defect complex of V_{Ga} and a donor in these reports, however the detail has not been clearly understood up to now. Since the background donor concentration in our samples is estimated to be extremely low (less than about 10^{15} cm^{-3}) and the 1.2 eV band observed in Fig. 2 is not observed in as-grown samples at all (in Fig. 1), we can not associate the 1.2 eV band in Fig. 2 with the previously reported 1.2 eV bands [20,21], which were reported to have correlation with the incorporation of donor impurities. Therefore it is reasonable to assign the 1.2 eV band observed in Fig. 2 to a new emission band, which is presumably be related with Be incorporation in conjunction with annealing effects.

There have been no reports on the appearance of the 1.35 eV broad band observed in Fig. 1, which disappears entirely after annealing, for all as-grown GaAs. Therefore the emission mechanism seems to be strongly related with defects introduced through Be incorporation during MBE growth and which disappear after annealing. Appearance of the weak and broad 1.0 eV band after annealing in GaAs (in Fig. 2) was reported in [17], where its emission mechanism was estimated to be a result of the V_{Ga}-V_{As} divacancy.

Since one of the most probable events during annealing in GaAs is the creation of a large number of V_{As}, the generation and relaxation mechanisms of the defects associated with the above deep PL emission bands can be explained as follows. If a large number of V_{As} are created during annealing, Be atoms in Ga sites (Be_{Ga}) can move to As sites and create Be atoms in As sites (Be_{As}), according to the relation ; $Be_{Ga} + V_{As} \rightarrow Be_{As} + V_{Ga}$, leaving a large number of vacancies in Ga sites. If there are Be atoms in interstitial sites, they can also move to either V_{As} sites or V_{Ga} sites which are created through the above mechanism during annealing, and finally occupy either Ga or As sites. Therefore the appearance mechanism of the 1.35 eV narrow band and its LO phonon replica (1.31 eV band) can be understood as a consequence of the formation of V_{Ga}, which is consistent with previous reports [17-19], through above model of Be site transformation from Ga sites to As sites. The appearance of the 1.0 eV band in Fig. 2, which is associated with V_{Ga}-V_{As} divacancy [17], is also consistent with above discussions.

On the basis of the above model, a novel and dominant defect, in addition to V_{Ga}, which can expected to appear after annealing is Be_{As}. Hence the Be_{As} or Be_{As} related defect complex such as Be_{As}-V_{Ga} are expected to result in the appearance of a 1.2 eV emission band seen in Fig. 2. In addition, one of the main defects which are associated with Be and can be be expected to disappear after annealing based on above model are interstitial Be atoms. Therefore it is logical to associate the emission mechanism of the 1.35 eV broad band in Fig. 1 with interstitial Be atoms. Reports on the numerical calculation of the electronic structure of Be doped in GaAs suggest that Be atoms in As sites can create both deep donor and acceptor levels, as well as Be atoms in interstitial sites can create deep donor level [22]. These results seem to support the above speculations of the Be site transformation mechanism. Thus we can conclude that the 1.2 eV band in Fig. 2 is associated with deep donor or acceptor level which might be created by Be_{As} defects, on the other hand, the 1.35 eV broad band in Fig. 1 is likely associated with deep donor level which might be created by Be interstitial atoms.

CONCLUSION

Annealing effects on deep level PL properties in Be doped MBE grown GaAs were studied for a wide range of p concentrations up to 2×10^{20} cm^{-3}. The appearance and disappearance of several strong and new PL emission bands were observed after annealing. The formation and annihilation mechanism of these deep levels after annealing were discussed based on the creation of V_{As} defects due to As evaporation, and resultant generation of V_{Ga}, Be_{As} defects and annihilation of interstitial Be atoms. PL emission bands related to deep impurity levels which presumably are associated with Be_{As} and Be interstitials were observed for the first time. Their photon energies were found to be equal to around 1.2 eV and 1.35 eV, respectively.

REFERENCES

1 Y. Makita, T. Nomura, M. Yokota, T. Matsumori, T. Izumi,
   Y. Takeuchi, and K. Kudo, Appl. Phys. Lett. 47, 623 (1985).
2. T. Nomura, Y. Makita, K. Irie, N. Ohnishi, K. Kudo, H. Tanaka,
   and Y. Mitsuhashi, Appl. Phys. Lett. 48, 1745 (1986).
3. Y. Makita, Y. Takeuchi, N. Ohnishi, T. Nomura, K. Kudo, H. Tanaka,
   H. C. Lee, M. Mori, and Y. Mitsuhashi, Appl. Phys. Lett. 49, 1184 (1986).
4. N. Ohnishi, Y. Makita, M. Mori, K. Irie, Y. Takeuchi, and S. Shigetomi,
   J. Appl. Phys. 62, 1833 (1987).
5. Y. Makita, M. Mori, N. Ohnishi, P. Phelan, T. Taguchi, Y. Sugitama,
   and M. Takano, Mat. Res. Soc. Symp. Proc. 102, 175 (1988).
6. N. Ohnishi, Y. Makita, H. Shibata, A. C. Beye, A. Yamada, and M. Mori,
   Mat. Res. Soc. Symp. Proc. 145, 493 (1989).
7 H. Shibata, Y. Makita, M. Mori, T. Takahashi, A. Yamada, K. M. Mayer,
   N. Ohnishi, and A. C. Beye, Inst. Phys. Conf. Ser. No.106 : Chapter 5
   (Proc. of the 16th International Sym. on GaAs and Related Compounds),
   p.245 (1990).
8. M. Ilegems, in The Technology ond Physics of Molecular Beam Epitaxy,
   edited by E. H. C. Parker (Plenum Press, New York, 1985) p.83.
9. M. Ilegems, J. Appl. Phys. 48, 1278 (1977).
10. K. Ploog, A. Fisher, and K. Kunzel, J. Electrochem. Soc. 128, 400 (1981).
11. G. B. Scott, G. Duggan, P. Wawson, G. Weimann, J. Appl. Phys. 52, 6888
    (1981).
12. N. Duhamel, P. Henoc. F. Alexandre, and E. V. K. Rao, Appl. Phys. Lett.
    39, 49 (1981).
13. P. K. Bhattacharya, H. J. Buhlmann, M. Ilegems, and J. L. Staehli,
    J. Appl. Phys. 65, 6931 (1982).
14. J. D. Wiley, in Semiconductor and Semimetals, Vol.10, edited by
    R. K. Willardson and Beer (Academic Press, New York, 1975) p.91.
15. D. C. Look, in Electrical Characterization of GaAs Materials and Devices
    (John Wily & Sons, 1989).
16. H. Shibata, Y. Makita, A. Yamada, N. Ohnishi, M. Mori, Y. Nakayama,
    A. C. Beye, K. M. Mayer, T. Takahashi, Y. Sugiyama, M. Tacano,
    K. Ishituka, and T. Matsumori, Mat. Res. Soc. Symp. Proc. 163, 121
    (1990).
17. L. L. Chang, L. Esaki, and R. Tsu, Appl. Phys. Lett. 19, 143 (1971).
18. P. K. Chatterjee, K. V. Vaidyanathan, M. S. Durschlag, and
    B. G. Streetman, Solid State Commun. 17, 1421 (1975).
19. W. Y. Lum and H. H. Wieder, J. Appl. Phys. 49, 6187 (1978).
20. E. W. Williams, Phys. Rev. 168, 922 (1968).
21. S. Shigetomi and T. Matsumori, Phys. Stat. Sol. 89, K79 (1985).
22. L. M. R. Scolfaro, E. A. Menezes, C. A. C. Mendonca, J. R. Leite, and
    J. V. M. Martins, Materials Science Forum Vols.65-66 (1990) pp.369.

ORIENTED CARBON PAIR DEFECTS STABILIZED BY HYDROGEN IN *AS-GROWN* GaAs EPITAXIAL LAYERS

Y.M. CHENG[1], M. STAVOLA[1], C.R. ABERNATHY[2], S.J. PEARTON[2]
[1]Physics Department, Lehigh University, Bethlehem, Pennsylvania 18015
[2]AT&T Bell Laboratories, Murray Hill, New Jersey 07974

ABSTRACT

We have studied the IR absorption of heavily carbon doped GaAs grown by metalorganic molecular beam epitaxy. A striking observation is that the hydrogen-stretching vibration of a C-related complex at 2688 cm^{-1} is strongly polarized along just one of the <110> directions in the (001) growth plane. This polarized C-H vibration is assigned to a defect complex that is aligned at the growth surface and then maintains its alignment as it is incorporated into the growing crystal. In a series of experiments, we have studied the annealing of the 2688 cm^{-1} band and its alignment and suggest that the defect complex consists of a C_{As}-C_{As} pair stabilized by hydrogen.

I. INTRODUCTION

The H-stretching vibrations of complexes that contain carbon and hydrogen in epitaxial GaAs have been studied by infrared (IR)[1-9] and Raman[6,10] spectroscopies. For heavily carbon-doped GaAs grown by metalorganic molecular beam epitaxy (MOMBE), H-stretching features for as-grown samples have been observed at 2636, 2643, 2651, and 2688 cm^{-1} (refs. 2-4). The band at 2636 cm^{-1} was assigned by Clerjaud et al.[1] to a C_{As}-H complex with H near the bond center between the C_{As} acceptor and a Ga nearest neighbor. The bands at 2643, 2651, and 2688 cm^{-1} have also been assigned to complexes that contain carbon and hydrogen but their structures are unknown.[2-4] In this paper, we report that the 2688 cm^{-1} band observed in heavily C-doped GaAs grown by MOMBE is strongly polarized along a particular <110> direction in the (001) growth plane. Our results give clues to the structure of the defect complex that gives rise to this absorption band. Annealing results allow us to propose a mechanism for the reorientation of the aligned complex.

II. EXPERIMENTAL PROCEDURES AND RESULTS

We have examined several heavily carbon-doped GaAs epitaxial layers that were grown by MOMBE on semi-insulating GaAs substrates. MOMBE growth was performed in a Varian gas-source Gen II MOMBE system with arsine (AsH$_3$) and trimethylgallium ((CH$_3$)$_3$Ga) source gases and He carrier gas.[11] The arsine gas was introduced through a low pressure Varian cracker to decompose the hydride. The growth temperature was 500°C. IR absorption spectra were measured with a Bomem DA3.16 Fourier transform infrared spectrometer equipped with an InSb detector. Several samples were annealed in H$_2$, D$_2$ or He ambients following growth. These samples were sealed in quartz ampoules with 2/3 atm of the annealing gas and annealed in a muffle furnace for 30 minutes. Following the anneal, the samples were quenched to room temperature in ethylene glycol.

273

A. Strongly polarized absorption band at 2688 cm⁻¹

In Fig. 1, are shown IR absorption spectra for a GaAs:C sample that were measured with light polarized parallel to the two different <110> directions in the (001) growth plane. The sample was in the as-grown state and had an acceptor concentration of 5×10^{20} cm⁻³. Absorption bands are observed at 2636, 2643, 2651, and 2688 cm⁻¹. The band at 2688 cm⁻¹ is strongly polarized along a particular <110> direction, while the other bands have equal intensity for both polarization directions. Samples from four different wafers grown by MOMBE were examined. For all of these samples, the IR absorption band at 2688 cm⁻¹ was found to be preferentially polarized along one of the <110> directions for the (001) growth plane.

It is unexpected to observe polarization effects along a particular <110> direction in the (001) growth plane because these directions are crystallographically equivalent in the bulk of the epilayer. However, similar polarization effects have been observed previously in MBE-grown GaAs epitaxial layers which showed the photoluminescence lines polarized along one of the <110> directions.[12-15] These luminescence lines were attributed to bound exciton recombination at defect pairs oriented parallel to a particular <110> direction.[12,13] The chemical identity of the defect complexes is unknown. We will present evidence later that the polarized absorption spectra we have observed in heavily C-doped GaAs are not due to the same defects observed previously in the luminescence spectra. However, the mechanism for the defect alignment during growth is likely to be similar. Skolnick et al.[15] noted that the <110> directions are not equivalent in the zinc-blende lattice *at the growth surface*. For an As-stabilized (001) surface, the bonds joining a (001) As plane to the underlying plane of Ga atoms are all in just one of the {110} planes, Thus the two <110> directions are distinguished at the growth surface even though they are equivalent in the bulk. Therefore, the defect complexes must be aligned in a particular {110} plane at the growth surface and then maintain their alignment as they are incorporated into the growing crystal. As long as the growth temperature is low enough that the defect does not reorient thereafter, the defect complexes will remain aligned in the as-grown sample.

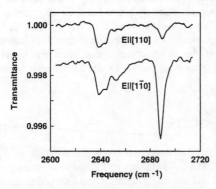

Figure 1. IR absorption spectra, measured near 4.2K, are shown for heavily carbon-doped GaAs grown by MOMBE with $N_A = 5 \times 10^{20}$cm⁻³. The spectra were measured with light polarized along [110] and [1$\bar{1}$0] directions in the (001) growth plane.

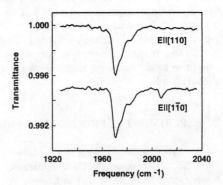

Figure 2. IR absorption spectra, measured near 4.2K, are shown for heavily carbon-doped GaAs with $N_A = 5 \times 10^{20}$cm⁻³ into which deuterium was introduced following growth. The spectra were measured with light polarized along [110] and [1$\bar{1}$0] directions in the (001) growth plane.

An MOMBE-grown sample with $N_A = 5 \times 10^{20}$ cm^{-3} was annealed in D$_2$ gas at 450°C for 30 min. Spectra for this sample are shown in Fig. 2. The bands in the H-stretching spectrum each have a corresponding band in the D-stretching spectrum following the anneal in D$_2$ gas.[3,4] We find that the 2688 cm^{-1} band's deuterium shifted counterpart at 2008 cm^{-1} is also strongly polarized along the same <110> axis as the corresponding H-vibration at 2688 cm^{-1}. The result that a strongly polarized D-stretching band is formed when D is diffused into the crystal following growth leads us to conclude that there is an underlying low symmetry defect that can bind H or D.

B. Annealing stability of the 2688 cm^{-1} center and its alignment

We have annealed samples from the wafer with $N_A = 5 \times 10^{20}$ cm^{-3} to explore the loss of alignment and dissociation of the complex that gives rise to the 2688 cm^{-1} band. One set of samples was annealed in sealed ampoules that contained 2/3 atm of H$_2$. Spectra are shown in Figs. 3a - 3c (solid curves). The intensity of the 2688 cm^{-1} center is not changed from its as-grown state by annealing in hydrogen at 520 °C. However, as the annealing temperature is increased, the preferential polarization of the 2688 cm^{-1} band disappears. Annealing in H$_2$ at temperatures higher than 620°C causes the intensity of the 2688 cm^{-1} center (the average of both polarizations) to decrease as the center is annealed away. These data are collected in Fig. 4 where the intensity and dichroism are plotted for the 2688 cm^{-1} band as a function of annealing temperature. It is seen that the dichroism is annealed away at 600 °C in an H$_2$ ambient and that the 2688 cm^{-1} band disappears at 700 °C.

A second set of samples, annealed in a He ambient, gave different results from those discussed above that were annealed in H$_2$. Polarized absorption spectra are shown in Figs. 3a - 3c (dashed curves) for samples annealed in sealed ampoules that contained 2/3 atm of He. The dichroism for the 2688 cm^{-1} band does not decay. Instead the polarized 2688 cm^{-1} band is annealed away as the temperature is increased without loss of alignment. (The C_{As}-H complex observed at 2636 cm^{-1} is not stable for any of these annealing temporaries in a He ambient.) This decay of the intensity of the 2688 cm^{-1} band upon annealing in the He ambient is also plotted as a function of

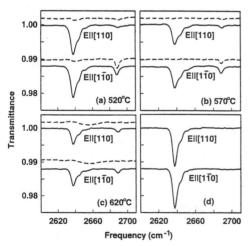

Figure 3. IR absorption spectra, measured near 4.2K, are shown for heavily carbon-doped GaAs grown by MOMBE with $N_A=5 \times 10^{20}$cm^{-3}. The spectra were measured with light polarized along [110] and [1$\bar{1}$0] directions in the (001) growth plane. In (a)-(c), spectra are shown for samples annealed in H$_2$ (solid curves) or He (dashed curves) at the temperatures indicated. In (d) spectra are shown for the sample that had been annealed in He at 620 °C [spectra shown dashed in (c)] and subsequently annealed in H$_2$ at 450 °C.

Figure 4. Annealing data for samples from the wafer with $N_A = 5 \times 10^{20}$ cm^{-3}. The solid triangles and dashed curve ($\blacktriangle$,---) are for the dichroism (right axis) and the open squares and solid curve ($\square$,—) are for the intensity of the 2688 cm^{-1} band for samples annealed in H$_2$. The open circles and solid curve (O, —) are for the intensity of the 2688 cm^{-1} band for samples annealed in He. The crosses and chained curve (+, —·—) are for the

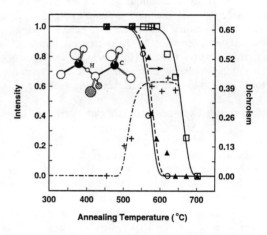

intensity of the 2688 cm^{-1} band for samples that had been annealed at 620 °C in He and were subsequently annealed in H$_2$ at the indicated temperatures for 30 min. In the inset is shown a $(C_{As})_2H$ complex with a bond-centered hydrogen atom between a close pair of C_{As} atoms. The atoms drawn as open circles are Ga atoms and the shaded circles are As atoms.

annealing temperature in Fig. 4. We find that the decay of the intensity of the 2688 cm^{-1} band for samples annealed in He, closely follows the loss of alignment for samples annealed in H$_2$. This suggests that during annealing in H$_2$ there is a dynamic equilibrium in which the 2688 cm^{-1} complexes dissociate and reform elsewhere with random orientation.

To further explore the disappearance of the 2688 cm^{-1} band upon annealing in He, we have reintroduced hydrogen into these samples with a subsequent anneal at 450 °C in H$_2$ gas. Absorption spectra are shown in Fig. 3d for a sample that had been annealed at 620 °C in He and subsequently annealed in H$_2$ at 450 °C. The 2636 cm^{-1} band due to the C_{As}-H center is regenerated by the introduction of H into the sample while the 2688 cm^{-1} band does not reappear. We conclude that the anneal in He not only causes H to dissociate from the 2688 cm^{-1} complex but also eliminates the center to which the H was attached.

The temperature where the center begins to lose its alignment is just above the MOMBE growth temperature of 500 °C. This result shows that the 2688 cm^{-1} center studied here is different from the aligned center previously observed by luminescence measurements for GaAs grown by MBE.[12,13,15] The MBE growth temperature was 620 °C which is higher than the temperature where the 2688 cm^{-1} center loses its alignment.

III. A MODEL FOR THE 2688 cm^{-1} CENTER AND ITS ANNEALING CHARACTERISTICS

A. Concentration of 2688 cm^{-1} centers: a $(C_{As})_2H$ model

The following model for the concentration of 2688 cm^{-1} centers and their annealing behavior is proposed. During growth or annealing in the presence of a source of hydrogen, C_{As}^- acceptors and passivated acceptors, $(C_{As}H)^0$, will be present. Because of the high C_{As} concentrations in our samples, there will be a substantial concentration of $(C_{As}H)^0$ - C_{As}^- next nearest neighbor pairs. If we assume the fraction of C_{As} acceptors that is complexed with hydrogen at the growth temperature is roughly 10% as is suggested by previous results on the passivation of heavily carbon-doped GaAs grown by MOMBE (ref. 3) and that there is a random distribution of $(C_{As}H)^0$ and C_{As}^- centers, then we can estimate the concentration of $(C_{As}H)^0$ - C_{As}^- pairs. The estimate from this model is very close to the concentration of 2688 cm^{-1} centers determined from the strength of the IR absorption.[16] Thus, a $(C_{As})_2H$ center can reasonably account for the concentration of observed 2688 cm^{-1} centers.

If we anneal samples containing $(C_{As}H)^0$ - C_{As}^- pairs in an ambient that does not contain H at temperatures near 600 °C, hydrogen will dissociate from the center and leave the sample. In this case, the Coulomb repulsion that results will cause the C_{As}^- - C_{As}^- pair to separate if the C_{As} atoms are sufficiently mobile. We estimate that the concentration of close pairs would be reduced by a factor of roughly 5 by the increased Coulomb repulsion when H leaves the sample.[16] This reduction of the concentration of close pairs is consistent with our observation that if H is reintroduced into a sample at 450 °C, following an anneal in He that causes the 2688 cm^{-1} band to disappear, then no 2688 cm^{-1} centers are formed even though C_{As}-H centers are created.

From our data we cannot determine where the H would be located in the $(C_{As})_2H$ complex we have proposed. We suggest that it is in a bond-centered position between one of the C_{As} atoms and the Ga atom that lies between the C atoms of the pair as shown in the inset of Fig. 4. In this case, the C atoms would lie along what we have taken to be a $[1\bar{1}0]$ direction which is consistent with the preferential polarization of the absorption at 2688 cm^{-1} along this direction.

B. Loss of alignment and the formation of 2688 cm^{-1} complexes

The loss of alignment of the 2688 cm^{-1} center is explained as follows. When a sample with 2688 cm^{-1} centers that have been aligned during growth is annealed in H_2 at a temperature above ≈ 525 °C, then $(C_{As})_2H$ centers will dissociate and reform elsewhere with random orientation so that an equilibrium in the reactions,

$$[(C_{As})_2H]^- \rightleftarrows (C_{As}H)^0 + (C_{As})^- \rightleftarrows 2(C_{As})^- + H^+ \tag{1}$$

will be maintained. This mechanism for the loss of alignment upon annealing in H_2 requires that if a highly carbon-doped sample, initially with no 2688 cm^{-1} centers, is annealed in H_2 at temperatures above 525 °C, then $(C_{As})_2H$ complexes will be formed. We have annealed several MOMBE grown samples with $N_A = 5 \times 10^{20}$ cm^{-3} in He at 620 °C for 30 min to eliminate the 2688 cm^{-1} center. Subsequently, we annealed these samples in 2/3 atm of H_2 for 30 min. The area of the 2688 cm^{-1} band vs. the annealing temperature is plotted in Fig. 4 . The 2688 cm^{-1} band begins to reappear at an annealing temperature near 500 °C and is not polarized as expected. In the context of our model, this corresponds to the temperature at which C_{As} becomes sufficiently mobile for $(C_{As})_2H$ complexes to be formed. The 30 min annealing treatments performed at 450 °C in H_2 to introduce H into heavily carbon-doped samples are performed at a sufficiently low temperature that carbon is immobile and $(C_{As})_2H$ complexes are not formed. When samples are annealed in H_2 or He above 525 °C, the carbon is sufficiently mobile for pair defects to reform elsewhere in a H_2 ambient or simply to separate in a He ambient.

IV. CONCLUSION

In summary, an absorption band at 2688 cm^{-1} observed in heavily carbon-doped GaAs grown by MOMBE from TMG and AsH$_3$ has been found to be preferentially polarized along a particular <110> direction in the (001) growth plane. Our results suggest that H is bound to a low symmetry defect and that this complex has been aligned at the growth surface and has maintained its alignment as the epilayer was grown. We have studied the annealing of the 2688 cm^{-1} center in H$_2$ and He ambients and have found that the center is annealed away for temperatures above 525 °C in the He ambient whereas in an H$_2$ ambient, it loses its alignment for the same annealing temperatures and is not annealed away until the temperature is increased further. Our results suggest that the 2688 cm^{-1} center is due to a $(C_{As})_2H$ center and that the hydrogen in the center controls the pairing and depairing of the C_{As} atoms.

Acknowledgments

The work performed at Lehigh University was supported by the U.S. Navy Office of Naval Research (Electronics and Solid State Sciences Program) under Contract No. N00014-90-J-1264.

References

1. B. Clerjaud, F. Gendron, M. Krause, and W. Ulrici, Phys. Rev. Lett. **65**, 1800 (1990).
2. D.M. Kozuch, M. Stavola, S.J. Pearton, C.R. Abernathy, and J. Lopata, Appl. Phys. Lett. **57**, 2561 (1990).
3. D.M. Kozuch, M. Stavola, S.J. Pearton, C.R. Abernathy, and W.S. Hobson, J. Appl. Phys. **73**, 3716 (1993).
4. M. Stavola, D.M. Kozuch, C.R. Abernathy and W.S. Hobson, *Advanced III-V Compound Semiconductor Growth, Processing and Devices*, edited by S.J. Pearton, D.K. Sadana, and J.M. Zavada, (MRS, Pittsburgh, 1992), p. 75.
5. K. Woodhouse, R..Newman, R. Nicklin and R..Bradley, J. Cryst. Growth **120**, 323 (1992).
6. J. Wagner, M. Maier, Th. Lauterbach, K.H. Bachem, M. Ashwin, R.C. Newman, K. Woodhouse, R. Nicklin, and R.R. Bradley, Appl. Phys. Lett. **60**, 2546 (1992).
7. K. Woodhouse, R.C. Newman, T.J. de Lyon, J.M. Woodall, G.J. Scilla, and F. Cordone, Semicond. Sci. Technol. **6**, 330 (1991).
8. K. Watanabe and H. Yamazaki, J. Appl. Phys., to be published.
9. R. Rahbi, B. Pajot, J. Chevallier, A. Marbeuf, R.C. Logan, and M. Gavand, J. Appl. Phys. **73**, 1723 (1993).
10. J. Wagner, M. Maier, Th. Lauterbach, K.H. Bachem, A. Fischer, K. Ploog, G. Mörsch and M. Kamp, Phys. Rev. B **45**, 9120 (1992).
11. C.R. Abernathy, S.J. Pearton, F. Ren, W.S. Hobson, T.R. Fullowan, A. Katz, A.S. Jordon, and J. Kovalchick, J. Cryst. Growth **105**, 375 (1990).
12. M.S. Skolnick, D.P. Halliday, and C.W. Tu, Phys. Rev. B **38**, 4165 (1988) and the references contained therein.
13. S. Charbonneau and M.L.W. Thewalt, Phys. Rev. B **41**, 8221 (1990) and the references contained therein.
14. L. Eaves and D.P. Halliday, J. Phys. C: Solid State Phys. **17**, L705 (1984).
15. M.S. Skolnick, T.D. Harris, C.W. Tu, T.M. Brennan, and M.D. Sturge, Appl. Phys. Lett. **46**, 427 (1985).
16. Y. Cheng, M. Stavola, C.R. Abernathy, S.J.Pearton, and W.S.Hobson, Phys. Rev. B (to be published)

HYDROGEN PASSIVATION EFFECTS IN HETEROEPITAXIAL InSb GROWN ON GaAs BY LPMOCVD

BYUENG-SU YOO, SANG-GI KIM and EL-HANG LEE
Electronics and Telecommunications Research Institute
P.O. Box 8, Daeduk Science Town, Daejeon, Korea

ABSTRACT

The effects of hydrogen plasma exposure upon electron Hall mobilities in InSb heteroepitaxial film grown on GaAs substrate have been investigated. After exposure to a hydrogen plasma at 250°C, the electron Hall mobility is significantly increased at low temperatures and the temperature dependence of the mobility is reduced. For the film with a broad x-ray rocking-curve width, 4 h-hydrogen plasma exposure can give rise to the enhancement of the mobility up to 6 times at low temperature. The mobility for the film with a narrow line width is enhanced around 1.5 times. These enhanced mobilities are nearly restored by 350°C rapid thermal annealing. The enhancement of the mobility due to hydrogenation is attributed to the satisfaction of the dangling bonds generated by the misfit dislocations.

INTRODUCTION

InSb heteroepitaxial films grown on wide band-gap materials have received increasing attention as a potential material for completely integrated long-wavelength infrared systems and for high-speed devices because of the narrowest band-gap and the highest electron mobility among the III-V compound semiconductor materials.[1,2] However, the large lattice mismatch between InSb epilayer and the wide band-gap substrates causes the deterioration of the characteristics of the epitaxial films.[3,4] The electron mobility at low temperatures is greatly reduced compared to that of a homoepitaxial film.

The incorporation of atomic hydrogen into semiconductors shows many beneficial effects on both electrical and optical properties. In compound semiconductors, hydrogen deactivates both shallow acceptor and donor impurities and also passivates the electrical activity of many defective and dangling bonds.[5,6]

In this work, we focused on hydrogen incorporation effects on the electrical properties of the InSb heteroepitaxial film grown on GaAs. The changes in the electron Hall mobility and carrier concentration due to hydrogen plasma exposure were investigated.

EXPERIMENTAL

InSb epitaxial layers were grown on semi-insulating (100) GaAs substrate misoriented 2° toward (110) using low-pressure metal-organic chemical vapor deposition (LPMOCVD). The growth system was a vertical reactor, equipped with a resistively heated, high speed rotating 5 in. molybdenum susceptor. Trimethylindium (TMIn) and trimethylantimony (TMSb) were used as indium and antimony sources, respectively. The growth pressure and temperature were 240

Torr and 475°C, respectively. The susceptor rotation speed of 325 RPM was used. These conditions result in nearly specular surface morphologies and highly uniform films over 2 in. wafer at the wide range of the V/III ratios. The details of the growth conditions were presented elsewhere.[7]

In this work, the epitaxial films of 1.5 ~ 2 μm thickness grown at the V/III ratios of 18 and 20 were mostly used. The x-ray diffraction for these films was measured in order to investigate the structural properties. Figure 1 shows the x-ray double crystal rocking curves for (400) diffraction. The full widths at half-maximum (FWHM) for the films of the V/III ratios of 18 and 20 are 470 and 894 arcsec, respectively.

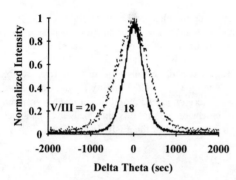

Fig. 1 X-ray double crystal rocking curves for the heteroepitaxial InSb films grown on GaAs. The line widths for the films grown at the V/III ratios of 18 and 20 are 470 and 894 arcsec, respectively.

Hydrogen plasma exposure was performed using capacitively coupled rf plasma system. The hydrogenation temperature was 250°C. The pressure was 0.3 Torr and H_2 flow rate was 42 standard cm^3 per minute. In order to avoid damages induced by the hydrogen plasma, the plasma power density was reduced to 0.05 W/cm^2.

Rapid thermal annealing (RTA) was carried out under a flow of N_2 at a pressure of 0.6 Torr. The Hall mobility and carrier concentration were measured using the van der Pauw Hall geometry. The current and magnetic field used for this measurement were 1 x 10^{-6} A and 5000 G, respectively.

RESULTS AND DISCUSSION

Since the InSb grown on GaAs system has a large lattice mismatch of 14 % between the epilayer and the substrate, the electrical and structural properties of the heteroepitaxial films are deteriorated compared to those of homoepitaxial films. The bulk-like temperature dependence of the mobility is observed at the thicknesses more than 30 μm.[8] Figure 2 shows the 77 K electron Hall mobilities as a function of the dislocation density for 1.5 ~ 2.0 μm thick films grown at various V/III ratios from 15 to 20. Most of the films have nearly specular surface morphologies.

The dislocation density is evaluated from the x-ray rocking-curve width. For this highly mismatched system, the broadening of the rocking-curve width can be determined almost by the mosaic spreading of InSb epitaxial layers. Using the model developed by Gay et al.[9], the dislocation density, N_{dl}, is given by

$$N_{dl} = (F^2-f^2)/9b^2, \qquad (1)$$

where F and f are the FWHM, in radians, of the epilayers and the monochromator crystal, respectively, and b is the magnitude of the Burgers vector. In this evaluation, the smallest Burgers vector, a/2(011), in the face-centered cubic structure produces the upper limit of the dislocation density. Because of the large lattice mismatch, the dislocation density is found to be larger than 10^{14} cm^{-2} at the interface between InSb epilayer and the GaAs substrate and rapidly reduces as the InSb epilayer grows. Thus, the dislocation density evaluated by the x-ray line width is the average value over the whole epilayer thickness.

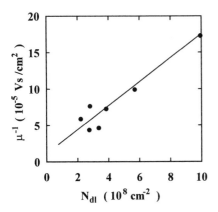

Fig. 2 Inverse 77 K Hall mobilities as a function of the dislocation density (N_{dl}) for the films grown at various V/III ratios. The solid line is a guiding line.

The 77 K electron mobilities are approximately inversely proportional to the dislocation densities, as shown in figure 2. If the electronic transport is limited to the scattering by dislocations and the electron mean free path is larger than the dislocation scattering length, the electron mobility will be inversely proportional to the dislocation densities. This result, therefore, suggests that the electron mobilities near 77 K are dominantly affected by the dislocation-induced scatterings for the InSb heteroepitaxial films of these thicknesses.

Figure 3 shows the temperature dependence of the electron Hall mobilities as a function of hydrogenation time for the film grown at the V/III ratio of 20. After the hydrogen plasma exposure, the electron mobilities are enhanced and the temperature dependence of the mobilities is reduced. At lower temperatures, the mobilities are increased up to 3 and 6 times for 1 and 4 h-hydrogenation, respectively. Exposure to H_2 gas was done in the same apparatus under the same conditions as hydrogenation, except no plasma was excited. The mobility change is negligible for the heat-treated film in H_2. These results indicate that the mobility enhancement by hydrogenation is not due to thermal annealing, but due to atomic hydrogen incorporation effects.

In the previous work, we reported that the temperature dependence of the mobility for this film at low temperatures could be explained by the charged dislocation scattering screened by free carriers.[10] Thus, the increase of the mobility after hydrogenation is attributed to the passivation of the defective and dangling bonds associated with the dislocations generated by the large lattice mismatch. The passivation of the defective and dangling bonds of the dislocations can reduce the charged dislocation scattering. Pearton et al. observed a reduction in electrical activity due to the dislocations by hydrogenation in Ge.[11] They mentioned the reduction of the surrounding lattice strain by the hydrogen binding to the dislocation as one of the possibilities of the deactivation of the dislocations. In this film, however, we found no changes of the x-ray line width by 1 h hydrogenation. Therefore, the strain reduction effect would be excluded for this passivation.

The electronic transport near room temperature is affected by LO phonon, acoustic phonon by deformation potentials and/or neutral dislocation scatterings as well as the charged dislocation scatterings. Accordingly, the enhancement of the mobility by hydrogenation near room temperature is small compared to low temperatures.

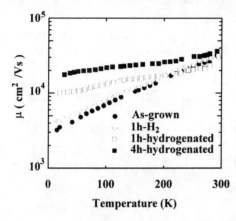

Fig. 3 Temperature dependence of the electron Hall mobilities for the film grown at the V/III ratio of 20. (●) as-grown, ( ) 250℃-annealed with a flow of H_2 for 1 h, ( ) hydrogen plasma-exposured for 1 h and (■) hydrogen plasma-exposured for 4 h.

Figure 4 shows the temperature dependence of the Hall carrier concentrations for the film of the V/III ratio of 20. For as-grown film, the carrier densities are slightly decreased with increasing temperatures and reach a minimum value of 6 x 10^{15} cm^{-3} at 140 K, suggesting that several impurities with different energy levels exist. The intrinsic carrier densities are dominant above 140 K. After hydrogenation, the carrier density at low temperatures is reduced and shows nearly a constant value up to 140 K.

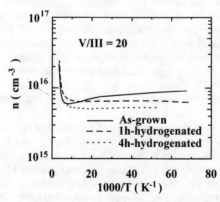

Fig. 4 Hall carrier concentrations as a function of temperature.

In GaAs, hydrogen incorporation is not only able to passivate the deep level defects, but also deactivates both shallow donors and acceptors.[5,6] Thus, the reduction of the carrier density is due to the deactivation of the shallow impurities by atomic hydrogen. And the constant carrier density at low temperatures is attributed to the deactivation of another deep impurities or defects which cause the decrease of the carrier density with increasing temperatures. However, the deactivation of the shallow impurities and the reduction of the impurity scattering can not completely explain the enhancement of the electron mobilities, as shown in figure 3.

Figure 5 shows the mobility of the film grown at the V/III ratio of 18. The film has a better crystal quality and a higher mobility than the film of the V/III ratio of 20, as shown in figures 1, 3 and 5. The temperature dependence of the mobility of this film is smaller than that of the V/III ratio 20 film. The temperature dependence of this film could not be explained by the charged dislocation scattering alone.[10] As the film quality is better, other scattering effects such as impurity and/or neutral dislocation scatterings contribute to the electronic transport. After hydrogenation, the mobility of this film is enhanced at low temperatures, but the amount of the mobility change is smaller than that of the worse quality film. Since the charged dislocation scattering is less dominant, the passivation effect of the dangling bonds could be small in the electronic transport. The change of the room temperature Hall mobility is negligible, because LO phonon, neutral dislocation and/or acoustic phonon by deformation potentials scatterings dominantly influence to the electronic transport at this temperature.

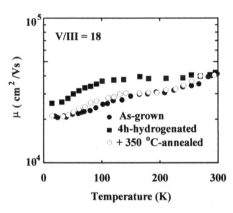

Fig. 5 Temperature dependence of the electron Hall mobilities for the film grown at the V/III ratio of 18. (●) as-grown, (■) hydrogen plasma-exposed for 4 h and (○) 350°C rapid thermal-annealed for 5 min after hydrogen plasma exposure for 4 h.

The enhanced mobility by hydrogenation is nearly completely restored via 350°C-RTA for 5 min. The carrier density of the hydrogenated film is also completely recovered to the value of as-grown film. The electrical activity of most passivated deep levels for GaAs and Si can be restored by thermal annealing above 400°C. In most of semiconductors, the wider is the band-gap of the material, the more thermally stable is hydrogen passivation of defects.[5] Since the band-gap of InSb is small, the passivated defective and dangling bonds could be easily restored.

In this work, we observed hydrogen passivation effects on electrical properties for the highly mismatched system of the InSb on GaAs. These results would be applicable to the passivation of another highly mismatched systems. According to Podor's calculation[12], the mobility enhancement of factor 6 for the V/III ratio 20 film indicates the passivation of the dangling bonds to factor 0.4. But this estimation comes from the assumption that the electron Hall mobility is limited only to the charged dislocation scattering and the changes of the mobility by hydrogenation are only due to the passivation of the dangling bonds generated by the dislocations.

The optical transmission spectrum at room temperature was measured using a Fourier transform spectrometer. However, no changes of the transmission spectrum by hydrogenation were observed. Further experiments using luminescence, which is sensitive to the defect density, are necessary in order to detect the quantitative passivation effects on optical properties.

SUMMARY

In summary, we investigated hydrogen passivation effects on the electronic transport for the InSb heteroepitaxial layers grown on GaAs by LPMOCVD. After hydrogenation, the electron Hall mobility is enhanced at low temperatures. The lower is the crystal quality of the film, the larger is the enhancement of the electron mobility. These electron mobility enhancements and carrier density changes are completely restored to as-grown values by 350°C RTA. The mobility is significantly enhanced at low temperatures where the charged dislocation scattering mainly affects the electronic transport, indicating that hydrogen passivates the dangling bonds of the dislocations generated by the large lattice mismatch.

ACKNOWLEDGMENT

We would like to thank J.-Y. Yi for helpful discussions and K.-Y. Lee for technical assistance. This work was supported by the Korea Telecom.

REFERENCES

1. R.M. Biefeld, J. Cryst. Growth, **75**, 255(1986).
2. P.E. Thompson, J.L. Davis, J. Waterman, R.J. Wagner, D. Gammon, D.K. Gaskill and R. Stahlbush, J. Appl. Phys. **69**, 7166(1991).
3. R.M. Biefeld and G.A. Hebner, J. Cryst. Growth, **109**, 272(1991).
4. Y. Iwamura and N. Watanabe, Jpn. J. Appl. Phys. **31**, L68(1992).
5. S.J. Pearton, J.W. Corbett and T.S. Shi, Appl. Phys. A, **43**, 153(1987).
6. N.M. Johnson, R.D. Burnham, R.A. Street and R.L. Thornton, Phys. Rev. B, **33**, 1102(1986).
7. M.A. McKee, B.-S. Yoo and R.A. Stall, J. Cryst. Growth, **124**, 286(1992).
8. A.J. Noreika, J. Greggi, Jr., W.J. Takei and M.H. Francombe, J. Vac. Sci. Technol. A, **1**, 558(1983).
9. J. Auleytner, X-ray Methods in the Study of Single Crystals (PWN-Polish Scientific Publishers, Warszawa, 1967) p. 152
10. B.-S. Yoo, M.A. McKee, S.-G. Kim and E.-H. Lee, Solid State Commun. **88**, 447 (1993).
11. S.J. Pearton and J.M. Kahn, Phys. Stat. Sol. A, **78**, K65(1983).
12. W. Zawadzki, in Handbook on Semiconductors, Vol. 1, edited by W. Paul (North-Holland, New York, 1982) p. 713.

DIPOLE RELAXATION CURRENT IN N-TYPE $Al_xGa_{1-x}As$

LUIS V.A. SCALVI*, L. OLIVEIRA** AND M. SIU LI**
*Departamento de Física, UNESP - Bauru, Cx. Postal 473, 17033 Bauru - SP, Brazil
**Departamento de Física e Ciência dos Materiais, Instituto de Física e Química de São Carlos - USP, Cx. Postal 369, 13560 São Carlos - SP, Brazil

ABSTRACT

Thermally stimulated depolarization current (TSDC) spectra are reported for $Al_xGa_{1-x}As$ alloys with direct and indirect bandgap for the first time. The experimental data reveal the presence of electric dipoles in both samples which are interpreted by using the negative-U DX center model. These data are fitted by a relaxation time distribution approach yielding average activation energies in close coincidence with the DX center binding energy value.

INTRODUCTION

In the Chadi and Chang's negative-U model for the DX center [1], the substitutional impurity traps two electrons concomitant with lattice distortion along the <111> direction becoming located at an interstitial site. Then, the DX center ground state is negatively charged (DX^- state) and charge balance assures that the same amount of d^+ centers (substitutional sites) are generated in the $Al_xGa_{1-x}As$ sample, suggesting the formation of electric dipoles. Although Chadi and Chang's model has been accepted by most DX center researchers, Maude et al [2,3] have used mobility data to support a positive-U model. O'Reilly[4] has argued that mobility can be better fitted by a negative-U model where d^+ and DX^- center are strongly correlated and a dipole-like picture is used to describe scattering by DX^- - d^+ pairs. If impurity donors are randomly distributed in the AlGaAs sample, the probability that d^+ and DX^- to be first neighbors is negligibly low. However, the extra electrostatic energy gained by placing a d^+ center close to a DX^- center is of significant magnitude at a large fraction of sites [4]. Then, trapout occurs such that formation of DX^- centers close to d^+ centers is energetically very favorable.

In this paper we report experimental evidence for the existence of electric dipoles in n-type $Al_xGa_{1-x}As$ of direct and indirect bandgap, which are fitted by a Havriliak-Negami relaxation time distribution approach [5]. The average activation energy for dipoles

Mat. Res. Soc. Symp. Proc. Vol. 325. ©1994 Materials Research Society

reorientation has approximately the same value of the DX center binding energy, which represents the most striking feature of our data.

EXPERIMENTAL PROCEDURE

We have used Si-doped molecular beam epitaxy (MBE) grown $Al_xGa_{1-x}As$ samples, 2 μm thick with the same structure as proposed by Chand et al [6], which has been published elsewhere [7]. The direct bandgap sample used is doped with $\cong 1x10^{18}$ cm^{-3} (sample I) and the indirect bandgap sample is doped with $\cong 5x10^{17}$ cm^{-3} (sample II). Aluminum composition of sample I is $x \cong 0.32$ and sample II, $x \cong 0.50$.

Thermally stimulated depolarization current (TSDC) is carried out as follows [7]: the sample is biased at room temperature (2.5 V) in darkness and cooled down to liquid He temperature, where the applied bias is removed. Subsequently the temperature is increased at a fixed rate, always in darkness, and the current is measured with the help of an electrometer and a recorder. No voltage is applied on the sample during the increase of temperature and no light is irradiated on the sample during the whole measurement.

TSDC RESULTS AND SIMULATION

Figure 1 shows the experimental TSDC spectrum for sample I. The temperature is increased at a rate of 0.081 K/sec and the measured thermally stimulated current presents a peak at 39 K. Considering that there is no light on the sample, no electron is photoexcited to the conduction band. The current seen in figure 1 can not be associated to an electronic current, which is evident by its order of magnitude, since the photoinduced current obtained for this sample is about 5 orders of magnitude higher than the TSDC current [7]. There is no known process which could be responsible for releasing few electrons and trapping them back at 39 K in the dark, since there is not enough thermal energy to overcome the DX center thermal emission barrier. We believe that the current peak seen in figure 1 is due to dipole reorientation. When the sample is biased at room temperature, the dipoles are oriented according to the electric field direction. Then the applied bias is removed at the lowest temperature. Increasing the temperature the oriented dipoles are allowed to reorient to their original positions, randomly distributed throughout the sample. Then an ionic dipole current is observed as shown in figure 1.

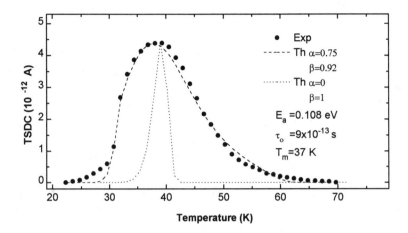

Figure 1. Thermally stimulated depolarization current of Si doped $Al_{0.32}Ga_{0.68}As$.

TSDC experimental curve is usually fitted by a single curve, showing Debye behavior [8,9], with a single relaxation time and activation energy. However our data are better fitted by an asymmetric relaxation time distribution, the type of distribution first proposed by Havriliak and Negami [5]. In this case TSDC current density is given by:

$$J(T) = \frac{Q_p}{\tau_o} \int_{-\infty}^{\infty} f(u) \exp\left\{-\left[\left(u + \frac{E_a}{kT}\right) + \frac{kT^2}{\tau_o bE_a} \exp\left[-\left(u + \frac{E_a}{kT}\right)\right]\right]\right\} \qquad (1)$$

Where Q_p is the polarization charge (area under the experimental curve), τ_o is the relaxation constant time from Arrhenius formula, k is the Boltzmann constant, b is the heating rate and E_a is the activation energy. The distribution function is given by:

$$f(u) = \frac{1}{\pi} \text{sen}\, \beta\theta \left\{1 + 2\cos(1-\alpha)\exp\left[-u(1-\alpha) + \exp\left[-2u(1-\alpha)\right]\right]\right\}^{-\frac{\beta}{2}} \qquad (2)$$

where: $$\tan(\theta) = \frac{\operatorname{sen} \pi(1-\alpha)}{\left\{ \exp[u(1-\alpha)] + \cos \pi(1-\alpha) \right\}} \qquad \text{and} \qquad u = \frac{E - E_a}{kT} \qquad (3)$$

The best fit to TSDC experimental curve is obtained with the continuous distribution of activation energy given by equation 2. The average activation energy is about 0.108 eV and the constant relaxation time is 9×10^{-13} sec, with a peak at 37 K. It is shown in figure 1 - dashed line. Debye single relaxation time ($\alpha = 0$, $\beta = 1$) gives a curve similar to the dotted line, using the same parameters as the average ones used for the distribution we get a curve with the same shape as the dotted line, but with a peak a 41 K. In order to obtain the dotted line of figure 1 we use a single constant relaxation time of 2.3×10^{-13} sec with the same activation energy ($E_a = 0.108$ eV) and get a peak at 39 K. As it can be seen the results are poor compared to the one obtained with Havriliak-Negami relaxation time distribution approach. Experimental data can also be fitted with single relaxation time approach. However it yields quite unreasonable parameters and must be disregarded. The obtained activation energy has the same value of the DX center binding energy what is the most striking feature of our data, suggesting a mechanism of electron transfer involving the DX center, which needs further investigation to confirm it.

TSDC spectrum is also reported for indirect bandgap material (sample II) as shown in figure 2 (solid circle). The heating rate is 0.076 K/sec and the current peak is at 46 K. Simulating the curves according to the relaxation time distribution approach (dashed line) yields an average activation energy of 0.124 eV and constant relaxation time of 9.7×10^{-13} sec.

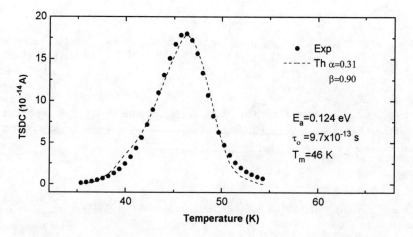

Figure 2. Thermally stimulated depolarization current of Si doped $Al_{0.5}Ga_{0.5}As$.

The data reported here are strong evidences that the electron freezeout at DX centers, which occurs when temperature is lowered down, leads to charged localized states, since neutral impurities would not be responsible for ionic dipole current. In Chadi and Chang's model there are three first neighbors equivalent positions for a d^+ center close to a DX^- center. Although the probability that a d^+ and a DX^- centers to be first neighbors is negligibly small, O'Reilly [4] has argued that for a large fraction of sites the d^+ center can be considered as a perturbation in the DX screening potential and then, the DX^- - d^+ pair has to be treated as a dipole-like picture. A strong point to support the hypothesis that the observed current comes from the relaxation of DX^- - d^+ pairs is the fact that no single relaxation time approach can be used to fit satisfactorily the experimental data. This is consistent with other points such as the random distribution of d^+ center around the DX^- charged impurity (generation of a dipole length distribution) and the random distribution of Al atoms around both charge states which should contribute to the local field potential.

Although there is persistent photoconductivity (PPC) in indirect bandgap $Al_xGa_{1-x}As$ in the range 80-100 K [10], the order of magnitude of the photoinduced current is lower than in direct bandgap material. Below approximately 60 K the PPC vanishes, because the hydrogenic state associated with the X valley is deep and capture electrons metastably [11,12]. There is no reason to believe that there is a barrier for electron capture by this X valley effective mass state, since there is no lattice relaxation involved. Then, the obtained TSDC current shown in figure 2 is apparently contradictory with our DX^- - d^+ dipole hypothesis since vanishing of PPC in indirect bandgap AlGaAs should rule out this possibility, because the negatively charged DX ground state would be destroyed. However the way TSDC is carried out, support O'Reilly hypothesis because there is no light irradiated on the sample during the whole measurement. Thus, when the temperature is lowered, DX^- state is generated in the range 100 -80 K, and at lower temperature the electrons remain trapped at this negatively charged state since there is no light to excite electrons from the DX center to the effective mass X valley state. The lower magnitude of TSDC signal for sample II compared to sample I is also consistent with this hypothesis, since the number of generated DX^- state above 60 K is lower in indirect bandgap material. The obtained average activation energy is also in fair agreement with the DX center binding energy [6].

CONCLUSION

Thermally stimulated depolarization current (TSDC) measurements clearly evidenced the presence of relaxing electric dipoles in n-type (Si doped) $Al_xGa_{1-x}As$ at low temperature in the dark. Experimental results are interpreted under O'Reilly's dipole scattering model. Experiments combining monochromatic light and dipole relaxation current are under investigation and will be fundamental to clearly determine whether observed currents come from DX^- - d^+ pairs or any other kind of dipole. The most striking feature of our data is that the obtained activation energy has a very close value to the DX center binding energy. A model to explain these features has also been worked out and shall be published opportunely.

ACKNOWLEDGMENTS

These samples were grown at Advanced Materials Laboratory of Oregon State University. One of us (Luis V. A. Scalvi) thanks Prof. John R. Arthur for the hospitality during his stay at O.S.U. We also acknowledge the partial financial support from brazilian agencies: Fapesp, CNPq, Finep and Capes.

REFERENCES

[1] D.J. Chadi and K.J. Chang, Phys. Rev. Lett. **61**, 873 (1988); Phys Rev. B **39**, 10063 (1989)

[2] D.K. Maude, L. Eaves, T.J. Foster and J.C. Portal, Phys Rev. Lett. **62**, 1922 (1989)

[3] D.K. Maude, J.C. Portal, L. Dmowski, T.J. Foster, L. Eaves, M. Nathan, M. Heiblum, J.J. Havris and R.B. Beal, Phys. Rev. Lett. **55**, 815 (1987)

[4] E.P. O'Reilly, Appl. Phys. Lett. **55**, 1409 (1989)

[5] S. Havriliak and Negami, Polymer **8**, 161 (1967)

[6] N. Chand, T. Hendersen, J. Klem, W. T. Masselink, R. Fisch, Y.C. Chang and H. Morkoç, Phys. Rev. B **30**, 4481 (1984)

[7] L.V.A. Scalvi, L. Oliveira, E. Minami and M. Siu Li, Appl. Phys. Lett. **63**, (in press)

[8] L. Oliveira, P. Magna, H.J.H. Gallo, E.C. Domenicucci and M. Siu Li, phys. stat. solidi (b) **171**, 141 (1992)

[9] W. van Weperen and H.W. den Hartog, Phys. Rev. B. **18**, 2857 (1978)

[10] L.V.A. Scalvi and E. Minami, phys. stat. solidi (a) **139**, 145 (1993)

[11] J.E. Dmochowski , L. Dobackewski, J.M. Langer and W. Jantsch, Phys. Rev. B **40**, 9671 (1989)

[12] L. Dobaczewski and P. Kaczor, Phys. Rev. B **44**, 8621 (1991)

PART V

Impurities in Semiconductors

OXYGEN DOPING OF GaAs DURING OMVPE
CONTROLLED INTRODUCTION OF IMPURITY COMPLEXES

Y. PARK*, M. SKOWRONSKI*, T.S. ROSSEEL**, AND M.O. MANASREH***
*Department of Materials Science and Engineering, Carnegie Mellon University, Pittsburgh PA 15213
**Oak Ridge National Laboratory, Oak Ridge TN 37831
***Wright Laboratory , WL/ELRA, WPAFB OH 45433

ABSTRACT

GaAs epilayers have been grown by Organo-Metallic Vapor Phase Epitaxy using dimethylaluminum methoxide as a dopant source. This compound contains a strong aluminum-oxygen bond which is thought to remain intact during low temperature deposition and result in the incorporation of Al-O as a complex. Incorporation of aluminum and oxygen was investigated by Secondary Ion Mass Spectroscopy as a function of growth conditions: growth temperature, growth rate, V/III ratio, reactor pressure and dopant mole fraction. High doping levels up to 10^{20} cm^{-3} (for both oxygen and aluminum) were achieved without degradation of surface morphology and/or precipitation of a second phase. Oxygen concentration is lower than that of aluminum for all investigated growth conditions but at low deposition temperatures oxygen/aluminum ratios approach 1, indicating that Al-O is incorporated as a pair. Infrared absorption measurements in the 600-1200 cm^{-1} range did not detect well known isolated oxygen localized vibrational modes (LVM). Also in layers grown at low temperatures the intensity of isolated aluminum LVM at 362 cm^{-1} is much smaller than the concentration obtained by SIMS. Both observations prove that oxygen not only is incorporated as an Al-O pair but remains bonded in the bulk of the layer. Low temperature photoluminescence measurements indicate that the Al-O complex is electrically active in GaAs, forms a deep level within the GaAs band gap, and serves as an efficient non-radiative recombination center. Near band edge luminescence intensity correlates well incorporation of oxygen. The Al-O pairs act as deep acceptors in GaAs and cause the compensation of shallow tellurium donors.

INTRODUCTION

The original goal of this investigation was twofold. First, it was believed for some time that starting aluminum compounds in OMVPE can be a source of serious oxygen contamination of aluminum alloys.[1-3] Aluminum and oxygen form an extraordinarily strong bond and any traces of oxygen in the growth chamber are likely to result in oxygen incorporation into the Al$_x$Ga$_{1-x}$As epilayer. To make things worse oxygen is known to form efficient non-radiative recombination centers in Al$_x$Ga$_{1-x}$As and acts as "luminescence killer".[4,5] For decades researchers have sought ways to reduce the oxygen level in arsine (by using gettering baffles[6], molecular sieves and/or eutectic Al-Ga-In bubblers[7]) and to eliminate leaks in OMVPE systems. After years of effort there was much improvement but still the optimum growth temperatures for Al$_x$Ga$_{1-x}$As are in the 700 ^{o}C range and are much higher than typical deposition temperatures of GaAs. This is the result of oxygen contamination which is known to incorporate more easily at lower growth temperatures. The most serious contamination source today are the traces of alkoxides in aluminum alkyls. Since dimethylaluminum methoxide and diethylaluminum ethoxide are fairly similar to corresponding alkyls in terms of physical properties they are extremely difficult to remove from sources used in OMVPE. Intentional doping with alkoxides as studied here can lead to better understanding of the oxygen chemistry and to purer sources.

Mat. Res. Soc. Symp. Proc. Vol. 325. ©1994 Materials Research Society

The second reason for interest in this topic is the possibility of using oxygen as a compensating defect in certain electronic devices. Residual oxygen has been occasionally used in order to produce a high resistivity $Al_xGa_{1-x}As$ by several groups.[8-10] However, these early attempts could not be incorporated into a reliable processing sequence because of the lack of control over oxygen content. In fact initial intentional oxygen doping experiments failed to introduce significant concentration of oxygen into the GaAs matrix.[11,12] The breakthrough in oxygen doping was achieved by an IBM group using dimethylaluminum methoxide as a doping compound.[13,14] This approach showed early promise in terms of compensation of silicon donors but a number of important issues remained unresolved. At this time it is not clear if the compensation is of a chemical (oxygen reacting with silicon on the growth surface) or electrical nature. We do not know the ionization energy of the dominant traps induced by oxygen doping and their acceptor or donor character. Also the atomic structure of defects and their thermal stability remain unknown.

In this paper we briefly review the data on growth chemistry and incorporation of oxygen into GaAs. In addition, insights into both the atomic structure of oxygen-related centers and their effect on layer properties are discussed. Possible device applications will also be mentioned.

Of particular importance are the initial results concerning the atomic structure of oxygen-related centers. It appears that the aluminum and oxygen are present in the GaAs matrix not in the form of isolated dopants but, at least at low growth temperatures, they remain closely bonded together in the volume of layer. In other words they form a complex. Apparently, alkoxides are not only a convenient agent for delivering aluminum and oxygen atoms to the growing crystal but the structure of the starting compound affects the final atomic structure of the center. This finding opens up interesting possibilities. Namely, it demonstrates that it is possible to design and create the multi-atom defect centers during the growth process. By selecting a dopant molecule and conducting the growth process at low temperatures some fragments of this molecule can be incorporated as complex defects. In principle, given the advances of a defect theory and atomic scale modeling we can predict properties of different complexes and than create the desired center. In some cases it should be possible to predict a center's properties based on currently available experimental data. The Al-O center appears to be one such complex. Oxygen, being one of the most electronegative elements, is capable of attracting extra electron and forms deep acceptor levels in most of semiconductors. As argued below, it retains this ability while bonded to aluminum in the GaAs matrix. Aluminum on the other hand provides a center with an incorporation route during growth and lowers its diffusion coefficient which explains high thermal stability of material.

EXPERIMENTAL PROCEDURES

The layers were grown by Organometallic Vapor Phase Epitaxy (OMVPE) in a horizontal reactor heated by infrared radiation. Substrates were (100) Cr-doped semi-insulating GaAs wafers misoriented 2° toward (110). Substrates were used as-received without cleaning or etching prior to growth and the native oxide layer was removed by thermal desorption at 750 ^{o}C for 10 minutes in an arsenic overpressure. Trimethylgallium (TMG), tertiarybutylarsine (TBA), and dimethylaluminum methoxide ($(CH_3)_2AlOCH_3$, purchased from Advanced Technology Materials Inc. (Danbury, CT)) were employed as starting compounds. In both liquid and gas phase (at low temperatures) DMAlMO forms a trimer with the ring structure[15,16] and has a melting temperature of 36 ^{o}C. TMG, TBA, and DMAlMO bubblers were maintained at -15 ^{o}C, 2 ^{o}C and 38 ^{o}C, respectively.

The oxygen and aluminum concentrations have been measured by Secondary Ion Mass Spectroscopy (SIMS) with the Cameca ims 4f microprobe. Because the samples charged to varying degrees depending upon concentration of dopants, data was collected using the auto-voltage mode. For all measurements, the primary ion beam was rastered over a 200x200 micron area and pressure in the sample analysis chamber was less than 3×10^{-8} Torr. The mass resolution was approximately 600 and the energy slit was typically set to 50 volts. Depth profiles of aluminum were collected using an 8 keV O_2^+ primary ion beam and a current ranging from 80 to 120 nA. This corresponds to a sputter rate of approximately 0.3 to 0.5 nm/s. Oxygen depth profiles were collected using a 14.5 keV Cs^+ primary ion beam and currents ranging from 30 to 50 nA. This corresponds to sputter rates of 0.5 to 1 nm/s. Samples were heated in the evacuated (5×10^{-7} Torr) dual-sample inlet system for approximately one hour prior to oxygen analysis to reduce surface moisture. The depth profiles were converted to concentration using tabulated sensitivity factors (RFS), referenced to either As or Ga. Accuracy is estimated to within a factor of two.

Photoluminescence measurements were performed with samples immersed in liquid helium (4.2 K). Excitation was provided by the 488 nm line of an Ar^+ laser and photoluminescence was analyzed by SPEX 1404 0.85 m double grating monochromator and detected by either S20 photomultiplier or North Coast liquid N_2-cooled Ge. Localized Vibrational Mode absorption measurements have been performed using Bomem DA3 Fourier Transform Infrared spectrometer with resolution of 0.1 cm^{-1}. Samples were mounted at the cold finger of continuous flow cryostat cooled with the flow of liquid helium. C-V profiles were obtained using Polaron profiler employing electrolytic etch.

RESULTS AND DISCUSSION.

Specular morphologies were obtained for GaAs grown between 500 °C and 700 °C and DMAlMO fractions between . Only for heavily doped layers (DMAlMO mole fraction in excess of 3×10^{-7}) and high temperatures the surfaces became hazy possibly indicating formation of the second phase. None of the growth experiments indicated existence of vapor phase reactions or long term changes inside the DMAlMO bubbler. One bubbler was in use in our laboratory now for about two years and we have not observed any changes in the vapor pressure which could be indicative of any kind of instability such as association reaction or disproportionation into alkyls and alkoxides.

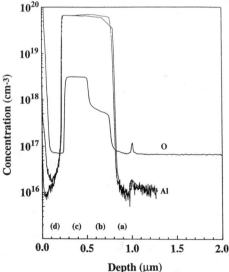

Fig. 1 Typical oxygen concentration profile obtained by SIMS.

A typical structure used in SIMS analysis was composed of several (4-6) layers of GaAs grown in different conditions such as DMAlMO mole fraction, growth rate, deposition temperature, V / III ratio, and reactor pressure. This method allowed for observations of relative changes of aluminum and oxygen concentration induced by change in growth conditions and eliminated the error of absolute concentration determination. A representative SIMS profile is shown in Fig. 1.

Two observations are apparent from the figure above. First, the concentration of intentionally introduced oxygen can be as high as 3×10^{18} cm^{-3} and is orders of magnitude higher than what was possible with either gaseous oxygen or water vapor doping. In fact the highest oxygen content achieved so far without degradation of the surface morphology was 8×10^{19} cm^{-3}. The second observation which is worth pointing out is the sharpness of interfaces. There are no long decay curves indicative of memory effects or significant interdiffusion at growth temperatures. The width of interfaces is limited by the resolution of the characterization method rather than by spread of doping profiles.

Incorporation of oxygen and aluminum.

Doping with DMAlMO allows for incorporation of both aluminum and oxygen into the layer.[17] However, the concentrations of aluminum and oxygen in GaAs are markedly different which implies that different incorporation mechanisms are active. Fig. 2 shows the dependence of the aluminum concentration on the growth temperature. The layers have been grown at atmospheric pressure with constant DMAlMO mole fraction of 3×10^{-7} in the gas phase, and at temperatures between 475 °C and 650 °C. It is evident that the aluminum concentration is independent of growth temperature above 575 °C and slightly decreases below 575 °C. Lack of dependence on growth temperature above 550 °C is characteristic of mass transport limited growth.

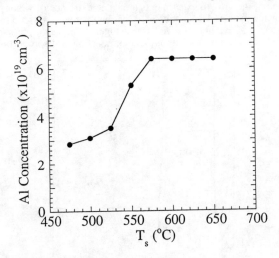

Fig. 2 Aluminum concentration in DMAlMO doped GaAs versus growth temperature.

The incorporation of oxygen is distinctly different. For all growth conditions the oxygen concentration was lower than that of aluminum. This should be considered together with the well know fact that oxygen does not incorporate into pure GaAs grown by OMVPE or MBE. It is either due to relatively weak bonds between oxygen and species on the GaAs growth surface or to the evaporation of volatile suboxides. The energy of the Al-O bond, on the other hand, is very high (120 kcal/mol) and aluminum bonding with arsenic on the GaAs surface can "pull" oxygen into the solid. In most growth conditions oxygen content is much below that of aluminum indicating that the Al-O bond can be severed. Fig. 3, for example, presents the ratio of oxygen and aluminum concentrations as a function of growth temperature. At temperatures above 500 °C, the (O)/(Al) ratio decreases exponentially with a characteristic energy of 2.0 eV as determined from the Arrhenius plot. It is worth noting that similar behavior in the high temperature range, with the same characteristic energy, was observed by Goorsky *et al.*[13] in low pressure (76 Torr) OMVPE growth performed using arsine rather than TBA. This is the activation energy related to breaking the Al-O bond. Since it is much smaller than the bond energy it is unlikely that the process responsible for oxygen removal is a simple bond cleavage. It probably involves interaction with other active species produced during pyrolysis of TBA and TMG such as atomic hydrogen. At this point, based on the kinetic data available in the literature, it is impossible to ascertain the pathways of DMAlMO decomposition. However, several features of this decomposition process are quite apparent. At temperatures below 500 °C the oxygen content saturates and approaches that of aluminum. The ratio (O)/(Al) is 1.2 and 0.9 at growth temperatures of 500 °C and 475 °C, respectively. This indicates that the oxygen content at low temperatures is limited by the flux of DMAlMO toward the substrate and its decomposition rather than by the reactions leading to removal of oxygen. Thus, at temperatures below 500 °C, we have possibly reached the limit of one oxygen atom incorporated per each aluminum, i.e. not only is oxygen incorporated only if bonded to aluminum but also each aluminum atom has a oxygen partner. Since the absolute concentrations determined by SIMS could be as much as a factor of two different from real values the complex in question could be AlO, Al_2O, or Al_2O_3.

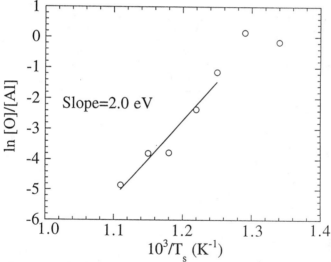

Fig. 3 Ratio of oxygen to aluminum concentration versus deposition temperature.

Also, based on the information contained in Figs. 2 and 3, one can eliminate one of the proposed mechanisms for oxygen removal. It has been suggested that lower oxygen content in $Al_xGa_{1-x}As$ grown at elevated temperatures is due to re-evaporation of volatile Al-O suboxides.[5] This contradicts the fact that the aluminum concentration remains constant at high deposition temperatures.

Localized Vibrational Mode Absorption Measurements

Because of their small mass, both aluminum and oxygen produce high frequency localized vibrations in GaAs. The centers due to isolated oxygen have their LVM modes at 730, and 715 cm[-1] (three charge states of the off-center substitutional oxygen O_{As}) and 845 cm[-1] (interstitial oxygen bonding with Ga and As atoms).[18-20] Aluminum is known to produce an LVM line at 361 cm[-1].[21] If one was to assume that the force constants in the Al-O complex are not drastically affected, such complex should produce two bands corresponding to vibrations of oxygen and aluminum located in the vicinity of LVM lines of isolated defects.

Optical absorption measurements have been performed in the 300-1500 cm[-1] range. No lines were detected in the energy range corresponding to vibrations of oxygen (600-1500 cm[-1]) although the oxygen concentration in investigated samples was more than an order of magnitude above the detection limit. This result proves that the percentage of oxygen atoms which are in the form of isolated centers characteristic of GaAs is less than 10%. In other words, almost all oxygen atoms form complexes with other species. This together with growth data suggest that oxygen not only is incorporated as an Al-O pair but that these pairs remain bonded in the volume of the epilayer. If this rationalization is true, than one should observe a corresponding change in the intensity of the aluminum LVM line. In layers deposited at temperatures in excess of 600 °C virtually all of the aluminum atoms will be present in the form of isolated substitutional Al_{Ga} centers. As the growth temperature decreases more and more, aluminum should retain its oxygen partner and as a result the intensity of the 361 cm[-1] LVM mode should decrease. The experimental spectra of layers grown at 600 °C and 475 °C are shown in Fig. 4(a) and Fig. 4(b), respectively. It is clear that the concentration of isolated Al_{Ga} decreased considerably with decreasing growth temperature. After normalization of spectra (taking into account different layer thicknesses and change of aluminum concentration as determined by SIMS) it was determined that aluminum present as isolated Al_{Ga} constitutes only 25% of the total number of Al atoms in the layer. Although the above data prove that most aluminum and oxygen atoms in GaAs epilayers doped with DMAlMO are present in the form of complexes, we have not observed any new absorption lines. At this point we cannot explain the lack of vibrational modes due to the Al-O complex.

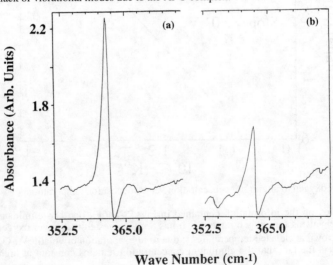

Fig. 4 Localized Vibrational Mode due to isolated Al_{Ga} in (a) layer deposited at 600 °C and (b) layer grown at 475 °C.

One of the consequences of the model above, namely that most of aluminum and oxygen are in the form of a Al-O complex, is the temperature behavior. It is expected that even strongly bonded pairs can dissociate during short term high temperature annealing. The preliminary results of annealing experiments are presented in Fig. 5.

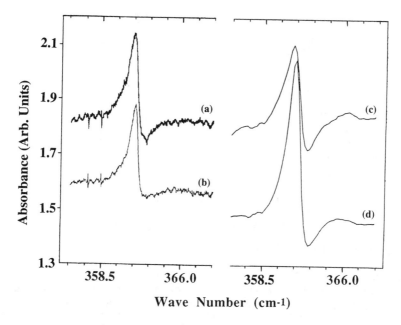

Fig. 5 Effects of annealing on the intensity of AlGa LVM absorption line. (a) As-grown layer deposited at $T_G = 500$ °C, (b) same layer as in (a) after annealing at 800 °C for 20 minutes, (c) as-grown layer (T_G=475 °C), and (d) the same layer as in (c) after 20 second anneal at 1000 °C.

The first pair of before and after samples (a) and (b) show the effect of annealing at 800 °C for 20 minutes. During annealing, the layer was protected by the flow of TBA. There is only a slight decrease of Al_{Ga} LVM intensity which is within experimental error. Apparently the complex is stable at 800 °C. Spectra (c) and (d) were obtained on layers grown at 475 °C and annealed in a rapid thermal annealing system (face to face anneal) at 1000 °C for 20 seconds. The annealing increased the concentration of isolated substitutional aluminum by a factor of two. This result is in agreement with our interpretation of atomic structure of Al-O defects.

Photoluminescence measurements.

Oxygen is known to produce deep states in semiconductors. In III-V semiconductor compounds the substitutional (in GaP) and off-center substitutional oxygen (in GaAs) results in states with ionization energies between 0.5 - 0.9 eV. Presence of deep traps in the semiconductor crystal leads to reduced carrier lifetime and luminescence efficiency. This effect is well documented for non-intentional oxygen contamination of $Al_xGa_{1-x}As$. Frequently, the decrease of near band gap luminescence is associated with the appearance of infrared luminescence bands characteristic of deep levels. Qualitatively the same effects were observed in layers doped with DMAlMO (Fig. 6).

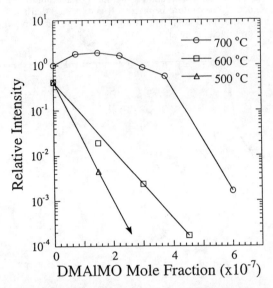

Fig. 6 Effect of DMAlMO doping and temperature on near-band-edge luminescence of GaAs.

All data points were normalized to the intensity of the undoped sample grown at 700 °C. The near-band-edge luminescence intensity of layers grown at 700 °C does not change significantly with DMAlMO doping up to a mole fraction of 3.75×10^{-7}. For higher flow rates, however, it decreases abruptly. This behavior correlates well with oxygen incorporation which is superlinear with DMAlMO mole fraction at high temperatures.[22] Layers grown at 500 °C and 600 °C show a rapid decrease of intensity with increased doping, the decrease being faster in layers grown at 500 °C. In fact, we were not able to detect luminescence from layers grown at 500 °C with mole fractions of 3×10^{-7} or above. Our best estimate is that the relative intensity in this layer is below 10^{-4} which is visualized by the arrow in Fig. 6. The dramatic decrease of luminescence intensity is a direct indication that oxygen-related centers behave as deep level traps and serve as efficient non-radiative recombination centers. This result, together with very high achievable doping concentrations, suggests that doping with alkoxides can produce material with optimized (carrier lifetime)$^{-1}$ x (carrier mobility) product which is required for ultrafast photoconductive switches.[23]

Besides a decrease of near band edge luminescence intensity, doping of GaAs with DMAlMO results in the appearance of several new luminescence peaks due to deep centers. Fig. 7 shows near infrared spectra of three layers grown at 700 °C with increasing DMAlMO mole fraction. The spectrum (a) (X(DMAlMO) = 3.75×10^{-7}) was obtained on the same sample as presented in Fig. 1(c). In addition to exciton and (e, C^0_{As}) peaks, there are four more peaks located at 1.470 eV, 1.450 eV, 1.425 eV, and 1.363 eV . In spectrum (b), corresponding to higher DMAlMO flow rate (X(DMAlMO) = 4.5×10^{-7}), the three deep peaks are dominant and a new even deeper PL band appeared at 1.115 eV. A layer grown with a DMAlMO mole fraction of 6.0×10^{-7} shows only a very weak line close to the band gap and a wide featureless PL band centered at 1.016 eV. All of the above bands most likely involve transitions between relatively deep levels and either band states or other impurities (i.e. donor - acceptor transitions). If this is the case, the distance between band edge and peak position should correspond to the deep level ionization energy. These energies are 72 meV, 97 meV, 150 meV, 410 meV, and 510 meV in order of increasing depth of the level and agree very well with values found by Bhattacharya *et al.*[24] in $Al_xGa_{1-x}As$ epilayers grown by OMVPE and intentionally contaminated with oxygen. They reported observation of deep level PL peaks shifted by 78 meV, 100 meV, 160 meV, and 400 meV from the band edge. The shape of our spectra is very similar to ones reported by Goorsky *et al.*[14] The number of observed PL bands indicates that doping with DMAlMO induces several different deep centers of, as of yet, unknown nature. Such behavior is to be expected since oxygen is known to easily form complexes with impurities and native defects in GaAs. It is also worth noting that one of the well known oxygen-related defects in GaAs, the off-site substitutional oxygen on arsenic site, has a level 150 meV below the conduction band[25,26] and could be responsible for one of the above PL bands. At high doping levels the intensity of all of the above transitions decrease even further and the only remaining peak is located at 1.5 μm (Dr. P.W. Yu, private information). This energy would correspond to the dominant level located close to the middle of GaAs band gap.

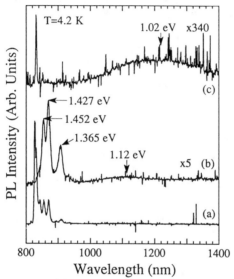

Fig. 7 Photoluminescence spectra of GaAs epilayers doped with DMAlMO.

Compensation of shallow donors.

Based on the characterization results of oxygen-doped GaAs and GaP, one could expect that Al-O pairs should produce a deep acceptor level in GaAs and compensate shallow donor states. This appears to be the case for oxygen contaminated $Al_xGa_{1-x}As$ which frequently turns out to be highly resistive if grown at low temperatures.[1,2,9,27] The same type of behavior was reported for GaAs layers intentionally doped with silicon providing n-type background carriers and dimethylaluminum methoxide.[13] For increasing DMAlMO flow rates the carrier concentration dropped and ultimately layers became fully depleted. It was not clear if the observed compensation was chemical (i.e., an Si-O defect) or electronic in nature. In order to resolve this issue we have deposited a series of GaAs layers doped with diethyltellurium and dimethylaluminum methoxide. Tellurium being from group VI of periodic table is not expected to form bond with oxygen and the probability of Te-O complexes is low. Layers co-doped with DMAlMO always exhibited lower carrier concentration measured either by Hall effect or C-V electrochemical profiler. Some of them heavily doped with oxygen were fully depleted. So far, the highest electron concentration which was compensated by Al-O was 2×10^{18} cm^{-3} i.e. DMAlMO induced doping can produce concentrations of deep acceptors in the 10^{18} cm^{-3} range. Further work on this aspect of alkoxide doping is in progress.

CONCLUSIONS

Results of SIMS analysis of oxygen and aluminum concentration in GaAs epilayers doped with dimethylaluminum methoxide indicate that oxygen is incorporated only when bonded to the aluminum atom. At low growth temperatures concentrations of both elements are approximately equal which suggest that most aluminum atoms are also accompanied by an oxygen partner. These conclusions are supported by infrared absorption measurements. The signature of isolated Al_{Ga} decreased at low growth temperatures and bands due to isolated oxygen have not been detected. Also, short term annealing at 1000 °C results in breaking up of the Al-O complex and an associated increase of Al_{Ga} LVM intensity. Pairs appear to be stable at temperatures up to 800 °C. Al-O pairs produce deep traps within the GaAs band gap which result in a decrease of near-band-gap luminescence by many orders of magnitude and the appearance of infrared luminescence bands. The doping induced traps are primarily of acceptor character capable of compensating tellurium donors. The concentration of Al-O traps can be as high as 10^{18} cm^{-3}.

Acknowledgments: This work was in part supported by NSF Grant No. DMR-9202683, AFOSR Grant No. AFOSR-91-0373 and by US Department of Energy, Office of Basic Energy Sciences, Division of Chemical Sciences under contract DE-AC05-84OR21400 with Martin Marietta Inc.

References:

[1]C.R. Abernathy, A.S. Jordan, S.J. Pearton, W.S. Hobson, D.A. Bohling and G.T. Muhr, Appl. Phys. Lett., **56**, 2654 (1990).

[2]M. Hata, H. Takata, T. Yako, N. Furuhara, T. Maeda and Y. Uemura, J. Crystal Growth, **124**, 427 (1992).

[3]T.F. Kuech, R. Potemski, F. Cardone and G. Scilla, J. Electron. Mater., **21**, 341 (1992).

[4]C.T. Foxon, J.B. Clegg, K. Woodbridge, D. Hilton, P. Dawson and P. Blood, J. Vac. Sci. Technol. B, **3**, 703 (1985).

[5]K. Akimoto, M. Kamada, K. Taira, M. Arai and N. Watanabe, J. Appl. Phys., **59**, 2833 (1986).

[6]D.W. Kisker, J.N. Miller and G.B. Stringfellow, Appl. Phys. Lett., **40**, 614 (1982).

[7]J.R. Shealy, V.G. Kreismanis, D.K. Wagner and J.M. Woodall, Appl. Phys. Lett., **42**, 83 (1983).

[8]B. Kim, H.Q. Tserng and J.W. Lee, IEEE Electron Devices Lett., **EDL-7**, 638 (1986).

[9]H.C. Casey, A.Y. Cho, D.V. Lang, E.H. Nicollian and P.W. Foy, J. Appl. Phys., **50**, 3484 (1979).

[10]H.C. Casey, J.S. McCalmont, H. Pandharpurkar, T.Y. Wang and G.B. Stringfellow, Appl. Phys. Lett., **54**, 650 (1989).

[11]D.S. Ruby, K. Arai and G.E. Stillman, J. Appl. Phys., **58**, 825 (1985).

[12]B. Lee, K. Arai, B.J. Skromme, S.S. Bose, T.J. Roth, J.A. Aguilar, T.R. Lepkowski, N.C. Tien and G.E. Stillman, J. Appl. Phys., **66**, 3772 (1989).

[13]M.S. Goorsky, T.F. Kuech, F. Cardone, P.M. Mooney, G.J. Scilla and R.M. Potemski, Appl. Phys. Lett., **58**, 1979 (1991).

[14]M.S. Goorsky, T.F. Kuech, P.M. Mooney, F. Cardonne and R.M. Potemski, Mat. Res. Soc. Symp., **204**, 177 (1991).

[15]R. Tarao, Bull. Chem. Soc. Japan, **39**, 725 (1966).

[16]D.A. Drew, A. Haaland and J. Weidlein, Z. anorg. allg. Chem., **398**, 241 (1973).

[17]Y. Park, M. Skowronski and T.M. Rosseel, Proc. Mat. Res. Symp., **282**, 75 (1993).

[18]J. Schneider, B. Dischler, H. Seelewind, P.M. Mooney, J. Lagowski, M. Matsui, D.R. Beard and R.C. Newman, Appl. Phys. Lett., **54**, 1442 (1989).

[19]H.C. Alt, Appl. Phys. Lett., **55**, 2736 (1989).

[20]H.C. Alt, Appl. Phys. Lett., **54**, 1445 (1989).

[21]O.G. Lorimor, W.G. Spitzer and M. Waldner, J. Appl. Phys., **37**, 2509 (1966).

[22]Y. Park, M. Skowronski and T.S. Rosseel, J. Cryst. Growth, (1993).

[23]A.C. Warren, N. Katzenellenbogen, D. Grischkowsky, J.M. Woodall, M.R. Melloch and N. Otsuka, Appl. Phys. Lett., **58**, 1512 (1991).

[24]P.K. Bhattacharya, S. Subramanian and M.J. Ludowise, J. Appl. Phys., **55**, 3664 (1984).

[25]M. Skowronski, S.T. Neild and R.E. Kremer, Appl. Phys. Lett., **57**, 902 (1990).

[26]H.C. Alt, Phys. Rev. Lett., **65**, 3421 (1990).

[27]R.H. Wallis, M.A.D. Forte-Poisson, M. Bonnet, G. Beuchet and J.P. Duchemin, Inst. Phys. Conf. Ser., **56**, 73 (1981).

DEEP LEVEL STRUCTURE OF SEMI-INSULATING MOVPE GAAS GROWN BY CONTROLLED OXYGEN INCORPORATION

J. W. HUANG AND T. F. KUECH
Department of Chemical Engineering, University of Wisconsin, Madison, WI 53706

ABSTRACT

Semi-insulating oxygen-doped GaAs layers have been grown by low pressure metalorganic vapor phase epitaxy (MOVPE) using aluminum-oxygen bonding based precursor diethyl aluminum ethoxide (DEALO). Resistivities of more than 2×10^9 Ω-cm at 294 K have been achieved. Deep level structure responsible for the high resistivity was investigated by deep level transient spectroscopy (DLTS) using DEALO and disilane co-doped GaAs p^+-n homojunction. Multiple deep level peaks were observed, and the relative peak heights were found to vary with the dopant concentrations. Major deep levels were electron traps with ionization energy between 0.75 and 0.95 eV below conduction band edge minimum. An activation energy of 0.81 eV was deduced from temperature-dependent resistivity measurement, and should be closely related to the major 0.75 eV peak in DLTS spectra.

INTRODUCTION

For GaAs-based metal-semiconductor field-effect transistor (MESFET) devices, semi-insulating (SI) substrates are generally used to reduce parasitic capacitance and eliminate device cross-talk. A number of significant problems of MESFET for both digital and analog circuit applications, however, are associated with the SI substrates, including the sidegating and backgating effects, light sensitivity, low output resistance, and low source-drain breakdown voltage [1]. A buffer layer providing higher resistivity and optical insensitivity should be inserted between the active layer and the substrate to alleviate or even eliminate these problems.

The controlled introduction of deep levels has long been recognized as a means of producing SI compound semiconductors. Controlled introduction of Cr and Fe into GaAs and InP respectively by metalorganic vapor phase epitaxy (MOVPE) was demonstrated to be able to serve as the desired buffer layer for device isolation [2, 3]. Development of an epitaxial growth technique for the controlled formation of SI materials has been the focus of several recent studies [1, 4, 5]. The choice of suitable chemical species for the fabrication of SI layers is complicated by the additional constraints on the diffusion coefficient and solubility of the impurity as well as the depth of the level produced. Oxygen has been recently investigated as a promising candidate for the MOVPE-based growth of SI GaAs. We have previously demonstrated that oxygen can be intentionally incorporated into GaAs through the independent introduction of aluminum-oxygen bonding based precursors dimethyl aluminum methoxide (DMALO, $(CH_3)_2AlOCH_3$) or diethyl aluminum ethoxide (DEALO, $(C_2H_5)_2AlOC_2H_5$) [4, 6]. These oxygen-doped epitaxial layers have specular surface morphology and high crystal quality. Oxygen concentration in excess of 10^{18} cm^{-3} was readily achievable, and oxygen-related deep levels were generated, compensating shallow silicon impurities. Deep level structure responsible for this observed compensation, however, was not fully investigated.

DMALO is a solid at room temperature (melting point is 35 $^\circ$C). High vapor pressure of DMALO in liquid form prohibits its use as a dopant in a conventional liquid bubbler. When employed in solid form, the source pick-up can sometimes vary in time due to the change of internal surface area. In contrast, DEALO is a commercially available liquid at room temperature (melting point is 2.5 ^{O}C) with suitable vapor pressure as a dopant, leading to a more reproducible oxygen incorporation [6].

This paper presents the deep level structure of DEALO and disilane co-doped MOVPE GaAs as determined by deep level transient spectroscopy (DLTS). Resistivity of DEALO-doped SI GaAs is also reported. An activation energy acquired from temperature-dependent current-voltage (I-V) measurement, which is related to the deep level structure responsible for the SI behavior, is compared to the DLTS spectra.

EXPERIMENTAL PROCEDURE

Samples were grown in a conventional horizontal low pressure (78 Torr) reactor [7]. TMGa and AsH$_3$ were used with H$_2$ as the carrier gas. Disilane and carbon tetrachloride were employed for n- and p-type doping respectively. Growth temperature was 600 ^{O}C and the V/III ratio (AsH$_3$/TMGa) was 40. The growth rate was held constant at 0.05μm/min., corresponding to a TMGa mole fraction of about 1.8×10^{-4} in the reactor.

For DLTS measurement, p^+-n homojunctions [8], which consisted of a 1 μm thick n-type region capped by a 0.5 μm p^+ layer doped with 4×10^{18} cm^{-3} carbon, were grown on Si-doped n^+ <100> substrates. These p^+-n junctions were fabricated by a standard lithography and liftoff process. 500 μm diameter Ge/Ni/Al dots were deposited and alloyed for ohmic contact on p^+ capping layer followed by mesa etching (fig. 1(a)). In/Sn ohmic contact was then alloyed to the n^+ substrate for back side ohmic contact. The n-type region was co-doped with DEALO and disilane to facilitate the investigation of oxygen-induced defects by DLTS. DLTS measurements, using double boxcar correlators, were performed over the temperature range of 77-425 K.

Resistivity measurements were performed on n-i-n structures [5], having a 2 μm DEALO-doped GaAs with two 0.5 μm n-type cladding layers doped with 2×10^{17} cm^{-3} silicon on Si-doped n^+ <100> substrates. Variable area (500, 1000, and 2000 μm square) mesa structures were fabricated by standard lithography and liftoff process to alloy Au/Ge/Ni/Ti/Au on the top n-layer for ohmic contact followed by mesa etching (fig. 1(b)). In/Sn ohmic contact was then alloyed to the n^+ substrate for back side ohmic contact. Keithley 617 electrometer was used as a picoammeter, and DC voltage was supplied by Keithley 230 voltage source. Resistivity was calculated from the ohmic region of current-voltage (I-V) curve. Temperature-dependent I-V measurements were performed over the range of 294-357 K.

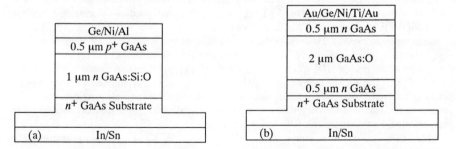

Figure 1. (a) DLTS sample : p^+-n homojunction. (b) I-V sample : n-i-n mesa structure.

RESULTS AND DISCUSSION

DLTS Study

Two sets of mole fraction conditions were used during the n-layer growth of p^+-n homojunction based on electrochemical capacitance-voltage (EC-V) profiling results from our previous work [6]. Concentrations of Si, Al, and O, as determined by secondary ion mass spectroscopy (SIMS) measurements [6], as well as EC-V profiling results are shown in table I. DLTS spectra for this pair of samples are illustrated in fig. 2. Multiple deep levels were observed, underlining the complexity of possible defect configurations due to the doping of DEALO and disilane. During the measurement, the quiescent reverse bias voltage across the diode was kept at -1.0 V, while the amplitude of the filling pulse was +1.5 V. Therefore both majority and minority carrier traps could be measured. These spectra indicated that no minority carrier (hole) traps were detected. The electron trap signatures are shown in fig. 3 and table II. The quiescent reverse bias voltage was also increased to -2.0 V while the amplitude of the filling pulse remained the same. It was found that relative peak heights and peak positions stayed unchanged. The former indicated an uniform distribution of oxygen-related deep traps, confirming the same finding from EC-V profile [6]. The latter showed that no field-assisted thermal emission of carriers by the traps was observed when the average electric field within the depletion region [8] was increased from 6×10^4 to 7×10^4 V/cm.

A comparison of the spectra in fig. 2 indicates that the principal spectral features, peaks D and E, are present in both samples. The Si concentration remained constant and the shift in the relative peak heights between these two samples should be related to the increased Al and O concentrations. There are two shifts in the spectra which can be noted. As the DEALO mole fraction was increased in the reactor, the peak labeled D, as well as the other lower energy peaks, increased relative to that of peak E. In our previous study [6], compensation was found to be directly related to the incorporation of Si as well as Al and O. We have tentatively assigned the peak E to be a defect associated with a Si-O based species. This assignment is based on both the previous observations of the presence of Si-related deep level [6] and the decrease in peak E relative to the remaining peaks as the Al and O contents were increased at a constant Si concentration. The remaining peaks of the DLTS spectra from these samples revealed a set of lower energy peaks (< 0.9 eV) which are associated with the incorporation of O.

The presence of multiple O-based peaks can be rationalized by the analogy to the case of DX center in low Al content $Al_xGa_{1-x}As$. In the case of the DX center attributed to the Si-based deep level, multiple emission peaks were observed in samples containing a low amount of Al [9]. These multiple emission energies are thought to be related to the shift in the local environment of the DX center. At low Al concentrations, there is a variation in the local environment on the second nearest neighbor shell (Si resides on a cation site) due to the random distribution of Al atoms on the cation sublattice. In the present case of O in GaAs co-doped with Al, there is a potential variation in the first nearest neighbor shell, giving rise to multiple emission peaks which may be well separated in energies. This coincides with the initial DLTS observation on DMALO-

Table I. DLTS Samples Description

Sample	[Si] (cm^{-3}) (SIMS)	[Al] (cm^{-3}) (SIMS)	[O] (cm^{-3}) (SIMS)	N_d-N_a (cm^{-3}) (with DEALO)	$\Delta(N_d$-$N_a)$ (cm^{-3})	Compen -sation
1	1.06×10^{17}	3.2×10^{17}	1×10^{15}	9.44×10^{16}	1.16×10^{16}	10.94%
2	1.06×10^{17}	6.5×10^{17}	1.5×10^{16}	5.44×10^{16}	5.16×10^{16}	48.68%

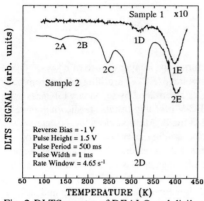

Fig. 2 DLTS spectra of DEALO and disilane co-doped GaAs.

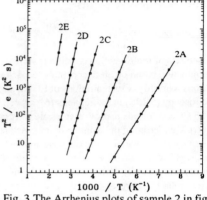

Fig. 3 The Arrhenius plots of sample 2 in fig. 2 showing the signatures of electron traps.

Table II. Deep Level Characteristics

Sample	N_d-N_a (cm^{-3}) (p-n Junction)	Trap	ΔE_t (eV)	σ (cm^2) (from intercept)	N_t (cm^{-3})	Total N_t (cm^{-3})
1	7.8×10^{16}	1D	0.72	7.88×10^{-14}	5.83×10^{13}	2.78×10^{14}
		1E	0.94	1.27×10^{-13}	2.20×10^{14}	
2	1.7×10^{16}	2A	0.24	1.54×10^{-15}	8.16×10^{13}	5.66×10^{15}
		2B	0.39	2.22×10^{-14}	5.17×10^{13}	
		2C	0.54	5.19×10^{-14}	4.81×10^{14}	
		2D	0.75	2.02×10^{-13}	3.19×10^{15}	
		2E	0.95	1.07×10^{-13}	1.85×10^{15}	

doped samples which indicated a variation in the DLTS spectra based on growth temperature and the Al-O concentration changes [4]. The presence of multiple O-related DLTS peaks complicates the defect structure but provides an interesting case where local environment of the defect can be independently altered by a growth variable.

During DLTS measurement, filling pulse time width was set to be 1 ms. This setting was verified to be enough to fill the trap up to more than 95% of complete filling. Total trap concentrations, however, can not account for all of the observed compensation from EC-V results (table I and II). The use of p^+-n junction in DLTS measurement explores only the upper half of the bandgap. There could be some other deep levels closer to valence band edge that were beyond the detection range. We have also noticed a change in N_d-N_a between EC-V results and C-V measurements on p^+-n junction, the former (table I) being higher than the latter (table II). While EC-V measurement is usually performed on the sample without any preparation, fabrication of p^+-n junction requires a lot of processing steps. Among these steps, thermal annealing of ohmic contact could be the most critical one in terms of affecting N_d-N_a. As N_d-N_a becomes smaller, the amount of detectable trap concentration will also be reduced. Thermal annealing could have changed the defect configurations or their charge states. More investigations are needed.

Temperature-Dependent Resistivity Measurement

Two samples of different DEALO mole fractions were grown and processed into variable-area *n-i-n* mesa structures. Typical I-V characteristics are shown in fig. 4. According to the theory of single carrier injection in solids [10], I-V curve will have three different current regimes (fig. 4) for a single set of traps with energy level lying above Fermi level. The currents within the first two regions have first and second order power law dependence on voltage, corresponding to ohmic and space-charge limited behaviors respectively. Resistivity can be calculated from the ohmic region. The third regime is marked by the sharp increase of current due to complete trap filling. The space-charge limited region, however, will be absent if the trap energy level is located below Fermi level. I-V curves in fig. 4 suggested that regions corresponding to an I-V second order power law relationship were relatively narrow. Assuming GaAs:O has only one major deep trap, this single level lying below Fermi level should be a more reasonable picture. A sublinear I-V dependence following ohmic regime was observed in fig. 4. Origins of these sublinear regions are still unknown. Trap-filled limited voltage (V_{TFL}), which is proportional to electrically active trap concentration [5, 10], was estimated to be 1 and 2 V for samples A and B in fig. 4. The B/A ratio of V_{TFL}, resistivity, as well as Al concentration are all at about 2. Since concentration of O is smaller than that of Al in both samples, electrically active defect structure in GaAs:O could be involving species like Al-O-Al, which has multiple Al atoms as the nearest-neighbors of substitutional O on As site.

Resistivity value more than 2×10^9 Ω-cm at 294 K was obtained (fig. 5). I-V measurements were performed on mesa structures of different areas since a small amount of surface conduction can shunt the very high bulk resistance of the GaAs:O layer. The results are illustrated in fig. 5, showing ohmic currents scaled with different mesa areas and giving the same resistivity. Bulk conduction is thus dominating over surface conduction in our samples. From the temperature-dependent resistivity data, an activation energy 0.81 eV was deduced from both samples, indicating a similar transport mechanism in both even though concentrations of Al and O are different. These transport measurements have clearly associated the temperature dependence of resistivity with the underlying deep level structure. The resistivity-derived 0.81 eV activation energy and the principal DLTS emission peaks described above are in close agreement. The

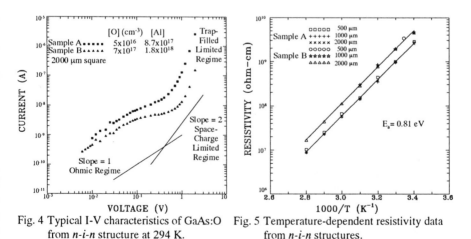

Fig. 4 Typical I-V characteristics of GaAs:O from *n-i-n* structure at 294 K.

Fig. 5 Temperature-dependent resistivity data from *n-i-n* structures.

complexity of the deep level structure, however, limits our ability to present a model involving the exact direct comparison between I-V and DLTS measurements.

The high resistivity in GaAs:O layer implies a very low net carrier concentration, which was confirmed by capacitance-voltage measurement at 1 MHz. Constant capacitance values (less than 0.5 % change) of 12, 55, 230 pF on 500, 1000, and 2000 μm squares on both samples in fig. 4 were obtained with bias ranging from -1 to 0 V. This is in good agreement with parallel plate capacitor model as the calculated electrical thickness matches the physical layer thickness (2 μm) [5]. The Fermi level should then be located near the middle of the bandgap, possibly no deeper than 0.75 or 0.81 eV below E_c. This is consistent with the picture of single deep trap in GaAs:O lying below Fermi level as described above.

CONCLUSION

Semi-insulating oxygen-doped GaAs layers have been grown by MOVPE using diethyl aluminum ethoxide (DEALO) as oxygen precursor. Resistivities of more than 2×10^9 Ω-cm at 294 K have been achieved. An activation energy of 0.81 eV, which is related to the deep level structure responsible for the high resistivity, was obtained from temperature-dependent resistivity measurement. The deep level structure was investigated by DLTS using DEALO and disilane co-doped GaAs p^+-n homojunction. Multiple deep level peaks were observed and the relative peak heights were found to change as dopant concentrations were varied. The major deep traps from DLTS spectra have ionization energy between 0.75 and 0.95 eV, and should be closely related to the 0.81 eV activation energy from resistivity measurement.

ACKNOWLEDGMENT

The authors would like to acknowledge the support of the National Science Foundation under DMR-9201558, the Naval Research Laboratory, and the Army Research Office.

REFERENCE

1. F. W. Smith, A. R. Calawa, C. L. Chen, M. J. Manfra, and L. J. Mahoney, IEEE Electron Device Lett. **9**, 77 (1988).
2. S. J. Bass, J. Crystal Growth **44**, 29 (1978).
3. J. A. Long, V. G. Riggs, and W. D. Johnston, Jr., J. Crystal Growth **69**, 10 (1984).
4 M. S. Goorsky, T. F. Kuech, F. Cardone, P. M. Mooney, G. J. Scilla, and R. M. Potemski, Appl. Phys. Lett. **58**, 1979 (1991).
5. Ph. Pagnod-Rossiaux, M. Lambert, F. Gaborit, F. Brillouet, P. Garabedian, and L. Le Gouezigou, J. Crystal Growth **120**, 317 (1992).
6. J. W. Huang, D. F. Gaines, T. F. Kuech, R. M. Potemski, and F. Cardone, presented at the 1993 EMC, Santa Barbara, CA, 1993 (to be published).
7. T. F. Kuech, D. J. Wolford, E. Veuhoff, V. Deline, P. M. Mooney, R. Potemski, and J. Bradley, J. Appl. Phys. **62**, 632 (1987).
8. P. J. Wang, T. F. Kuech, M. A. Tischler, P. Mooney, G. Scilla, and F. Cardone, J. Appl. Phys. **64**, 4975 (1988).
9. P. M. Mooney, J. Appl. Phys. **67**, R1 (1990).
10. M. A. Lampert and P. Mark, Current Injection in Solids (Academic Press, New York, 1970).

PROTON IRRADIATION DAMAGE IN Zn AND Cd DOPED InP

George C. Rybicki* and Wendell S. Williams**
*Photovoltaics Branch, NASA Lewis Research Center, Cleveland, OH 44135
**Department of Materials Science and Engineering, Case Western Reserve University,
Cleveland, OH 44106

ABSTRACT

Deep Level Transient Spectroscopy (DLTS) was used to study the defects introduced in Zn and Cd doped Schottky barrier diodes by 2 MeV proton irradiation. The defects H3, H4 and H5 were observed in lightly Zn doped InP, while only the defects H3 and H5 were observed in more heavily Zn doped and Cd doped InP. The defect activation energies and capture cross sections did not vary between the Zn and Cd doped InP.

The concentration of the radiation induced defects was also measured. The introduction rate of the defect H4 in the lightly Zn doped InP and the introduction rate of the defect H3 in the heavily Zn and Cd doped InP were about equal, but the introduction rate of the defect H5 varied strongly among the three types of material. The introduction rate of H5 was highest in the heavily Zn doped InP but the lowest in the heavily Cd doped InP, even though they were doped comparably. As a result, the total defect introduction rate was lowest in the highly Cd doped InP.

The results can be interpreted in terms of the models for the formation and annealing of defects, and by the different diffusion rates of Zn and Cd in InP.

INTRODUCTION

The superior radiation resistance of InP solar cells has previously been demonstrated, and has been associated with low defect introduction rates, rapid defect annealing rates and the strong effects of doping on these factors.[1,2]

However, InP solar cells are not immune from radiation damage. The irradiation induced defects associated with the degradation of InP solar cells have been identified by DLTS. In p-type InP, three majority carrier traps dominate the DLTS spectra - H3, at $E_v + 0.28$ eV and H4, at $E_v + 0.37$ eV, both of which are associated with phosphorus displacements [3,4], and H5, at $E_v + 0.57$ eV, which is related to an indium displacement.[4] The introduction rate of the defect H4 has been shown to decrease with increasing carrier concentration, possibly due to self annealing [5], but the introduction rate of H5 has been shown to increase with increasing carrier concentration, suggesting that it may be a point defect/impurity complex.[5]

The annealing rate of the defects in p-type InP is found to increase with increasing carrier concentration.[5] H4 is thought to be eliminated in a reaction with a dopant atom, while H5, whose concentration increases upon annealing, forms through a reaction with an impurity atom. These observations have led to the development of models for defects and defect reactions in p-type InP [5], but with the exception of the brief comparison of introduction rates and annealing rates of H3 and H4 in Zn and Cd doped InP by Sibile et al.[4], the models have not been tested by the substitution of different p-type dopants. In this work the defects and their properties in Cd doped InP are studied and compared with those in Zn doped InP.

Mat. Res. Soc. Symp. Proc. Vol. 325. ©1994 Materials Research Society

EXPERIMENTAL

Schottky barrier diodes were fabricated on Zn doped InP wafers with carrier concentrations of 2.5 x 10^{16} cm^{-3} and 2.5 x 10^{17} cm^{-3}, and on Cd doped InP with a carrier concentration of 2.5 x 10^{17} cm^{-3}. I-V, C-V and DLTS analyses of the samples were used to verify carrier concentration and diode quality; no deep levels were detected in the unirradiated diodes.

The Schottky barrier diodes were irradiated at room temperature with 2 MeV protons generated by the Van de Graaff accelerator in the Physics department at Case Western Reserve University. The beam current was limited to 2 nanoamperes to avoid heating of the samples.The three types of diodes were irradiated to the empirically determined fluence required to produce a good DLTS spectrum. The lightly Zn doped samples were irradiated to a fluence of 2 x 10^{12} cm^{-2}, the heavily Zn doped to a fluence of 2 x 10^{13} cm^{-2} and the Cd doped samples to a fluence of 5 x 10^{13} cm^{-2}.

After irradiation, the Schottky diodes were analyzed by DLTS. The concentration of irradiation induced defects, their energy levels, capture cross sections and annealing rates were measured.

RESULTS

DLTS spectra from the lightly Zn doped, the heavily Zn doped and heavily Cd doped InP diodes are shown in Figures 1, 2 and 3. The spectra all contain two major peaks, and a small additional peak appears on the side of the lower temperature peak in the lightly Zn doped InP sample.

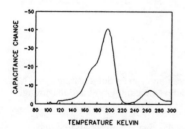

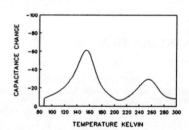

Figure 1.   DLTS spectrum of proton-damaged lightly Zn doped InP.

Figure 2.   DLTS spectrum of proton-damaged heavily Zn doped InP

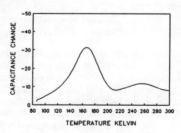

Figure 3. DLTS spectrum of proton-damaged heavily Cd doped InP

The energy levels of the traps associated with the peaks were determined and corrected for the effect of the electric field in the depletion region on emission of carriers from the deep levels. Corrections were applied for the Frenkel-Poole effect, which is the lowering of the trap potential barrier for carrier emission due to the electric field, and for the emission of carriers by phonon assisted tunneling out of the trap levels.[6] The capture cross section, σ, of the defects was measured directly by the pulse-width variation technique rather than inferred from the intercept of the activation energy plot.[7] The activation energy and capture cross section data appear in Table 1.

	Lightly Zn doped	Heavily Zn doped	Heavily Cd doped
ΔE H3 eV		0.27 ± 0.04	0.29 ± 0.04
ΔE H4 eV	0.34 ± 0.05		
ΔE H5 eV	0.59 ± 0.06	0.56 ± 0.02	0.59 ± 0.03
σ H3 cm^2		$9.4 \times 10^{-19} \pm 0.08$	$1.2 \times 10^{-18} \pm 0.01$
σ H4 cm^2	$1.7 \times 10^{-17} \pm 0.6$		
σ_0 H5 cm^2	$1.3 \times 10^{-19} \pm 0.2$	$3.0 \times 10^{-21} \pm 0.5$	$4.9 \times 10^{-21} \pm 1.2$
ΔE_σ eV (H5 only)	0.07 ± 0.01	0.03 ± 0.01	0.04 ± 0.01

Table I Activation energies and capture cross sections of defects in proton irradiated InP, ΔE is the activation energy, σ is the capture cross section, and ΔE_σ is the activation energy associated with the variation of the capture cross section of H5 with temperature.

The energy level of the trap associated with the major peak in the lightly Zn doped InP was $E_v + 0.34$ eV. This is consistent with the value reported for H4 .[3] The capture cross section, σ, of H4 was 1.7×10^{-17} cm^2, much less than the value of 8×10^{-16} cm^2 reported in the literature.[3] The values of capture cross section in the literature however are commonly estimates from the intercept of the activation energy plots and are often greatly in error.[8] By contrast, the energy levels of the trap associated with the lower temperature peaks in the heavily Zn and Cd doped InP were lower, $E_v + 0.27$ and $E_v + 0.29$ eV respectively. These values are consistent with the values reported for H3.[3] The capture cross sections measured for H3 in heavily Zn and Cd doped InP were 9.4×10^{-19} cm^2 and 1.2×10^{-18} cm^2 respectively, again much smaller than the value of 6×10^{-16} cm^2 reported for the intercept method.[3] The capture cross section of this level is also an order of magnitude smaller than that for H4 in lightly Zn doped InP, further supporting the identification of the trap in the heavily Zn and Cd doped InP as H3 and not H4. The activation energy and capture cross section of the traps, though, was the same for comparably Cd and Zn doped InP.

The energy level of the trap associated with the peak at higher temperatures was $E_v + 0.59$ eV in the lightly Zn doped InP, $E_v + 0.56$ eV in the heavily Zn doped InP and $E_v + 0.59$ eV in the Cd doped InP, all consistent with the value for H5. [3] The capture cross section of H5 varied with temperature following an Arrhenius relationship with $\sigma_0 = 1.3 \times 10^{-19}$ cm^2 and $\Delta E_\sigma = 0.07$ eV in the lightly Zn doped InP, $\sigma_0 = 3 \times 10^{-21}$ cm^2 and $\Delta E_\sigma = 0.03$ eV in the heavily Zn doped InP, and $\sigma_0 = 4.9 \times 10^{-21}$ cm^2 and $\Delta E_\sigma = 0.04$ eV in the heavily Cd doped InP. These values of the capture cross section are much smaller than the value of 6×10^{-15} cm^2 estimated by previous authors using the intercept method.[3] The activation energy of H5 is thus not dependent on the type or concentration of dopant. Although the values of σ_0 are smaller in the highly doped InP, it must be remembered that these values are the intercept of the σ vs. inverse temperature plot and that the activation energies for the variation of the capture cross section also vary between the three types of material. The result is that the actual capture cross section of H5 varied little between the three materials.

The concentration of the defects, N_t, in the three types of samples was measured and corrected for incomplete trap filling near the Schottky contact.[9] Due to band bending near the Schottky contact some of the traps will always be below the Fermi level; the occupancy of these traps is not changed during the DLTS experiment. The corrected irradiation induced defect concentrations appear in Table 2. The introduction rate was calculated from the defect concentration assuming a linear relation between the defect concentration and the particle fluence.

	Lightly Zn doped	Heavily Zn doped	Heavily Cd doped
N_t cm^{-3} H3	$2.1 \times 10^{14} \pm 0.4$	$5.2 \times 10^{15} \pm 0.45$	$10.1 \times 10^{15} \pm 1.6$
N_t cm^{-3} H4	$4.6 \times 10^{14} \pm 0.6$		
N_t cm^{-3} H5	$2.5 \times 10^{14} \pm 0.2$	$5.2 \times 10^{15} \pm 1.4$	$7.6 \times 10^{15} \pm 1.9$
ϕ cm^{-2}	2×10^{12}	2×10^{13}	5×10^{13}
$dN_t/d\phi$ cm^{-1} H3	105 ± 20	257 ± 22	213 ± 38
$dN_t/d\phi$ cm^{-1} H4	229 ± 29		
$dN_t/d\phi$ cm^{-1} H5	116 ± 34	271 ± 70	151 ± 33

Table II   Defect concentrations and introduction rates in proton irradiated InP. N_t is the defect concentration, ϕ is the particle fluence and $dN_t/d\phi$ is the introduction rate in defects per incident proton.

H4 was present only in the lightly Zn doped InP. The defect introduction rate for H4 was 229 per 2 MeV proton. H3 was present in all three materials. The introduction rates for H3 were 105, 257 and 213 per 2 MeV proton in the lightly Zn doped, heavily Zn doped and heavily Cd doped InP respectively. The introduction rate of H3 did not vary significantly between the heavily Zn and Cd doped InP, but was higher in the heavily Zn doped InP than it was in the lightly Zn doped InP. This result is in contrast with the result of Sibile et al.[4] who found both H3 and H4 in electron irradiated, lightly Zn and Cd doped InP, and also found that the introduction rate of H3 was lower in Cd doped InP than in comparably Zn doped InP.

By contrast the introduction rate of H5 varied significantly in the three materials. The introduction rates for H5 were 116, 271 and 151 per 2 MeV proton in the lightly Zn doped InP, the heavily Zn doped InP and heavily Cd doped InP respectively. The introduction rate of H5 was significantly lower in the Cd doped InP than in the Zn doped InP even though the doping concentration was identical. The result of the lower introduction rate of H5 in Cd doped InP is that the total introduction rate of defects is lower in the heavily Cd doped InP than in comparably Zn doped InP.

The thermal annealing rates of the defects were measured, while empty of holes, under 3 volts reverse bias, at temperatures from 350-425 K. The data appear in Table III.

The annealing rate of H4 was measured in the lightly Zn doped InP. The annealing was a reasonably good fit to first order kinetics. The annealing rate constant was determined from a least squares fit to the data, and an activation energy for the annealing of 1.35 ev was determined from an Arrhenius plot of the annealing data.

The annealing of H3 was a reasonable fit to first order kinetics, and the rate constants were determined by a least squares fit to the data. Initial annealing of H3 in the lightly Zn doped InP produced a dramatic growth of the H3 peak until the concentration was comparable to that of H4,

	Lightly Zn doped	Heavily Zn doped	Heavily Cd doped
A H3 350K			$9.8 \times 10^{-5} \pm 0.1$
A H3 375K	$5.2 \times 10^{-5} \pm 1.5$	$2.1 \times 10^{-4} \pm 0.2$	$6.1 \times 10^{-4} \pm 0.1$
A H3 400K	$1.0 \times 10^{-3} \pm 0.4$	$1.2 \times 10^{-3} \pm 0.1$	$6.0 \times 10^{-3} \pm 0.5$
A H3 425K	$1.2 \times 10^{-2} \pm 0.1$	$1.1 \times 10^{-2} \pm 0.3$	
A H4 375 K	$8.4 \times 10^{-5} \pm 0.1$		
A H4 400K	$1.0 \times 10^{-3} \pm 0.4$		
A H4 425K	$1.1 \times 10^{-2} \pm 0.1$		

Table III Annealing rates of defects in proton irradiated InP.

but further annealing caused a reduction of the concentration. An activation energy for the annealing of 1.49 eV was determined from an Arrhenius plot of the annealing rate constants. The annealing of H3 in the more heavily doped materials occurred without the initial increase in concentration observed in the lightly Zn doped InP, and activation energies of 1.10 ev in the heavily Zn doped InP and 1.01 eV in the heavily Cd doped InP were determined from Arrhenius plots of the annealing data.

The annealing rate constant for H3 was not strongly dependent on dopant concentration in the Zn doped InP but was 5 to 6 times higher in the Cd doped InP. The annealing rate constants at 400K, for H3 in all materials in this work are considerably less than the rate constant 1.25×10^{-2} s^{-1}, measured by Sibile, for H3 at 400 K in electron irradiated InP.[10] The lower annealing rate observed here suggests that proton irradiation damage may be more difficult to anneal than electron radiation damage.

The concentration of H5 increased on annealing in all three materials. The annealing of H5 was observed in all three materials but the kinetics were complicated and no reaction order or rate constant could be determined. The growth rate of H5 was highest in the heavily Cd doped material and lowest in the heavily Zn doped material.

DISCUSSION

From the results it can be seen that, although the type of defects vary between the lightly Zn and heavily Zn and Cd doped InP, there is no difference between the defect type, activation energy or capture cross section in comparably Zn and Cd doped InP. The difference between the lightly Zn doped and heavily Zn and Cd doped InP, ie. the absence of H4 in the heavily doped samples, can be understood by considering the effect of doping on the annealing rate of H4. The effect of the carrier concentration on the properties of defects in Zn doped InP has been studied by Yamaguchi and Ando.[5] They showed that the measured introduction rate of H4 was much lower in highly Zn doped InP when irradiated at room temperature, but that it was independent of carrier concentration when irradiation was carried out at 100 K. The observed introduction rate of H4 thus was affected by self annealing of the defect, and the annealing rate increases rapidly with increasing carrier concentration. In the present work H4 was observed only in lightly Zn doped InP for this reason. H4 was annealed out very rapidly due to the high carrier concentration in the heavily Zn and Cd doped materials. This result is in good agreement with Yamaguchi's work in which he found that above a doping concentration of 1×10^{17} cm^{-3} in Zn doped InP, the introduction rate of H4 had fallen to the point where H5 was the major defect.[5] The activation energy of the annealing of H4 was measured to be 1.35 eV, in good agreement with 1.32 eV measured by Sibile[11].

The defect H3 was found in all three materials, but the introduction rate was lowest in the lightly Zn doped InP. Within the experimental uncertanty, the concentration of H3 did not vary between the heavily Zn and Cd doped InP. The annealing rate of H3 varied significantly between the three materials: it was highest in the Cd doped InP and lowest in the lightly Zn doped InP. The activation energies for annealing in the heavily Zn and Cd doped InP were 1.09 and 1.01 eV, in reasonable agreement with the 1.2 eV measured by Sibile.[11] The activation energy for annealing of H3 in the lightly Zn doped InP was much higher, 1.47 eV, the measurement of the annealing rate in this case may have been complicated by the initial growth of H3. Nevertheless the annealing rate of H3 was higher in the Cd doped InP than in comparably Zn doped InP. A possible reason for this observation may be the larger size and slower diffusion of the Cd atom. During irradiation of InP, Zn or Cd interstitials are created; the slower diffusion of Cd than Zn [12] would lead to a slower loss of Cd interstitials than Zn interstitials at sinks. If the annealing of H3 involved the dopant interstitial, a higher annealing rate would be observed in Cd doped InP. The elimination of the related defect H4 by an interaction with the dopant atom has been suggested by Yamaguchi.[5]

The introduction rate of H5 was highest in the heavily Zn doped InP, lower in the lightly Zn doped InP and lowest in the heavily Cd doped InP. The introduction rate of H5 has been observed to increase with increasing doping concentration by Yamaguchi, who identifies H5 as an impurity/defect complex. The fact that the introduction rate is lower in Cd doped InP than in comparably Zn doped InP, may again be due to the slower diffusion of the Cd atom. If H5 is a dopant/defect complex the slower diffusion of Cd could limit its formation and thus its introduction rate. The annealing rate of H5, although of a complex order, was slowest in the highly Zn doped InP, faster in the lightly Zn doped InP, and fastest in the Cd doped InP. This may also be a result of the slow diffusion of Cd limiting the room temperature formation rate of H5: a larger concentration of the unreacted constituents for H5 may be available in the Cd doped InP, and thus on annealing it grows at the most rapid rate.

CONCLUSIONS

The energy levels and capture cross sections of irradiation induced defects in InP are not affected by the substitution of Cd for Zn. However, the introduction rate of defects is lower in heavily Cd doped InP than in comparably Zn doped InP. The annealing rate of defects is also higher in the Cd doped InP. These results suggest that the effect of dopant type on defects in p-type InP is limited to the rate of formation and annealing reactions, as the energy levels and capture cross sections are not affected by the substitution of Cd for Zn. Nonetheless the effect of Cd doping is to reduce the defect introduction rate, and thus Cd doped InP may allow the fabrication of even more highly radiation resistant InP solar cells.

REFERENCES

1. M. Yamaguchi, C. Umera, and A. Yamamoto , Jpn. Appl. Phys., 23, 302 (1984).
2. I. Weinberg, C.Swartz and R.Hart, Proc. of 18th IEEE PVSC., 723, (1985).
3. A. Sibile and J. Bourgoin, Appl. Phys. Lett. 41, 956 (1982).
4. A. Sibile, J. Suski and M. Gilleron, J. Appl. Phys, 60, 595 (1986).
5. M. Yamaguchi and K. Ando, J. Appl. Phys., 60, 935 (1986).
6. S. Messenger, R. Walters and G. Summers, J.Appl. Phys., 71, 4201 (1992).
7. A. Telia, B. Lepely and C. Michael , J. Appl. Phys., 69, 7159 (1991).
8. Deiter Schroder, Semiconductor Materials and Device Characterization,
       (John Wiley, NY, 1990), p.310.
9. Y. Zohta and M. Watanabe, J. Appl. Phys, 53, 1809 (1982).
10. A. Sibile and J.Suski, Phys. Rev. B., 31, 5551 (1985).
11. A. Sibile, Phys. Rev. B., 35, 3929 (1987).
12. N. Chand and P. Huston, J. Elec. Mater., 11, 37 (1982).

THEORY OF RARE-EARTH IMPURITIES IN SEMICONDUCTORS

C. Delerue and M. Lannoo
Institut d'Electronique et de Microélectronique du Nord
U.M.R. C.N.R.S. 9929
Département Institut Supérieur d'Electronique du Nord
41 boulevard Vauban, 59046 Lille CEDEX France

The theory of rare-earth impurities in semiconductors is reviewed. Results of Green's function calculations are presented. The particular case of Ytterbium (Yb) in InP is analyzed in detail. We show that the optical transitions of Yb in InP and their Zeeman splitting can be interpreted by a small coupling between quasi degenerate 5d and 4f states. It also explains the observed lifetimes. This model is coherent with the attribution of the E_c-30 meV level to the Yb^{3+}/Yb^{2+} acceptor level as proposed recently. We detail the mechanism which allows the energy transfer between a free exciton and the rare-earth impurity. The conditions required for an efficient luminescence are analyzed.

I - INTRODUCTION

The introduction of rare-earth impurities in semiconductors is of technological interest for their potential optical properties. Rare-earth doped semiconductors can emit light under optical or electrical excitation. The emission is characterized by lines corresponding to intra-4f-shell transitions. These lines are very sharp because 4f states remain atomic-like, i.e. the coupling with other valence states is small. However, this coupling is essential because it allows the transfer of energy between free excitons and 4f internal excitations which explains the luminescence of rare-earth impurities in semiconductors. The mechanism of the energy transfer is not yet definitely understood [1]. The aim of this paper is to discuss the theory of rare-earth impurities. In a first part, we present results obtained using a tight binding Green's function calculation. Chemical trends of rare-earth impurities are analyzed and compared to other independent studies. In the following parts, we concentrate on the case of Ytterbium in InP for which the experimental data are the most important. We show that the attribution of the E_c-30 meV level to the Yb^{3+}/Yb^{2+} acceptor level and a small 4f-5d coupling coherently explain the Zeeman splitting of the luminescence lines, the radiative lifetime, the delocalization of the 4f excited level and the mechanism of the energy transfer.

II - CHEMICAL TRENDS

In this section, we summarize the main results of a tight binding calculation of the electronic structure of substitutional rare-earth impurities in semiconductors. Details have already been published elsewhere [2]. For most of the free rare-earth atoms, the valence electronic configuration can be written as $4f^{n+1}6s^2$. As mentioned already, the 4f orbitals are very contracted on the atom and therefore can be considered in a first approximation as a frozen core. In contrast, the 6s shell is very extended. In the free atom, the 5d shell is usually empty but in compounds - in rare-earth metals for example -, 5d states participate to the bonding and some 4f electrons can be transferred to the 5d shell. Note that 5d orbitals are more extended than 3d states of transition metal atoms for comparison.

In our tight binding framework, the rare-earth atom is represented by 4f, 5d and 6s orbitals and

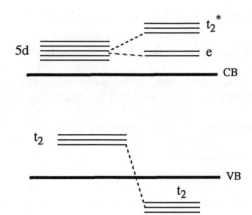

Fig. 1: level structure in the defect molecule model.

atoms of the semiconductor by s and p orbitals. The calculation is done for the three possible frozen 4f configurations, $4f^{n-1}$, $4f^n$ and $4f^{n+1}$. As the neutral valence population of rare-earth atoms is n+3, these correspond to 4+, 3+ and 2+ oxidation states because it is usually assumed that electrons of the 5d and 6s shells are transferred to the more electronegative neighbor atoms. The 6s orbitals are so extended that they strongly interact with s,p states of the neighbors. In consequence, the 6s density is equally shared between the valence and conduction band leading to a $6s^1$ configuration [2]. So it we are left with the problem of the 5d states which is quite similar to the one of transition metal impurities [3].

Figure 1 summarizes the situation for the 5d states of rare-earth impurities at substitutional cation sites. 5d states transform like e and t_2 in T_d symmetry. Doubly degenerate e states remain nearly uncoupled (they are uncoupled in a simple molecular model where only d atomic orbitals and sp^3 hybrid states on the neighbor atoms are taken into account [3]). $t_2(5d)$ states interact with the t_2 combinations of sp^3 states on the neighbor atoms. This leads to triply degenerate bonding t_2 and antibonding t_2^* states. In the case of transition metal impurities, e or t_2^* states can be in the forbidden gap [3]. For rare-earth impurities, we obtain no gap levels in most cases. The reason is twofold: first, the 5d level is very high in energy and secondly, the coupling between d states and s,p neighbor states is ~ 2.5 times larger for rare-earth impurities than for transition metal impurities [2]. The t_2 level is always in the valence band (at least in the 3+ oxidation state) and the e and t_2^* states are always in the conduction band. The charge state of the defect is entirely determined by the number of 4f electrons. For example, a rare-earth in a 3+ oxidation state is neutral in a III-V compound, negatively and positively charged respectively in IV and II-VI semiconductors.

We have also calculated the number of 4f electrons corresponding to the stable configuration. This requires the knowledge of the ionization levels $E_{m,m-1}$

$$E_{m, m-1} = E(m) - E(m-1) \tag{1}$$

where E(m) is the total energy of the impurity in the $4f^m$ configuration. The position of the Fermi level E_F with respect to the ionization level determines the stable configuration. If E_F is lower than $E_{m,m-1}$ the $4f^{m-1}$ configuration is the stable one otherwise it is $4f^m$. We have calculated

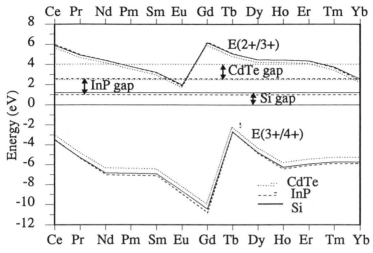

Fig. 2: trends in the E(2+,3+) and E(3+,4+) levels for different semiconductors.

the levels for m = n and m = n+1 which can be more conveniently written E(3+/4+) and E(2+/3+) making reference to the oxidation states. Results are plotted on Figure 2 for Si, InP and CdTe. These can be plotted on the same diagram because we have verified that the trends along the rare-earth series are identical within 0.2 eV irrespective of the host except for a rigid shift of the band structure. The same kind of result has been obtained for transition metal impurities in semiconductors. It is due to the efficient screening in the d shell of such systems [4].

We see on Figure 2 that the E(3+/4+) level is always in the valence band and the E(2+/3+) level is in the conduction band except for few exceptions. It means that the 3+ state is the stable one. In the case of Sm, Yb, Tm and Eu, the 2+ state is possible. The comparison with experimental data is not really meaningful because the lattice position of the impurity is usually not known. However, we can mention that in III-V materials, one observes Er^{3+} in InP [5], GaAs [6][7] and GaP [6]; Nd^{3+} in GaAs [6] and GaP [6]; Tm^{3+} in GaAs [6] and Pr^{3+} in GaP [8], Yb^{3+} in GaAs [9], GaP [9] and in InP [10][11]. Our results agree with these observations. Note however that our Yb(2+/3+) level in InP is practically degenerate with the bottom of the conduction band. In II-VI semiconductors, Nd^{3+}, Gd^{3+}, Er^{3+} and Eu^{2+} have been found in CdTe in accordance with our calculation [12][13]. Yb is seen in the 3+ state in CdTe while we predict a 2+ state. We must mention that we do not expect an accuracy of our levels better than 0.5 eV even if we believe that Fig. 2 gives the correct trends along the series. This uncertainty is due in particular to the importance of correlation effects in the electronic structure of 4f states. This is a very complex problem and these correlation effects cannot presently be calculated accurately even using the best local density techniques.

Recently, using these results and the analogy with transition metal impurities, Langer has shown that the ionization energies follow a unique curve along the rare-earth series in a class of isovalent semiconductors provided that the band structures are shifted by their natural band offsets [14]. Therefore, the experimental knowledge of one ionization level of one rare-earth impurity allows to position accurately the ionization levels for the whole rare-earth series in isovalent semiconductors. For example, he has found that our E(2+/3+) level should be shifted upwards by 0.6 eV in the II-VI compounds [14].

Recently, results of a local density calculation of Er in silicon at substitutional, interstitial and hexagonal interstitial sites have been published [15]. The 4f shell is also treated as a frozen core since the coupling of the 4f orbitals with the host is small [16]. The total energy is determined for different oxidation states. It is found that the 3+ oxidation state is always the stable one. The 6s density is shared between the valence and conduction bands and the 5d atomic state is high in the conduction band for the substitutional impurity. All these results are in agreement with ours. The tetrahedral interstitial site is found to be the stable one. We must mention however that the luminescent Erbium in silicon seems to be a complex with oxygen [16][17].

III Ytterbium in InP

Yb in InP has been extensively studied because it can be easily introduced in the material and because there is experimental evidence that it occupies a substitutional cation site [11][18]. It gives a strong luminescence at ~ 1.23 eV which is interpreted as the transition between the spin-orbit split 4f levels $^2F_{5/2} \rightarrow {}^2F_{7/2}$ of Yb^{3+} [1][11] (Yb^{3+} has 13 electrons in the 4f shell which corresponds to one hole in the closed shell). The mechanism which allows the energy transfer from a free exciton to the 4f shell is still unclear. It must obviously imply some coupling between the 4f orbitals and the host. Therefore we need to go beyond the model of the frozen 4f core. Similarly, a study of the Zeeman splittings of the luminescence lines shows that the excited state $^2F_{5/2}$ must be delocalized by 25% [11] (the ground state $^2F_{7/2}$ has an orbital reduction factor of only 5%). This large delocalization has to be explained. In the following, we analyze these points in the light of the tight binding model described above.

III.1 Origin of the electron and hole traps

The knowledge of the electronic structure of Yb in the bandgap of InP is of prime importance to interpret the luminescence. An acceptor level has been measured at E_c - 30 meV [19]. Its origin is still a matter of controversy. With Electron Paramagnetic Resonance, the signal of Yb^{3+} is observed both in p-type and n-type materials [10]. It was deduced that the acceptor level is not the Yb(2+/3+) level [10]. However, very recently, Bohnert et al. have suggested that inhomogeneities

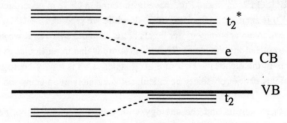

Fig. 3: the polarizability of the core of the RE pushes traps towards the bandgap.

in the sample could explain the observation of Yb^{3+} in n-type samples and that in fact the acceptor level would be the Yb(2+/3+) level [20]. Therefore, other experimental studies are necessary to conclude. A hole trap at $E_v + 50$ meV is also reported [21]. Since our calculation does not predict any level in the bandgap, the origin of these traps must be discussed.

As mentioned already, our calculation gives a Yb(2+/3+) level very close to the bottom of the conduction band. Therefore the assignation of the acceptor level to Yb(2+/3+) is possible taking into account the accuracy of our results [20].

Another possibility is to suppose that the traps correspond to 5d derived levels. In effect, the e(5d) level could be the acceptor level and the t_2 level the hole trap (our calculation gives for Yb^{3+} the t_2 level at ~ $E_v - 0.5$ eV, e at ~ $E_c + 1.5$ eV and t_2^* at ~ $E_c + 4.0$ eV). Recently, we have shown that the lattice distortion could not explain the presence of the observed levels [22]. In the same paper, we have discussed in detail the influence of the polarizability on the 5d levels. We find that the core of the rare-earth atom is very polarizable. The interesting point is that this leads to a potential attractive for both an electron and a hole. Therefore, the electron and hole levels are pushed towards the bandgap (see Figure 3). This effect is not included in our tight binding calculation (it would be present only in a GW calculation). Using a simple model [22], we obtain that the e(5d) level shifts from $E_c + 1.5$ eV to $E_c + 0.2$V and the t_2 level from $E_v - 0.5$ eV to $E_v - 0.2$ eV for Yb^{3+} in InP. Therefore the core polarizability can possibly push 5d states into the bandgap. However, our calculation is too simple and a more accurate treatment is necessary to conclude.

III.2 Delocalization, Zeeman splittings of the 4f states and detailed model of the energy transfer

In this part, we want to study the physical origin of the large delocalization of the $^2F_{5/2}$ state. Before discussing this point, let us summarize the present models proposed for the excitation of the luminescence. The situation is described on the total energy diagram of Figure 4. The mechanisms proposed in the literature can be decomposed into different steps [1][20][24]. First, an electron-hole pair is created in the semiconductor. Then, the electron is captured on the acceptor level at $E_c - 30$ meV (AE). The defect is now negatively charged and can capture a hole by Coulomb attraction (BE on Figure 4). Then the energy is transferred to the 4f shell leaving an excited $Yb^{3+}(^2F_{5/2})$ impurity. Finally, the relaxation $^2F_{5/2} \rightarrow {}^2F_{7/2}$ occurs by emission of a photon. As already discussed, in refs. [1][10] the acceptor level is an isovalent state while in ref. [20], it is the Yb(2+/3+) level.

We have shown in a previous paper that the delocalization of 25% in the excited state $^2F_{5/2}$ can be qualitatively interpreted using simple perturbation theory [23]. In the frozen core approximation, the $^2F_{5/2}$ and $^2F_{7/2}$ states are pure 4f states. If we allow some perturbation V not included in this scheme, the orbital reduction factor will be proportional to V/Δ where Δ is some average distance in energy between the 4f state and the other states with higher energy. From Figure 4, we see that the $^2F_{5/2}$ state is much closer to the other states - in particular to the continuum - than the $^2F_{7/2}$ state. Therefore, the delocalization must be more important in the $^2F_{5/2}$ state than in the $^2F_{7/2}$ state.

Obviously, if the previous model explains the delocalization, it does not give the nature of the coupling V. We have also recently proposed a simple model of the electronic structure of Yb which explains both the delocalization and the Zeeman splittings of the 4f states [22]. Here we

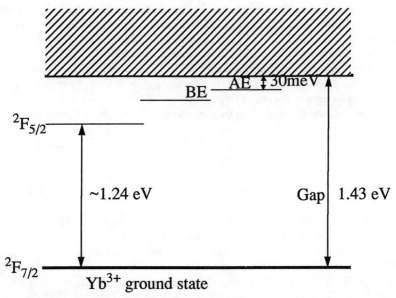

$^2F_{5/2}$

$^2F_{7/2}$

BE —— AE $\updownarrow$ 30meV

~1.24 eV

Gap | 1.43 eV

Yb^{3+} ground state

Fig. 4: Total energy diagram characteristic of Yb in InP

describe this model and we try to connect it with the total energy diagram of Fig. 4 (this comparison was not performed in ref. [22] because only preliminary results were given). It starts from the tight binding model described in section I but with the introduction of the coupling of the 4f orbitals to the host states. Because Yb^{3+} has 13 electrons in the 4f shell corresponding to one hole in the closed shell, the problem is treated using one-electron theory. We have only to remember that going from a one-electron energy diagram to a total energy diagram, we have to invert all the energies since we are dealing with a hole.

The coupling of the 4f states can have two origins. A first one is the coupling with the s,p orbitals of the nearest neighbors. In ref. [23], we have shown that this is very small. A second origin is the crystal field. The crystal field leads to two kinds of perturbations. First, there are terms within the 4f states which simply split the levels. This does not induce a delocalization of the 4f states. Secondly, the crystal field couples 4f states to states of different atomic multiplets with similar representations. In our case, there can be a coupling between 4f and 5d states. As seen in section I, the 5d orbitals are actually spread into the t_2 bonding and t_2^* antibonding states (Fig. 1). Therefore, there is a coupling between 4f states of t_2 symmetry with the t_2 and t_2^* states (see Fig. 5).

The coupling of the 4f states must be very small otherwise the observed $^2F_{5/2} \rightarrow {}^2F_{7/2}$ transition energy would not be close to the one in the free Yb^{3+} ion. This is due to the screening of the crystal field by the $5s^2 5p^6$ shell whose maximum is outside the 4f shell [25]. The estimation of the effect of the crystal field is not easy. In consequence, we have fitted to the experimental data on the Zeeman splittings of the luminescence lines [11]. Details of the calculation are given in the Appendix.

The results are summarized on Figure 5. Small letters represent one-electron states. The main

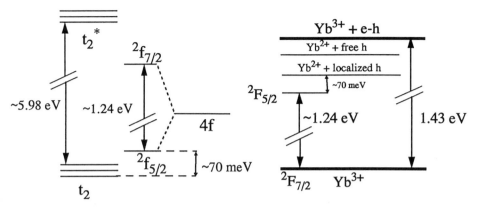

Fig. 5 (left): electronic structure in a one-electron model including the coupling between 4f and 5d states obtained by fitting the Zeeman splittings of the luminescence lines. Fig. 6 (right): equivalent total energy diagram when Yb(2+/3+) is the acceptor level and the hole is localized on the t_2 bonding level.

results of the fit are that we obtain the $^2f_{5/2}$ level very close to the t_2 bonding level and the final average splitting between the $^2f_{5/2}$ and t_2 levels is of the order of 70 meV (in fact, there is only a coupling between states of $\Gamma 8$ symmetry in the double group representation and levels are also split by the crystal field: see Appendix for details). Therefore, even if the interaction between these two levels is small (~ 35.3 meV), the delocalization of some $^2f_{5/2}$ states can reach 20% on the t_2 bonding states. In contrast, since the $^2f_{7/2}$ states are far in energy from the t_2 and t_2^* states, they remain localized on the 4f shell. We conclude that the delocalization of the $^2f_{5/2}$ state and the Zeeman splittings of the luminescence lines [11] are interpretable by a quasi degeneracy and a small coupling of the $^2f_{5/2}$ and t_2 levels. This model also explains the $\Gamma 6$-$\Gamma 8$ level reversal observed within the $^2f_{5/2}$ state [11] (see Appendix).

If the one-electron model explains well some experimental observations, it is essential to verify how it is compatible with the total energy diagram of Figure 4. We need to look at the nature of the coupling of the $^2F_{5/2}$ state with states of higher energies (BE, AE and the continuum on Figure 4). We analyze the two possible situations for the acceptor level (AE) at E_c - 30 meV discussed in section III.1. In the first model, the acceptor level is not the Yb(2+/3+) level, i.e. it is an electron trap on a level different from the 4f level (isoelectronic trap) [1][10]. After capture of a hole, the defect corresponds to a bound exciton (BE on Figure 4). In that case, the many-electron wave functions of the $^2F_{5/2}$ and BE states (or AE + a free hole in the valence band) differ by two single particle states. Therefore, their coupling cannot be described by a one-electron matrix element as described above. Of course, there could be one-electron couplings with states in resonance in the continuum but the difference in energy between these states and the $^2F_{5/2}$ state should be at least equal to 1.43 - 1.24 - 0.03 = 0.160 eV which is too large compared to the 70 meV splitting obtained from the fit of Zeeman lines (it is 0.160 eV because there is actually a continuum above the AE level on Figure 4 corresponding to an electron on the electron trap and a free hole in the valence band). We conclude that the model of the isoelectronic trap is not compatible with the one-electron model.

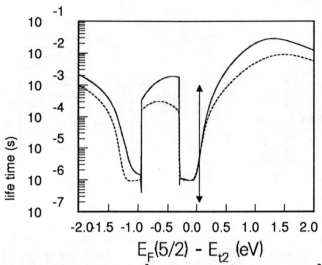

Fig. 7: lifetime for the excited state $^2F_{5/2}$ as function of the position of the $^2f_{5/2}$ level with respect to the t_2(5d) level (continuous line T = 30 K, dashed line T = 305 K). The arrow corresponds to the case of Yb as obtained from the fit of the luminescence lines.

The second model for the acceptor level is the 4f level Yb(2+/3+) (Figure 6) [20]. The bound exciton (BE) corresponds to Yb^{2+} with a hole on some bound state [20]. In that case, the $^2F_{5/2}$ and BE many-electron wave functions only differ by one single particle state. The coupling between the two is just the one-electron matrix element between a 4f state and the hole bound state. Therefore it coincides with our one-electron coupling parameter V if the hole bound state is the t_2 bonding state. The difference between the $^2F_{5/2}$ and BE levels would be equal to ~ 70 meV. Since the difference in energy between the $^2F_{5/2}$ and AE levels is equal to ~ 160 meV (Figure 4 and 6), it means that the binding energy of the hole would be of the order of 160 - 70 = 90 meV. It is a quite large value which corresponds to an already deep level. This would be compatible with a t_2 bonding level ~ 90 meV above the valence band in the one-electron model. An important point is that it is the t_2 ionization level for Yb^{2+}. So we have done again the Green's function calculation for Yb^{2+}. We obtain that the t_2 bonding level is in the bandgap, close to the top of the valence band. This is due to the fact that the 5d energy is higher for Yb^{2+} than for Yb^{3+}. Therefore the model of Yb(2+/3+) at E_c - 30 meV plus a hole trap for Yb^{2+} [20] is supported by our calculation.

In the last model, the mechanism of energy transfer is quite simple (see Fig. 4 and 5). After excitation of an electron-hole pair, the electron is trapped on the Yb(2+/3+) acceptor level. The impurity is in the oxidation state Yb^{2+} and is negatively charged. Then the hole can be captured on the t_2 bonding level which is relatively deep in the bandgap (~ E_v + 90 meV). The hole is easily transferred non radiatively to the $^2f_{5/2}$ level because there is a small coupling between the quasi degenerate $^2f_{5/2}$ and t_2 states (Fig. 5). Finally, the hole relaxes to the $^2f_{7/2}$ states by emission of a photon. This gives a coherent microscopic model for the mechanism proposed by Bohnert et al. [20].

III.3 Optical matrix elements for Yb in InP

With the one-electron wave functions calculated with the Hamiltonian described above, we have computed the optical matrix elements using atomic orbitals [26]. Lifetimes are obtained from the Fermi Golden rule assuming a Boltzmann distribution in the excited states [27]. The essential point is that optical matrix elements are non vanishing only between 4f and 5d atomic orbitals of the rare-earth. We calculate a lifetime between 6 µs and 2 µs depending on the temperature. This is in agreement with the experimental lifetime of the luminescence of 10 µs [28]. We interpret this quite short and unusual lifetime for a rare-earth impurity to the delocalization of the $^2f_{5/2}$ states on t_2 5d states. Indeed, for another rare-earth impurity, the $^2f_{5/2}$ and t_2 levels have no reasons to be quasi degenerate. On Figure 7, we plot the lifetime when we shift the $^2f_{5/2}$ level with respect to the t_2 level. We see that very quickly, the lifetime is of the order of 1-10 ms, in agreement for example with the estimated lifetime of ~ 1 ms for Er in InP [28].

Very recently, Bohnert et al. [20] have observed using time-resolved photoluminescence another broad emission of light between ~ 1.31 eV and ~ 1.25 eV connected to Yb in InP which they attributed to a donor-acceptor like transition. Assuming a donor D at a distance R away from the Yb ion, such transition would correspond to

$$D^0 + Yb^{3+}\left({}^2F_{5/2} \right) \rightarrow D^+ + Yb^{2+} + h\nu\,(R) \tag{2}$$

and would be in competition with the "normal" transition

$$Yb^{3+}\left({}^2F_{5/2} \right) \rightarrow Yb^{3+}\left({}^2F_{7/2} \right). \tag{3}$$

In the following, we show that the efficiency of the transition (2) is negligible compared to (3) and propose an alternative explanation. We have just seen that the transition (3) is slightly allowed because there is some 5d mixing in the $^2f_{5/2}$ states. The transition (2) corresponds to the transition of an electron from a donor state to the very localized 4f state. In the most favorable case, the probability of transition for process (2) is the one of process (3) multiplied by the weight of the donor envelope function on the rare-earth atom. This factor is equal to

$$\frac{1}{8\pi}\left(\frac{c}{a}\right)^3 e^{-2\frac{R}{a}} \tag{4}$$

where R is the distance between the rare-earth impurity and the donor, a is the effective Bohr radius and c is the cubic lattice constant of InP. Eqn. (4) gives a factor 1.25 x 10^{-5} for R = 0 and 2.04 x 10^{-6} for R being the average distance between donor atoms for a donor concentration N = 5. 10^{17} cm^{-3} used in ref. [20] (R ~ $(3/4\pi N)^{1/3}$). We see that in any case the process (3) is strongly favored and (2) is unlikely.

Instead of transition (3), we propose that the observed emission corresponds to the relaxation of a hole from the t_2 bonding level to the $^2f_{7/2}$ level (see Figure 5). The probability of emission is of the same order than the $^2f_{5/2}$ - $^2f_{7/2}$ transition since $^2f_{5/2}$ and t_2 states are mixed (the probability is even higher). The energy of the transition would be ~ 1.24 + 0.07 = ~ 1.31 eV (Figure 5). With a broadening and a Franck-Condon shift induced by a possible electron-phonon coupling occurring in the t_2 states, this could explain the photon emission between ~ 1.31 eV and ~ 1.25 eV. Of course other studies are necessary to confirm this interpretation.

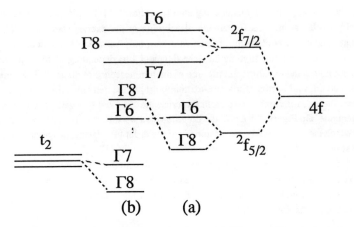

Fig. 8 : electronic structure in the defect molecule model. The case (a) only includes the effect of the crystal-field within 4f states. Case (b) includes the effect of the crystal-field within 4f states and the coupling V to the 5d states (mainly to the t_2 bonding state).

IV CONCLUSION

We have presented a Green's function calculation of the electronic structure of substitutional rare-earth impurities in semiconductors. We have shown that the luminescence of Yb in InP can be explained by the quasi degenerescence of the $^2f_{5/2}$ and $t_2(5d)$ levels and by the assignation of the acceptor level to the Yb(2+/3+) level as recently proposed by Bohnert et al. [20].

APPENDIX: Detailed model of the Zeeman splittings of the luminescence lines

Here we give details on the calculation of the coupling between 5d and 4f states. It can be simplified using group theory. In T_d symmetry, f states transform like a_1 (non degenerate), t_2 and t_1 states (triply degenerate) while d states behave like e and t_2. First, the crystal field splits states of different representations, i.e. a_1, t_2 and t_1 4f states have no longer the same energies. Secondly, as discussed previously, it couples the 5d and 4f states. In the present case, only the 5d and 4f states of t_2 symmetry can be coupled. Let us call V this coupling. If we define by α^2 the localization of the t_2 bonding state on the 5d orbitals, the localization of the $t_2{}^*$ antibonding state is approximately equal to 1- α^2 (strictly speaking, this is only valid in the molecular model [3]). Therefore the coupling of the 4f orbitals of t_2 symmetry with the t_2 and $t_2{}^*$ states is respectively equal to αV and $(1-\alpha^2)^{1/2}$V. The 4f states are also split by the spin-orbit coupling $\lambda_f\left(\vec{l}\cdot\vec{s}\right)$.

We write the Hamiltonian matrix in the basis of the 4f, t_2 and $t_2{}^*$ spin states. The e(5d) states are not included because they do not interact with 4f states. The energy of the t_2 bonding state is taken as reference. The diagonal terms of the matrix are E_{t_2}, $E_{t_2{}^*}$ the energies of the t_2 and $t_2{}^*$ 5d states and E_{ft2}, E_{fa1}, E_{ft1} the energies of the 4f states respectively of t_2, a_1 and t_1 symmetry (they are different because of the crystal field). For convenience, we write: $E_{fa1} = E_{ft2} + \Delta_{t2a1}$ and $E_{ft1} = E_{ft2} + \Delta_{t2t1}$. The difference $E_{t_2{}^*} - E_{t_2} = 5.98 eV$ and $\alpha^2 = 0.38$ are obtained from the Green's function

326

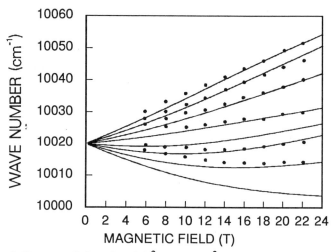

Fig. 9: Zeeman splitting of the $\Gamma 8(^2f_{5/2}) \rightarrow \Gamma 6 \, (^2f_{7/2})$ transition in high magnetic fields for H// <100>.

calculation for Yb^{3+}. V, E_{ft2}, Δ_{t2a1}, Δ_{t2t1} are taken as free parameters. The Zeeman Hamiltonian $\beta\left(g_0\vec{s}+\vec{l}\right) \cdot \vec{H}$ is added when necessary ($\vec{H}$ is the magnetic field). The Hamiltonian is diagonalized numerically. Note that the method is slightly different from the one already published [22] because we do not take into account the spin-orbit coupling within the 5d states for simplicity. Its inclusion does not change the results in a significant way.

Results are summarized on Figure 8. From group theory, we know that $^2f_{5/2}$ transforms into $\Gamma 8$ + $\Gamma 6$, $^2f_{7/2}$ into $\Gamma 8 + \Gamma 7 + \Gamma 6$ and t_2 (or t_2^*) into $\Gamma 8 + \Gamma 7$ in the double point group representations. Parameters are adjusted to get a perfect fit for all the observed transitions (Figure 5 of ref. [11]) and a correct fit of the Zeeman splittings of the $\Gamma 8(^2f_{5/2}) \rightarrow \Gamma 6 \, (^2f_{7/2})$ transition [11] (see Figure 9). We obtain V = 57.3 meV, Δ_{t2a1} = 12.4 meV and Δ_{t2t1} = 11.8 meV. The important parameter is actually the coupling between the t_2 and $^2f_{5/2}$ states because they are quasi degenerate (αV = 35.3 meV).

It is interesting to look at the effect of the crystal field on the 4f shell alone. This can be done easily in our calculation by just imposing V = 0 leaving the other parameters unchanged. We obtain the spectrum of Fig. 8 (a). The order of the $\Gamma 8$-$\Gamma 6$ levels is the one predicted by a simple point charge model of the crystal field [11]. We see that the introduction of the coupling V leads to an interaction of the $\Gamma 8$ states of t_2 and $^2f_{5/2}$ levels. This pushes the $\Gamma 8(^2f_{5/2})$ to higher energies, above the $\Gamma 6(^2f_{5/2})$ (Case (b) on Fig. 8). Therefore **the unusual $\Gamma 6$-$\Gamma 8$ level reversal observed within the $^2f_{5/2}$ state [11] is due to the coupling to the t_2(5d) bonding state.**

REFERENCES

[1] K. Takahei, Proceedings of the 21th International Conference on the Physics of Semiconductors, ed. by Ping. Jiang and Hou-Zhi Zheng, World Scientific Publishing, Beijing (1992).
[2] C. Delerue and M. Lannoo, Phys.Rev.Lett. 67, 3006 (1991).
[3] C. Delerue, M. Lannoo and G. Allan, Phys.Rev.B 39, 1669 (1989).
[4] C. Delerue, M. Lannoo and J.M. Langer, Phys.Rev.Lett. 61, 199 (1988).

[5] B. Lambert, A. Le Corre, Y. Toudic, C. Lhomer, G. Grandpierre and M. Gauneau, J. Phys. Condensed Matter **2**, 479 (1990).

[6] M. Ennen and J. Schneider, in Proceedings of the Thirteenth International Conference on Defects in Semiconductors, Coronado, CA, 1984, edited by L.C. Kimerling and J.M. Parsey (Metallurgical Society of AIME, New York, 1985), Vol. **14a**, p. 115.

[7] H. Ennen, J. Wagner, H.D. Mueller and R.S. Smith, J. Appl. Phys. **61**, 4877 (1987).

[8] V.A. Kasatkin, F.P. Kesumanly and B.E. Samorukov, Fiz. Tekh. Poluprovodn. **15**, 616 (1986) [Sov.Phys.Semicond. **15**, 352 (1981)].

[9] H. Ennen, G. Pomrenke and A. Axmann, J.Appl.Phys. **57**, 2182 (1985).

[10] B. Lambert, Y. Toudic, G. Grandpierre, A. Rupert and A. Le Corre, Electron. Lett. **24**, 1446 (1988).

[11] G. Aszodi, J. Weber, Ch. Uihlein, L. Pu-Lin, H. Ennen, U. Kaufmann, J. Schneider and J. Windscheif, Phys.Rev.B **31**, 7767 (1985).

[12] R. Boyn, Phys.Stat.Solidi **148**, 11 (1988).

[13] R.S. Title, in Physics and Chemistry of II-VI compounds, edited by M. Aven and J.S. Prener (North-Holland, Amsterdam, 1967).

[14] J.M. Langer, Proceedings of the 17th International Conference on Defects in Semiconductors, Gmunden, Austria, to be published (1993).

[15] M. Needels, M. Schlüter and M. Lannoo, Phys.Rev.B, **47**, 15533 (1993).

[16] J. Michel, J.L. Benton, R.F. Ferrante, D.C. Jacobson, D.J. Eaglesham, E.A. Fitzgerald, Y.H. Xie, J.M. Poate and L.C. Kimerling, J. Appl.Phys. **70**, 2672 (1991).

[17] D.L. Adler, D.C. Jacobson, D.J. Eaglesham, M.A. Marcus, J.L. Benton, J.M. Poate and P.H. Citrin, Appl.Phys.Lett. **61**, 2181 (1992).

[18] A. Kozanecki and R. Grotzschel, J. Appl.Phys. **68**, 517 (1990).

[19] P.S. Whitney, K. Uwai, H. Nakagome and K. Takahei, Appl.Phys.Lett. **53**, 2074 (1988).

[20] G. Bohnert, J. Weber, F. Scholtz and A. Hangleiter, Appl.Phys.Lett. **63**, 382 (1993).

[21] D. Seghier, T. Benyattou, G. Bremond, F. Ducroquet, J. Gregoire, G. Guillot, C. Lhomer, B. Lambert, Y. Toudic and A. Le Corre, Appl.Phys.Lett. **60**, 983 (1992).

[22] C. Delerue and M. Lannoo, Proceedings of the 17th International Conference on Defects in Semiconductors, Gmunden, Austria, to be published (1993).

[23] M. Lannoo and C. Delerue, in the Proceedings of the MRS Spring Meeting (San Francisco), Symposium on Rare-Earth Doped Semiconductors, Volume 301 (Ed. G.S. Pomrenke, P.B. Klein, D.W. Langer), p. 385 (1993).

[24] B.J. Heijmink Liesert, M. Godlewski, A. Stapor, T. Gregorkiewicz, C.A.J. Ammerlaan, J. Weber, M. Moser and F. Scholz, Appl.Phys.Lett. **58**, 2237 (1993).

[25] B.G. Wybourne, Spectroscopic Properties of Rare Earths, Interscience Publishers, John Wiley and Sons, Inc., New-York (1965).

[26] F. Herman and S. Skillman, Atomic Structure Calculation, Prentice Hall, New York (1963).

[27] D.L. Dexter, in Solid State Physics, ed. by F. Seitz and D. Turnbull (Academic, New York, 1958), Vol. **6**, p. 360.

[28] C. Lhomer, F. Clérot, B. Lambert, B. Sermage and B. Deveaud, Proceedings of the 21th International Conference on the Physics of Semiconductors, ed. by Ping. Jiang and Hou-Zhi Zheng, World Scientific Publishing, Beijing (1992).

[29] There is a misprint in ref. [22]. The parameter V obtained by fitting the luminescence lines is equal to 75.1 meV instead of 121.8 meV (the calculation in ref. [22] included the spin-orbit coupling in the 5d shell contrarily to the present paper).

Rh: A DOPANT WITH MID-GAP LEVELS IN InP AND InGaAs AND SUPERIOR THERMAL STABILITY

H. Scheffler, B. Srocka, A. Dadgar, M. Kuttler, A. Knecht, R. Heitz, D. Bimberg,
J.Y. Hyeon*, and H. Schumann*
Institut für Festkörperphysik, Technische Universität Berlin, Hardenbergstraße 36,
10623 Berlin, Germany
*Institut für Anorganische und Analytische Chemie, Technische Universität Berlin,
Straße des 17. Juni 115, 10623 Berlin , Germany

ABSTRACT

We investigate the influence of Rh-doping on electrical properties of InP and InGaAs. Deep level transient spectroscopy carried out on Rh-doped as well as undoped reference samples reveals a single Rh-related deep level in both semiconductors. While a Rh-related electron trap in InGaAs is situated at E_C-0.38eV a Rh-related hole trap is found to exist in InP approximately 0.73 eV above the valence band edge. The internal reference rule for 3d transition metal deep levels is found to be valid for these 4d transition metal levels: their energy is constant across the InGaAs/InP heterojunction. We conclude these levels to be the single acceptor states of Rh substitutionally incorporated on cation sites.

INTRODUCTION

Transition metals (TMs) with energy levels near the middle of the band gap are of particular importance for the fabrication of semi-insulating (s.i.) III-V semiconductor layers needed for high speed electronic and optoelectronic devices. Since nominally undoped III-V semiconductors are commonly n-type, highly resistive layers are most simply obtained by doping with deep acceptors compensating the net shallow donors. In the case of InP, generally Fe is used for this purpose. However, Fe redistributes strongly at elevated temperatures during epitaxial growth or device processing steps. Redistribution and accumulation at interfaces often leads to degradation of device performance. Thus, there is a need for thermally more stable compensators. Because of their larger ion radii the 4d and 5d transition metals are predicted to fulfil the thermal stability demand and recently for Zr (4d) and Hf (5d) in InP a superior thermal stability compared to Fe could indeed be demonstrated [1].

While all 3d TMs introduce deep levels into the band-gap of InP, close to nothing is known about the electrical properties of the 4d and 5d TMs so far. On the one hand calculations carried out by Makiuchi et al. [2] lead to the prediction of 4d and 5d TM deep acceptor and donor levels, most of them situated energetically slightly above the corresponding levels of the isovalent 3d elements. On the other hand with deep level transient spectroscopy (DLTS), optical absorption, and electron paramagnetic resonance spectroscopy no TM-specific signals could be detected in n-type InP doped by the Liquid Encapsulated Czochralski (LEC) technique with the 4d ions Mo, Ru, Rh, and Pd [3]. In contrast to the latter study we report here on DLTS measurements on *both* n-type and p-type Rh-doped materials. Our purpose was to scan the whole band gap of InP and InGaAs to clearly detect possibly existing Rh-related levels as majority carrier emitting traps. We find new Rh-related levels for both InP and InGaAs.

Mat. Res. Soc. Symp. Proc. Vol. 325. ©1994 Materials Research Society

CRYSTAL GROWTH AND DIODE FABRICATION

All investigated InP samples are grown by Metal Organic Chemical Vapour Deposition (MOCVD) at a reduced pressure of 20mbar using a substrate temperature of 640°C. Doping with Rh was achieved with a new metalorganic compound especially synthesized for this purpose. It is cyclooctadienyl(cyclopentadienyl)rhodium. The $In_{0.53}Ga_{0.47}As$ layers are grown by Liquid Phase Epitaxy (LPE) using a conventional sliding boat technique. Growth from In-solution is carried out at a temperature of 635°C. For Rh-doping 3N rhodium is added to the growth solution giving for concentrations up to 0.5at% Rh in the growth solution mirrorlike epitaxial layers.

For samples to be investigated by DLTS a two-layer-sequence is grown on conductive InP:S or InP:Zn substrates. In these samples the first (bottom) layer is of same conductivity type as the substrate while the top one is of opposite type and highly doped ($\approx 10^{18}cm^{-3}$). Thus, an one-sided abrupt p/n-junction is formed. Two kinds of samples are fabricated containing either Rh-doped or undoped bottom layers, the latter ones being needed as reference samples for the interpretaion of DLTS spectra. p/n-junction mesa diodes are fabricated by photolithography and wet chemical etching.

EXPERIMENTAL RESULTS

During MOCVD growth InP can be doped by Rh in concentrations as high as $5 \times 10^{17} cm^{-3}$, as is determined by secondary ion mass spectroscopy (SIMS). A typical profile is shown in fig.1. For LPE grown InGaAs layers, however, the Rh-concentration is always below the SIMS detection limit of about $5 \times 10^{15} cm^{-3}$ in this case. All Rh-doped epitaxial layers as well as reference samples exhibit mirrorlike surfaces. Compared to undoped reference samples the X-ray rocking curves of Rh-doped layers are not broadened indicating perfect crystalline quality. By light microscopy inspection (lateral resolution of about 2μm) we find no indication for the formation of precipitates.

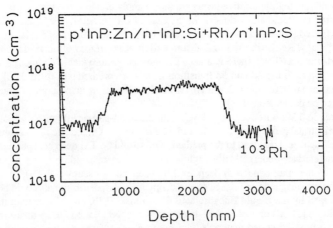

Fig.1: SIMS depth profile of a Rh-doped p^+InP:Zn/n-InP:Si+Rh structure grown by MOCVD on an n-type InP:S substrate. The detection limit is about 1×10^{17} Rh/cm^3 for this measurement.

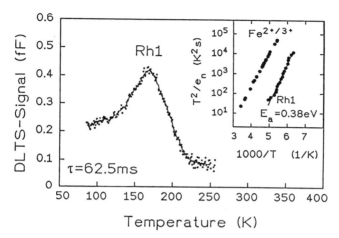

Fig.2: DLTS spectrum of Rh-doped n-type InGaAs for an emission time constant of τ=62.5ms. In the inset the thermal emission signature of the Rh1 level is compared to that of the Fe-acceptor level in InGaAs.

Rh-implanted InP samples are annealed at temperatures up to 750°C for 20min. Compared to the original implantation-profile after annealing no Rh-diffusion towards the bulk is observed by SIMS. Since Fe is known to diffuse under these annealing conditions [4] Rh is found to exhibit superior thermal stability compared to Fe.

Comparative Hall effect measurements on Rh-doped and reference samples give free carrier concentrations of about 1×10^{16}cm^{-3} not significantly influenced by doping with Rh for the highly Rh-doped InP-layers. Thus, in InP most of the Rh seems to be electrically inactive. However, Rh-related deep levels are detected in smaller concentrations in both InP and InGaAs by DLTS, which is many orders of magnitude more sensitive than Hall-effect measurements.

As seen in fig.2 the DLTS spectra recorded from n-type Rh-doped InGaAs layers show a majority carrier emission peak labelled Rh1. This level is not present in undoped reference samples. In addition, from p-type Rh-doped InGaAs layers no hole emission signals are observed. Thus, the peak Rh1 is caused by an electron trap. The Arrhenius-plot of the thermal electron emission rate is depicted in the inset of fig.2. The apparent activation energy is $E_a^{Rh1}=(0.38\pm0.03)$eV. This value is close to the activation energy of the Fe^{2+}/Fe^{3+} acceptor level. Although iron is known to be a major impurity in the Rh-source material used for the LPE growth, it is obvious from the inset of fig.2 that the level Rh1 is different from the Fe^{2+}/Fe^{3+} level we observed in Fe-doped InGaAs. Furthermore, unlike Fe^{2+}/Fe^{3+} this signal is not detected in hole emission on p-type Rh-doped layers. Thus, the Rh1 level seems to be directly related to Rh. Taking alloy broadening effects into account its concentration is calculated to $N_{Rh1}=8\times10^{13}$ cm^{-3}.

For InP a comparison of typical DLTS spectra obtained on various Rh-doped and reference samples is depicted in fig.3. While n-type MOCVD grown Rh-doped layers do not contain any electron traps in concentrations larger than 1×10^{13}cm^{-3}, in p-type material a quite intensive DLTS peak labelled Rh-A is observed for all Rh-doped samples. Its peak intensity corresponds to a deep level concentration of 8×10^{14}cm^{-3} for the MOCVD grown InP:Cd+Rh sample containing

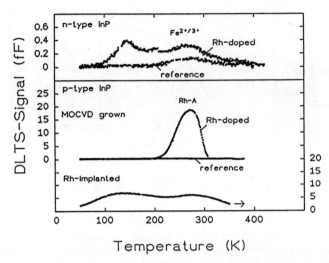

Fig.3: Comparison of DLTS spectra obtained on various n-type and p-type Rh-doped InP samples grown by MOCVD. The emission time constant is $\tau=62.5$ms.

1×10^{17}Rh/cm^3 measured by SIMS. In addition, the Rh-A level is also observed in an InP:Zn layer implanted with Rh ions. No hole traps having concentrations larger than 1×10^{13}cm^{-3} could be detected on InP:Cd reference layers.

The Arrhenius-plots of the thermal hole emission rate from the Rh-A level measured on three different Rh-doped layers are plotted in fig.4. The thermal hole emission signature of Rh-A is found to shift towards lower temperatures with increasing shallow acceptor concentration N_A. This behavior is typical for an electric field induced enhancement of emission rates. Thus, the electric fields present in the test zone of space charge layer during DLTS measurements is calculated from Capacitance-Voltage (C-V) measurements taking into account the λ-distance (crossover-point of Fermi level with deep level position). The electric field effect results in a dramatic reduction of the apparent activation energy with increased electric field ranging from $E_a=0.73$eV for a field of $F=5\times10^4$V/cm to $E_a=0.55$eV for $F=1.6\times10^5$V/cm. The energetic position of the Rh-A level within the band-gap of InP corresponds to the zero-field activation energy. In consequence, this level is situated in a distance of at least 730meV above the valence band edge.

For all investigated InP samples the Rh-A level concentration does not exceed 8×10^{14}cm^{-3}, which is much smaller than the Rh-concentration detected by SIMS (about 10^{17}cm^{-3}). Nevertheless, this level is almost certainly Rh-related, since Rh-A is observed in MOCVD grown Rh-doped samples as well as in Rh-implanted InP while it is absent in reference samples. In addition, this level can not be caused by 3d TMs introduced as contaminations of the Rh-precursor used for the MOCVD growth, since all hole traps formed by 3d TMs are known to have very different thermal emission signatures as is obvious from fig.4. Furthermore, for the Rh-implanted sample the presence of 3d TMs is very unlikely. However, a hole trap H5 known in literature [5] exhibits a thermal emission signature similar to Rh-A (see fig.4). Whether it is the same as Rh-A observed in our Rh-doped InP is unclear yet.

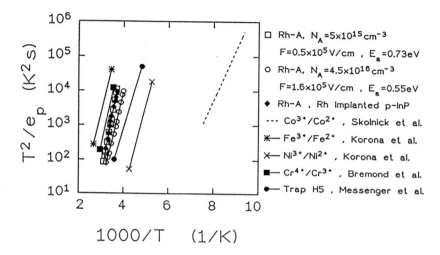

□ Rh-A, N_A=5x10¹⁵cm⁻³
  F=0.5x10⁵V/cm , E_a=0.73eV
○ Rh-A, N_A=4.5x10¹⁶cm⁻³
  F=1.6x10⁵V/cm , E_a=0.55eV
◆ Rh-A , Rh Implanted p-InP
--- Co³⁺/Co²⁺ , Skolnick et al.
✳— Fe³⁺/Fe²⁺ , Korona et al.
✕— Ni³⁺/Ni²⁺ , Korona et al.
■— Cr⁴⁺/Cr³⁺ , Bremond et al.
●— Trap H5 , Messenger et al.

Fig.4: Thermal hole emission signatures measured on Rh-doped InP compared with those of hole traps already known in the literature.

CONCLUSIONS

A correlation between the energy of isovalent transition metal levels within the semiconductor and the binding energy of the free TM ion valence electrons has been postulated [3]. Thus, in a semiconductor the level position should shift upwards going down in one column of the periodic table of elements from 3d to 4d and 5d. A similar trend can be deduced from calculations carried

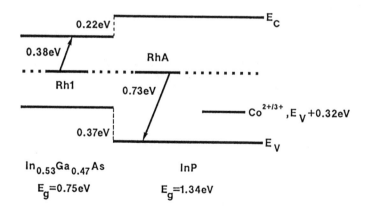

Fig.5: Schematical band diagram of a InGaAs/InP heterostructure with the Rh-related levels Rh1 in InGaAs and Rh-A in InP depicted.

out by Makiuchi et al. [2]. Co being isovalent to Rh introduces a deep acceptor level at $E_V+0.32eV$ in InP [6]. From this position the energetic level of the Rh-related hole trap Rh-A in InP is shifted by 410meV towards the conduction band (fig.5). Thus, we tentatively relate the Rh-A level to the acceptor state of isolated Rh incorporated substitutionally on a cation site, i.e. Rh^{2+}/Rh^{3+}.

Fig.5 shows schematically the band diagram of an InP/InGaAs heterostructure with the band discontinuities taken from reference 7 and the energy positions of Rh-A in InP as well as Rh1 in InGaAs plotted. The internal reference rule [8], which implies that the energetic position of a **3d** TM level is constant across the interface, is obviously valid for these **4d** TM-related levels. Thus, we conclude for *both* semiconducturs these observed Rh-related levels to be the deep acceptor states of Rh.

At first sight it seems to be abnormal that the Rh levels are *hole* traps in InP and *electron* traps in InGaAs. However, a similar behavior is already known in the literature: the Fe deep acceptor level acts as a hole trap in GaAs [9] while in InP it is a recombination center, which can be observed by DLTS as an electron [10] or a hole [11] emitting level. Thus, for TM deep levels situated around the middle of the band-gap a change of the semiconductor host can indeed cause a change of the *trap character*, too.

ACKNOWLEDGMENTS

Parts of this work were supported by the Deutsche Forschungsgemeinschaft and the Telecom FTZ. We would like to thank A. Näser for some SIMS measurements.

REFERENCES

[1] A. Knecht, M. Kuttler, H. Scheffler, T. Wolf, D. Bimberg, and H. Kräutle; Nuclear Instruments and Methods in Physics Research **B80/81** , p.683 (1993)

[2] N. Makiuchi, T.C. Macedo, M.J. Caldas, and A. Fazzio; Defect and Diffusion Forum **62/63**, p.145 (1989)

[3] G. Bremond, A. Nouailhat, G. Guillot, Y. Toudic, B. Lambert, M. Gauneau, R. Coquille and B. Devaud; Semicond. Sci. Technol. **2**, p.772 (1987)

[4] H. Ullrich, A. Knecht, D. Bimberg, H. Kräutle, W. Schlaak; J. Appl. Phys. **70**, p.2604 (1991)

[5] S.R. Messenger, R.J. Walters, G.P. Summers; J. Appl. Phys. **71**, p.4201 (1992)

[6] M.S. Skolnick, R.G. Humphreys, P.R. Tapster, B. Cockayne, and W.R. MacEwan; J. Phys. C: Solid State Physics **16**, p.7003 (1983)

[7] N. Baber, H. Scheffler, A. Ostmann, T. Wolf, and D. Bimberg; Phys. Rev. **B45**, p.4043 (1992)

[8] J.M. Langer, C. Delerue, M. Lannoo, and H. Heinrich; Phys. Rev. **B38**, p.7723 (1988)

[9] M. Kleverman, P. Omling, L-A. Ledebo, and H.G. Grimmeiss; J. Appl. Phys. **54**, p.814 (1983)

[10] M. Sugawara, M. Kondo, T. Takanohashi, and K. Nakajima; Appl. Phys. Lett. **51**, p.834 (1987)

[11] K. Korona, K. Karpinska, A. Babinski, and A.M. Hennel; Acta Physica Polonica, Vol. A77, p.75 (1990)

[12] G. Bremond, G. Guillot, B. Lambert, Y. Toudic; 5th Conference on Semi-insulating III-V Materials, Malmö, Sweden ; IOP Publishing Ltd (1988), p.319

METASTABLE DEFECTS AND STRETCHED EXPONENTIALS

DAVID REDFIELD and RICHARD BUBE
Stanford University, Department of Materials Science and Engineering, Stanford, CA 94305

ABSTRACT

Metastability of defects in semiconductors provides the basis for (1)An integrated picture of defects that includes the DX center in crystalline III-V compounds, the dangling-bond defect in amorphous Si, and "self-compensation" effects in II-VI compounds. The unifying physical property is the ability of many foreign atoms to have two nearly equal-energy sites, one with a shallow electronic level and the other with a deep level. (2)Elucidation of stretched-exponential kinetics in these materials, including the first demonstrable microscopic model for them. This stretched-exponential model is a distribution of independent response times.

INTRODUCTION

Metastability in the properties of defects in semiconductors has been observed with increasing frequency in recent years. Its consequences include a number of puzzling effects and instructive explanations. These have opened new understandings of the properties of a variety of defects and of electron-lattice interactions. The purposes of this paper are (1)to interrelate several such defects in different semiconductors, and thus help unify their underlying models, and (2)to use the special kinetics properties of these defects to elucidate the stretched exponentials (SEs) that typify their time dependences. This latter has broad implications since there seems to be no other case in which a microscopic explanation for a SE can be so well documented, although SEs are very widely observed.

The central physical properties of metastable defects are their now-established localization and the associated "deep" electronic energy levels. In this context, "deep" implies that observation temperatures are not too high. A major consequence of localization is strong coupling of the electrons in defect states to the lattice, even in materials like most semiconductors which are not strongly ionic. This in turn causes important lattice relaxation effects when defects change their electronic state. These reconfigurations are in fact responsible for the observed metastabilities by forming energy barriers to transitions between the ground and metastable states, although purely electronic models had been proposed earlier. They also offer the only explanation for states with negative effective electron correlation energies. Generally these transitions involve capture or release of a charge carrier, so the material properties are significantly different when the defects change state. One notable example of this is persistent photoconductivity in homogeneous materials.

A key means of characterizing properties of such defects has been measurement and analysis of the kinetics of transitions that may be induced either optically or thermally, between their ground and metastable states. The commonly observed SEs are helping in the formulation of physical models; as confidence in these models has grown it has become possible to invert this reasoning to infer the first microscopic explanation for some SEs themselves.

Two specific metastable defects in very different materials are treated here for these purposes: the DX center in III-V compounds and the dangling-bond defect in hydrogenated amorphous silicon (a-Si:H). The recent status of research in the DX center can be found in the proceedings of the *International Symposium on DX-Centres and other Metastable Defects in Semiconductors.*[1] A somewhat comparable overview for a-Si:H appears in *Amorphous Silicon Materials and Solar Cells.*[2] In addition, the properties of these metastable defects are closely connected to a major limitation in dopability of the wide-gap II-VI compounds like ZnSe that has restricted their application in devices. The DX center has been so well characterized that, following more than ten years of confusion, there is now considerable agreement on interpretation of its properties, so it is useful as a prototype for discussion of other metastable defects.

Mat. Res. Soc. Symp. Proc. Vol. 325. ©1994 Materials Research Society

THE DX CENTER IN III-V COMPOUNDS

Properties

A few selected properties from the large literature on the DX center are summarized here to serve as the basis for the following discussion. Its understanding began with the convincing interpretation of the temperature dependence of capture cross sections for carriers by Henry and Lang in terms of multi-phonon emission processes,[3] and the subsequent proposal by Lang and Logan of a large-lattice-relaxation (LLR) model.[4] These widely accepted (although incomplete) descriptions of the DX center have also formed the basis for a broader recognition of the role of LLR in semiconductor defects; now metastable behavior in defects is generally associated with LLR. An introductory discussion summarizing these relationships and the related configuration-coordinate diagrams has been given by Baraff.[5]

The DX center is known to be associated with donor atoms; a common example is Si on a Ga site in GaAs, although the DX center does not appear unless GaAs is subjected to high pressures or alloyed with AlAs to at least 22%. A fundamental characteristic of the DX center consistent with the LLR picture is that the threshold energy for optical excitation is much larger than the thermal excitation energy. But perhaps the most dramatic metastable effect of the DX center is persistent photoconductivity in which a light-induced increase in conductivity at low temperatures can persist for exceedingly long times in the dark. This is now known to be caused by an energy barrier separating the metastable and ground states due to large configuration changes. Measurements of the kinetics of decay of this condition at various temperatures show that the barrier height is significantly larger that the binding of an electron in the state — *i.e.*, the energy that is inferred from, say, a temperature-dependent Hall measurement, which is of course an equilibrium measurement. This illustrates another general feature of LLR: the difference between results of equilibrium and rate measurements.

The time dependence of such decays in AlGaAs alloys has been found to be non-exponential in character, although it approaches simple exponential behavior as the composition approaches either binary compound, GaAs or AlAs.[6] This was a key observation in the identification of the nature of the center because it showed that the local environment of the donor atom is critical, and companion pressure data showed that the properties were not purely electronic. It was later shown that the mysterious "non-exponential" behavior is in fact a SE.[7] The form of a SE for the number of defects N decaying from an initial value of N_0 is

$$N(t) = N_0 \exp[- (t / \tau_0)^\beta] \tag{1}$$

where β is the stretch parameter and is less than one. Obviously, as β approaches one, Eq. (1) approaches a simple exponential; it was also shown in Ref. 7 that β is proportional to temperature.

We interpret the above-mentioned alloy effects on the kinetics as reductions in the value of β from one in the binary compounds as the 50-50 alloy is approached. These features will be discussed below in a more general framework.

Physical Model

After years of uncertainty, the origin of the DX center is now quite well accepted to be the displaced-atom model of Chadi and Chang.[8] There a Si donor atom, for example, is found to have two possible sites with nearly the same energy: the usual substitutional, four-fold-coordinated site with sp^3 wave functions; and an alternative site displaced along a (111) direction toward an interstitial position. In the displaced site the donor is three-coordinated and the wave functions of the bonds become primarily sp^2 in character. For the substitutional site, the lowest-energy electronic state is a normal shallow-donor state with a large orbit, whereas for the displaced site it is well localized, leading to a deep level. This localized character is responsible for the major role of lattice relaxation in the energetics of the deep state, and the resulting metastability. In GaAs the energy of the displaced state is about 0.2 eV above that of the substitutional state, but pressure or alloying can reverse the order of energies. Then the deep-level state is the ground state and there is an energy barrier of about 0.33 eV (depending on conditions) that inhibits return of a trapped electron to the shallow state, and there is also a barrier to capture from a band state into the DX state. This latter is the explanation for persistent photoconductivity.

Chadi and Chang found that the deep state has a negative effective correlation energy, so two electrons must change state each time (there is continuing disagreement over this aspect of the model). They also extended their calculations to dopants in II-VI compounds with informative results.[9] In these materials metastability is reported less often (one case is in ZnCdSe[10]), but the calculations show that for acceptors in ZnSe the ground state is the displaced-atom state with a deep electronic level. The lack of metastability observations is probably due simply to the fact that the energy of the substitutional state is too high for ready access. This deep-level character of the acceptor state offers an explanation for the difficulty in achieving p-type conductivity in ZnSe with most dopants, another important practical problem. This explanation promises to replace the older "self-compensation" model for doping limitations in these materials; that model seems to be discredited now for other reasons.

THE DANGLING–BOND DEFECT IN a-Si:H

Properties

Again summarizing some properties,[2] it must be recognized that details are much less well known in a-Si:H because of the amorphous structure and difficulties in making reproducible materials. Nevertheless, the density and local bonding are nearly the same in this material as in crystalline Si, and the dominant defect has been identified as the Si dangling bond (DB) by its EPR signal. (Note that such a dangling Si bond does not occur in bulk Si.) Metastability is its other clear property, and what is called the defect is a localized center in its metastable state; the center cannot be detected when in its ground state. The defect can be "created" (*i.e.* the center can be excited to its metastable state) by light or other excitation, and annealed thermally. In this case, however, there is considerable evidence that "light-induced" transitions are caused by the recombination or capture of photoexcited carriers, rather than by direct photon absorption. One unusual feature is that the threshold energy for optical excitation is about equal to the activation energy for anneal.

The kinetics for generation and anneal of this defect are now known to be SEs, with the form

$$N(t) = N_S - (N_S - N_0) \exp[-(t/\tau_0)^\beta] \qquad (2)$$

where N_S is the eventual steady-state value which is determined by the excitation rate and temperature.[11,12] When N_S is less than N_0, this describes anneal, and *vice versa* for generation. Clearly Eq. (1) is just a special case of Eq. (2) for $N_S = 0$. For anneal, β is often proportional to temperature,[13] although some contrary evidence exists. The kinetics of defect generation has been studied extensively, with uncertainty still present in the temperature dependence of β. These studies have shown that τ is determined principally by the light intensity I and that $\tau \approx I^{-2}$, agreeing with the prediction[11] that $\tau \approx I^{-1/\beta}$, since observed values of β are about 1/2. Notice that increased intensity does not increase the range of values of N, as had been erroneously thought, but instead it shortens the time in which their generation occurs.

<u>Physical Models</u>

Uncertainty over the origin of the DB in a-Si:H remains today; in fact there is still disagreement over whether the origin of this defect is extrinsic or intrinsic to the material. We have described some of the elements of this debate previously,[14] and have favored the extrinsic picture, with foreign atoms one likely source of the centers that can form the DBs (structural imperfections such as small voids are other extrinsic possibilities). For this reason and because of the many similarities in the properties of the DB in a-Si:H and the DX center, we proposed[15,16] a model for the DB defect that is based on the Chadi and Chang model of the DX center. For undoped material this model assumes the presence of some foreign atoms (there are large quantities of O, C, and N in all a-Si:H) that may have two possible sites, as shown in the upper part of Fig. 1. Note that, although a foreign atom D is the source of the defect, the result is a dangling bond of the host Si. The lower part of Fig. 1 shows a two-level configuration-coordinate diagram that we associate with these states. Experiments indicate that the energy difference between the minima is $\Delta E \approx 0.2$ eV, and the barrier for anneal out of the metastable state is $E_2 \approx 1$ eV.

We believe that existing data on metastable defects in these various materials and their models summarized here provide a unifying picture of these various phenomena.

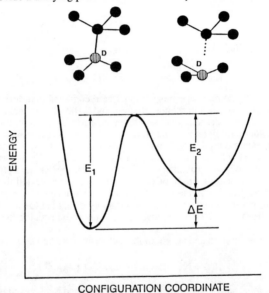

Figure 1. Proposed atomic model[15,16] for the metastable defect in a-Si:H (above) and its configuration-coordinate diagram (below).

STRETCHED EXPONENTIALS

Whether or not these physical models of the defects are correct, they do not by themselves explain the stretched-exponential form for the kinetics nor the properties of the stretch parameter β. Moreover, SE kinetics are observed in very many other material systems including dielectrics, polymers, and glasses; there are essentially no accepted theories for these materials.[17] Nevertheless, the available data cited here appear to offer the first microscopic explanation for SE kinetics.

In virtually all attempts to explain SEs in other materials, a central property that has been invoked is strong interactions among different contributing parts of a material. That is, the behavior in any one part of a material is correlated with the behavior in other parts. A vital feature of the metastable defects in semiconductors that distinguishes them from most of the other materials exhibiting SEs is their independence from each other. This independence is assured by two facts: the defects are typically dilute ($\approx$10 ppm) and localized (as evidenced by their large lattice relaxations). For example, the identifying EPR signal of the Si dangling bond in a-Si:H would be distorted by interaction effects.

The crucial evidence that permits interpretation of the SE here comes from the detailed observations of what was then called "non-exponential" transients in kinetics of DX centers in AlGaAs alloys.[6] As mentioned above, the occurrence of simple exponentials in the binary compounds GaAs and AlAs makes it clear that the SE results from minor variations of the local environments of the DX centers caused by the presence of alloy atoms in their neighborhood. This has been confirmed experimentally in a variety of ways. We now interpret these transients as being generally SEs, but with values of β that are controlled by the alloy composition because that determines the probability of having one or more minority-constituent atoms near a DX center, thus altering its local environment. All the centers with a single alloy atom nearby (actually second neighbors) form one subset with the same properties, but those with two (or none) form another subset with common properties that are different from the first, *etc.*

When the defects change their state, the transient response of each subset is characterized by a simple exponential because it has a well-defined time constant τ set by its physical parameters. The observed SE behavior is the sum of these various contributions, each weighted by the number of members of the set $f(\tau)$. Analysis of SEs in terms of such sums in the continuous limit has shown, as in Fig. 2, that wider distributions of $f(\tau)$ produce lower values of β.[18] Thus the general

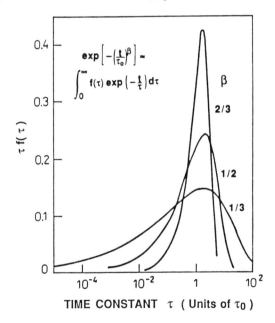

$$\exp\left[-\left(\frac{t}{\tau_0}\right)^\beta\right] = \int_0^\infty f(\tau)\exp\left(-\frac{t}{\tau}\right)d\tau$$

Figure 2. Probability distribution of time constants τ forming a stretched exponential whose effective time constant is τ_0.

picture of SEs in these materials is that they are caused by a distribution in the values of physical parameters that determine the time constants of response to changes.

This picture of a SE as a weighted sum of independent contributors has been widely known as a possibility, although in most other materials it has been rejected because they are believed to be strongly interacting. But there is strong evidence for the independence of these localized defects in semiconductors, and for this interpretation of SEs for the DX center in AlGaAs. For these reasons we proposed this interpretation also for the SEs in a-Si:H and found that the temperature dependence of β and the observed distribution of activation energies for anneal could be inferred naturally from this interpretation.[19]

ACKNOWLEDGMENT

This work was supported in part by the Electric Power Research Institute.

REFERENCES

1. *DX-Centres and other Metastable Defects in Semiconductors*, Ed. by W. Jantsch and R. Stradling (Hilger, Bristol 1991).
2. *Amorphous Silicon Materials and Solar Cells*, Ed. by B. Stafford, Amer. Inst. Phys. Conf. Proc. **234**, (1991).
3. C. Henry and D. Lang, Phys. Rev. B **15**, 989 (1977).
4. D. Lang and R. Logan, Phys. Rev. Lett. **39**, 635 (1977).
5. G. A. Baraff, in *Proc. 14th Intl. Conf. on Defects in Semiconductors*, Ed. by H. Von Bardeleben (Trans Tech, Switzerland, 1986) p. 377.
6. E. Calleja, *et al*, Appl. Phys. Lett. **49**, 657 (1986).
7. A. Campbell and B. Streetman, Appl. Phys. Lett. **54**, 445 (1989)
8. J. Chadi and K. Chang, Phys. Rev. B **39**, 10063 (1989). See also Ref. 1.
9. J. Chadi and K. Chang, Appl. Phys. Lett. **55**, 575 (1989).
10. A. Dissanayake, *et al*, Phys. Rev. B **45**, 13996 (1992).
11. D. Redfield and R. Bube, Appl. Phys. Lett. **54**, 1037 (1989).
12. R. Bube and D. Redfield, J. Appl. Phys. **66**, 820 (1989).
13. J. Kakalios, *et al*, Phys. Rev. Lett. **59**, 1037 (1990).
14. D. Redfield and R. Bube, J. Non-Cryst. Solids **137-138**, 215 (1991).
15. D. Redfield and R. Bube, Phys. Rev. Lett. **65**, 464 (1990).
16. D. Redfield, Mod. Phys. Lett. **5**, 933 (1991).
17. For example, see the conference proceedings in J. Non-Cryst. Solids **131-133** (1991).
18. A. Plonka, *Time-Dependent Reactivity of Species in Condensed Media*, (Springer-Verlag, Berlin 1986).
19. D. Redfield, in *Amorphous Silicon Technology-1992*, Matl. Res. Soc. Symp. Proc. **258**, 34 (1992).

HYDROGENATION OF GALLIUM NITRIDE

M. S. BRANDT,* N. M. JOHNSON,* R. J. MOLNAR,** R. SINGH,** AND T. D. MOUSTAKAS**

* Xerox Palo Alto Research Center, 3333 Coyote Hill Road, Palo Alto CA 94304
** Department of Electrical, Computer, and Systems Engineering, Boston University, Boston, MA 02215

ABSTRACT

A comparative study of the effects of hydrogen in n-type (unintentionally and Si-doped) as well as p-type (Mg-doped) MBE-grown GaN is presented. Hydrogenation above 500°C reduces the hole concentration at room temperature in the p-type material by one order of magnitude. Three different microscopic effects of hydrogen are suggested: Passivation of deep defects and of Mg-acceptors due to formation of hydrogen-related complexes and the introduction of a hydrogen-related donor state 100 meV below the conduction band edge.

INTRODUCTION

The effects of hydrogen on the electronic and vibrational properties of various semiconductors have been studied extensively [1]. In elemental semiconductors, one observes, e.g., the passivation of deep defects (dangling bonds) in amorphous silicon, resulting in a decrease of the subgap absorption, or the passivation of dopant atoms such as boron or phosphorous, resulting in a decrease of the carrier concentration. In recent years, these studies have been extended to compound semiconductors, but have been mostly restricted to the GaAs/AlAs system and InP. With respect to the possible effects of hydrogen wide-bandgap III-V and II-VI semiconductors like GaN and ZnSe have only recently received attention. This interest has been triggered by an intriguing difficulty to achieve p-type doping in films grown with one particular growth technique. In both the II-VI and III-V systems it is found that metal-organic chemical vapor deposition (MOCVD) is unable to produce p-type material, while molecular-beam epitaxy appears to easily grow p-type samples. This behaviour has been linked to the presence of hydrogen in the gas phase during MOCVD growth, which might lead to the formation of acceptor-hydrogen complexes. Indeed, in the case of N-doped p-type ZnSe, the observation of N-H local vibrational modes in MOCVD material strongly supports this assumption [2].

In this contribution, we report on continuing studies of the effects of hydrogen in both n-type (unintentionally doped, sometimes called autodoped, and Si-doped) and p-type (Mg-doped) GaN. In p-type samples grown by MOCVD, acceptor dopants have to be activated by either a low-energy electron beam irradiation or by a thermal annealing step [3]. Again, the formation of acceptor-hydrogen complexes under the abundant presence of hydrogen has been suggested as the origin for the necessity of a post-growth process. We therefore use MBE-grown GaN and deliberately introduce hydrogen by remote-plasma hydrogenation to study the effects of hydrogen in this III-V compound.

SAMPLE GROWTH AND ANALYTICAL PROCEEDURES

Wurtzite GaN epilayers were grown on (0001) sapphire substrates by ECR-assisted MBE [4]. Gallium and the dopant elements (Mg and, for intentional n-type doping, Si) were evaporated

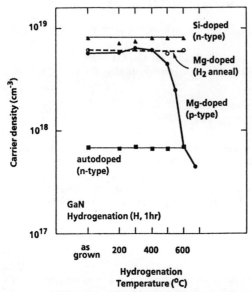

Figure 1: Dependence of the carrier concentration on hydrogenation temperature in epitaxial layers of GaN.

from conventional Knudsen cells, while nitrogen radicals were produced by passing molecular nitrogen through an ECR source at a pressure of 10^{-4} Torr. A microwave power of 35 W or higher was used for the growth of semi-insulating GaN films. Prior to the growth of the GaN epilayer, the substrates were exposed to the N-plasma for 30 min to form an AlN layer. A further GaN buffer layer was then grown at 500°C followed by the high-temperature growth of the actual film at 800°C at a growth rate of 200-250 nm/hr. Hydrogenation was performed with a remote microwave plasma operating at 2 Torr. This technique excludes effects due to charged particle bombardment of the sample or its illumination from the plasma. A high temperature sample holder allowed hydrogenation at temperatures up to 675°C. The films were electrically characterized in a standard Hall-effect apparatus, with the samples abrasively etched into a clover-leaf shape. For ohmic contacts Au was deposited on p-type GaN and Al on n-type material. Depth profiles were determined from secondary ion mass spectroscopy (SIMS), with a Cs^+ primary ion beam for the detection of hydrogen/deuterium and Si and an O^- primary beam for Mg. Calibration of the SIMS data was achieved with Mg and deuterium-implanted GaN reference samples. To increase the sensitivity, the GaN samples for SIMS were treated with a plasma containing deuterium instead of hydrogen. The photoluminescence measurements were performed with a pulsed N_2 laser as the excitation source (3.678 eV).

RESULTS

Figure 1 shows the carrier concentration at room temperature for three differently doped samples subjected to isochronal (1hr) hydrogenation. The n-type samples are completely uneffected by the hydrogenation, however the p-type sample shows a significant decrease in the hole concentration after hydrogenation at $T \geq 500$°C. The mobility in all three cases remained unchanged after hydrogenation at typically 50 cm^2/Vs (autodoped), 10 cm^2/Vs (Si-doped) and 0.3 cm^2/Vs (Mg-doped). A control experiment on the p-type sample is included in Fig. 1, where the microwave plasma was switched off during the post-growth treatment demonstrating that exposure

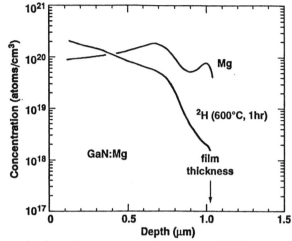

Figure 2: Depth profiles of hydrogen and magnesium in p-type GaN determined by SIMS.

to molecular hydrogen at temperatures up to 600°C does not affect the hole concentration.

Secondary-ion mass spectrosopy is used to verify the introduction of hydrogen into the material. In addition, together with the carrier concentrations determined in Fig. 1 the measurement of the dopant concentrations allows the determination of the doping efficiencies in Si- and Mg-doped GaN. Figure 2 shows the H- and Mg-depth profiled for p-type GaN. An average Mg-concentration of $10^{20}cm^{-3}$ is found, which indicates a very high doping efficieny of about 10%. The average hydrogen concentration is about the same as the Mg-concentration. The slight decrease in H-concentration up to a depth of 0.7μm could be due to a diffusion process. The steep decrease in the concentration, however, coincides with the doping inhomogeneity visible in the Mg-profile at 0.9μm. The high Mg-concentration at the substrate interface might be due to outdiffusion of Mg from doped GaN layers deposited on the substrate holder during previous depositions.

A similar comparison of the dopant and deuterium depth profiles is given for Si-doped n-type GaN in Fig. 3. The average Si-concentration is about $4 \times 10^{20}cm^{-3}$, the doping efficiency of several percent is therefore slightly lower than in the p-type material. Two hydrogen profiles are included in Fig. 3. It becomes obvious that after hydrogenation at 600°C, the hydrogen concentration levels at $2 \times 10^{20}cm^{-3}$, independent of the local Si concentration, which is found to vary throughout the sample.

The hydrogen concentration, after the same passivation step, incorporated in the unintentionally (autodoped) material is about one order of magnitude lower than in Si-doped material (Fig. 4), and has a constant concentration profile. Comparing the hydrogen concentrations in the three differently doped samples we find that the hydrogen incorporation is not just limited by factors like solubility, but does indeed depend on the type and concentration of the dopant present.

Further information on the effects of hydrogen in GaN were obtained from photoluminescence (PL). In Fig. 5, a comparison is shown between the photoluminescence of Mg-doped and autodoped samples, both before and after hydrogenation at 600°C. The two dominant PL-lines in the as-grown samples are the exciton line at 3.44 eV and the Mg-acceptor related line at 3.25 eV, with its two phonon replicas. After hydrogenation, these lines are still observed in the respective samples, but a quantitative comparison of signal heights is difficult. In the autodoped sample, the hydrogenation led to a complete quenching of the defect related line at 2.3 eV [5]. In addition to these known PL lines, a new PL line at 3.35 eV appears after hydrogenation in both samples, which has not been previously reported in GaN.

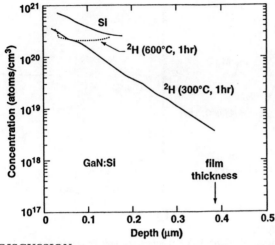

Figure 3: Depth profiles of hydrogen and silicon in n-type GaN determined by SIMS. The hydrogen profiles have been obtained on two identical samples treated at different passivation temperatures.

DISCUSSION

A similar hydrogen-related study on p-type GaN has already been performed by Nakamura and coworkers [6]. However, there are various significant differences to the results reported here. Nakamura used MOCVD grown samples, which were LEEBI-treated to obtain p-type conductivity. In our case, the MBE gown films are p-type without any need for a post-growth treatment. Nakamura used an NH_3 ambient for hydrogenation and had to rely on the thermal decomposition of the molecules on the surface of the film to obtain atomic hydrogen. Our experiments, in which the atomic hydrogen is produced in a remote plasma, therefore show that the hydrogenation temperature of 500°C is indeed typical for the material, and not just for the NH_3 cracking process. Details of the photoluminescence results also differ. Nakamura et al. saw an increase in a possibly defect-related PL band centerd at 1.65 eV after their ammoniation treatment. No indication for such a PL line is visible in our spectra. The remote hydrogen plasma used in our hydrogenation technique therefore does not lead to defect formation detectable in PL. In fact, the quenching of the defect-related line at 2.3 eV for the n-type sample (Fig. 5) shows that hydrogen can passivate this deep defect state, most probably by a complex formation process. We can conclude that in the experiments presented here, no compensation due to deep defect formation takes place. The observed changes in the hole concentration therefore appear to arise from compensation by shallow donor states created by the incorporation of hydrogen, or by a Mg-H complex formation.

A donor-like state related to the introduction of hydrogen would indeed be consistent with both the Hall data and the photoluminescence experiments. Under this assumption, the position of the new PL line 100 meV below bandgap would indicate the energy level of this new donor state. One has then to rationalize that the introduction of hydrogen into the n-type samples does not change the effective electron concentration, as visible in Fig. 1. The SIMS data reveal about $10^{19} cm^{-3}$ and $10^{20} cm^{-3}$ hydrogen in the autodoped and in the Si-doped samples, respectively, after hydrogenation, thereby giving an upper limit for the additional donor concentration in these samples. Only 1/100 of these would be thermally activated at room temperature, leading to an added electron concentration well below the Hall concentrations measured for these samples which would be difficult to detect.

The presence of Mg-H complexes, which would passivate the acceptors in contrast to compensation as discussed above, would be shown by the observation of a local vibrational mode of such

344

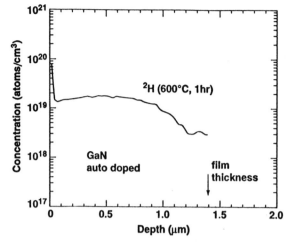

Figure 4: Depth profiles of hydrogen in unintentionally (autodoped) n-type GaN determined by SIMS.

a complex, as has been achieved for the N-H-complex in ZnSe. Preliminary Raman and infrared studies on the samples studied here indeed show such modes in the vicinity of 2200cm^{-1} when Mg concentrations above 10^{19}cm^{-3} are present in the samples [7]. No indication for Si-H, C-H and N-H vibrational modes has been found, however. Details of these results will be subject to a future publication.

The observation of both the new PL line in n- and p-type samples and of the LVM of Mg-H complexes in p-type material suggests that hydrogen can have various effects on the properties of GaN. Indeed, in the p-type sample one has to conclude from vibrational spectroscopy and the PL measurements that both effects, the compensation and the passivation, are present, although from the relative intensities of the PL line at 3.35 eV in the p-type and autodoped samples one might expect that the hydrogen-donor state concentration is small in the p-type material. However, the chemistry of hydrogen in GaN is clearly a challenging subject [8], which will require considerably more experimental and theoretical work.

SUMMARY

We have presented a comparative study of the effects of hydrogen in n- and p-type MBE-grown GaN. We find that the hole concentration in Mg-doped samples can be significantly reduced by hydrogenation above 500°C. The electron concentration in autodoped and Si-doped material is unaffected by hydrogenation. The observation of a new photoluminescence line at 3.35 eV in both n- and p-type samples suggests that hydrogen has a donor-like state in GaN. The defect-related PL line at 2.3 eV can be effectively quenched by hydrogenation, which indicates defect-hydrogen complex formation. As a third effect of hydrogen in GaN, Mg-H complexes are observed in vibrational spectroscopy.

ACKNOWLEGEMENTS

The authors are pleased to acknowledge J. Ager and W. Götz for the preliminary results from Raman spectroscopy and infrared absorption spectroscopy. They also thank C. Van de Walle

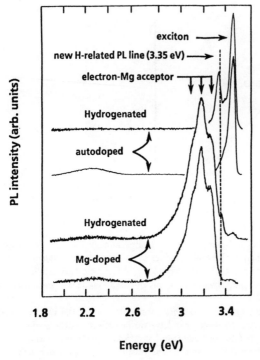

Figure 5: Photoluminescence spectra of autodoped and Mg-doped GaN, both before and after hydrogenation.

and E. E. Haller for helpful discussions and J. Walker and S. Ready for technical assistance. The work at Xerox was supported by AFOSR under contract F49620-91-C-0082 and at Boston University by ONR under contract N0014-92-j-1436. One of the authors (M.S.B.) acknowledges partial support from the Alexander von Humboldt-Stiftung.

REFERENCES

[1] *Hydrogen in Semiconductors*, edited by J. I. Pankove and N. M. Johnson (Academic, San Diego, 1991).

[2] J. A. Wolk, J. W. Ager III, K. J. Duxstad, E. E. Haller, N. R. Taskar, D. R. Dorman, and D. J. Olego, Appl. Phys. Lett. **63**, 2756 (1993).

[3] H. Amano, M. Kito, K. Hiramatsu, and I. Akasaki, Jpn. J. Appl. Phys. **28**, L2112 (1989).

[4] T. D. Moustakas and R. J. Molnar, Mat. Res. Soc. Conf. Proc. **281**, 753 (1993).

[5] J. I. Pankove and J. A.Hutchby, J. Appl. Phys. **47**, 5387 (1976).

[6] S. Nakamura, T. Mukai, M. Senoh, and N. Iwasa, Jpn. J. Appl. Phys. **31**, L139 (1992), S. Nakamura, N. Iwasa, M. Senoh, and T. Mukai, Jpn. J. Appl. Phys. **31**, 1258 (1992).

[7] J. Ager and W. Götz, private communication.

[8] J. A. Van Vechten, J. D. Zook, R. D. Horning, and B. Goldenberg, Jpn. J. Appl. Phys. **31**, 3662 (1992).

Optical Spectroscopy of a
Nitrogen-Hydrogen Complex in ZnSe

J. A. Wolk* and J. W. Ager III

Lawrence Berkeley Laboratory, Berkeley, California, 94720

K. J. Duxstad and E. E. Haller

*Department of Materials Science and Mineral Engineering, University of California at Berkeley and
Lawrence Berkeley Laboratory, Berkeley, California, 94720*

N. R. Taskar, D. R. Dorman, and D. J. Olego

Philips Laboratories, Briarcliff Manor, New York, 10510

ABSTRACT

We have observed two local vibrational modes related to H bonded to N acceptors in ZnSe samples grown by metal organic vapor phase epitaxy. The modes have been seen in both infrared and Raman spectroscopy. The new mode seen at 3194 cm^{-1} is assigned to an N-H stretching vibrational mode and the mode found at 783 cm^{-1} is tentatively assigned to an N-H wagging vibrational mode. Polarized Raman spectroscopy was used to determine that the symmetry of the defect complex is C_{3v}, which implies that the H atom is in either a bonding or anti-bonding position.

I) INTRODUCTION

The attempt to develop blue lasers has led to a vigorous effort to achieve high doping levels in thin films of ZnSe, which has a bandgap of 2.73 eV. While it is relatively simple to dope ZnSe n-type, it has proven to be extremely difficult to achieve high levels of p-type doping. The difficulty in doping ZnSe both heavily n-type and p-type is actually a problem encountered when working with all wide-band gap semiconductors. An additional example is ZnTe, which can only be doped p-type. The origin of this behavior is not agreed upon, but several explanations have been suggested.[1-3] In the case of ZnSe, the difficulties doping samples p-type have been partially overcome using N as the impurity. N sits on the Se site where it acts as an acceptor. Recently, net acceptor concentrations as high as 10^{18} cm^{-3} have been achieved using molecular beam epitaxy (MBE).[4] It would be preferable, though, to grow such layers using metal organic vapor phase epitaxy (MOVPE), which is a much less expensive technique. Unfortunately, attempts to grow heavily doped p-type layers by MOVPE have proven unsuccessful. One possible explanation for this lack of success is that the N acceptors are passivated by H present during the growth process. The possible sources of H in the growth process include 1) the use of NH_3 as the source of N, 2) the use of H_2 as a carrier gas and 3) the decomposition products of the Zn and Se precursor molecules. In this paper we present spectroscopic evidence that H is bonding to the N impurities, implying that it is extremely likely that the proposed passivation is actually taking place.

present address Simon Fraser University, Burnaby, British Columbia, Canada V5A1S6

Mat. Res. Soc. Symp. Proc. Vol. 325. ©1994 Materials Research Society

II) EXPERIMENTAL DETAILS

The ZnSe layers used for this study were grown by photo-assisted MOVPE on (100) semi-insulating GaAs substrates using a Hg lamp with an illumination intensity of 50 mW/cm^2 and Dimethyl-Zinc and Dimethyl-Selenium as the precursors. The growth temperature was 350 °C and the layers were 2.5-3 μm thick. The layers containing nitrogen were delta doped with NH$_3$ using flow modulation epitaxy.[5] (Our samples are not those referred to in reference 5). Photo-assisted MOVPE is used because the illumination generates free carriers near the growth surface which aid in the cracking of Dimethyl-Zinc and Dimethyl-Selenium[6,7]. This allows the growth to take place at a lower temperature, thereby increasing the sticking coefficient of the N. Even though Secondary Ion Mass Spectroscopy (SIMS) measurements revealed N concentrations above 3×10^{18} cm^{-3}, capacitance-voltage measurements showed the net active acceptor concentrations to be less than 10^{15} cm^{-3}.

Infrared absorption spectra were obtained using a Digilab 80-E Fourier transform spectrometer. Spectra were taken at 9 K with a resolution of 0.5 cm^{-1} and at 300 K with a resolution of 1 cm^{-1}. A Ge:Cu photoconductor was used as the detector. ZnSe epilayers grown without nitrogen doping were used as reference samples.

Room temperature Raman scattering measurements were performed in a pseudobackscattering geometry. The focused beam from an argon-ion laser (514.5 nm) was incident on the sample at 65 to the normal. The laser power was 120 mW and the spot diameter was 10 μm. The backscattered light was analyzed with a single grating monochromator and detected using a microchannel plate photomultiplier . A holographic notch filter was used to suppress elastically scattered light. Polarized spectra were corrected for differences in grating efficiency by calibration with a white light source. The spectral resolution was 2 cm^{-1}. The polarization geometry is defined with respect to the sample surface ([100]): the x, y, and z axes are parallel to the [100], [010], and [001], respectively, while the y' and z' axes are parallel to [01$\bar{1}$] and [011], respectively.

III) RESULTS AND DISCUSSION

We first discuss our infrared absorption results, shown in Figure 1. Peaks are seen at 3194 cm^{-1} and 783 cm^{-1} at 9K. Both of these peaks are assigned to vibrational modes of a N-H complex, as we now discuss. We assign the peak at 3194 cm^{-1} to the N-H stretching vibrational mode based on the following arguments. First, we reiterate that the reference samples for these studies were undoped ZnSe layers. We also took absorption spectra of the undoped layer using the substrate as a reference and did not see either peak. Second, we confirmed that our samples have a high concentration of N using Secondary Ion Mass Spectroscopy (SIMS), which revealed nitrogen and hydrogen concentrations above 3×10^{18} cm^{-3}. Considering the typical strength of infrared absorption by local vibrational modes, it is unlikely that an LVM could be observed in a 2-3 μm thick layer unless the associated impurities were present in concentrations of at least 10^{18} cm^{-3}. The SIMS results also demonstrated that the concentration of the N impurities was correlated with that of the H. Third, the frequency is reasonable for a N-H bond. The frequency of the N-H vibration in the ammonium molecule is 3444 cm^{-1},[8] which is roughly 8% higher than the mode we observe. It is known empirically that many X-H (where X represents an impurity) LVM frequencies are several percent lower in the lattice than in a free molecule.[9] This is quite reasonable since some of the electron density of the impurity-hydrogen bond is probably present in a weak bond between the impurity and the nearest neighbor host lattice atom. The N-H LVM frequency was observed to be 2886 cm^{-1} in GaP[10] and a line at 3079 cm^{-1} in GaAs has been tentatively identified as being due to a N-H bond.[10] Finally, only N-H, C-H, and O-H bonds have been found to have LVMs in this frequency range in semiconductors.[8] Since the sample was not intentionally doped with O or C, it is improbable that they are present in concentrations approaching 10^{18} cm^{-3}.

Polarized Raman spectroscopy was used to probe the symmetry of the defect. As seen in Figure 2, the N-H stretch vibration is most intense in the $X(Z'Z')\overline{X}$ and $X(ZZ)\overline{X}$ geometries, is weaker in the $X(ZY)\overline{X}$ geometry, and is not observed in the $X(Z'Y')\overline{X}$ geometry.[11] Although the $X(ZY)\overline{X}$ peak is small, it is reproducible and is observed in both samples where the peak is seen in infrared absorption. Selection rules for Raman scattering at tetrahedral, trigonal, orthorhombic, and monoclinic centers may be obtained in a straightforward manner.[12] The results shown in Figure 2 are only consistent with a defect of C_{3v} symmetry, which is the symmetry commonly observed for other H-passivated donor and acceptor impurities which have been studied.[13] All other modes of a trigonal center, and modes of lower point group symmetry, predict a finite scattering intensity in the $X(Z'Y')X$ geometry.

The two most probable models for the passivated complex are illustrated in Figure 3. The correct model must have C_{3v} symmetry and have the H atom bonded to the N acceptor. Drawing an analogy with C passivation in GaAs,[14,15] the H atom could be bonded to the N atom in a bonding direction, with

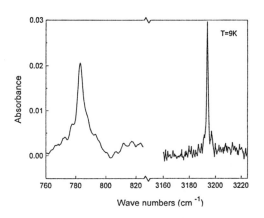

Fig. 1. Infrared absorption spectra of ZnSe:N showing the two N-H local vibrational mode peaks. T = 9K.

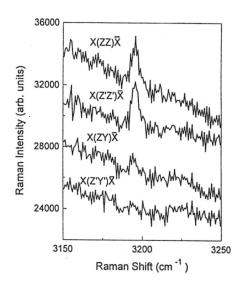

Fig. 2. Polarized Raman measurements of the 3194 cm^{-1} local vibrational mode. T = 300K. The polarization geometry is defined with respect to the sample surface ([100]). Spectra are offset vertically for clarity.

349

the bond to the nearest neighbor Zn atom being very weak. The other possibility is that the H atom is bonded to the N atom in an anti-bonding direction, but this geometry has not previously been observed for acceptor passivation.

We tentatively assign the peak at 783 cm^{-1} to the wag mode of the N-H complex. The relative strengths of the 783 cm^{-1} and 3194 cm^{-1} peak are the same in both samples we observed, suggesting they are related to the same defect. As seen in Figure 4, in polarized Raman scattering the 783 cm^{-1} peak is observed for the four scattering geometries described above, which is consistent with the E symmetry expected for a wag mode. The frequency of this peak is too high to be identified with a stretching or wagging vibration of the Zn-N or Se-N bonds.

We finally discuss one additional possibility for the identification of the two LVM peaks we observe. Our samples were initially characterized by photoluminescence spectroscopy carried out at 7K. The spectra were dominated by donor-acceptor pair transitions and did not show any excitonic transitions. The donor-acceptor pair transitions were seen to occur at longer wavelengths than in samples grown under similar conditions with N concentrations of approximately 1×10^{18} cm^{-3}. It has been suggested that this difference in emission wavelength could be due to the presence of nitrogen related compensating defect complexes, such as a V_{Se}-Zn-N_{Se} complex.[16] Since these complexes have not been

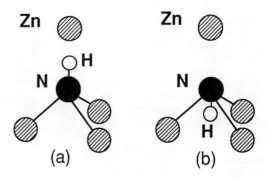

Fig. 3. Possible models for H-passivation of the N acceptor with a) the H atom in a bonding site and b) the H atom in an anti-bonding site.

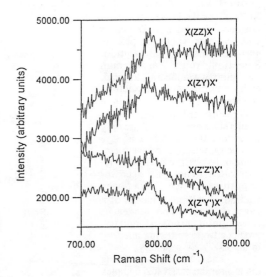

Fig. 4. Polarized Raman measurements of the 783 cm^{-1} local vibrational mode. T = 300K. The polarization geometry is defined with respect to the sample surface ([100]). Spectra are offset vertically for clarity.

positively identified, we cannot completely exclude the possibility that we are observing LVMs of H bonded to a N atom which is part of such a defect. However, given the high symmetry of the defect we observe and the narrowness of the LVM peaks (which become even narrower upon annealing of the samples), the most reasonable explanation for our data is that we are observing LVMs of H bonded to an uncomplexed substitutional N impurity.

In order to determine whether or not the N-H bond could be broken by annealing, the samples were capped with SiO_2 and heated to 650 and 750 °C. Figure 5 shows infrared absorption spectra of the wag mode for unannealed and annealed samples. While the

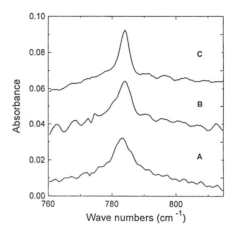

Fig. 5. Infrared absorption spectra of the N-H stretch mode in a) an unannealed sample b) a sample annealed at 650 °C c) a sample annealed at 750 °C.

LVM peak does narrow, the area of the peak does not change, implying that there is no change in the concentration of the N-H centers. The narrowing of the peaks is likely to be due to a relaxation of strain present in the epilayer. A similar result is seen for the high frequency stretch mode.

The positions and FWHM of the observed peaks at 9K and 300K are listed in Table 1. We note the unusual feature that the frequency of the 783 cm^{-1} peak increases with increasing temperature. This behavior is not presently understood.

Table I. Temperature dependence of frequency and FWHM of N-H LVM peaks

Temperature (K)	N-H Wag Mode (?)		N-H Stretch Mode	
	Peak Position (cm^{-1})	FWHM (cm^{-1})	Peak Position (cm^{-1})	FWHM (cm^{-1})
9	783.0	3.8	3193.6	1.5
300	789.3	9.0	3192.7	3.0

IV) CONCLUSION

In conclusion, using infrared absorption and polarized Raman scattering, we have discovered two new local vibrational modes in ZnSe:N,H. The higher frequency peak at 3194 cm^{-1} is assigned to the stretch mode of a N-H bond, while the lower frequency peak at 783 cm^{-1} is tentatively assigned to the N-H wag mode. The Raman polarization scattering results are only consistent with a C_{3v} symmetry for the defect, which implies that the H atom is either in a bonding or an anti-bonding position. This study confirms that H forms a complex with N acceptors in ZnSe, implying that the presence of H during the growth process of MOVPE ZnSe plays a role in the problems encountered in achieving high p-type doping levels in this material.

ACKNOWLEDGMENTS

The authors wish to acknowledge J. N. Heyman, W. Walukiewicz, and N. M. Johnson for helpful discussions, J. W. Beeman for constructing our Ge:Cu photoconductor, and K. Jensen and J. Soohuh for providing samples. This work was supported in part by the director, Office of Energy Research, Office of Basic Energy Sciences, Materials Science Division, of U. S. DOE under contract No. DE-AC03-76SF00098, in part by the US NSF under contract DMR-9115856, and in part by the U. S. Air Force under contract F49620-91-C-0082.

REFERENCES

[1] G. Mandel, Phys. Rev., 134, A1073 (1964).

[2] D. B. Laks, C. G. Van de Walle, G. F. Neumark, and S. T. Pantelides, Phys. Rev. Lett., 66, 648 (1991).

[3] W. Walukiewicz, to be published in Proceedings of the 1993 Spring Meeting of the Materials Research Society.

[4] J. Qiu, J. M. Depuydt, H. Cheng, and M. A. Haase, Appl. Phys. Lett., 59, 2992 (1991).

[5] N. R. Taskar, B. A. Khan, D. R. Dorman, and K. Shahzad, Appl. Phys. Lett., 62, 270 (1993).

[6] Sz. Fujita and Sg, Fujita, J. Crystal Growth 117, 67 (1992).

[7] A. Yoshikava and T. Okamoto, J. Crystal Growth 117, 107 (1992).

[8] J. Chevalier, B. Clerjaud, and B. Pajot, in Semiconductors and Semimetals, eds. J. I. Pankove and N. M. Johnson, (Academic Press, Inc., Boston, 1991), Vol.34, p. 447.

[9] B. Clerjaud, D. Cote, F. Gendron, W.-S. Hahn, M. Krause, C. Porte, and W. Ulrici, in Proceedings of the 16th International Conference on Defects in Semiconductors, edited by G. Davies, G. G. De Leo, and M. Stavola, (Trans Tech Publications, Switzerland, 1992), p.563.

[10] B. Pajot, C. Song, and C. Porte, Proceedings of the 16th International Conference on Defects in Semiconductors, edited by G. Davies, G. G. De Leo, and M. Stavola, (Trans Tech Publications, Switzerland, 1992), p.581.

[11] In the notation [a(b,c)d], a (d) refers to the propagation vector of the incident (scattered) light, while b (c) refers to the polarization vector of the incident (scattered) light.

[12] M. Cardona, in Light Scattering in Solids II, edited by M. Cardona and G. Güntherodt, (Springer-Verlag, Berlin, 1982), p.19.

[13] Semiconductors and Semimetals, eds. J. I. Pankove and N. M. Johnson, (Academic Press, Inc., Boston, 1991) Vol.34.

[14] B. Clerjaud, F. Gendron, M. Krause, and W. Ulrici, Phys. Rev. Lett., 65, 1800 (1991).

[15] R. Jones and S. Öberg, Phys. Rev. B, 44, 3673 (1991).

[16] I. S. Haukssan, J. Simpson, S. Y. Wang, K. A. Prior, and B. C. Cavenett, Appl. Phys. Lett. 61, 2208 (1992).

CHARACTERIZATION OF DEFECTS IN N-TYPE 6H-SiC SINGLE CRYSTALS BY OPTICAL ADMITTANCE SPECTROSCOPY

A.O. EVWARAYE*, S.R. SMITH**, and W.C. MITCHEL
Wright Laboratory, Materials Directorate, WL/MLPO, Wright-Patterson Air Force Base, Ohio 45433-7707.

ABSTRACT

Optical admittance spectroscopy is a technique for measuring the conductance and capacitance of a junction under illumination as a function of the wavelength of the light and the frequency of the measuring AC signal. For the first time, this technique is applied to characterize deep defect levels in 6H-SiC:N. Nitrogen is a donor atom in 6H-SiC which substitutes for carbon in three inequivalent sites (h, k_1, k_2), giving rise to n-type conduction. Deep defect levels attributible to transition metal impurities have been identified in 6H-SiC:N by optical admittance spectroscopy.

INTRODUCTION

The unique physical and electronic properties of silicon carbide make it an ideal choice for high-power, and high-operating temperature devices, and devices that are radiation resistant. The frequency of device operation depends on the minority carrier lifetime, which, in turn, is controlled by deep traps that act as recombination centers. Furthermore, the development of semi-insulating materials requires the introduction of a trap near mid gap in sufficient quantities to restrict the minority carrier lifetime to very small values.

Nitrogen is introduced into the growth ambient when it is de-adsorbed from the graphite components of the growth chamber. It is consequently incorporated into the SiC lattice at a hexagonal site(h) and two cubic sites(k_1,k_2) and gives rise to n-type conduction. Nitrogen donors have been studied by electron paramagnetic resonance(EPR)[1-5], photoluminescence(PL)[6-8], Hall effect and infrared(IR) absorption[9], and Admittance Spectroscopy[10].

In order to electrically characterize traps that lie near mid gap in wide band gap semiconductors(e.g., 6H-SiC) by thermal techniques such as Deep Level Transient Spectroscopy(DLTS), or Admittance Spectroscopy, it is necessary to raise the temperature of the specimen to high

Mat. Res. Soc. Symp. Proc. Vol. 325. ©1994 Materials Research Society

temperatures. This requires a hot stage and great care in the choice and application of the metallization. A method that has the sensitivity of the capacitance techniques, but that does not require the use of high temperatures would be useful. Therefore, we have studied the deep levels in n-type(N-doped) 6H-SiC by using Optical Admittance spectroscopy of Schottky diodes.

DISCUSSION

Admittance spectroscopy was first introduced by Losee to study deep traps in compound semiconductors[11]. The details of the technique have been worked out by him and others[12,13]. The optical variation of this technique was introduced by Vincent, et al.[12], and further developed by Duenas, et al.[14]. Figure 1 is a schematic of the Optical Admittance Spectroscopy experiment.

Whereas thermal admittance spectroscopy detects the effect of thermal emission of carriers from deep centers on the conductance of a diode, optical admittance spectroscopy detects the effect of carriers that have been optically excited from trap level(s) to the conduction band. Figure 2 shows a schematic representation of a Schottky diode on n-type material containing one deep level. When the trap level crosses the Fermi level the additional charges that are generated(a thermal process) cause a change in the conductance and capacitance of the diode. During an optical experiment the temperature is held below the point where the trap level crosses the Fermi level.

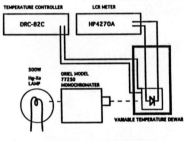

FIGURE 1. Schematic diagram of the optical admittance spectroscopy experimental apparatus.

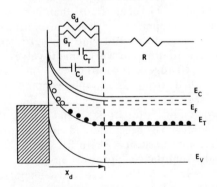

Figure 2. Schematic of a band diagram for a Schottky barrier on n-type material containing one deep trap.

Traps introduce into the measurement, a conductance G_T, given by[12]

$$G_T = \frac{e_n \omega^2}{e_n + \omega^2} \frac{N_T}{n} A \left(\frac{\varepsilon q N^+}{2V} \right)^{\frac{1}{2}},$$

and the measured capacitance in the presence of traps is given by

$$C = C_0 + C_T = \left(1 + \frac{N_T}{n} \left(\frac{e_n \omega^2}{e_n + \omega^2} \right) \right) A \left(\frac{\varepsilon q N^+}{2V} \right)^{\frac{1}{2}},$$

where e_n is the emission rate, n is the number of free carriers in the bulk, N_T is the number of deep traps, N^+ is the number of ionized deep centers, ω is the measurement frequency, ε is the bulk dielectric coefficient, A is the area of the diode, and q is the electronic charge.

Clearly, when N+ is changed by the process of illumination, both C and G_T change, and in the same manner. Hence, the photo-capacitance and the photo-conductance curves have the same shape.

Monochromatic light is focused on the diode and the wavelength of the light is scanned by the use of a monochromater. Thus, when light of the proper energy is incident on the diode, a carrier may be excited from the trap level to the conduction band by the absorption of the photon. The presence of these extra carriers in the conduction band produces a change in the conductance of the diode which can be detected using a capacitance/conductance meter, hence, we see peaks in the conductance corresponding to the energy of the absorbed photons. When the wavelength (energy) of the light is not of the proper value to excite carriers, the conductance is lower.

Schottky diodes were fabricated on n-type SiC that was first oxidized and etched to yield a clean, ordered, surface. Actual metallization consisted of sputtering Ni onto n-type material and annealing at 900 C for five minutes in forming gas to produce ohmic contacts. Schottky diodes were fabricated by evaporating Al dots 600μm in diameter onto the other side of the wafer. The diodes that were formed by this process were evaluated by examining the C^{-2} vs V behavior. The capacitance and conductance were measured using an HP4270A Multifrequency LCR Meter, operated in the high resolution mode. The measurements reported here were all made at a frequency of 100 kHz.

RESULTS

Five peaks were seen in the spectrum obtained from the SiC specimens, as shown in figure 3. The energies of the transitions indicated by the peaks are shown in the figure. Consideration of the transition energies rules out the possibility that any of these peaks are related to nitrogen in any other manner than a complex of some sort. The band-to-band transition for 6H-SiC is clearly seen along with three peaks that are attributable to deep levels in SiC arising from transition metal impurities. A fifth peak is visible in the spectrum and has been identified as a transition metal impurity using DLTS, and IR absorption(figure 4); these results are reported elsewhere.[10] Figure 5 shows a spectrum from a specimen having a different nitrogen concentration and no transition metal IR signature. Two of the levels are clearly seen, while the level at 1310 nm is seen as a shoulder to the middle peak.

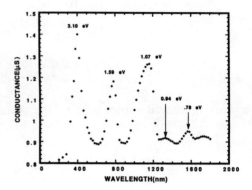

Figure 3. Optical Admittance spectrum for a specimen having N_D-N_A = 6.4 X $10^{15} cm^{-3}$.

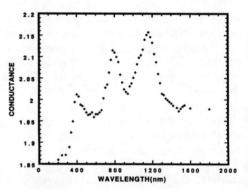

Figure 4. IR Absorption spectrum of transition metal, V, in 6H-SiC.

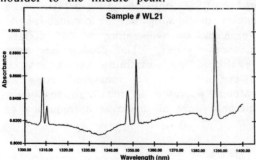

Figure 5. Optical Admittance spectrum for a specimen having N_D-N_A = 1.7 X $10^{17} cm^{-3}$.

We have shown that optical admittance spectroscopy is a valuable tool for the characterization of wide band gap semiconductors, specifically nitrogen doped 6H-SiC. We have determined both the band-to-band transition in this material, and the energies of the three transition metal impurity levels.

ACKNOWLEDGEMENTS

We would like to acknowledge the work of Mr. Paul Von Richter, Mr. Robert V. Bertke, and Mr. Gerald Landis in preparing the specimens for these experiments. The work of SRS was supported by Air Force contract no. F33615-91-C-5603.
*Visiting Scientist, Permanent address, Physics Department, University of Dayton, Dayton, Ohio 45469-2314, **University of Dayton Research Institute, Dayton, Ohio 45469-0178.

REFERENCES

1. H. H. Woodbury And G. D. Ludwig, Phys. Rev **124**, 1083(1961)
2. E. N. Kalabukhova, N. N. Kabdin, and S. N. Lukin, Sov. Phys. Solid State **29**, 1461(1987)
3. E. N. Kalabukhova, N. N. Kabdin, S. N. Lukin, and t. L. Petrenko, Sov. Phys. Solid State **30**, 1457(1988)
4. O. V. Vakulenko and V. S. Lysyi, Sov. Phys. Solid State **30**, 1446(1988)
5. Lyle Patrick, Phys. Rev. **127**, 1878(1962)
6. W. J. Choyke and L. Patrick, Phys. Rev. **127**, 1868(1962)
7. D. R. Hamilton, W. J. Choyke, and L. Patrick, Phys Rev. **131**, 127(1963)
8. P. J. Dean and R. L. Hartman, Phys Rev. **B5**, 4911(1972)
9. W. Suttrop, G. Pensl, W. J. Choyke, R. Stein, and S. Leibenzeder, J. Appl. Phys. **27**, 3708(1992)
10. A. O. Evwaraye, S. R. Smith, and W. C. Mitchel, J. Appl. Phys. to be published.
11. D. L. Losee, J. Appl. Phys. **46**, 2204(1975)
12. G. Vincent, D. Bois, and P. Pinard, J. Appl. Phys. **46**, 5173(1975)
13. W. D. Oldham and S. S. Naik, Solid State Electron. **15**, 1085(1972)
14. S. Duenas, M. Jaraiz, J. Vicente, E. Rubio, L. Bailon, and J. Barbolla, J. Appl. Phys. **61**, 2541(1987)

PART VI

Defects in Low Temperature Grown Semiconductors

DEFECTS IN LOW–TEMPERATURE–GROWN MBE GaAs

DAVID C. LOOK
University Research Center, Wright State University, Dayton, OH 45435

ABSTRACT

Defect concentrations in molecular beam epitaxial (MBE) GaAs range from 10^{12} to 10^{20} cm^{-3} as the growth temperature is lowered from 600 to 200 ºC; however, very high quality layers can be grown over this whole range. The dominant defect is the As antisite, but there is also good evidence for As interstitials and gallium vacancies. The particular form of the As antisite center in low–temperature (LT) MBE GaAs is not known at this time, but it is definitely not EL2, because both the thermal activation energy and the electron capture cross section differ significantly. However, other features, such as the EPR spectrum and metastable–to–normal recovery kinetics are identical to those of EL2. The donor (As antisite) to acceptor ratio seems to hold at about one order of magnitude as growth temperature is varied from $200 - 400$ ºC; thus, the Fermi level stays near mid–gap over this whole range. However, hopping conduction among the As antisite centers is strong for samples grown between 200 and 300 ºC and keeps the material from being semi–insulating, while for those grown between $350 - 480$ ºC, the resistivity is greater than 10^7 Ω cm. The annealing dynamics are particularly interesting and include such features as the mobility going through a sharp maximum at an annealing temperature of 400 ºC for a layer grown at 200 ºC. The donor and acceptor concentrations can be determined both by Hall effect and absorption measurements as the layer is annealed up to 600 ºC. Above 550 ºC, large precipitates are formed. The relative roles of the precipitates and point defects in influencing compensation, lifetime, and device characteristics are a source of much controversy and will be discussed.

INTRODUCTION

In the early days of molecular beam epitaxial (MBE) growth of GaAs, emphasis was placed on developing pure, defect–free material. Although it is likely that many workers explored the limits of their machines, particularly in substrate growth temperature T_g, and As$_4$/Ga beam–equivalent pressure (BEP), most of the published work dealt with T_g in the region $580 - 600$ ºC, which produced the best material [1]. However, in 1978, Murotani et al [2] demonstrated that growth at 400 ºC produced a highly resistive layer which was useful as a buffer for a GaAs metal–semiconductor field–effect transistor (MESFET). Even with that impetus for further low–temperature (LT) growth studies, only a few investigations were reported in the next few years [3–5], and they mainly dealt with the increase in deep centers and decrease in doping activation as T_g was lowered. Then, in 1988, Smith, Calawa, and colleagues [6] showed that a MESFET buffer layer grown at 200 ºC, and annealed at 600 ºC, had a remarkable ability to eliminate cross–talk between closely spaced devices, an important quality for successful integrated circuit (IC) production. This work spurred strong activity in the LT–MBE GaAs (henceforth simply called LT GaAs) area, and to date, $400 - 500$ papers have been published [7,8]. The reasons for the effects discovered by Smith and Calawa, as well as for other useful effects such as a very short recombination time, involve extremely high concentrations of point defects, in some cases more that 10^{20} cm^{-3}. Furthermore, after anneal, large As precipitates are formed [9], and many workers believe that the precipitates can also affect electrical, optical, and device properties [10].

In this review, we will mainly concentrate on the point defects and how they move and annihilate at high temperatures. We will consider the electrical, optical, and magnetic–resonance properties and show how these properties change during annealing. The effects on devices will also be discussed.

NON–STOICHIOMETRIC GROWTH

The samples first discussed by Smith, Calawa and co–workers [6] were grown at 200 °C, and this has been the most popular growth temperature for LT GaAs samples since then. Such samples have a high excess of As, much of it in interstitial or split–interstitial positions [11], which tends to expand the lattice, as shown in Fig. 1. Other workers have suggested even higher percentages of excess As at a given temperature, but it must be remembered that such low growth temperatures are difficult to measure by the usual non–contacting temperature sensors, so that values assigned by various laboratories may differ by as much as several tens of degrees. As might be expected, another important factor in determining the As content is the As_4/Ga BEP.

Besides filling interstitial positions, the excess As can also go on Ga sites, forming As antisites (As_{ga}), or it can be accommodated by Ga vacancies (V_{ga}). However, beyond a critical growth thickness the lattice can evidently not continue to absorb the excess As as point defects, and large "pyramidal" defects are formed [12]. For a BEP of 10, and a growth temperature of 200 °C, the critical thickness above which these large defects are formed is about 3 μm [13]. It is important to use single–crystal layers for the most accurate measurements of many of the quantities discussed in the present work.

IDENTIFICATION OF POINT DEFECTS

A single–crystal LT GaAs layer grown at about 200 °C contains very large concentrations of deep donors ($\sim 10^{20}$ cm^{-3}) and acceptors ($\sim 10^{19}$ cm^{-3}) [14]. Since background impurity levels are in the $10^{15} - $ cm^{-3} range, these donors and acceptors must be pure defects. The samples are nearly always n–type; in fact, to our knowledge, the only report of p–type material up to now was in a very early paper, and was evidently never repeated [15]. (In that case, a Hall mobility could be measured only

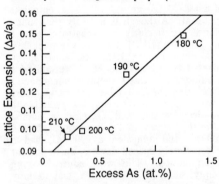

Fig. 1. Dependence of the lattice parameter expansion on the excess As content for MBE GaAs layers grown at 180, 190, 200, and 210 °C. (From Ref. 13.)

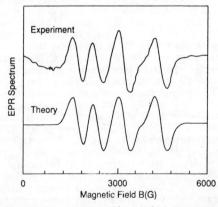

Fig. 2. Experimental (T=4K) and theoretical EPR spectra of a 15–μm–thick MBE GaAs layer grown at 200 °C. (From Ref. 17.)

between 100 and 200 K, and outside of that range, the sample was too resistive to measure. Since the mobilities were also very low, $1 - 100$ cm^2/V s, the data would have to be considered questionable.) Recently, we have found p–type conduction in a sample grown at 300 ºC and furnace annealed at 450 ºC (but not at 400 or 500 ºC !) for 10 min., or rapid thermally annealed for 10 s at 850 ºC; these acceptor data will be discussed in more detail after consideration of the donors, about which we know considerably more.

As Antisites

The only defect which has been firmly identified in any type of GaAs is the As antisite, As_{ga}. However, it is usually difficult to claim with certainty whether the As_{ga} is isolated or is complexed with another defect, say at a second–nearest–neighbor position. (Even the famous and widely studied EL2 suffers this uncertainty!) Thus, although it is known that the dominant donor in LT GaAs contains As_{ga}, the exact form of the donor is not known.

As first presented by Kaminska et al [16], the electron paramagnetic resonance (EPR) spectrum of LT GaAs contains the classic four–line pattern known to belong to As_{ga}. A comparison of experiment and theory shown in Fig. 2 (from another paper [17]) strongly supports this view. Unfortunately, the paramagnetic state of As_{ga} is As_{ga}^+, so that a concentration determination from EPR gives only the (+)–state concentration, which is the same as the acceptor concentration N_a in a sample which has a small concentration of free carriers (i.e., one in which the Fermi level E_f is near midgap). The data of Fig. 2 are consistent with $[As_{ga}] = N_a = 3$ x 10^{18} cm^{-3}, while independent EPR measurements of 200–ºC samples grown in other laboratories give similar values, 5 x 10^{18} [16] and 7 x 10^{18} cm^{-3} [18], respectively.

The donor and acceptor concentrations can also be determined from optical absorption measurements. Consider the near–IR absorption spectrum of 300–ºC GaAs shown in Fig. 3. The shape is nearly identical to that of EL2, so that it seems reasonable to apply the photoionization coefficients already known for EL2 [19]. That is, the absorption coefficient α as a function of wavelength λ should be given by

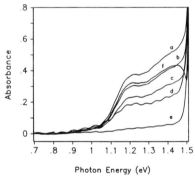

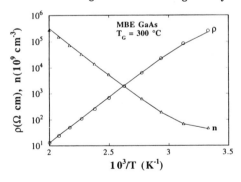

Fig. 3. IR absorption spectra (corrected for substrate absorption) of GaAs:Be layers grown at 300 ºC. a. 10^{17} cm^{-3} doped layer, b. 10^{16} cm^{-3}, c. 10^{18} cm^{-3}, d. 10^{19} cm^{-3}, e. 10^{18} cm^{-3} doped layer after 30 min. white–light illumination. f. Photoquenching absorption spectrum of 10^{17} cm^{-3} doped layer given by the ratio of spectra taken before and after illumination.

Fig. 4. The measured resistivity ρ and carrier concentration n (actually 1/eR) vs. T^{-1} for an MBE GaAs layer grown at 300 ºC. The solid lines are theoretical fits.

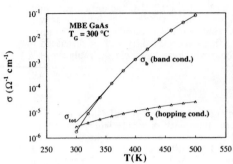

Fig. 5. Hopping conductivity σ_h and band conductivity σ_b vs. T^{-1} for an MBE GaAs layer grown at 300 °C. The solid lines are theoretical fits.

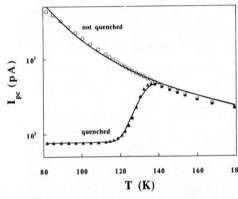

Fig. 6. The photocurrent vs. T for an MBE GaAs layer grown at 400 °C. The quenched data occur after IR illumination at 82 K. The solid lines are theoretical fits. (From Ref. 23.)

$$\alpha(\lambda) = \sigma_n(\lambda)[As_{ga}^0] + \sigma_p(\lambda)[As_{ga}^+]$$

where $\sigma_n(\lambda)$ and $\sigma_p(\lambda)$ are the photoionization cross–sections for electrons and holes, respectively [19]. By measuring α at two different wavelengths (say, 1.1 and 1.2 μm), both $[As_{ga}^0]$ and $[As_{ga}^+]$ can be determined. Such an analysis for 200–°C GaAs gives a total $[As_{ga}]$ of about 1×10^{20} cm^{-3}, and an ionized concentration, $[As_{ga}^+] \simeq 1 \times 10^{19}$ cm^{-3} [14]. These values should correspond to N_d and N_a, respectively, as long as no donors shallower than As_{ga} are important. Note that the value of N_a as determined by absorption is within error of that determined by EPR. Discussion concerning the identity of the acceptors will be deferred to a later section of this paper.

There is at least one other way to quantitatively determine the donor and acceptor concentrations – electrical measurements, including the Hall effect and conductivity [20]. For growth temperatures less than about 300 °C, hopping conduction between the As_{ga} centers is stronger than the usual band conduction, at least for room–temperature electrical measurements [21]. The reason for this is a strong wavefunction overlap, since the estimated (hydrogenic) radius of an 0.65–eV center is about 9 Å, while the average distance between centers of density 10^{20} cm^{-3} is only about 13 Å. Thus, the conductivity σ and Hall coefficient R of these systems must be modeled with both band ("b") and hopping ("h") components [20,21]:

$$\sigma = \sigma_b + \sigma_h$$
$$R = \frac{R_b\sigma_b^2 + R_h\sigma_h^2}{(\sigma_b + \sigma_h)^2}$$

Since σ_b, σ_h, R_b, and R_h, are known [21] functions of N_d, N_a, and E_d (the donor activation energy), we can fit the temperature dependences of σ and R to determine N_d, N_a, and E_d. Such fits of $\sigma(=\rho^{-1})$ and $n(\simeq e^{-1}R^{-1})$ for 300–°C material are shown in Fig. 4 and give values $N_d \simeq 3 \times 10^{18}$ cm^{-3}, $N_a = 2 \times 10^{17}$ cm^{-3}, and $E_d \simeq 0.65$ eV (at T = 0). It is interesting that the hopping conduction σ_h is stronger than the band conduction σ_b at a measurement temperature T_m of 300 K, but quickly becomes weaker at higher temperatures, as shown in Fig. 5. For 200–°C material, on the other hand, σ_h is much greater than σ_b even at $T_m = 500$ K [21].

The value of E_d for LT GaAs, which is 0.65 ± 0.01 eV for 300–, 350–, 400–, and 450–°C material, is quite interesting because it is significantly less than the value of E_d

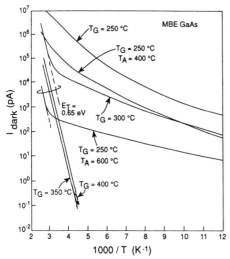

Fig. 7. Dark current vs. T^{-1} for MBE GaAs layers grown at various T_g's and annealed at various T_a's.

for EL2, which is 0.75 ± 0.01 eV. Thus, the LT donor is not identical to EL2, and there may be doubt in the mind of some that the donor is even related to As_{ga}. To prove the latter relationship, we appeal to quenching experiments involving illumination by IR (say, 1.1–μm) light which is known to transform As_{ga}, or EL2, from a normal state to a metastable state at temperatures below 120 K [22]. In the metastable state, the donor level is very deep [22] and both band and hopping conduction are greatly diminished. Recovery to the normal state occurs between 120 and 140 K, and the original conduction also returns. Photoconduc–tivity, photoluminescence, absorption, EPR, and other phenomena are affected the same way. In Fig. 6 we show photocurrent vs. T for both quenched and unquenched material grown at 400 ºC, a temperature at which hopping conduction σ_h is negligible. The solid lines are fits from an essentially rigorous analysis [23] which gives an electron capture barrier of 0.040 eV for the unquenched data, and a metastability recovery barrier of 0.26 eV for the quenched data. The values of these same two parameters for EL2 are 0.075 eV and 0.26 eV, respectively. The identical values of the metastability barriers leaves no doubt that As_{ga} is <u>involved</u>, while the difference in the capture barriers shows, again, that the two centers are not identical. A summary of our knowledge, including the EPR results, is given in Table 1 of Ref. 23 for the donors in 300 − 450–ºC material.

For 200 − 300–ºC GaAs, the comparison of the dominant donor with respect to EL2 is not as clear, especially for as–grown material. The reasons are at least two–fold: (1) σ_h is so strong that the fitted magnitudes of σ_b are inaccurate, making determinations of N_d, N_a, and E_d, poorer than those for 300 − 400–ºC material; and (2) the As_{ga}– related centers do not quench, or at least only a small fraction ($\sim$ 10%) of them do. The lack of quenching may be related to the strain present in as–grown 200–ºC layers, or even more likely to the strong recombination paths which compete with the metastability transition path. In any case, we cannot, with certainty, assign the value $E_d = 0.65$ eV to 200–ºC GaAs; it may be as high as 0.75 eV. The situation is illustrated in Fig. 7, in which it is seen that strong band conduction is not present at any temperature (at least up to 400 K) for as–grown 250–ºC material. However, by annealing at 600 ºC, a short section of activated behavior is seen at the highest measurement temperatures, while the much shallower hopping conduction is still dominant at lower temperatures. Although the slope of the activated portion is listed as 0.65 eV, this cannot be considered a firm number.

In spite of these uncertainties for <u>as–grown</u> 200– or 250–ºC GaAs, once the material is annealed above 500 ºC the situation becomes clearer; e.g., the deep donors in annealed material can be quenched. (We cannot establish unambiguously whether the quenchable donors are microscopically different than the main body of original donors, or whether they are the same but just in a different environment because of the anneal.) Interestingly enough, one of the quenched parameters in this case is the hopping conduction itself, as shown in Fig. 8. The recovery curve, shown in Fig. 9 is again well fitted with a metastability barrier of 0.26 eV and a prefactor of 3 x 10^8 s^{-1}, very close to

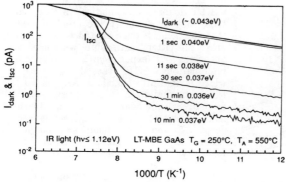

Fig. 8. Current vs. T for an MBE GaAs layer illuminated by IR light at 82 K for various times (I_{dark} represents zero time.) Slopes of the lines are given in eV. (From Ref. 27.)

the respective EL2 values. Thus, As_{ga} is also involved in the donor defects of 200 – 300–°C GaAs.

We now return to the acceptors, which have been studied much less. From the known As–rich stoichiometry, the most obvious guess for the dominant acceptor would be the Ga vacancy, V_{ga}. Indeed, a diffusion model based on V_{ga} centers has been presented [24] and positron–annihilation experiments have been interpreted to involve V_{ga} defects [25]. Also, we have recently seen p–type conduction in material grown at 300–°C and annealed at 450 °C, and the activation energy is about 0.5 eV, certainly within range of the several charge–state transitions theoretically postulated to occur for V_{ga} [26]. However, these assertions are all somewhat speculative at this point, and the acceptor defect may be more complicated than a simple, isolated V_{ga}. For example, the dominant thermally stimulated current peak in LT GaAs, occurring at 140 K and having an activation energy of 0.27 eV, has been tentatively assigned to V_{ga}– As_{ga} [27], and a sharp zero–phonon line in photoluminescence is believed to arise from an exciton bound to a V_{ga}– As_i center [28]. Again, all of these assignments require further confirmation.

ANNEALING EFFECTS

The annealing dynamics of the donors and acceptors in LT GaAs are quite interesting. In Fig. 10, we show the results of two independent experiments [21]: Hall–effect measurements, which determine N_d and N_a, and absorption measurements, which determine $[As_{ga}]$ and $[As_{ga}^+]$. As can be seen, there is a clear correlation between N_d and $[As_{ga}]$, and between N_a and $[As_{ga}^+]$. Above T_a = 450 °C, all quantities decrease, as might be expected. However, the most interesting behavior is between 400 and 450 °C; in this range N_a first drastically decreases, and then drastically increases. This behavior is consistent with that of 300–°C material, in which N_a increases enough at 450 °C that the sample actually becomes p–type. Models have been presented for this phenomenon, but it must be stated that our understanding is not yet complete.

Another very interesting result of annealing, at least above 500 °C, is the formation of large As precipitates [9]. Table 1 gives densities and average diameters as a function of T_g, with T_a = 600 °C in all cases. Some of the larger precipitates are composed of metallic As, and thus could act as buried Schottky barriers which could compensate free carriers and influence recombination times of excess carriers [10]. Thus, it has been proposed by some (and rebutted by others) that the As precipitates are responsible for the semi–insulating nature of annealed LT GaAs, and also for the extremely fast recombination times. Recent papers on these and other issues have appeared in a special December 1993 issue of the Journal of Electronic Materials, and discussion can also be found in Ref. 8.

366

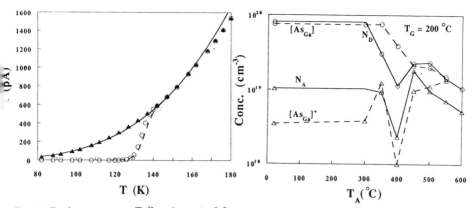

Fig. 9. Dark current vs. T (heating rate 0.3 K/s) for an MBE GaAs layer grown at 200 °C, and annealed at 500 °C. The lower curve (dashed line) follows 5 min. of IR illumination at 82 K. The lines are theoretical fits.

Fig. 10. The donor (N_d), acceptor (N_a), total antisite [As_{ga}], and charged antisite [As_{ga}^+] concentrations vs. annealing temperatures T_a for an MBE GaAs layer grown at 200 °C. (From Ref. 21.)

Table 1. Density, average diameter, and volume fraction of As precipitates as a function of growth temperature T_g. All samples were annealed at 600 °C for 1 hr. (Data from Ref. 9.)

T_g (°C)	Density (10^{16} cm^{-3})	Diameter (Å)	Volume frac.(x 10^{-3})
225	4.3	74	3.410
250	3.9	55	0.951
275	1.7	61	0.635
300	0.78	76	0.556
325	0.52	46	0.077
350	0.39	29	0.016

DEVICES

The interest in LT GaAs and other III–V compounds has been largely driven by device applications, so a few remarks on this subject are in order. The very first paper on the subject (i.e., in the "modern era", 1988 –) showed that device crosstalk (called "backgating" and "sidegating") as well as light sensitivity could be eliminated in MESFET devices by growing the device layer on a buffer layer which itself was grown at 200 °C and annealed at 600 °C [6]. This discovery greatly enhanced the prospect of GaAs ICs. Then it was found that this same type of LT layer could be used as a gate insulator in a MESFET device (LT layer on top of the active layer) [29,30], and in fact, a record for power at 1 GHz was established [30]. In a similar application, an LT layer on top of a simple MESFET device, with the gate and ohmic metals exposed, has been shown to passivate the surface [31,32] and increase the breakdown voltage [33]. Finally, the very short recombination time (~ 100 fs) in LT GaAs has led to a record speed for photoconductive switches [34,35]. Some of these applications have been commercialized, but most of them will require much more developmental work. The detailed explanations for the unusual device properties remain an area of much speculation and argument, particularly in the relative roles of point defects and As precipitates.

SUMMARY

LT–grown GaAs contains extremely large concentrations of donor and acceptor point defects, roughly 10^{20}, 10^{18}, and 10^{17} cm^{-3} for materials grown at 200, 300, and 400 °C, respectively. The dominant donor defects are $As_{ga}-$ related, and the acceptor defects are probably $V_{ga}-$ related, although this latter assertion is not as definite. Also present are As interstitials, or dimers, of even higher concentrations ($\sim 10^{21}$ cm^{-3}) in 200–°C material. Upon annealing at 500 °C or higher, the point–defect densities decrease, and large As precipitates are formed. The relative roles of the point defects and precipitates in influencing the optical, electrical, and device properties of LT GaAs have received much attention but are still controversial.

ACKNOWLEDGMENTS

I wish to thank my many associates at Wright Patterson Air Force Base and Wright State University who have contributed greatly to the material which went into this review. They include T.A. Cooper, J.E. Ehret, K.R. Evans, Z–Q. Fang, J.T. Grant, J.E. Hoelscher, M.O. Manasreh, M. Mier, J.T. Prichard, G.E. Robinson, J.R. Sizelove, C.E. Stutz, E.N. Taylor, D.C. Walters, H. Yamamoto, and P.W. Yu. I also wish to thank C.W. Litton and G.L. McCoy for their leadership, support, and encouragement of the LT GaAs work. Outside colleagues who have contributed significantly to my own understanding include A.R. Calawa, Z. Liliental–Weber, M. Melloch, F.W. Smith, E.R. Weber, and G. Witt. Finally, I am grateful to Robin Heil for typing this manuscript. Support was received from the U.S. Air Force under Contract F33615–91–C–1765.

REFERENCES

1. H. Sakaki, in R.J. Malik (ed.), III–V Semiconductor Materials and Devices (North Holland, Amsterdam, 1989) p. 217.
2. T. Murotani, T. Shimanoe, and S. Mitsui, J. Crystal Growth 45, 302 (1978).
3. C.E.C. Wood, J. Woodcock, and J.J. Harris, Inst. Phys. Conf. Ser. No. 45, 29 (1979).
4. R.A. Stall, C.E.C. Wood, P.D. Kirchner, and L.F. Eastman, Electronics Lett. 16, 171 (1980).
5. G.M. Metze and A.R. Calawa, Appl. Phys. Lett. 42, 818 (1983).
6. F.W. Smith, A.R. Calawa, C.–L. Chen, M.J. Manfra, and L.J. Mahoney, IEEE Electron Device Lett. EDL–9, 77 (1988).
7. G.L. Witt, A.R. Calawa, U. Mishra, and E. Weber (eds.), Mater. Res. Soc. Symp. Proc. 241 (1992).
8. D.C. Look, Thin Solid Films 231, 61 (1993).
9. M.R. Melloch, N. Otsuka, J.M. Woodall, A.C. Warren, and J.L. Freeouf, Appl. Phys. Lett. 57, 1531 (1990).
10. A.C. Warren, J.M. Woodall, J.L. Freeouf, D. Grischkowsky, D.T. McInturff, M.R. Melloch, and N. Otsuka, Appl. Phys. Lett. 57, 1331 (1990).
11. K.M. Yu, M. Kaminska, and Z. Liliental–Weber, J. Appl. Phys. 72, 2850 (1992).
12. Z. Liliental–Weber, W. Swider, K.M. Yu, J. Kortright, F.W. Smith, and A.R. Calawa, Appl. Phys. Lett. 58, 2153 (1991).
13. Z. Liliental–Weber, Mat. Res. Soc. Symp. Proc. Vol. 241, 101 (1992).
14. D.C. Look, D.C. Walters, M. Mier, C.E. Stutz, and S.K. Brierley, Appl. Phys. Lett. 60, 2900 (1992).
15. M. Kaminska, E.R. Weber, Z. Liliental–Weber, R. Leon, and Z.U. Rek, J. Vac.

Sci. Technol. B. 7, 710 (1989).

16.   M. Kaminska, Z. Liliental–Weber, E.R. Weber, T. George, J.B. Kortright, F.W. Smith, B.–Y. Tsaur, and A.R. Calawa, Appl. Phys. Lett. 54, 1881 (1989).

17.   H.J. von Bardeleben, M.O. Manasreh, D.C. Look, K.R. Evans, and C.E. Stutz, Phys. Rev. B 45, 3372 (1992).

18.   K. Krambrock, M. Linde, J.M. Spaeth, D.C. Look, D. Bliss, and W. Walukiewicz, Semicond. Sci. and Tech. 7, 1037 (1992).

19.   P. Silverberg, P. Omling, and L. Samuelson, Appl. Phys. Lett. 52, 1689 (1988).

20.   D.C. Look, D.C. Walters, M.O. Manasreh, J.R. Sizelove, C.E. Stutz, and K.R. Evans, Phys. Rev. B 42, 3578 (1990).

21.   D.C. Look, D.C. Walters, G.D. Robinson, J.R. Sizelove, M.G. Mier, and C.E. Stutz, J. Appl. Phys. 74, 306 (1993).

22.   J. Dabrowski and M. Scheffler, Phys. Rev. B 40, 10391 (1989).

23.   D.C. Look, Z–Q. Fang, and J.R. Sizelove, Phys. Rev. Lett. 70, 465 (1993).

24.   D.E. Bliss, W. Walukiewicz, K.T. Chan, J.W. Ager III, S. Tanigawa, and E.E. Haller, Mat. Res. Soc. Symp. Proc. Vol. 241, 93 (1992).

25.   D.J. Keeble, M.T. Umlor, P. Asoka–Kumar, K.G. Lynn, and P.W. Cooke, Appl. Phys. Lett. 63, 87 (1993).

26.   M.J. Puska, J. Phys.: Condens. Matter 1, 7347 (1989).

27.   Z–Q. Fang and D.C. Look, Appl. Phys. Lett. 61, 1438 (1992).

28.   P.W. Yu, D.C. Reynolds, and C.E. Stutz, Appl. Phys. Lett. 61, 1432 (1992).

29.   L.–W. Yin, Y. Hwang, J.H. Lee, R.M. Kolbas, R.J. Trew, and U.K. Mishra, IEEE Electron Device Lett. 11, 561 (1990).

30.   C–L. Chen, F.W. Smith, B.J. Clifton, L.J. Mahoney, M.J. Manfra, and A.R. Calawa, IEEE Electron Device Lett. 12, 306 (1991).

31.   Y. Hwang, W.L. Yin, J.H. Lee, T. Zhang, R.H. Kolbas, and U.K. Mishra, Abstracts of the 1990 Electronic Materials Conference (TMS, Warrendale, PA, 1990) p. 15.

32.   D.C. Look, C.E. Stutz, and K.R. Evans, Appl. Phys. Lett. 57, 2570 (1990).

33.   C–L. Chen, L.J. Mahoney, M.J. Manfra, F.W. Smith, D.H. Temme, and A.R. Calawa, IEEE Electron Device Lett. 13, 335 (1992).

34.   M.Y. Frankel, J.F. Whitaker, G.A. Mourou, F.W. Smith, and A.R. Calawa, IEEE Trans. on Electron Devices ED–37, 2493 (1990).

35.   T. Motet, J. Nees, S. Williamson, and G. Mourou, Appl. Phys. Lett. 59, 1455 (1991).

STOICHIOMETRY RELATED PHENOMENA
IN LOW TEMPERATURE GROWN GaAs

M. Missous and S. O'Hagan
Department of Electrical Engineering and Electronics and Centre for Electronic Materials
University of Manchester Institute of Science and Technology
PO Box 88, Manchester M60 1QD, England, UK

ABSTRACT

The growth of GaAs at low temperatures (LT-GaAs) at or below 250 °C, under standard Molecular Beam Epitaxy (MBE) growth conditions, usually results in a massive incorporation of excess As in the lattice which then totally dominates the electrical and optical characteristics of the as grown material. We report on new phenomena associated with the growth of GaAs at 250 °C and we show, for the first time ,data on highly electrically active doped material. By careful control of the growth conditions, namely As_4/Ga flux ratios, material in which total defect concentrations of less than 10^{17} cm^{-3}, well below the huge 10^{20} cm^{-3} that is normally obtained in LT-GaAs, can be achieved thereby demonstrating that high quality GaAs can in effect be grown at extremely low temperatures.

INTRODUCTION :

Ever since the first report by Smith et al[1] on the beneficial effects of growing GaAs buffer layers by MBE at low temperature (~ 200-250 °C) on the characteristics of FETs, extensive efforts aimed at investigating the structural, optical and electrical properties of such materials have been maintained in the last 3 to 4 years.

The ability to grow near perfect GaAs by MBE at standard growth temperatures of greater than 550 °C , without the need for precise control of the As to Ga flux ratio as long as there is excess As overpressure, has been the reason behind the great success of MBE in growing high quality GaAs. The situation changes dramatically however as the growth temperature is lowered to 250 °C or below.The growth of nominally undoped low temperature GaAs (LT-GaAs) under "normal" growth conditions results in highly defective and strained material with $\Delta a/a$ ~ 1 x 10^{-3} and defect concentrations of the order of 10^{20} cm^{-3} range[2] .The kinetics of growth at such low temperatures bears little or no resemblance to its high temperature counterpart in so far as the excess arsenic incorporation leads to generation of a high concentration of point defects in the GaAs lattice. Summarised , the statistics of defects in undoped un-annealed LT-GaAs grown near 250 °C , is as follows :

(i)   ~ 2-5 x 10^{19} cm^{-3} of neutral antisites $[As_{Ga}]^{0}$
(ii)   ~ 5 x 10^{18} cm^{-3} of positively charged antisites $[As_{Ga}]^{+}$
(iii) ~ 2-6 x 10^{20} cm^{-3} As interstitial $[As_i]$
(iv)   ~ 1-5 x 10^{18} cm^{-3} Ga vacancies $[V_{Ga}]$

It is the interaction between these various defects that leads to heavily compensated and barely conductive GaAs when grown at low temperatures.In the work reported here, our aim was to test whether these very large concentrations of defects were intrinsic, by virtue of the low growth temperatures used, or whether they can in fact be controlled somehow. Most studies on un-annealed LT-GaAs ,use MBE growth conditions that are more appropriate for high temperature growth, namely an As₄/Ga beam equivalent pressure (BEP) ratio in the 10 to 20 which in terms of fluxes translates into ratios of As:Ga from 3 to 7 approximately. These values are deemed necessary to introduce the excessive defects in LT-GaAs. In this report, we investigate the exact role of the group V to group III BEP ratio on the incorporation of excess arsenic and the associated structural and electrical properties of both doped and undoped LT-GaAs.

EXPERIMENTAL :

All the samples studied here were grown in an all solid source VG V90H MBE system with 4 inch growth capability.The details of the growth conditions have been reported elsewhere and will not be repeated here[3]. The As₄/Ga BEP ratios used varied from 6, which was more than enough to ensure growth at 600 °C , to 3 where growth could not be sustained at this very same temperature(ie the layers were highly arsenic deficient). The arsenic species used throughout was the tetramer As₄ generated from a solid arsenic source.

The growth temperature for all the samples was kept at 250 °C. The growth rate used was kept at 1 μm/h and the layer thicknesses varied from 1 to 3 μm for both the undoped and doped samples. The dopant concentrations were varied from 1 x 10¹⁸ to 1 x 10¹⁹ cm⁻³ for the "n-type" dopant Si.

All the layers were characterised in their as-grown state , using double X-ray crystallography, Hall effect measurements and optical absorption, care was taken not to heat them to temperatures greater than the growth temperature.

RESULTS AND DISCUSSION :

(i) The effect of As/Ga BEP on structural properties of LT GaAs :

It is well known that excess arsenic incorporation in LT GaAs gives rise to extra diffraction peaks in X-ray diffraction measurements[4] . The extra peak or peaks are thought to emanate from the epilayer which is strained by the excess arsenic incorporated in interstitial sites[2] .In this section the effect of varying the BEP is investigated with particular attention given to the amount of excess arsenic incorporated The BEP ratios were varied from 6 , a "typical" value , to 3 which is close to the minimum required at 250 °C to still grow under As stable conditions.

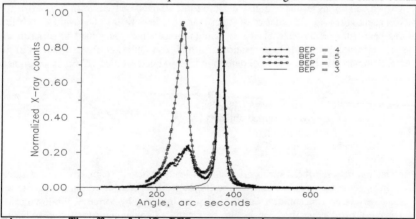

Figure 1 The effect of As/Ga BEP on excess As incorporation. Layer thickness is 1μm, except for the case with BEP = 6 , where it is 3 μm.

Figure 1 shows the double crystal X-ray rocking curves of the various layers grown. It is quite clear that the effect of the BEP ratio is a very small one until one gets very close to stoichiometric conditions. TABLE I summarises the results on the different layers and it is seen that the amount of excess arsenic is almost constant until one gets to a flux ratio of close to 1. The excess arsenic concentrations were extrapolated from the data of Man Yu et al[5] on deviation from stoichiometry of As and corresponding changes in lattice parameters on similar structures. Concomitantly, we have also measured the amount of the neutral $[As_{Ga}]°$ using infrared absorption and the values extracted (see TABLE I) are again seen to vary little with BEP and are equal to $\approx 2 \times 10^{19}$ cm^{-3} which is comparable to what has been reported in the literature[6] for growth at 250 °C. The signal from the stoichiometric layer was indistinguishable from the background.

TABLE I :   Excess As incorporated as a function of BEP

LAYER	BEP	As$_i$ (X-ray) (cm^{-3})	$[As_{Ga}]°$ (cm^{-3})
#410	6	1.88×10^{20}	1.8×10^{19}
#417	5	1.57×10^{20}	2.1×10^{19}
#418	4	1.62×10^{20}	1.5×10^{19}
#431	3	below detection limit	below detection limit

(ii) Stoichiometry in LT-GaAs :
        It is clear from figure 1, that a decrease in BEP ratio from 6 to close to 3 leads to the complete removal of the second x-ray peak (and presumably strain in the lattice) meaning a decrease in the incorporation of excess arsenic.
Because the density of the latter at the growth temperature used here (250 °C) seems to be a very weak function of the arsenic overpressure used during growth (except for the almost abrupt disappearance of the second x-ray peak at a BEP of ~3),we have shown recently [3,] in a detailed experiment, that a very small deviation from stoichiometry is sufficient to lead to excess As incorporation in quantities almost the same as when high flux ratios were used. Therefore at a given growth temperature there is a solid solubility limit to the amount of the excess As incorporated almost irrespective of BEP, as long as it is higher than the minimum required for stoichiometric conditions.
The overall conclusion of this section is therefore that growth of undoped stoichiometric GaAs, while possible at low temperatures, is nevertheless very much more demanding than at higher temperatures. This is of little consequence for the majority of work on LT GaAs since the object there is indeed to make use of the excess defect concentrations to achieve high resistivity through post-growth annealing. In our case where the driving force is the growth of electrically active material , the consequences are far more serious.

(iii) The effect of doping on the structural characteristics in LT GaAs :

        Figure 2 shows the double crystal X-ray rocking curves of two Si doped layers with concentrations of 1×10^{18} cm^{-3} and 1×10^{19} cm^{-3} at a relatively high BEP value of 6.The striking feature in the diagram is the disappearance of the second x-ray peak for the most heavily doped layer. We have in fact shown elsewhere[3] that this situation prevails irrespective of the BEP values for a doping of 1×10^{19} cm^{-3}.

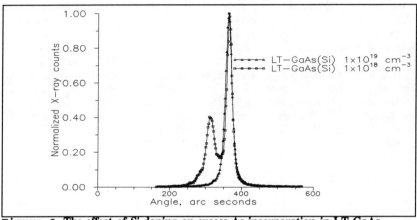

Figure 2 The effect of Si doping on excess As incorporation in LT-GaAs

The consequence of this is that the excess As incorporation is now almost completely inhibited. During doping with Si, the competition between Si and As for the Ga site leads to preferential incorporation of Si at Ga sites and this is indeed borne out in the near band gap infra-red absorption measurements[7] where the concentration of the neutral As_{Ga} related defect was found to decrease by almost an order of magnitude. The removal of the huge excess arsenic interstitial concentration ($\tilde{\ } 2 \times 10^{20}$ cm^{-3}) is far more difficult to understand however.

A point worth emphasising is that the growth of GaAs at these low temperatures might be governed by the Fermi-level effect ie the charge state of the various species will determine the Fermi level in the system which in turn will have a profound effect on the incorporation of the various species.

The effects of dopants on the growth of semiconductors is an increasingly reckoned phenomena especially at high concentrations (usually $\geq 1 \times 10^{19}$ cm^{-3}).In MBE, the effect of tin segregation[8] was one of the first manifestations of dopant effects during MBE growth. Later on , anomalous Be diffusion was discovered while fabricating HBT[9] structures . In our case, the phenomena we observe which is that of *stoichiometry induced growth,* seems to also conform to the same general rule. Indeed doping the LT GaAs to 1×10^{18} cm^{-3} leads to the re-appearance of the second x-ray peak and hence the excess arsenic incorporation (see figure 2).

(iv) Electrical properties :
    TABLE II below summarises the room temperature Hall effect measurements.
TABLE II   Electrical properties of Si-doped LT-GaAs

LAYER	[Si] (cm^{-3})	BEP	n (cm^{-3})	μ(cm^2/Vs)
# 332	1×10^{19}	6	1.9×10^{17}	364
# 419	1×10^{19}	4	1.6×10^{18}	540
# 324	1×10^{19}	3	1×10^{19}	989
# 327	1×10^{18}	4	2.9×10^{12}	727
# 325	1×10^{18}	3	6×10^{17}	2200

Starting with the 1 x 10^{19} cm^{-3} Si doped layers first, it is apparent that even for the highest BEP used here, 6, the layer was still conducting and of n-type conductivity with a mobility high enough to suggest free carrier conduction rather than hopping conduction even though the layer is clearly heavily compensated. As the BEP falls towards the stoichiometric value, the layer becomes suddenly extremely conducting , with all the Si (1 x 10^{19} cm^{-3}) activated and with a mobility of ~ 1000 cm^2/V.sec. These values are comparable to the best that has ever been achieved at such high concentrations[10] but at much higher temperatures (420 °C).This result can be understood in terms of both the reduction of excess arsenic at close to a BEP of 3 and also because near stoichiometric conditions, the concentration of V_{Ga} is predicted to fall from ~ 9 x 10^{18} cm^{-3} for growth under As-rich conditions to below ~5x10^{17} cm^{-3} for Ga-rich conditions[11] at a Si doping level of 1x10^{19} cm^{-3}.

The case of the lower doped sample (1 x 10^{18} cm^{-3}) shows again that by growing near stoichiometric conditions, 60% activation of the dopant is achieved with a fairly high mobility (2200 cm^2/V.sec). The effect of increasing the As overpressure is however much more dramatic leading to a total collapse of the carrier concentration (see TABLE II) but in line with the X-ray and optical absorption data .

CONCLUSIONS

We have investigated the structural and electrical properties of as-grown undoped and Si doped LT-GaAs as a function of As/Ga BEP ratio. In undoped LT-GaAs, the amount of excess As is almost independent of As/Ga BEP ratio, a small deviation from stoichiometric conditions is sufficient to lead to excess As being incorporated.

For doped materials, and under high As overpressure conditions, a new phenomena is seen concerning the effect of doping of LT-GaAs by Si in the 10^{19} cm^{-3} region which leads to the complete removal of the interstitial component, which contributes some few 10^{20} cm^{-3} atoms , and also to the reduction in the concentration of the As_{Ga} related defects by almost an order of magnitude. Doping with Si ,at 1 x 10^{19} cm^{-3}, always leads to n-type conductivity with free carrier rather than hopping conduction. Under near stoichiometric conditions, state of the art GaAs doped with Si is obtained , we believe, for the first time at as low a temperature as 250 °C using conventional MBE growth techniques. The implications from these measurements are that even at temperatures as low as 250 °C, the dopants are incorporated substitutionally and are electrically active.

Finally the question of the fundamental lower temperature limit to the growth of high electrical quality GaAs material has to be addressed. In the light of the results presented here, it is clear that the fundamental condition that has to be fulfilled at low temperatures is that of stoichiometry. It is clear that by supplying "stoichiometric" beams of As and Ga, high quality layers with defect concentrations lower than 10^{17} cm^{-3} can be produced, and with careful control of the BEP ratio , these could be further reduced. In conclusion, we would like to propose the defects-temperature-flux ratio phase diagram of Figure 3 as a possible venue to explain the increase in the defect concentrations as MBE growth is performed at temperatures lower than the standard of 550-600 °C while still keeping higher As/Ga flux ratios. The challenge set by this phase diagram is whether we can control flux ratios accurately enough near stoichiometric conditions to achieve low defect densities in low temperature GaAs.

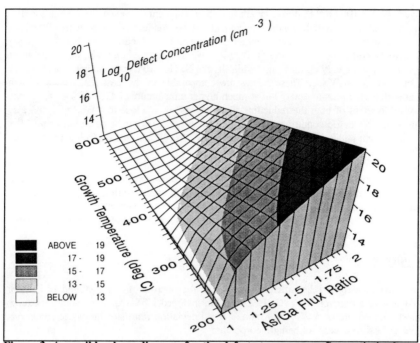

Figure 3. A possible phase diagram for the defects-temperature-flux ratio in the MBE growth of GaAs.

ACKNOWLEDGMENTS :
We gratefully acknowledge the many useful comments and discussions with Professors A.R. Peaker and K.E. Singer and Drs. M.R. Brozel and W.S. Truscott

REFERENCES

1. F.W. Smith , A.R. Calawa, M.J. Manfra and L.J. Mahoney, IEEE Electron Device Lett. 9(2) , 77 (1988)
2. M. Kaminska, E.R. Weber, Z. Liliental-Weber, R. Leon and Z.U. Rek, J.Vac.Sci.Technol.B7,710 (1989)
3. M.Missous and S. O'Hagan, J.Appl.Phys,*In press*
4. M. Kaminska, Z.Liliental-Weber, E.R. weber and T. George, Appl.Phys.Lett.54,1881 (1989)
5. K. Man Yu, M. Kaminska and Z. Liliental-Weber, J.Appl.Phys. 72(7),2850, (1992)
6. M.O. Manasreh,D.C. Look,K.R. Evans and C.E. Stutz, Phys.Rev.B 41,10272 (1990)
7. S.O'Hagan and M. Missous, to be published in J.Appl.Phys.
8. C.E.C Wood and B.A. Joyce, J.Appl.Phys. 49,4854 (1978)
9. P.M. Enquist, L.M. Lunardi, D.F. welsh, G.W. Wicks, J.R. Shealy, L. F. eastman and A.R. Calawa, Ins.Phys.Conf.Ser. 74,599 (1985)
10. M. Ogawa and T. Baba, Jap.J.Appl.Phys.,24(8),L572 (1985)
11. T.Y. tan, H.M. Tou and U.M. Gosele, Appl.Phys.A 56,1(1993)

IMPROVEMENT OF THE STRUCTURAL QUALITY OF GaAs LAYERS GROWN ON Si WITH LT-GaAs INTERMEDIATE LAYER

Zuzanna Liliental Weber, H. Fujioka,* H. Sohn,* and E.R. Weber*

Center for Advanced Materials, Lawrence Berkeley Laboratory 62/203, University of California, Berkeley, CA 94720, *Department of Materials Science, University of California, Berkeley, CA 94720

ABSTRACT

Superior electrical and optical quality of GaAs grown on Si with an inserted low-temperature (LT)-GaAs buffer-layer was demonstrated. Photoluminescence intensity was increasing and leakage current of Schottky diodes build on such a structure was decreasing by few orders of magnitude. These observations were correlated with structural studies employing classical and high-resolution transmission electron microscopy (TEM). Bending of the threading dislocations and their interaction was observed at the interface between a cap GaAs layer and the LT-GaAs layer. This dislocation interaction results in the reduction of dislocation density by at least one order of magnitude or more compared for the GaAs layers with the same thickness grown on Si. The surface morphology of the cap GaAs layer is improving as well.

INTRODUCTION

The growth of GaAs on Si (GaAs/Si) has attracted much attention because it combines large and mechanically strong Si substrates with the properties of a GaAs layer. These advantages make GaAs/Si suitable for substrates of low cost GaAs integrated circuits (ICs). The GaAs/Si epitaxy technique is also promising for monolithic integration of III-V devices with Si devices. However, this technology poses several severe problems which have to be solved such as, poor crystal quality caused by the relatively large differences in lattice constants and thermal expansion coefficients, and poor surface morphology. A large number of studies have concentrated on these problems. These works include the two step growth technique, the use of a strained layer superlattice, and thermal annealing [1-3]. These techniques do improved crystal quality, but the density of residual threading dislocation is still much higher than in bulk GaAs wafers. The rough surface of GaAs/Si is also a problem, especially for heterojunction devices which require almost atomically flat interfaces.

Recently it was demonstrated that the use of GaAs buffer layers grown at substrate temperatures of as low as 240 °C during the growth of GaAs/Si leads to smoother surface topography and improved electrical and optical properties [4]. This low temperature grown GaAs (LT-GaAs) buffer layer is also expected to help device isolation, because it is highly resistivity $(1.7 \times 10^7 \, \Omega cm)$. LT-GaAs was initially developed as a buffer layer to eliminate the side-gate effect of MESFET [5]. LT-GaAs layers are slightly conductive in their as-grown states but become highly resistive upon post-growth-annealing at 600°C. Hence, the growth of a standard active GaAs layer on top of the LT layer causes the layer to become highly resistive. As-grown LT-GaAs displays a dilated lattice constant ($\Delta a/a$ of 0.1%) and contains approximately 1-2 atomic % excess As [6,7]. Such layers contain about $10^{20}/cm^3$ As_{Ga}-related defects, and about $5 \times 10^{18}/cm^3$ of these are ionized. While the excess As remains unchanged even after a 600°C anneal, the deep level density and disorder are observed to decrease monotonously with increasing annealing temperatures. A 600°C anneal also results in As precipitates approximately 4-6 nm in diameter of a volume density of 10^{17}/cm^3, which accounts for almost all of the excess As present before the annealing.

In this paper, we will describe TEM study of how this LT-GaAs layer interacts with the threading dislocations which come from the GaAs/Si hetero-interface.

EXPERIMENTAL PROCEDURE

Two samples with different LT-GaAs buffer thickness were grown (0.5 μm and 0.07 μm). A sample without LT-GaAs was also grown for comparison. The total thickness of the samples was 3 μm for all samples. 2" Si wafers with (100) 4° off toward the [011] orientation were used for all the experiments to avoid the formation of anti-phase domains. The samples were degreased in organic cleaning solution first. Just before loading the substrate into the MBE chamber, the substrate was then cleaned using several cycles of oxidation by HCl/H$_2$O$_2$ followed by removal of the silicon dioxide by HF. All the samples were grown using a Varian GEN II MBE system. After having been transferred to the growth chamber, the samples were heated up to about 850°C by a direct radiation method for 15 minutes to remove surface oxide. The substrate temperature for GaAs growth was chosen initially to be 350°C for ten minutes and then increased to the normal growth temperature (600°C). This technique is known as "two step growth", which results in a smooth surface [1]. The growth rate for the normal GaAs was set at about 1μm/hr. After the growth of 1μm of normal GaAs, the substrate temperature was reduced to a nominal temperature of 230°C for the growth of the LT-GaAs buffer. On the top of this LT-GaAs buffer layer, another normal GaAs layer was grown. To investigate the electrical properties of the top layer which would be used as the active device region, in-situ n-type doping was done using a Si effusion cell. No special treatments, such as heat cycle annealing, were applied so as not to confuse the effect of the LT-GaAs layer in this study.

At this stage the samples were removed from the growth chamber and divided into two parts; one for transmission electron microscopy (TEM) and photo luminescence (PL) and the second part for electrical studies.

For TEM studies two types of thin foils were prepared: one for plan-view observations and the second for cross-section studies. To prepare plan-view thin foils the sample was dimpled from the substrate side to leave a 30 μm thick layer at the top of the sample. This was followed by chemical thinning from the substrate side or ion milling from the same side. For cross-sectional observation the samples were cut along both [110] and [$\bar{1}$10] directions and glued layer to layer side together. This was followed by mechanically thinning from both sides of the samples, followed by ion milling from both sides.

After taking the samples from the MBE chamber, gold Schottky diodes were fabricated for electrical studies. Both Schottky and ohmic contacts were formed on the top layer because LT-GaAs is highly resistive. Just before each metallization process, an HCl treatment was performed to remove the native oxide. After evaporation of Ni/AuGe through a mask, a rapid thermal anneal was performed at 350°C for 15 seconds in forming gas which contained 10 % H$_2$ and 90 % N$_2$. Au Schottky electrodes were evaporated after confirmation of the ohmic behavior of the Au/Ni electrode. Since Au Schottky electrodes readily react with GaAs, no thermal treatment was done after Au evaporation.

RESULTS AND DISCUSSIONS

TEM results

High resolution studies of cross sectioned samples showed no large difference in the misfit dislocation density or their arrangement between the Si and the GaAs buffer layer. For the sample cut along the 4° deviation of the interface from exact (001) was observed and edge 90° dislocations separated from each other by about 91Å were present (Fig. 1a). The dislocation cores in a majority of cases were located at the interface but some of them were into the substrate about 1-3 monolayers. This may be related to charge compensation at the interface as discussed by Harrison [8]. For the projection in the perpendicular direction the misorientation from (001) is not observed as expected (Fig. 1b). At this interface mostly partial dislocations were observed with associated stacking faults. This anisotropy of dislocation formation at the interface can be related to the polar structure of GaAs.

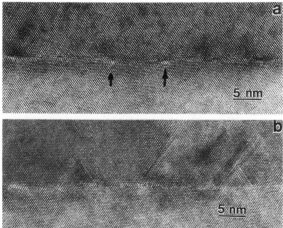

Fig. 1. a) High-resolution image of the Si/GaAs interface. Note 4° tilt along [110] and location of the dislocation cores. Dislocations marked by arrows are located in Si. All misfit dislocations shown here are edge type; b) The same interface along [$\bar{1}$10]. Note many partial dislocations and formation of stacking faults.

At a distance of 0.5 μm from the interface the density of dislocations was in the range of 10^{10}cm^{-2}. These dislocations interact with each other and their density decline two orders of magnitude before reaching the top layer due to annihilation and recombination reactions. Some of the remaining a/2<110> dislocations tend to lie along <110> directions but 27 % of them were lying along [$\bar{1}1\bar{2}$] or [$1\bar{1}\bar{2}$] which in projection is parallel to [$\bar{1}$10] in the interface plane. These dislocations are all inclined extending from top to bottom surface of the foil in plan-view images. The contrast of the <112> oriented dislocation was very strong for the [220] diffraction vector perpendicular to the dislocation line. These are perfect edge dislocations with their Burgers vector along [220]. They are straight with symmetrical black-black alternating with white/white contrast in the direction perpendicular to the diffraction vector [Fig. 2a]. For the perpendicular diffraction vector only occasionally were dislocations found along [$\bar{1}\bar{1}2$] or [112]. This again shows asymmetry of dislocation formation during the layer growth. It is probable the dislocations tend to align along <112> because this results, during growth of the layer, in the shortest possible length of dislocation per unit of layer thickness that still lies on a {111} plane.

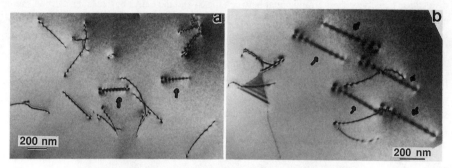

Fig. 2. a) Plan-view of the top layer of GaAs grown on Si. Note that dislocations marked by arrows lie along [$\bar{1}1\bar{2}$] or [$1\bar{1}\bar{2}$]. b) Plan-view of the top layer of GaAs grown on Si with LT-GaAs intermediate layer. Note relative increase of <112> type of dislocations in this layer.

For the GaAs on Si with the intermediate 0.5 μm thick (or 0.07 μm thick) LT-GaAs layer a very interesting phenomenon was observed. Starting with the same number of dislocations as in the previous case (without LT layer) dislocations propagated through the layer with no bending at the lower interface probably due to the fact that that the lattice mismatch 0.1% between buffer GaAs layer and LT layer is not large enough to cause significant strain in the layer. However, very strong bending of dislocations was observed at the upper interface between the LT-GaAs layer and the GaAs top layer (Fig. 3).

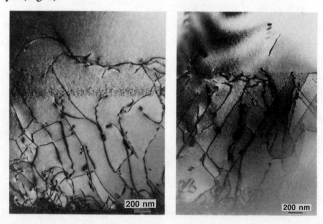

Fig. 3. Two examples of cross-sectioned samples with the LT-GaAs intermediate layer. Note drastic bending of threading dislocations at the upper interface between LT-GaAs layer and the cap GaAs layer and the formation of As precipitates in the LT-layer traping the dislocation mobility.

This probably shows that both dislocations that are formed during growth and those that form from thermal strain during cooling down from the growth temperature have difficulty in moving within the LT layer because of As precipitates which decorate the dislocations within this LT layer. This trapping of dislocations leads to bending of dislocations immediately above the LT layer and therefore to further annihilation and consequently fewer dislocations propagating toward the surface within the top layer. The growth of LT-GaAs layers with a cap GaAs layer can be treated as a thermal cycling which leads to the dislocation bending and their interaction.

Very nonuniform distributions of dislocations were observed in plan-view samples obtained from these layers. There were some areas with similar dislocation densities to those layers without LT-GaAs but there were other areas without any dislocations. It is very difficult to obtain good statistics from TEM concerning average dislocation densities. However, rough estimation suggests a factor of one order of magnitude fewer dislocations due to the LT-GaAs layer. A similar behavior was observed with the thinner (0.07μm) LT layer. Another difference was that in the samples with LT layers the relative number of <112> oriented dislocations was 10% higher compared to layers grown without LT layers.

Electrical properties

From C-V measurements it was confirmed that the Si doping was homogeneous and that the concentrations were almost identical for all the samples. The doping concentration and Schottky barrier height are estimated to be $1.3 \times 10^{16} cm^{-3}$ and 0.91eV, respectively. Fig. 4a shows I-V characteristics for Schottky diodes with and without LT buffer layers. The use of a LT buffer drastically reduces the leakage current. This is clear evidence for improvement in electrical properties by use of the LT-GaAs buffer because these diodes were processed simultaneously. The sample without LT intermediate layer did show lower series resistance in the forward bias region. This is due to the current through the Si substrate. The activation energies associated with the leakage current at -1V for the samples with and without LT-GaAs were estimated to be 0.36eV and 0.05 eV, respectively, from temperature dependence measurements. This result implies that the leakage current for the diode without LT-GaAs is dominated by a mechanism involving tunneling phenomena thorough deep levels.

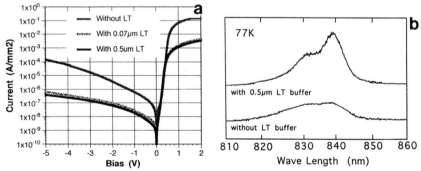

Fig. 4. a) I-V characteristics of Au Schottky diodes build on GaAs grown on Si with and without of intermediate LT-GaAs layer. Note decreasing of the leakage current in the samples with LT-GaAs layers; b) Photoluminescence intensity obtained from the GaAs grown on Si with and without of intermediate LT-GaAs layer.

Photoluminescence results

The layers with inserted LT-GaAs buffers have much higher intensity of the photoluminescence line observed at 839 nm at 77 K compared to the layers without LT buffers (Fig. 4b). This improvement can be attributed to the slightly lower density of dislocations as well as a probable lower density of point defects in these layers. It is possible that a LT-GaAs layer also acts as a barrier for Si outdiffusion and prevents introduction of other point defects into the layer. It was as well observed that amount of strain in the layers without LT buffer and with the LT buffer have similar strain, since energy position of PL line did not change.

CONCLUSIONS

Superior electrical and optical quality of GaAs grown on Si with an inserted LT-GaAs buffer-layer was demonstrated. These observations were correlated with structural studies employing classical and high-resolution TEM. The interface between the Si and GaAs was much cleaner than in samples studied in the past [9]. Large anisotropies in the types and distributions of dislocations were observed in these layers. Prominent bending of the threading dislocations was observed on the upper interface between the LT-GaAs layer and the cap GaAs layer. This bending was attributed to decoration and pinning of dislocations by As precipitates at the top of the LT-GaAs layer. The LT-layer also appeared to shield the top layer from accumulation of point defects. An increase in the fraction of dislocations lying along $[\bar{1}1\bar{2}]$ or $[1\bar{1}\bar{2}]$ was also observed in the top layer of samples with an inserted LT-GaAs layer.

ACKNOWLEDGMENT

This research was supported by AFOSR-ISSA-90-0009, through the U.S. Department of Energy, under Contract No. DE-AC03-76SF00098 and AFOSR-88-0162. The use of the Facility of the National Center for Electron Microscopy in Lawrence Berkeley Laboratory supported by U.S. Department of Energy under Contract No. DE-AC03-76SF00098 is greatly appreciated. Z.L.W. wants to thank W. Swider for very careful and successful TEM sample preparation.

REFERENCES

1. M.Akiyama, Y.Kawarada, and K.Kaminishi, Jpn. J. Appl. Phys. **23**, L843 (1984).
2. T.Soga, S.Hattori, S.Sakai, M.Takeyasu, and M.Umeno, Electron. Lett. **20**, 916 (1984).
3. N.Chand, R.People, F.A.Baiocchi, K.W.Wecht, and A.Y.Cho, Appl. Phys. Lett. **48**, 1815 (1986).
4. H.Fujioka, H.Sohn, E.R.Weber, A.Verma, J. Elect.Mat. (1993) in print.
5. F.W.Smith, A.R.Calawa, C.I.Chen, and M.J.Mahoney, IEEE Electron Device Lett. **9**,77 (1988).
6. Maria Kaminska, E.R.Weber, Z.Liliental-Weber, and R.Leon, Jour. Vac. Sci. Tech. B**7**, 710 (1989)
7. Z. Liliental-Weber, Mat. Res. Soc. Proc. vol. **241**, 101 (1992).
8. W.A. Harrison, E.A. Kraut, J.R. Waldrop, and R.W. Grant, Phys. Rev. B**18**, 4402 (1978).
9. Z. Liliental-Weber, Mat. Res. Soc.Symp. Proc. vol. **148**, (1989) p. 205.

PRESSURE RATIO (P_{As}/P_{Ga}) DEPENDENCE ON LOW TEMPERATURE GaAs BUFFER LAYERS GROWN BY MBE

M. LAGADAS[1], Z. HATZOPOULOS[1], M. CALAMIOTOU[2], M. KAYIAMBAKI[1] and A. CHRISTOU[3]
[1] Foundation for Research and Technology-Hellas,
Institute of Electronic Structure and Laser, Heraklion, Crete, Greece
[2] University of Athens, Department of Physics, Athens, Greece
[3] University of Maryland, Department of Materials Engineering, College Park, MD, USA.

ABSTRACT

We have investigated the influence of the pressure ratio (P_{As4}/P_{Ga}) on the structural and electrical properties of GaAs layers grown at 250°C by MBE. SEM photographs have revealed smooth surfaces for $P_{As4}/P_{Ga} \geq 15$ and Double crystal X-ray rocking curves have shown an increase on the lattice mismatch $\Delta a_r/a$ of the L.T. grown layers and high crystalline quality. Resistivity has not been affected by the different values of P_{As4}/P_{Ga}. n-GaAs epilayers grown on top of L.T. buffer layers have their mobility decreased and the electron trap density increased as revealed by Hall and DLTS measurements.

INTRODUCTION

Low Temperature GaAs layers grown by MBE have been attracting attention in recent years because they are useful as buffer layers in FET's for eliminating sidegating effects [1]. X-ray and PIXE measurements in the as-grown material revealed that these layers have increased lattice constant in the direction of growth and excess As concentration up to 1.3%. Lattice mismatch and As concentration depends on the growth temperature [2]. Electron and hole traps have been revealed by DLTS [3], TSC [4] and TEES [5]. In annealed L.T. grown GaAs layers the excess As forms precipitates whose dimensions and density depends on the growth temperature, the annealing temperature [6,7] and the use of As_4 or As_2 [8].

Up to now no work has been done on the effect of B.E.P. (Beam Equivalent Pressure) P_{As4}/P_{Ga} on the crystalline quality and the resistivity of L.T. GaAs layers. In this work we have investigated the effect of different B.E.P. on lattice expansion, resistivity, surface morphology and electron traps of GaAs layers grown at 250°C.

EXPERIMENT

All samples were grown in a VG V80H MBE system. L.T. GaAs layers have been grown using tetramer arsenic source As_4, at a substrate temperature T=250°C, on Indium bonded molyblocks. In order to have the same growth temperature in different runs which is so important for reproducibility of the low temperature only two molyblocks were used, with identical thermal behaviour as can be recognized from oxide desorption at 580°C.

The substrate temperature was determined by a thermocouple reading. At the actual temperature of 580°C the thermocouple reads in the neighbourhood of 750°C. The growth temperature of 250°C was determined by extrapolation.

For resistivity measurements (100) n^+(Si-doped=$2x10^{18}cm^{-3}$) GaAs substrates (ϱ_c=1.7x10^{-3}ohm.cm) were used, while for X-rays, DLTS and Hall measurements (100) S.I. GaAs were used. All substrates were decreased followed by etching in a 60°C solution of 5:1:1 (H_2SO_4:H_2O_2:H_2O) and then outgassed for 1 hr. at 350°C in the preparation chamber before introducing them into the deposition chamber. The growth rate was 1μm/hr. and the B.E.P. P_{As4}/P_{Ga} was varied from 10 to 40 in L.T. grown layers as measured by a nude ion-gange. After the oxide desorption an undoped buffer layer of 2500Å GaAs (n^+ doped GaAs for resistivity measurements) was grown at 580°C. The substrate temperature was then ramped down to 250°C and 1μm L.T. GaAs was grown. For DLTS (Hall) measurements an additional layer of 1.5μm (5000Å) of $5x10^{16}cm^{-3}$ ($2x10^{17}cm^{-3}$) n-GaAs Si-doped was grown on the top of L.T. GaAs at 580°C, while for resistivity measurements 5000Å of $2x10^{18}$ n^+-GaAs was grown. No additional layers were grown when the substrate temperature was ramped.

EXPERIMENTAL RESULTS AND DISCUSSION

Surface morphology of L.T. GaAs layers as viewed by SEM was smooth for all samples with BEP≥15. For BEP=10 there was a roughness due to ~1μm hillocks in <01̄1> direction. This roughness has also been observed in samples grown at 200°C [16]. Annealing the samples at 580°C for 1 hr. has shown no change on the surface morphology.

X-ray rocking curves were measured on a double axis diffractometer (Bede 150) in the (+, -) non dispersive mode, using a GaAs reference crystal and CuKa radiation. The (004) symmetric and the (115) asymmetric Bragg reflections were measured.

All annealed samples at 580°V for 1 hr. have shown one Bragg peak, with FWHM=11-13 arcsec, indicating that the layers had the same lattice constant as the substrate and that their crystalline quality was very good. In contrary, samples grown at 200°C have shown a much broad (400) Bragg peak with an asymmetric tail on the low-angle side after annealing at 580°C [12].

For the as-grown samples two Bragg peaks were observed. By measuring the (115) asymmetric reflection at low and high angle of incident configuration we have found that the epilayers were grown coherently at the substrate. Fig. 1 shows the relative increase of the relaxed lattice constant of $\Delta a_r/a$ as a function of flux ratio R=J_{As4}/J_{Ga} for different samples grown at 250°C. The increase of the relaxed lattice parameter $\Delta a_r/a$ of the non-stoichiometric GaAs epilayer has been calculated from the experimental vertical lattice increase of $\Delta a_\perp/a$ [12]. We estimated the flux ratio R from the B.E.P. P_{As4}/P_{Ga} [10].

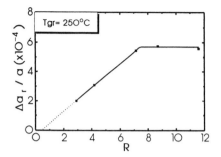

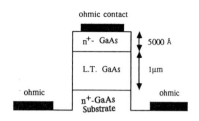

Fig. 1. Relaxed lattice expansion $\Delta a_r/a$ of as-grown L.T. GaAs at different flux ratios $R = J_{As4}/J_{Ga}$ for $T_{gr}=250°C$.

Fig. 2. Cross section of n-i-n structure for resistivity measurement of L.T. GaAs grown at $T_{gr}=250°C$. The growth temperature of n^+ epilayer was 600°C.

$\Delta a_r/a$ increase linearly with respect to R (P_{As4}/P_{Ga}) up to a value of 7.3 (25) where a saturation in $\Delta a_r/a$ occurs at the value of $(5.6\pm0.2)\times10^{-4}$. The FWHM of the epilayers is 22-24 arcsec indicating very good crystalline quality comparable with the FWHM of a simulated (004) intrinsic Rocking Curve of a perfect 1µm layer which is 20arcsec. Moreover rocking curves from the as-grown L.T. layers have shown well defined interference fringes indicating a high level of structural perfection. The FWHM also did not depend on B.E.P.

The sticking coefficient of As_4 depends on the substrate temperature [9] while the lattice mismatch is analogous with excess As presented in the L.T. GaAs layer [2], so the increase in $\Delta a_r/a$ is due to the excess As adsorption. The linear part of the curve intersects the R axis at 0.5 where stoichiometric growth and no difference in $\Delta a_r/a$ is expected. Neverhteless, growth of L.T. GaAs layers with B.E.P.=7 resulted in milky GaAs layers, indicating shortage of As_4 during growth. This is due to incomplete As_4 desorption at low growth temperature [9].

The saturation in $\Delta a_r/a$ indicates that there is a saturation of adsorption of excess As in the GaAs lattice. This saturation in adsorption of excess As have also been observed by RHEED intensity measurements in the absence of Gallium flux [10]. Experiments are in progress in order to investigate the $\Delta a_r/a$ and R saturation values on different growth temperatures.

For resistivity measurements typical AuGe/Ni/Au deposition was used for ohmic contact on n^+-GaAs followed by RTA at 410°C for 20 secs. A pattern with 200µm and 500µm dots diameter was used for the n-i-n structure (Fig. 2). To be sure that the thickness of L.T. GaAs layer did not affect the resistivity measurements we grew two samples with 1µm and 5µm in the i-layer at 250°C. The two samples appeared to have the same resistivity. For all n-i-n structure the L.T. GaAs layers were annealed at 580°C for 30 mins during the growth of the top 5000Å n^+-GaAs for ohmic contacts.

The I-V characteristics of n-i-n structure consisted of a linear region and a breakdown region at 30 Volts. The resistivity measurements were performed under d.c. current in the linear region of the I-V characteristic up to 10 Volts. The measured resistivity ϱ_c=3x10$^7\Omega$.cm (±2x10$^7\Omega$.cm) was the same for all the L.T. GaAs samples with B.E.P.=10-40. Annealing the samples at 650°C for 1 hr. in As$_4$ overpressure had no influence in the resistivity. The high resistivity of the L.T. GaAs layers is attributed to the density and the size of As precipitates which are formed during annealing. In accordance to the model proposed by Warren et. al. [11] for n$^+$-GaAs grown at low temperature, there is a depletion radius of ~200Å around the As-precipitates. This means that a precipitate density of ~6x10^{16}cm^{-3} is enough to deplete the L.T. GaAs transferring it into a semi-insulating. At T_{gr}=250°C and B.E.P.=10 (the lower B.E.P. we used) precipitate density is 10^{17}-10^{18} [8], higher than the limit of the 6x10^{16}cm^{-3}. In higher B.E.P. excess As adsorption can just increase precipitate size and/or density but does not affect the resistivity.

Table I

Trap	Ea(meV)	σ_c(cm^2)
1	875	4x10^{-13}
2	710	2.8x10^{-11}
3	350	1.5x10^{-14}

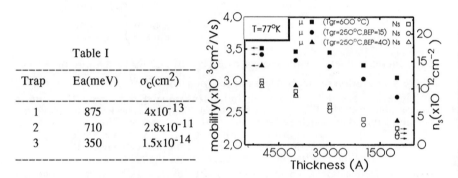

Fig. 3. Mobility and Carrier density at 77°K of n-GaAs epilayer grown on the top of L.T. GaAs buffer.

Fig. 3 shows mobility and carrier density of the n-GaAs epilayer as a function of n-GaAs thickness, extracted from Hall measurements at 77K, for 3 different samples: two samples with L.T. buffer GaAs grown at 250°C with B.E.P.=15, 40 and one conventional sample for reference growth at 600°C, B.E.P.=10. In samples with L.T. GaAs the mobility and the carrier density is lower that in reference sample.

n-epilayers grown on conventional and L.T. GaAs buffers have been compared by capacitance DLTS. The LTB samples revealed 3 common electron traps, summarized in table I, while a trap (Ea=605meV, σ_n=1x10^{-17}cm^2), tentatively identified as M6 [18] was observed on the conventional sample.

Trap 2 concentration increases as B.E.P. changes from 15 to 40 (Fig. 4) while an Arrhenius plot indicates trap 1 is EL2-like (Fig. 5).

Concentration of trap 1 and trap 2 in the epilayer increases with reverse bias (Fig. 6) up to the pinch-off voltage of -0.45V. Therefore trap concentration increases towards the buffer-epilayer interface.

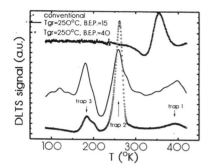

Fig. 4. Capacitance DLTS with rate window 80/sec and $V_{rev.}=-0.38V$

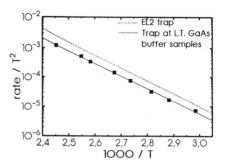

Fig. 5. Arrenius plot of the common trap1 appeared at samples with L.T. GaAs buffer and the EL2 for comparison.

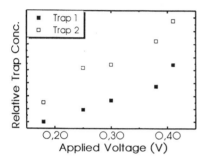

Fig. 6. Variation of trap concentration by different applied voltages in sample with L.T. GaAs buffer grown at 250°C, B.E.P.=40.

Before the growth of the n-GaAs epilayer the samples were annealed at 580°C for 5 minutes and recovery in (2x4) RHEED reconstruction was observed. We expect smooth interface between L.T. GaAs and n-GaAs layer [14], so the mobility degradation can not be attributed in interface roughness.

In L.T. layers there are native defects like V_{Ga}, As_{Ga} [13] which can diffuse up to the n-GaAs epilayer [15]. The observed decreased in the carrier density by decreasing the thickness of n-GaAs epilayers in samples with L.T. buffer is due to higher density of deep traps (Fig. 6). Diffusion of native defects and traps are responsible for the observed decrease in mobility. These defects act as scattering centers, resulting in lower mobility [17] in n-GaAs epilayer in sample with L.T. buffer.

CONCLUSION

We have studied the influence of the B.E.P. on the structural and the electrical properties on L.T. GaAs layers grow at 250°C by MBE. Lattice expansion was linearly increased up to B.E.P.=25 and saturation in Lattice expansion observed for B.E.P.≥25. Variation of B.E.P. had no influence on the resistivity of the annealed material. By using L.T. GaAs buffer, extra electron traps have been introduced in n-GaAs epilayers. Trap concentration increased by increasing B.E.P. For good surface morphology, high mobility and low trap density of n-GaAs epilayers B.E.P. around 15 is suggested.

REFERENCES

1. F.W. Smith et. al. IEEE Electron Devices Lett. **EDL-9**, 77 (1988).
2. Liliental-Weber et. al. Appl. Phys. Lett. **58**, 2153 (1991).
3. R.A. Puechner et. al. J. Cryst. Growth **111**, 43 (1991).
4. W.S. Lan et. al. J.J. Appl. Phys. **30**, L1843 (1991).
5. Z.C. Huang and C.R. Wie in Low Temperature (LT) GaAs and Related Materials, edited by G.L. Witt, R. Calawa, U. Mishra and E. Weber (Mater. Res. Soc. Proc. **241**, Boston, 1991) pp. 63-68.
6. M.R. Melloch N. Otsuka, K. Mahalingam, A.C. Warren, J.M. Woodall and P.D. Kirchner in Low Temperature (LT) GaAs and Related Materials, edited by G.L. Witt, R. Calawa, U. Mishra and E. Weber (Mater. Res. Soc. Proc. **241**, Boston, 1991) pp. 113-124.
7. O. Mahalingam et. al. J. Vac. Sci. Technol. B **10**, 812 (1992).
8. M.R. Melloch et. al. J. Cryst. Growth **111**, 39 (1991).
9. T.M. Brennan et. al. J. Vac. Sci. Technol. A **10**, 33 (1992).
10. Kun-Jing Lee et. al. J. Appl. Phys. **73**, 3291 (1993).
11. A.C. Warren et. al. J. Appl. Phys. **57**, 1331 (1990).
12. M. Calamiotou et. al. Solid. State Comm. **87**, 563 (1993).
13. D.C. Look et. al. Appl. Phys. lett. **60**, 2900 (1992).
14. A. Srinirasan et. al. J. Vac. Sci. Technol. B **10**, 835 (1992).
15. Isao Ohbu et. al. J.J. Appl. Phys. **31**, L1647 (1992).
16. M. Lagadas et. al. J. Cryst. Growth **127**, 76 (1993).
17. W. Walukiewicz et. al. Appl. Phys. Lett. **58**, 1638 (1991).
18. D.V. Lang et. al. J. Appl. Phys. **47**, 2558 (1976).

FEMTOSECOND PROBE-PROBE TRANSMISSION STUDIES OF LT-GROWN GaAs NEAR THE BAND EDGE,

H. B. RADOUSKY, A. F. BELLO, D. J. ERSKINE, L. N. DINH, M. J. BENNAHMIAS, M. D. PERRY, T. R. DITMIRE AND R. P. MARIELLA JR.,
Lawrence Livermore National Laboratory, Livermore, CA 94550.

ABSTRACT

We have studied the near-edge optical response of a LT-grown GaAs sample which was deposited at 300 ^{0}C on a Si substrate, and then annealed at 600 ^{0}C. The Si was etched away to leave a 3-micron free standing GaAs film. Femtosecond transmission measurements were made using an equal pulse technique at four wavelengths between 825 and 870 nm. For each wavelength we observe both a multipicosecond relaxation time, as well as a shorter relaxation time which is less than 100 femtoseconds.

INTRODUCTION

Low temperature grown GaAs[1-2] has been of considerable interest in the last few years due to the sub-picosecond recombination times. This increased interest is due both to the possibilities of utilizing the faster response times in electronic devices,[3-8] as well as understanding the basic physics underlying the faster response.[9] In many cases, the faster times are attributed to the presence of As impurities in the annealed samples, though fast times have also been observed in amorphous, unannealed samples.[10] In this paper we report femtosecond transmission spectra on a free standing LT-GaAs film initially grown on a silicon substrate.

EXPERIMENTAL DETAILS

Solid state lasers with short pulses comparable to CPM dye lasers have been recently developed based on Ti-Sapphire.[11-12] In addition to the simplicity of use compared to dye lasers, they provide orders of magnitude higher average power and are tunable over a broad range of frequencies. The laser used in these experiments can provide sub-100 fs pulses with 600 mW average power and is tunable in the current

389

configuration from 800 to 900 nm. This wavelength regime brackets the GaAs band gap at 870 nm.

When focused in a GaAs sample, these optical pulses can promote electrons from the valence band to the conduction band, non-destructively creating highly non-equilibrium conditions. In our experiment, we have utilized a probe-probe version[13-14] of the standard pump-probe technique. The beam is split into two equal parts, with one pulse delayed with respect to the second. The reflectivity or transmitivity is then measured as a function of the delay time. The second pulse probes the dynamical process at a given time delay relative to the first. In this way, the time averaged optical properties map out, as a function of delay, the relaxation over the first several picoseconds following excitation. The symmetrical nature of the probe-probe technique

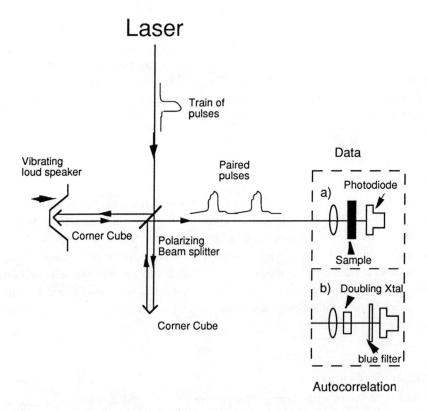

Figure 1 - Experimental set-up for the equal-pulse transmission measurements.

facilitates resolving relaxation times that are comparable or shorter than the pulse width. As a separate feature of the setup, a vibrating speaker is used to sweep through the relevant delays between the pulses many times each second. This allows for both ease in finding the signal, as well as faster collection times. It is analogous to taking a photoluminescence spectra using a scanning spectrometer and a photomultiplier tube, as opposed to using a multichannel analyzer. Both the pump-probe and probe-probe configurations can be used with the vibrating speaker. The diagram for the experimental setup is illustrated in Figure 1.

FILM GROWTH AND CHARACTERIZATION

The GaAs films were grown by MBE on Si substrates. The substrate temperature was 300 °C. Following growth, the film was annealed for 10 minutes at 600 °C. It is expected that defects, such as arsenic precipitates, are introduced by the low temperature growth and subsequent annealing.[7-10] The advantage of growth on Si is that this substrate is easy to etch away, leaving a free standing GaAs film , which was 3 microns thick. This thickness corresponds to one optical absorption length at 825 nm, which is the appropriate dimensions for this experiment. The sample was masked prior to etching away the substrate, so that a series of 3-micron, free standing GaAs windows was created.

The GaAs film was characterized by room-temperature photoluminescence, which is shown below in Fig. 2. The film was excited by green 514 nm light from an Ar ion laser. The peak of the spectrum is at 870 nm, corresponding to a gap energy of 1.43 eV. This indicates that the GaAs gap is essentially unchanged from bulk GaAs.[15] (The second peak near 900 nm is likely due to shallow acceptors, not the deep levels associated with the arsenic precipitates.) This would imply that we should see optical absorption even if the excitation energy is equal to or slightly greater than the gap energy. This is indeed what is observed in this sample.

RESULTS AND DISCUSSION

Prior to the availability of Ti-Sapphire lasers, the femtosecond studies of GaAs have relied on laser pump pulses near 2 eV in energy from CPM dye lasers.[13-14,16-17] The new laser sources are now providing the opportunity to study the fast optical response of GaAs using pump beams centered near the band gap energy of 1.43 eV.[18] This is in contrast to studies which use 2 eV pump beams and probe beams with energies close to the band edge.[16-17]

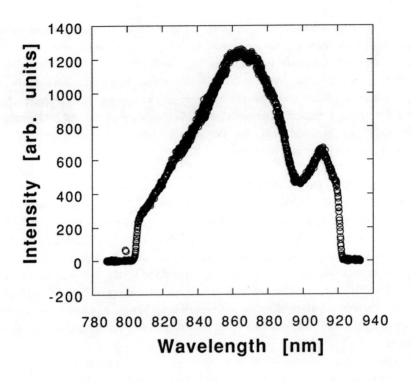

Figure 2. Photoluminescence for the LT-grown GaAs sample.

We show in Figure 3 the transmission correlation peak (TCP)[14] for the LT-GaAs sample taken at 870 nm. Spectra was also obtained at wavelengths of 825, 840, 855 nm. For all four wave lengths we observe two distinct lifetimes. A multipicosecond lifetime ($\approx$ 8 ps) is found to be consistent with literature values for other GaAs films grown at 300 C.[7] This time is take to be the recombination time, and is dependent on the recombination centers which have been introduced into the sample by the low temperature growth. The short lifetime is found to be less than 100 fs, which is consistent with values for pure GaAs.[13-14] Further results will be discussed in detail in a separate publication.[19] It would appear that growth of LT - GaAs on Si substrates

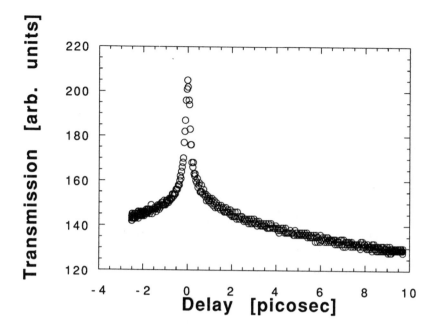

Fig. 3   TCP for LT-grown GaAs at 870 nm.

gives results similar to growth on other substrates, and provides a convenient means to create free standing films and windows.

ACKNOWLEDGMENTS

This work was performed at Lawrence Livermore National Laboratory under the auspices of the U.S. Department of Energy under contract number W-7405-ENG-48.

REFERENCES

[1]F. W. Smith, A. R. Calawa, C. L. Chen, M. J. Manfra and L. J. Mahoney, IEEE Electron Device Lett. 9, 77 (1988).

[2]F. W. Smith, H. Q. Le, V. Diadiuk, M. A. Hollis, A. R Calawa, S. Gupta, M. Frankel, D. R. Dykaar, G. A. Mourou, and T. Y. Hsiang, Appl. Phys. Lett. 54, 890 (1989).

[3]F. W. Smith, H. Q. Le, M. Frankel, V. Diadiuk, M. A. Hollis, D. R. Dykaar, G. A. Mourou and A. R. Calawa, OSA Proceedings on Picosecond Electronis and Optoelectronics, ed. T. C. L. Gerhard Sollner and D. M. Bloom, 4, 176 (1989).

[4]S. Gupta, S. L. Williamson, J. F. Whitaker, Y. Chen and F. W. Smith, Laser Focus World, July, 1992.

[5]S. Gupta, M. Y. Frankel, J. A. Valdmanis, J. F. Whitaker, G. A. Mourou, F. W. Smith and A. R. Calawa, Appl. Phys. Lett. 59, 3276 (1991).

[6]S. Gupta, J. F. Whitaker, S. L. Willaimson, G. A. Mourou, L. Lester, K. C. Hwang, P. Ho, J. Mazurowski and J. M. Ballingall, J. Electronic Material, in press.

[7]H. H. Wang, J. F. Whitiker, A. Chin, J. Maxurowski and J. M. Ballingall, J. Electronic Materials, in press.L. F. Lester, K. C. Hwang, P. Ho, J. Mazurowski, J. M. Ballingall, J. Sutliff, S. Gupta, J. Whitaker and S. L. Willaimson, IEEE Photonics Tech. Lett. (in press).

[8]M. Y. Frankel, J. F. Whitiker, G. A. Mourou, F. W. Smith and A. R. Calawa, IEEE Transaction on Electronic Devices, 37 (1990).

[9]X. Q. Zhou, H. M. van Driel, W. W. Ruhle, Z. Gogolak and K. Ploog, Appl. Phys. Lett 61, 3020 (1992).

[10]A. C. Warren, J. M. Woodall, J. L. Freeouf, D. Grischkowsky, M. R. Mellolch and N. Otsuka, Appl. Phys. Lett. 57,1331 (1990).

[11]W. S. Pelouch, P. E. Powers and C. L. Tang, Optics Letters 17, 1070 (1992).

[12]B. Proctor and F. Wise, Appl. Phys, Lett. 62, 470 (1993).

[13]C. L. Tang and D. J. Erskine, Phys. Rev. Lett. 51, 840, (1983).

[14]A. J. Taylor, D. J. Erskine and C. L. Tang, Appl. Physics Letters, 43, 989 (1983).

[15]I. L. Spain, M. S. Skolnick, G. W. Smith, M. K. Saker and C. R. Whitehouse, Phys. Rev. B 43, 14091 (1991) and references therein.

[16]C. J. Stanton, D. W. Bailey adn K. Hess, Phys. Rev. Lett. 65, 231 (1990). and references therein.

[17]T. Gong, W. L. Nighan and P. M. Fauchet, Appl. Phys. Lett. 57, 2713 (1990). and references therein.

[18]J. F. Whitaker, Materials Science and Engineering B (in press) and reference therein.

[19]A. F. Bello, D. J. Erskine, H. B. Radousky, L. N. Dinh, M. J. Bennahmias, M. D. Perry, T. R. Ditmire AND R. P. Mariella Jr., unpublished.

COMPOSITIONAL MODULATIONS AND VERTICAL TWO-DIMENSIONAL ARSENIC-PRECIPITATE ARRAYS AND IN LOW TEMPERATURE GROWN AL$_{0.3}$GA$_{0.7}$AS

K.Y. Hsieh[*], Y.L. Hwang[**], T. Zhang[***] and R.M. Kolbas[***]
[*]Institute of Materials Science and Engineering ,National Sun Yat-Sen University ,Kaohsiung, Taiwan, R.O.C.
[**]LSI logic, 1601 McCarthy Blvd M.S. B-142 Milpitas CA 95035 U.S.A.
[***]Department of Materials Science and Engineering North Carolina State University, Raleigh, NC 27695-7911 U.S.A.

ABSTRACT

Compositional modulations and arsenic precipitates in annealed Al$_{0.3}$Ga$_{0.7}$As layers which were grown at a low substrate temperature (200° C) by molecular beam epitaxy (MBE) were studied by transmission electron microscopy (TEM). These layers were used as surface layer which were applied on metal-insulator-semiconductor (MIS) diode. The planar and cross sectional TEM micrographs reveal that compositional modulations occurred when the thickness of LT AlGaAs was over 1500Å. The wavelength of the modulations varies between 100-200 Å and the direction of the modulation is along [011]. The arsenic precipitates were formed after annealed and the distribution of them followed the compositional modulation. Vertical two dimensional arsenic-precipitates arrays were arranged in the low aluminum constitute region. These novel microstructures result from the strain-induced spinodal decomposition and the arsenic precipitates redistribution process.

INTRODUCTION

It has been reported that GaAs epilayer structures grown at low substrate temperature by molecular beam epitaxy (MBE) and subsequently annealed have a great potential development in electrical[1,2] and optical device application[3,4]. The 1%-2% of excess arsenic which is incorporated into the film[5] form into arsenic precipitates after high temperature annealed. The formation of arsenic precipitates in the LT GaAs layers results in changes in material electrical properties from a conductive (10 ohm-cm) layer to a highly resistive (10^7 ohm-cm) insulators[6]. It is found that the distribution of the arsenic precipitates were strong effected by the presence of the AlGaAs/GaAs interface[7-8] and the type of the dopants[9]. The precipitates prefer to accuminate in the GaAs side of the heterojunction rather than form in the AlGaAs side of the heterojunction[7-8] and the donors enhance the precipitation while acceptors suppress the precipitation.[9]. The **in plane** two dimensional arsenic precipitate arrays were observed along the heterojunction interface and in the high concentration dopant (donor) region..

In this paper, we presented the recent study on arsenic precipitates in LT-AlGaAs/AlAs/n-GaAs heterojunction grown by MBE . The compositional modulations were found in the annealed LT AlGaAs layer and the direction of these modulation was along [011]. Numerous new AlGaAs/GaAs heterojunctions which lie along the growth direction were formed .The arsenic precipitates passed the redistribution process were arranged into the low Al composition region. A unique **vertical** two-dimensional arsenic-precipitate arrays were appeared. This is the first observation that the **compositional modulation** occurred in LT GaAs-AlGaAs system and **vertical** two dimensional arsenic-precipitate arrays were formed in LT AlGaAs.

EXPERIMENT

The layered structures used in this study were grown in a Varian 360 MBE system on undoped LEC(001) GaAs substrate. The substrate temperatures were measured by a standard Varian

26%Re-5%Re thermocouple arrangement in contact with the indium-bonded substrate mount. Except for the LT AlGaAs layer, the growth temperatures for the other active layers were normal in the range of 580° C to 640° C. The ratio of the group V to group III beam equivalent pressure for all AlGaAs and GaAs layers was approximate 10. The arsenic source was the tetramer As4. Prior to the deposition of the LT compound layer, the substrate temperature was decreased from a higher temperature to 200° C. Subsequently, if an *in situ* anneal for the as-grown LT compound layer was required, the substrate temperature was then increased to the annealing temperature. The *in situ* anneal was conducted at 600° C for 10 to 20 minutes under an arsenic overpressure of about 6×10^{-6} Torr. Precipitate microstructures in annealed samples were examined by TEM observations using a JEOL 200CX operated at 200 KV. Cross-sectional specimens were prepared by ion thinning at low temperature (77 K). Planar specimens were thinned from the substrate side with mechanically polishing and etched with a solution of bromine in methanol.

RESULTS AND DISCUSSION

The layer structure was schematically shown in Fig. 1. The structure of all the samples which had been analyzed in this study basically is similar. The only difference between each other is the thickness variation of AlAs and the low temperature grown layer. The LT AlGaAs is applied as the surface layer in a MESFET structure and the AlAs is played as a barrier which prevented the arsenic related defects diffusing into the channel. Fig. 2(a) and Fig. 2(b) are (200) dark field images of the sample 2-0113A which consists of a 2100Å AlGaAs surface layer, a 50Å AlAs barrier layer and a 2000Å n-GaAs channel of bulk doping level $5.36 \times 10^{17} cm^{-3}$. The sample were *in situ* anneal at 600° C for 10 minutes and 20 minutes respectively. As seen in the images, all the arsenic precipitates were stayed in AlGaAs region. The thin AlAs layer between the LT compound and n-channel totally blocked out the arsenic precipitates away from the channel. The distribution of the arsenic precipitates in these two sample showed different map. For the sample that annealed in 10 minutes, the arsenic precipitates were uniform distribution(Fig. 2(a)). While the sample annealed in 20 minutes (Fig. 2(b)), from the TEM micrographs it can be told that the density of arsenic ppts is less and the distribution of the arsenic precipitates can roughly be divided into three regions. At the region A, part of the arsenic precipitates accumulated, called the precipitate accumulation zone (PAZ), along the LTAlGaAs/AlAs interface. An area (Zone B) that with a thickness of around two hundred angstroms away from the heterojunction has less arsenic precipitates, call precipitate depletion zone (PDZ). At area C, it can be seen that a contrast modulation was occurred. The direction of the modulation was almost normal with the growth direction and the wavelength of this modulation was around 100-200Å. The lattice mismatch between in GaAs and Alas end-member is small ($\sim 1.3 \times 10^{-3}$ AlAs has the smaller unit cell); and structure factor contrast tends to dominate, particularly when {200} imaging conditions are used (the {200} structure factor reflects the difference in atomic scattering factors of the constitute III-type and V-type atoms, and is thus very weak in GaAs, but relatively strong in AlAs), therefore we can eliminate the suspicious of that the contrast modulation caused by strain contrast factor. This result indicates that the contrast modulation was due to the compositional modulation. The high intensity area showed that contained high Al composition. The boundary between white bands and black bands is clear only at the top of the LT AlGaAs layer. Once the bands extended to the heterojunction, the boundary could not be recognized . It has been reported that the arsenic ppts prefer to be formed in low Al constitute region than in high Al composition region due to the difference of in precipitation/matrix interfacing energy.[8] Since the compositional modulations created a

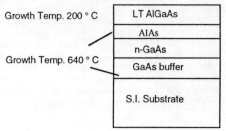

Fig. 1 Cross section of the Film structure

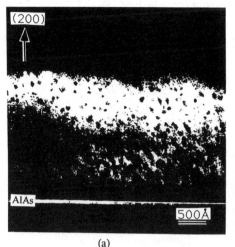 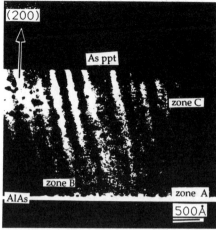

(a)   (b)

Fig. 2 TEM images of the cross sections of a GaAs MESFET structure capped with an in situ annealed LT AlGaAs surface layer, (a) annealing for 10 minutes, (b) annealing for 10 minutes. Note that the As ppts are well confined within the LT AlGaAs layer by a 50Å thick AlAs barrier layer. The compositional modulations were occurred in the LT compound region and the distribution of the As ppts followed the compositional modulation. The annealing time is longer, the phenomena of the composition modulation is more obvious.

numerous GaAs/AlGaAs heterojunction interface lay along the growth direction , thus the arsenic ppts were confined inside the black bands (GaAs). A vertical two dimensional arsenic precipitates were formed. The planar view TEM was applied to reveal the compositional modulation

Fig. 3 Plan view TEM image of a GaAs MESFET. Numerous short , weav bands ran diagonally on the micrograph. As ppts were formed in the dark region. This result confirmed that the composition modulation occurred after annealed.

phenomena. The image (Fig. 3) showed that numerous short white wavy bands (length:1000-2000Å;width:100Å) were oriented in a zigzag manner approximately along the $[01\bar{1}]$ direction. Again, it can been seen that the arsenic ppts were located in the outside of the white bands.(high Al composition region). Fig 4. is a 3-D schematic diagram of this layer structure which combined with the results of the cross section and planar view TEm images. It is clear to show that the compositional modulation is occurred only at $(01\bar{1})$ plane.

These results are different with that those reported by Melloch et al.[9] In their report, the compositional modulation (quantum well structure) was made artificially during the growth process and the arsenic ppts were formed in plane along with the heterojunction. The two dimensional arsenic ppt arrays were paralleled with the interface. In this case, the modulations occurred spontaneously. The LT AlGaAs layer that before high temperature annealed was examined with TEM. The image revealed that it was homogeneous. There had no clusters existed. The compositional modulation was taken place in the LT AlGaAs region after heat treatment. The longer annealing time was employed, the phenomena of compositional modulation were more clear. In addiation , the arsenic ppts were accumulated in the low Al composition region and formed a vertical two dimensional arrays.

The phenomena of the compositional modulation have been found in several III-V alloys[10-12.] In their results, the modulations did not result from a decomposition of an homogeneous layer, but occur during the growth of the layer itself, and subsequent frozen in this state. According to the Stringfellow's theoretic calculation[13], the interaction parameter Ω and enthalpy of mixing ΔH(mix) of ternary and quaternary III-V compound semiconductor alloys, both are positive. Phase Diagram calculations for these alloy system have hence predicted the existence of miscibility gaps in many cases at low temperatures and also the possibility that

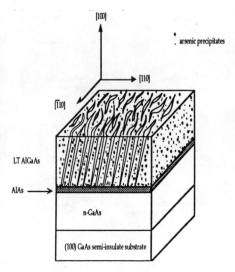

Fig. 4   A 3-D schematic diagram of microstructure in LT AlGaAs

Fig. 5 Images of the cross section of GaAs MESFET structure capped with an in situ annealed LT AlGaAs layer (4000Å). Note that the polycrystalline started at 1700Å and the amount of As ppt reduced near the poly region. The As ppts formed two dimensional arrays paralleled with the growth direction

some of these alloy systems may be unstable to have phase segregation by spinodal decomposition. Miscibility gaps have been experimentally observed for some alloy system[14-16] and the occurrence of spinodal decomposition in epitaxial layers of a few of these alloys has also been reported[17-20.] Since there is no experimental or theoretic data to support that the GaAs-AlAs system has a miscibility gap existed in such low temperature region (200° C), we can assume that the compositional modulation happened in this experiment is related to the spinodal decomposition. If the above the assumption is right, the next question is what the factor triggered the spindol decomposition happened. As it mentioned before, there is 1%-2% of excess arsenic in the LT GaAs layer, therefore we can expect that there must have excess arsenic in the LT AlGaAs layer. The arsenic ppts should be existed after the sample annealed and these assumption has been verified. Initially, the arsenic ppt uniformly appeared inside the LT AlGaAs during annealed. After the arsenic ppts kept coarsening, the strain which caused by the lattice parameter difference between the ppt and matrix also kept increasing. When the total strain reach to certain amount, the strain trigger the spinodal composition happened in order to release the extra energy to prevent the dislocation appeared. This strain induce ordering process has been observed by Hsieh et al[21.] Once there have phase modulation occurred, the arsenic ppts redistribution process then followed. The total strain depends on the total amount arsenic precipitates. The thickness of the LT AlGaAs will be the key factor to determine the modulation happened or not. We did observe that there is no compositional modulation occurred when the thickness of LT AlGaAs is below 1000Å. In addition, the AlAs barrier layer also plays a important role in this phenomena. The AlAs barrier stopped the arsenic precipitates out diffusion. It let the As ppts be confined in the LT AlGaAs region and it is allowed that the strain can be accumulated continuously to reach the threshold value, then triggered the ordering process. One of samples (not shown here) showed that if the AlAs barrier was replaced with a LT GaAs, the wavelength of the modulation increased and the compositional modulations would not be observed obviously. Combined with "strain induced modulation" and "arsenic redistribution process", it can be successfully to explain this experimental results.

In an early report, an unannealed LT GaAs layers grown 200° C has a larger lattice constant (0.1%) than that of the normal GaAs. This result suggested that the LT compound has a thickness limited before the epilayer turn into a polycrystalline. Fig. 5 is a (220) dark field images of the sample 2-0117A which consists of a 4000Å LT AlGaAs surface layer and a 200Å AlAs barrier layer. The sample was also *in situ* annealed at 600° C for 10 minutes. As seen from the image, the top area of the LT AlGaAs has become a polycrystalline and the arsenic precipitate were formed in discontinuous lines that were almost normal to the AlAs layer. It is also seen in this image that the regions between the boundary of the polycrystalline and the single crystal has arsenic ppts free and the contrast of this region is similar to that of the ALAs layer. These results indicate that this region has a high Al composition. Again, this image shows that a compositional modulation and vertical two dimensional arsenic-precipitate arrays were occurred in LT AlGaAs layer.

CONCLUSION

In summary, we have presented novel microstructures of arsenic precipitates which formed in LT AlGaAs/AlAs/n-GaAs layered structures grown at low substrate temperatures by MBE. It is shown that compositional modulation occurred when the thickness of LT AlGaAs was over 1500Å. The distribution of arsenic precipitates were followed the compositional modulation. Vertical two-dimensional arsenic precipitates were formed in the low aluminum constitute region. This is the first report that the compositional modulation was observed in an homogeneous LT AlGaAs layer after annealed. All these micrstructures can result from the strain-induced spinodal decomposition related mechanism and arsenic ppts redistribution process.

ACKNOWLEDGMENT

The author wish to thank Dr. K.C. Hsieh for his experimental assistance and many useful discussion.

REFERENCES

[1] F.W. Smith, A.R. Calawa, C. Lee, Mantra, and L.J. Mahoney, IEEE Electron. Device Lett. **EDL-9**, 77 -80 (1989)
[2] M.R. Melloch, D.C. Miller, and B.Das, , Appl. Phys. Lett. **54** . 943-945 (1989)
[3] A.C. Warren, N. Katzenellenbogen, D. Grischkowsky, J.M. Woodall, M.R. Melloch, and N.Otsuka, , Appl. Phys. Lett **58** 1512-1514 (1991).
[4] A.C. Warren, J.H. Burroughes, J. M. Woodall, D.I. McInturff, R.T. Hodgon, and M.R. Melloch, IEEE Electron Device Lett. **EDL-12**, 527-529 (1991)
[5] M. Kaminska, Z. Lilliental-Weber, E.R. Weber, T. George, J.B. Kortright, F.W. Smith, B.Y. Tsaur, and A.R. Calawa, , Appl. Phys. Lett. **54,** 1881-1883 (1989).

[6] A.C. Warren, J.M. Woodall, J.L. Freeout, D.Grischkowsky, D.T. McInsurff, M.R. Melloch, and Otsuka, , Appl. Phys. Lett. **57** . 1331-1333 (1990)

[7] K. Mahalingam, N. Otsuka, M.R. Melloch, J.M. Woodall and A.C. Warren, , J.Vac. Sci. Technol. B **10**. 812-814 (1992)

[8] K. Mahalingam and N.Otsuka, M.R. Melloch and J.M. Woodall, , Appl. Phys. Lett. **60**, .3253-3255 (1992).

[9] M.R. Melloch, N. Otsuka, K. Mahalingam, C.L. Chang, P.D. Kirchner, J.M. Woodall and A.C. Warren, , Appl. Phys. Lett. **61**, .177-179 (1992).

[10] P. Henoc, A.Izrael, M.Quillec and H.Launois, Appl. Phys. Lett.. **40**, .963-965 (1982).

[11] M.J. Treacy, J.M. Gibson and A.Howie, Phil. Mag. A **51**, 389 (1985).

[12] T.S. Kuan, T.F. Kuech, W.I. Wang, and E.L. Wilkie, Phys. Rev. Lett, **54**, 201 (1985).

[13] G.B. Stringfellow,J.Cryst. Gorwth, **27**, 21 (1974).

[14] E.K. MUller and J.L. Rchards, J. Appl. Phys. **35**, 1233 (1964).

[15] J.R. Pessetto and G.B. Stringfellow, J. Cryst. Growth. **62**, 1 (1982).

[16] M.Quillec, C. Daguet, J.L. Benchimol and H. Launois, Appl. Phys. Lett. **40**, 325-327 (1982)

[17] H. Launois, M.Quillex F.Glas and M.J. Treacy, GaAs and Related Cpds, Albuquerque 1982 Inst. Phys. Conf. Ser.**65** 537 (1983).

[18] J.H. Chiu, W.T. Tsang, S.N. G. Chu, J.Shan and J.A. Ditzenberger, Appl. Phys. Lett. **46**, 408-410 (1985).

[19] A.G. Norman and G.. R. Booker, ,J.Appl. Phys. Vol 57, pp4715- 1985

[20] S. Mahajan, B. V. Dutt, H. Temkin, R.J. Cava and W.A. Bonner, J.Cryst. Growth. . **68**. 589 (1984)

[21] K.C. Hsieh, J.N. Baillargeon, and K.Y. Cheng, Appl. Phys. Lett. **57** 2244-2246 (1990).

CLUSTERS AND THE NATURE OF SUPERCONDUCTIVITY IN LTMBE-GaAs

N.A.BERT*, V.V.CHALDYSHEV*, S.I.GOLOSHCHAPOV*, S.V.KOZYREV*,
A.E.KUNITSYN*, V.V.TRET'YAKOV*, A.I.VEINGER*
I.V.IVONIN**, L.G.LAVRENTIEVA**, M.D.VILISOVA**, M.P.YAKUBENYA**
D.I.LUBYSHEV***, V.V.PREOBRAZHENSKII***, B.R.SEMYAGIN***
*   Ioffe Physico-Technical Institute, 194021 St.Petersburg, Russia;
**   Siberian Physico-Technical Institute, 634050 Tomsk, Russia;
***   Institute of Semiconductor Physics, 630090 Novosibirsk, Russia.

ABSTRACT

The structure and properties of GaAs layers grown by molecular-beam epitaxy at low temperature (150-250 °C) have been studied. The samples were found to contain up to 1.5 at.% extra As, which formed nano-scale clusters under annealing. The dependences of the excessive As concentration and As-cluster size and density on the growth and annealing conditions were established. LT-GaAs layers were found to have high electrical resistivity, however, our investigations of microwave absorption in a weak magnetic field revealed a characteristic signal usually attributed to the superconducting phase. It has been proved that this microwave absorption is unlikely to be due to either the arsenic clusters in LT-GaAs films or indium in the substrate, as it was assumed previously. We suggest a new hypothesis that the superconducting phase in LT-GaAs is Ga nanoclusters formed on the growth surface.

INTRODUCTION

GaAs grown by molecular-beam epitaxy (MBE) at low temperature (less than 300 °C) has received a great deal of attention during the last few years (see, for example, [1-3]). The main feature of this material is high arsenic excess, far outside the homogeneity region. Annealing forces extra-As to form nano-scale clusters. High resistivity and short carrier lifetime make this material useful for buffer layers, high-speed photodetectors and other applications.

An unexpected phenomenon observed in LT-GaAs is the existence of a superconducting phase recently reported by Baranowski et al [4]. The presence of this phase can be detected using characteristic microwave absorption dependent on a weak magnetic field. The authors [4] attributed the superconducting phase to arsenic clusters or arsenic planes. However, Li et al [5] have lately reported that the superconducting phase may be formed in LT-GaAs due to indium often used to sold the substrates to a holder in MBE-systems.

In this paper we report the results on the structure and properties of LT-GaAs samples prepared under various conditions. The superconducting phase in these samples has also been investigated. We have checked the hypotheses mentioned above and suggest a new one accounting for the existence of the superconducting phase in LT-GaAs films.

EXPERIMENTAL

The LT-GaAs layers were grown in a dual-chamber "Katun" MBE-system on standard undoped semi-insulating 2-inch GaAs substrates with (100) orientation. The growth rate was 1 μm/h and the layer thickness 1 μm. The growth temperatures and As pressures for the layers investigated are given in Table I. The substrate temperature was controlled by the thermal radiation of a Ta-heater. Indium was not used to sold the substrate to the holder. The crystal growth was assessed by reflected high-energy electron diffraction (RHEED).

Table I.

N	P_{As} (10^{-5} Pa)	T_s (°C)	Θ (sec)	$a_l - a_s$ (10^{-4}Å)	[As]–[Ga] (at.%)
94	1.1	150	37	85	1.2
93	1.1	200	31	60	1.1
95	1.1	250	23	20	<0.3
96	1.8	150	47	83	1.4
97	1.8	200	29	65	1.0
98	1.8	250	27	27	0.4

After the growth, a part of the grown samples was annealed at the temperatures $T_a = 400$, 500 or 600 °C under As over pressure $P_{As} = 1.8 \times 10^{-5}$ Pa immediately in the growth chamber for 15 minutes.

Electron-probe microanalysis (EPMA), X-ray diffraction (XRD), transmission electron microscopy (TEM), electron microscopy of surface replica were used to characterize the microstructure of LT-GaAs layers. Special care was taken to improve the sensitivity of EPMA that allowed us to decrease the total absolute error down to 0.3 atomic %. Electrical and photoluminescense (PL) study were also performed.

In order to reveal the superconducting phase, the LT-GaAs films were investigated by magnetic-field-modulated microwave absorption (FMMA) in a weak magnetic field over the temperature range 2–50 K. The FMMA signal measured was the field derivative of microwave absorption dP/dH.

RESULTS AND DISCUSSION

During the layer growth a net of streaks of the main reflections was observed in the HEED pattern that allowed us to draw the conclusion about a layer-by-layer growth mechanism, atomically smooth surface and a high crystalline quality of the LT-GaAs being grown. As-stabilized conditions were realized at all the temperatures used. The intensive diffuse scattering, increasing with lowering growth temperature as well as with increasing layer thickness, was the main feature of the HEED pattern. This phenomenon seems to be due to the high concentration of extra arsenic on the growth surface and to its capture into the growing layer.

Electron microscopy of surface replica has shown a specific relief of the growth surface. The growth pits with submicron dimensions (80–120 nm) and with a density of 10^8–10^9 cm^{-2} were also detected.

The lattice constant of as-grown layers (a_l) measured by XRD exceeded that of the GaAs substrate (a_s) and increased rapidly with decreasing growth temperature. Simultaneously, the FWHM of the rocking curves (Θ) rose. The values of $a_l - a_s$ and Θ as well as the concentration of extra As determined by EPMA are given in Table I. The amount of extra As is in good agreement with the data obtained previously [2]. The maximum concentration of excessive arsenic is 1.5 at.% and it reduces sharply with increasing growth temperature. The change in the As molecular flux has a weak influence on the excessive arsenic concentration in the epitaxial layer. Consideration of LT-GaAs lattice constant measurements taking into account the Ga and As tetrahedral radii and tetrahedral interstice size show excessive As to be primarily situated at the interstitials. The concentration of antisite defects As_{Ga} seems to be far less than that of the interstitials As_i and is 5×10^{18}cm^{-3} in LT-GaAs grown at 200 °C according to EPR data [2].

The annealing caused the lattice constant and X-ray rocking curve width of the layers to approach that of the substrate. Fig. 1 shows the dependences of $a_l - a_s$ on annealing temperature. The data indicate an improvement of the crystal perfectness of the GaAs matrix under annealing. TEM study showed that excessive As formed nano-scale clusters in the annealed samples (Fig. 2).

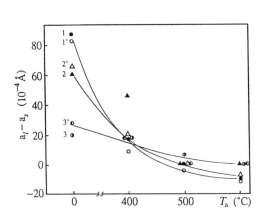

Fig. 1. Difference between lattice parameters
for LT-GaAs films and GaAs substrate
vs. annealing temperature. Growth conditions:
T_s (°C): 1,1'—150, 2,2'—200, 3,3'—250;
P_{As} (10^{-5} Pa): 1–3—1.1, 1'–3'—1.8.

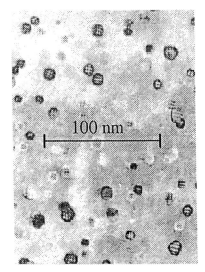

Fig. 2. Bright-field electron microscope
micrograph of As clusters
in sample 94 (T_s = 150 °C, T_a = 600 °C).

At least at high annealing temperature (500 and 600 °C) the clusters had a crystal structure indicated by the moire fringes in the TEM image and extra spots around (220) and (400) reflections in the [110] directions at selected area diffraction pattern. At the same growth temperature 150 °C, the average cluster size increased from 2 to 7 nm when the annealing temperature rose from 500 to 600 °C. Simultaneously, the cluster density reduced from 5×10^{17} to $(6-8) \times 10^{16} cm^{-3}$. Comparing the density and size of As clusters with the concentration of extra As from the EPMA data, one can find a complete precipitation of extra As at annealing temperature 600 °C and only partial one at lower T_a. When the growth temperature rose from 150 to 200 and up to 250 °C, i.e. the initial amount of extra As dropped, the cluster size in the samples annealed at 600 °C reduced from 7 to 5 and down to 4 nm while the cluster density remained approximately the same for $T_s = 150$ and 200 °C and decreased slightly for $T_s = 250$ °C. In addition to the As clusters, there were other defects in LT GaAs layers, in particular, dislocations of the edge type as well as defects causing the spotted dark contrast in the plane-view image. The defect density was evaluated to be of the same order of magnitude as that of the surface defects observed through replicas.

All the LT-GaAs samples (both as-grown and annealed) were found to have high electrical resistivity ($\rho > 10^4$ Ω cm) which was not measured precisely because of the shunt resistance of the thick (330 μm) SI–GaAs substrate.

The PL investigations at 4.2 K showed a weak carbon-related line, intensity of which increased with the annealing temperature in agreement with the conclusion on the improvement of the crystal perfectness. A specific feature of the spectra was the absence of excitonic lines. This fact supports the As-cluster model accounted for electronic properties of LT-GaAs [3].

It is usually assumed that the FMMA signal with characteristic dependences on the temperature and magnetic field indicates the existence of a superconducting phase in the samples [6, 7]. Most of the samples investigated showed superconductor-like FMMA signals (Fig. 3). The signal magnitude were found to vary strongly from one sample to another.

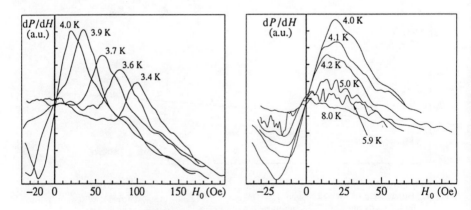

Fig. 3. Temperature dependence of field-modulated microwave absorption signal
for sample 95 ($T_s = 150$ °C, $T_a = 600$ °C).
The measurement temperatures are shown at the curves.

404

The superconducting transition temperature was estimated to be about 10 K. The following features of the signal were also detected: a hysteresis in a microwave absorption upon reversing the field-sweep direction; increasing noise as the temperature approaches T_c; in dependence of the signal on the sample orientation with respect to the external magnetic field. This features as well as the characteristic shape, temperature and magnetic field dependence of the FMMA signal seem likely to indicate a superconducting phase in LT-GaAs samples in accordance to the previous results [4].

It has been suggested in [4] that this phase in LT-GaAs can be due to As-rich plane structures or arsenic clusters, although arsenic does not exhibit superconducting properties in the bulk form. In order to check this hypothesis, we have compared the microwave absorption data with the growth and annealing conditions and with the results of the structure study. No correlations of the excess As concentration and structure including the precipitation stage, cluster size and cluster concentration with microwave absorption have been detected, although the FMMA signal varied by up to 2 orders of magnitude and qualitatively in some cases. The large number (24) of samples investigated and the gradual appropriate change in the crystal structure under the growth and annealing condition variations allow us to conclude that the microwave absorption signal in LT-GaAs samples is unlikely to be due to clusters or other defects formed by extra-As.

Recently Li et al. have shown [5] that the FMMA signal can be due to indium clusters formed in the GaAs substrate during the epitaxy and subsequent annealing if indium was used to sold the substrate to a holder in the MBE system. Indium clusters were assumed to have the superconducting transition temperature higher then bulk In ($T_c = 3.4$ K).

It should be pointed out that we have not used indium in our MBE system. The EPMA and PL study indicated that indium content in our substrates was less than 2×10^{18} cm^{-3}. Moreover, we reproduced for some substrates the heat treatment procedures similar to that for the low-temperature MBE growth followed by the annealing. The microwave absorption measurements showed no superconductor-like signal in the substrates without epitaxial layer. Thus, we have proved that the characteristic microwave absorption in our LT-GaAs samples is not due to indium in the substrates.

To understand the nature of the FMMA signal, one should be take into account the absence of clear regularity in the signal type and magnitude variations. This fact allows us to suppose that the microwave absorption may be due to specific defects formed at the initial growth stage and retained in the substrate-layer interface or moving with the growth surface. We observed a fairly high concentration of such kind of defects near the surface by surface replica and transmission electron microscopy. It should be pointed out that these defects are not a specific feature of our samples or low-temperature MBE. The growth defects of such kind have been observed earlier in vapour-phase epitaxy films [8]. These defects are 1–100 nm gallium particles doped with different impurities and are liquid at growth temperature.

405

It is well known that gallium is a type-I superconductor with a critical temperature T_c=1.1 K. The capture of different impurities in the liquid gallium particles during the crystal growth can result in an increase of T_c. The existence of the FMMA signal up to 10 K can be also attributed to the small size of gallium particles. Such an increase in T_c was earlier observed for Sn clusters [9].

CONCLUSION

We have proved that the superconductor-like microwave absorption is unlikely to be due to either arsenic clusters in LT-GaAs films or indium in the substrate, as it was assumed previously. We suggest that the superconducting phase in LT-GaAs is Ga nanoclusters formed on the growth surface. The hypothesis suggested for the microwave absorption nature in LT-GaAs accounts for the results on the structure, property and microwave absorption reported here and in the previous papers [4, 5].

REFERENCES

1. F.W.Smith, A.R.Calawa, Chang-Lee Chen, M.J.Mantra, L.J.Mahoney.
    IEEE Electron Devices Lett. **9**, 77 (1988)

2. M.Kaminska, Z.Liliental-Weber, E.R.Weber, T.George, J.B.Kortright, F.W.Smith, B.-Y.Tsang, A.R.Calawa. Appl.Phys.Lett. **54**, 1831(1989)

3. M.R.Melloch, K.Mahalingam, N.Otsuka, J.M.Woodall, A.C.Warren.
    J.Cryst.Growth **111**, 39 (1991)

4. J.M.Baranowski, Z.Liliental-Weber, W.F.Yau, E.R.Weber. Phys.Rev.Lett. **66**, 3079 (1991)

5. Y.K.Li, Y.Huang, Z.Fan, C.Jiang, X.B.Mei, B.Yin, J.M.Zhou, J.C.Mao, J.S.Fu, E.Wu.
    J.Appl.Phys. **71**, 2018 (1991)

6. A.S.Kheifets, A.I.Veinger. Physica C. **165**, 491 (1990)

7. V.F.Masterov, A.I.Egorov, N.P.Gerasimov, S.V.Kozyrev, I.L.Likholit, I.G.Saveliev, A.F.Fyodorov, K.F.Shtel'makh. Pis'ma JETP. **46**, 289 (1987)

8. L.G.Lavrentieva,I.V.Ivonin,L.M.Krasilnikova, M.D.Vilisova.
    Kristall und Technik. **15**, 683 (1980)

9. L.Giaver, H.R.Zeller. Phys.Rev.Lett. **20**, 1504 (1968)

# PART VII

Defects in Bulk and
Epitaxial Semiconductors

PHOTOLUMINESCENCE IMAGING OF III-V SUBSTRATES AND EPITAXIAL HETEROSTRUCTURES

W. JANTZ, M. BAEUMLER, Z.M. WANG* AND J. WINDSCHEIF
Fraunhofer-Institut für Angewandte Festkörperphysik, D-79108 Freiburg, FRG

ABSTRACT

The characterization of III-V compound semiconductor substrates and epitaxial layers with photoluminescence imaging is reviewed. The luminescence patterns of semi-insulating GaAs are dominantly determined by the concentration and distribution of nonradiative recombination centers, as shown by comparison with spectroscopic temperature and lifetime topography of photoexcited carriers. Wafers fabricated with various growth and annealing procedures are evaluated. Presently available informations on nonradiative centers in GaAs are summarized and discussed. The correlation of luminescence, absorption and resistivity topograms of InP substrates shows various interrelated influences of the Fe acceptor distribution. High resolution luminescence images of growth induced, strain induced and substrate induced defects in epitaxial heterostructures are obtained. The generation of relaxation dislocations in pseudomorphic layers is influenced by growth parameters, layer structures, layer doping and also by substrate properties. Nonradiative recombination center patterns replicate the arrangement of threading dislocations in the substrate.

INTRODUCTION

The increasing complexity of micro- and optoelectronic device fabrication has stimulated the development of characterization techniques satisfying the need for detailed control of substrate materials, epitaxial layers and fabrication processes. An important and widely used method is photoluminescence imaging (PLI), including photoluminescence topography (PLT) and photoluminescence microscopy (PLM).

PLT of III-V compound semiconductors was introduced about 8 years ago [1-3]. It is now firmly established as an indispensable diagnostic tool. Already the first PLT experiments, designed modestly to map the near bandgap intensity I_{PL} without spectral resolution at room temperature, immediately yielded important and surprising new informations, convincingly demonstrating the diagnostic value of PLT to characterize III-V semiconductors. In due course, improvements including mapping at low temperature with spectral selection were realized [4, 5]. These features increased both the range of applications and the information content of luminescence images. Today, PLT has been installed in many academic and industrial laboratories. Commercial systems developed for routine production control are available. More recently, spectrally selective PLM at cryogenic temperature [6] has been introduced as an easy-to-use tool to study growth-induced defects in compound semiconductor epitaxial layers.

EXPERIMENTAL DETAILS

Fig. 1 shows schematic diagrams of our PLI systems. In the PLT setup the focussed Ar$^+$ laser excitation is scanned in two dimensions across the stationary sample and the entire or a spectrally selected part of I_{PL} emitted from the sample is recorded. Excitation and collection optics are scanned simultaneously such that areas of 75 mm diameter can be imaged. The PLM approach is complementary to PLT because the entire imaged sample area (between 200 and 750 μm diameter) is illuminated with incoherent light and I_{PL} is recorded laterally selective with a video camera. Measurements were taken at room temperature, 77K and 2K (PLT only).

Mat. Res. Soc. Symp. Proc. Vol. 325. ©1994 Materials Research Society

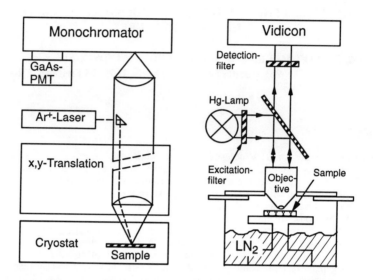

Fig. 1.   Schematic representation of the photoluminescence topography (left) and photoluminescence microscopy systems.

The characteristic distribution of an optically detectable defect and its direct, inverse or nonexisting correlation with patterns of other material properties is helpful to identify defect structures and defect interactions. On the other hand, the excellent (generally carrier diffusion limited) lateral resolution of PLM allows to image fine details of growth-induced, strain-induced and substrate-induced structural and chemical defects.

For the analysis of bulk GaAs properties commercial state-of-the-art wafers were used throughout. Material obtained from different vendors, grown with Liquid Encapsulated Czochralski (LEC) and Vertical Gradient Freeze (VGF) procedures, subject to ingot and wafer annealing, was available. Various epitaxial heterostructures were investigated, including thick (μm) GaAs layers, GaAs and InGaAs quantum well (QW) and multi-QW layers enclosed between AlGaAs and GaAs barriers. As usual, these test layers were separated from the substrate by buffer and superlattice layers and covered by a GaAs cap layer. Relaxation dislocations were observed in InGaAs QW and multi-QW structures similar to those used for the fabrication of laser diodes.

INVESTIGATION OF SUBSTRATES

The first PLT images of compound semiconductors were obtained with semi-insulating GaAs substrates [1, 2]. Such measurements are now an important part of routine optical wafer characterization (for a recent survey, see e.g. [8]). The capability of PLT to reveal bulk material properties is immediately seen from Fig. 2, showing, for LEC and VGF grown material, the well known cellular dislocation density structure and the precise correlation between the pattern of dislocations as measured by X-Ray topography and the pattern of near bandgap I_{PL}. Such images yield detailed informations that are well known and intensively exploited, but for this very reason worthwhile to be summarized here, including:

- the cellular dislocation structure itself, imaged with PLT faster and less expensively as compared to X-Ray topography,
- the distribution of optically active centers,
- the dependence of macroscopic and microscopic material homogeneity on growth techniques,
- the redistribution mechanisms of annealing procedures,
- the magnitude and inhomogeneity of photoexcited carrier lifetime,
- an estimate of the magnitude and homogeneity of surface recombination.

As an example how PLT can be used to characterize defect-induced material properties, we discuss results of spectroscopic lifetime topography [4, 7]. Briefly summarized, this method exploits the fact that photoexcited electrons relax, via LO phonon generation and electron scattering, to the bottom of the conduction band. Thus, prior to recombination, a thermalized electron gas is generated

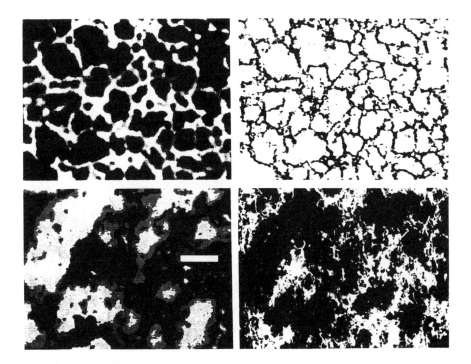

Fig. 2.   Near bandedge photoluminescence intensity topograms (left) and corresponding X-Ray scattering topograms (right) of LEC (top) and VGF (bottom) grown semi-insulating GaAs. Areas of high luminescence appear bright. The LEC X-Ray topogram was obtained in the reflection geometry, the VGF topogram in the transmission geometry. Therefore, areas of high dislocation density appear dark in the top, but bright in the bottom X-Ray topogram. Hence, luminescence intensity and dislocation density are anticorrelated in LEC, but correlated in VGF material. These interrelations, however, are not generally true, depending e. g. on the annealing history. The white bar indicates 1 mm.

with a temperature T_e that can be measured optically by recording the exponential high energy tail of the band-to-band recombination radiation. T_e itself is correlated with the carrier lifetime τ, because during this time the electron gas is cooling via TO phonon generation and other relaxation processes.

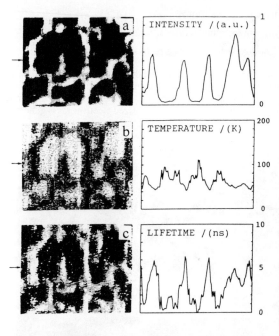

Fig. 3.   Topographies and linescans of luminescence intensity I_{PL}, temperature T_e and and lifetime τ of photoexcited electrons.

Experimentally an anticorrelation between I_{PL} and T_e is found. It is intuitively clear and corroborated by theory [7] that T_e and τ are also anticorrelated, such that finally I_{PL} and τ are directly correlated. In Fig. 3 quantitative topograms of I_{PL}, T_e and τ are given to demonstrate these well-defined interrelations. It follows that the dominant recombination process determining I_{PL} is nonradiative. Hence, the strong variations of I_{PL} in Fig. 2 indicate a corresponding concentration fluctuation of a nonradiative recombination center (NRRC).

These defects, omnipresent in bulk GaAs, are not yet unambiguously identified. Nevertheless, it is necessary to control, homogenize and reduce their concentration. T_e distributions semi-quantitatively describe the concentration and fluctuation of NRRC. The experimental approach described above is essentially independent of parameters that are difficult to control, such as optical excitation and detection efficiency. It is, therefore, well suited to compare wafers fabricated with different growth procedures and/or annealing treatments.

T_e has been evaluated topographically within 5x5 mm^2 sample areas. The data are plotted in histographic form in Figs 4 (a-c). A broad distribution peaked a high T_e indicates high, inhomogeneous NRRC distribution, whereas a shift towards lower peak T_e and a narrowing of the distribution is indicative for a reduction and homogenization of NRRC concentration. As a reference, each diagram reproduces the T_e distribution of a high-quality epitaxial layer grown with Metal Organic Vapor Phase Epitaxy (MOVPE). It is seen that substrate material is inferior to the epitaxial layer material and varies rather widely with respect to the position and width of the T_e distribution. Fig. 4a shows as-grown and various ingot annealed LEC materials. Evidently the annealing efficiency can be characterized quite sensitively. This is also the case for wafer-annealed material, as shown in Fig. 4b. However, the two overlapping distributions, representing wafers of the same ingot and subject to the same annealing procedure, show that the thermal treatment does not altogether eliminate material inhomogeneities. For comparison, Fig. 4c shows T_e histograms of VGF wafers. No information about annealing was available, but similarities to Figs 4a, b suggest that both as-grown and annealed material has been evaluated. The data were collected during the last two years, indicating continous and substantial improvements of the wafer quality as characterized by T_e topography.

412

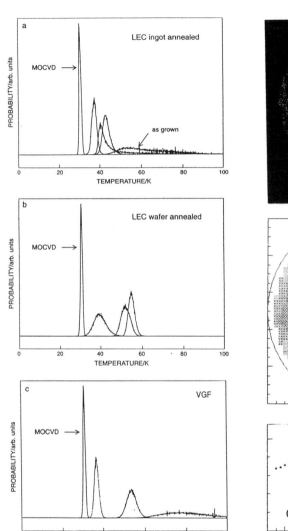

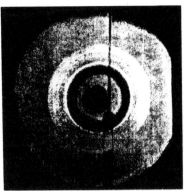

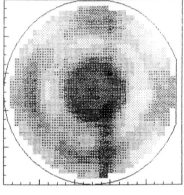

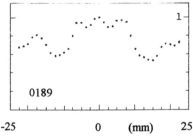

0189

-25   0   (mm)   25

Fig. 4.   Histograms of electron temperature for various semi-insulating GaAs wafer materials: (a) as grown and ingot annealed LEC, (b) wafer annealed LEC, (c) VGF.

Fig.5.   Photoluminescence (I_{PL}) topogram, resistivity (ρ) topogram and ρ linescan (normalized to the center value) of a two inch semi-insulating InP wafer. Bright areas in the ρ topogram indicate low ρ, hence I_{PL} and ρ are anticorrelated.

The diagnostic value of PLT is further increased by comparative analysis, using other topographic measurements of the same sample area. Fig. 5 shows PL and resistivity (ρ) topograms of a semi-insulating InP substrate [9]. It is seen that I_{PL} and ρ are strictly anticorrelated. These and similar topographic correlations (not shown here) between ρ and Fe^{2+} absorption [10] and between I_{PL} and T_e [11] demonstrate that

- fluctuations of the Fe concentration can reliably be measured with the convenient, nondestructive PLT technique

- fluctuations of ρ are dominantly caused by fluctuations of the Fe concentration,

- I_{PL} variations are dominantly caused by competing recombination via the deep Fe center.

As mentioned above, the chemical and structural identity of NRRC in GaAs is not yet established unambiguously, in spite of numerous investigations [12-16]. However, we believe that the following statements are now well established:

1. The NRRC causes the pronounced variations of I_{PL} shown in Fig. 2 [7, 10]. Hence its concentration tends to fluctuate with a pattern reproducing the dislocation density distribution.

2. The average concentration and the concentration fluctuation of NRRC depend on the crystal growth process and can be varied reversibly by annealing procedures [14, 17]). Hence the chemical species forming the defect can be accommodated in the GaAs matrix in different forms.

3. The NRRC is not EL2, as demonstrated e.g. by annealing procedures that homogenize EL2 while I_{PL} remains inhomogeneous [5, 18].

4. The NRRC does not contribute to the electrical compensation balance, as demonstrated e.g. by annealing procedures that homogenize ρ while I_{PL} remains inhomogeneous [16, 18].

5. Numerous efforts to identify a recombination center generating luminescence at below-bandgap energy suitable to explain the I_{PL} pattern, i.e. exhibiting an anticorrelated order-of-magnitude variation, were futile.

6. Bulk GaAs contains excess As on the order of 10^{18} cm^{-3}, incorporated inhomogeneously as precipitates, antisites and (probably) interstitials [19].

7. The center responsible for near bandgap, so-called reverse contrast (RC) absorption is closely related to NRRC [15]. RC absorption requires an optically active center with an energy level close to one of the band edges.

In general it is tacitly assumed that an optically active center also influences the electrical material properties. Therefore, statement 4 has thus far been circumvented by assuming that NRRC does contribute to the compensation, but does occur in such low concentration that this contribution is not significant. Conversely, the carrier capture cross section of the center is assumed to be high enough that it can, nevertheless, dominate all other optical recombination processes. These permissible, but suspiciously complicated assumptions are becoming progressively more difficult to maintain as compensation is now well controlled even at very low concentration levels of the contributing defects. Material with Carbon concentrations down to 3×10^{14} cm^{-3} can be grown semi-insulating with predictable resistivity ρ [20].

We, therefore, postulate that NRRC does *not* participate in compensation, i.e. does not supply or accept electrons to ionize other acceptors or donors (as e.g. EL2 ionizes the carbon acceptor in n-type semi-insulating GaAs). Such a center must have an ionization energy substantially exceeding that of EL2. An acceptor level very close to the conduction band edge E_c or a donor level very close to the valence band edge does not contribute to the compensation balance, even if incorporated in high concentration (statement 4). It would also satisfy statements 5 and 7, i.e. account for the probable non-existence of I_{PL}-competing below-bandgap recombination radiation and for the RC absorption. An acceptor level close to E_c has indeed been proposed recently to explain photoconductivity effects related to RC absorption [21]. These manifold interrelations are suggesting that the defects causing nonradiative recombination and reverse contrast absorption are not only closely related [15], but identical.

Of course, a defect structure with the required physical properties must still be identified and confirmed experimentally. Acceptor states close to E_c are, for instance, provided by local fluctuations of E_c. Taking into account all seven statements listed above, we suggest that local fluctuations of excess As exist, that these local fluctuations are generating local fluctuations of E_c which, in turn, cause the I_{PL} variations shown in Fig. 1 and the corresponding RC absorption patterns [15, 22].

Neither suggestion, probably not even their combination, is new. Nevertheless we believe that the arguments collected above should stimulate a detailed reconsideration of the presented issues. One possibly fruitful observation could be that strong below-bandgap absorption and low PL efficiency are prominent features of GaAs grown with Molecular Beam Epitaxy (MBE) at low temperature, known to contain excess As. This observation, if applicable to bulk GaAs, would explain the correlations between increased excess As, increased below-bandgap absorption and reduced I_{PL} quite naturally.

INVESTIGATION OF EPITAXIAL LAYERS

PLI systems with cryogenic cooling and spectral selection allow a detailed analysis of epitaxial layer systems to control and optimize deposition processes (MOVPE and MBE). These investigations may be subdivided into assessments of macroscopic homogeneity, such as thickness and composition of ternary and quaternary layers [22, 23], and of microscopic defects [24]. Consonant with the focus of this symposium on defects we shall concentrate on topics belonging to the second area, however emphasizing that the evaluation of the geometric and chemical homogeneity of epitaxial heterostructures, in particular laser structures, are an increasingly important task of spectroscopic PLT in our laboratory and elsewhere.

For luminescence investigations of local defects, requiring the highest possible lateral resolution, PLM is best suited. Defect structures usually are visible because I_{PL} is locally reduced. Less frequently, bright defects are observed. We shall discuss typical defects encountered in state-of-the-art epitaxial layers, induced by deposition processes, layer strain and substrate imperfections [24-27].

Fig. 6a shows the PLM image of an oval defect, frequently generated in MBE layers. While this rather pronounced growth irregularity in general is also visible with conventional (Nomarski) microscopy, the PLM image directly qualifies the resulting optical degradation and identifies finer details such as dislocation lines starting from the defect.

It is well known that the strain of layers grown on substrates or layers with a different lattice constant can be accommodated coherently only up to a critical thickness l_c. In layers with thickness $l > l_c$ strain relaxation dislocations (RD) are generated. Precise assessment of l_c is desireable, in particular for laser structures. RDs are intolerable material degradations that must securely be avoided. However, a desired device performance (e.g. laser emission wavelength) often requires to approach l_c as close as possible. This controversial situation has stimulated detailed studies how RD generation depends on parameters other than lattice mismatch and layer thickness [25].

Fig. 6b shows RD patterns in a MBE grown $In_xGa_{1-x}As$ layer with x = 0.2 and l= 20 nm. With an appropriate detection filter (see Fig. 1), only luminescence emitted by the InGaAs layer is recorded. The image shows that l is just beyond l_c for the given In content. For slightly higher l, the RD density increases dramatically. A particularly interesting feature in Fig. 6b is the different density of horizontal and vertical [110] RDs (α and β dislocations in the usual notation). The preferred generation of α dislocations leads to a peculiar asymmetry and temperature dependence of the sheet resistivity, satisfactorily interpreted in terms of depletion tubes surrounding the RDs [28].

Fig. 6b already indicates that a RD enters the layer at some particular point and propagates along a specific crystallographic direction, such that generally the majority of RDs cross the entire field of view. More information about RD generation points and propagation directions are presented in Figs. 6(c, d). Fig. 6c, showing an InGaAs layer with precisely l=l$_c$, reveals a lateral pattern of defects

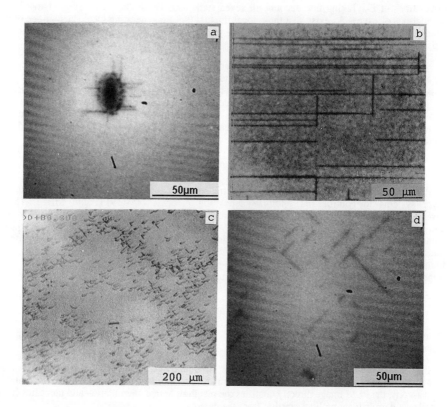

Fig. 6.   Defect structures observed in MOVPE and MBE heterostructures containing AlGaAs, InGaAs and GaAs layers. For details see text.

similar to the cellular structure shown in Fig. 2 [27]. The correlation demonstrates that RD generation, in spite of buffer, superlattice and barrier layers, can nevertheless be stimulated by substrate threading dislocations. This finding has decisive bearing on the choice of substrates and growth procedures used to fabricate laser heterostructure layers.

A typical PLM image of a laser structure containing 5 In$_x$Ga$_{1-x}$As quantum wells with x = 0.35 and l = 5.7 nm embedded between GaAs barriers is shown in Fig. 6d. Again the first appearance of RDs is an indicator that the critical thickness l$_c$ is realized. Few and still rather short [100] RD lines are seen to extend from the points where the crystal disturbance is generated [25, 26]. Such images serve as a sensitive indicator to evaluate the beneficial or adverse influence of growth temperatures, growth interruptions, increased number of otherwise identical quantum wells, barrier or quantum well doping, confinement layers and postgrowth annealing. For instance, laser structures identical to that shown in Fig. 6d, with the exception of the number of QWs, turned out to behave quite differently with respect to RDs. The structure with 6 QWs showed much more, the structure with 3 QWs no RDs. Last but not least, the importance of substrate quality, in this case the superiority of VGF wafers for deposition of laser structures, was clearly demonstrated.

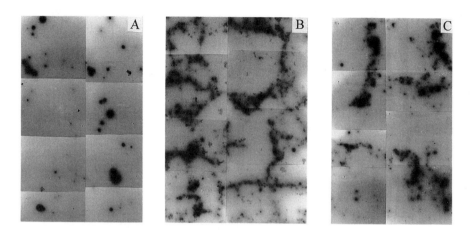

Fig. 7.   Substrate-correlated defects observed in an epitaxial GaAs layer grown with MBE on different GaAs wafers: (A) VGF, (B) ingot annealed LEC, (C) wafer annealed LEC.

As a final example how substrate properties can be "inherited" by epitaxial layers, Fig. 7 shows PLM images of a MBE test structure grown on VGF, ingot annealed LEC and wafer annealed LEC substrates [26]. The luminescence is emitted by a 1.2 μm GaAs layer embedded between buffer and barrier layers. In each case very prominent patterns of reduced I_{PL} (829 nm) are observed, clearly replicating the dislocation density patterns of the respective substrates. This correlation has previously been established for ingot annealed LEC wafers by directly comparing the PLM image with the underlying portion of the substrate after structural etching [24]. Fig. 7 demonstrates that an adverse influence of substrate inhomogeneites on the optical layer quality can thus far only gradually be controlled by using innovative (wafer annealed LEC and VGF) substrate material.

Of course, it is most appealing to imagine even more detailed interrelations between substrate inhomogeneities and defects in epitaxial layers, e.g. postulating diffusion of excess As, preferentially along dislocations, from the substrate into the layer. Such mechanisms would further emphasize the importance of the issues discussed in Section 3. Clearly some intricate problems must still be solved in order to supply a substrate quality ideally suited for the deposition of defect-free epitaxial layers.

ACKNOWLEDGEMENTS

The MOVPE and MBE samples were grown by K. H. Bachem, K. Köhler, E.C. Larkins and H. Riechert (Siemens). The X-ray topographies were supplied by N. Herres, the InP wafer and luminescence topogram (Fig. 5) by J. Völkl. We also acknowledge technical support by J. Forker, K.H. Schmidt and R. Stibal as well as helpful discussions with U. Kaufmann. Part of this work has been supported by the Bundesministerium für Forschung und Technologie under Contracts NT 2775 and TK 0578.

REFERENCES

* present address: Institute of Semiconductor Physics, Beijing, PR of China.

1. H.J. Hovel and D. Guidotti, IEEE Trans. Electron. Devices ED-32, 2331 (1985).
2. W. Wettling and J. Windscheif, Appl. Phys A40, 191 (1986).
3. S.K. Krawczyk, M. Garrigues and H. Bouredoucen, J. Appl. Phys. 60, 392 (1986).
4. Z.M. Wang, J. Windscheif, D.J. As and W. Jantz, Inst. Phys. Conf. Ser. 112, 191 (1990).
5. T.W. Steiner and M.L.W. Thewalt, Semicond. Sci. Technol. 7, A16 (1992).
6. Z.M. Wang, F. Scholz, H. Schuster and K. Streubel, J. Crystal Growth 97, 560 (1989).
7. Z.M. Wang, J. Windscheif, D.J. As and W. Jantz, J. Appl. Phys. 73, 1430 (1993).
8. K. Schohe, F. Krafft, C. Klingelhöfer, M. Garrigues, S.K. Krawczyk and J. Weyher, Mat. Science Eng. B20, 121 (1993).
9. J. Völkl, H.J. Wolf, P. Eckstein, G. Wittmann, R. Teichler, R. Stibal and W. Jantz, presented at the Semi-insulating III-V Materials Conference, Ixtapa 1992, unpublished.
10. W. Jantz, R. Stibal, J. Windscheif, F. Mosel and G. Müller, Semi-Insulating III-V Materials, edited by C.J. Miner, W. Ford and E.R. Weber (Inst. Phys. Publishing, Bristol 1993) 171.
11. P. Wellmann, S. Waldmüller and A. Winnacker, presented at DGKK Arbeitskreis, Göttingen 1993, unpublished.
12. K. Leo, W.W. Rühle and N.M. Haegel, J. Appl. Phys. 62, 3055 (1987).
13. H.CH. Alt, M. Müllenborn and G. Packeiser, Proc. 6th Conf. Semi-Insulating III-V Materials, edited by A.G. Milnes and C.J. Miner (Adam Hilger, Bristol, 1990), 309.
14. E. Molva, P. Bunod, A. Chabli, A. Lombardot, S. Dubois and F. Bertin, J. Crystal Growth 103, 91 (1990).
15. S. Tüzemen, L. Breivik and M.R. Brozel, Semicond. Sci. Technol. 7, A36 (1992).
16. M. Müllenborn and H.Ch. Alt, J. Appl. Phys 69, 4310 (1991).
17. S. Martin, Defect Control in Semiconductors, edited by K. Sumino (Elsevier, Amsterdam 1990), 795.
18. W. Jantz, R. Stibal, J. Windscheif and J. Wagner, Applied Surface Science 50, 480 (1991).
19. O. Oda, H. Yamamoto, M. Seiwa, G. Kano, T. Inoue, M. Mori, H. Shimakura and M. Oyake, Semicond. Sci. Technol. 7, A215 (1992).
20. B. Weinert, private communication.
21. M.R. Brozel, Proceedings of the LEOS'92 Confererence, (Boston, 1992) 321.
22. C.J. Miner, Semicond. Sci. Technol. 7, A10 (1992).
23. D.J. As, S. Korf, Z.M. Wang, J. Windscheif, K.H. Bachem and W. Jantz, Semicond. Sci. Technol. 7, A27, 1992.
24. Z.M. Wang, M. Baeumler, W. Jantz, K.H. Bachem, E.C. Larkins and J.D. Ralston, J. Crystal Growth 126, 205 (1993).
25. E.C. Larkins, M. Baeumler, J. Wagner, G. Bender, N. Herres, M. Maier, W. Rothemund, J. Fleißner, W. Jantz, J.D. Ralston, G. Fleming and R. Brenn, presented at the 26th Intern. Symposium on GaAs and Related Compounds, Freiburg 1993, to be published in Mat. Res. Soc. Symp. Proceedings.
26. M. Baeumler, E.C. Larkins, K.H. Bachem, D. Bernklau, H. Riechert, J.D. Ralston and W. Jantz, presented at the 5th Intern. Conf. on Defect Recognition and Image Processing in Semiconductors and Devices, Santander 1993, to be published in Semicond. Sci. Technol.
27. Z.M. Wang, M. Baeumler, W. Jantz, K.H. Bachem, E.C. Larkins and J.D. Ralston, Proceedings of the LEOS'92 Confererence, (Boston, 1992) 319.
28. P. Hiesinger, T. Schweizer, K. Köhler, P. Ganser, W. Rothemund and W. Jantz, J. Appl. Phys. 72, 2941 (1992).

PROPERTIES OF AN ARRAY OF DISLOCATIONS
IN A STRAINED EPITAXIAL LAYER

TONG-YI ZHANG
Department of Mechanical Engineering
Hong Kong University of Science and Technology
Clear Water Bay, Kowloon, Hong Kong

ABSTRACT

The energy of a general array of dislocations in an epitaxial layer is formulated and expressed in terms of per unit area of interface. Several limiting cases are used to verify the solution and the results are compared to other independent treatments. The critical thickness required for the generation of an isolated dislocation is found by solving for the layer thickness which corresponds to a zero value of the formation energy. The critical dislocation density at a given thickness is also determined. An additional work required for sequential generation of dislocations in an epitaxial layer arises from dislocation-dislocation interaction and has to be expressed in terms of per unit length of dislocation line. The work hardening effect is found to increase sharply with decreasing distance between the fresh and the pre-existing dislocations once the distance falls below approximately twenty times the layer thickness. The additional work achieves the level of the self energy of an isolated dislocation when the distance between the fresh and the nearest pre-existing dislocation is comparable to twice the layer thickness.

INTRODUCTION

Recently, Freund[1,2] and Willis and co-workers[3-6] developed rigorous treatments for the stability of dislocation arrays. However, some inconsistencies remain in the self energy of a dislocation array in a strained epitaxial layer. In particular, one does not recover appropriate solutions for special types of arrays in the limit where the thickness of the epitaxial layer approaches infinity. In this case, the dislocation array becomes a dislocation wall. For example, the self energy of a wall of pure screw dislocations approaches zero in Willis et al.'s solution[3,4], while it is easily shown that the appropriate limit is infinity. Similarly, the self energy of an array of pure edge dislocations with their Burgers' vector perpendicular to the interface differs from that for a pure tilt boundary[7] by a factor of "e" in the log term when the solution of Freund is used[2].

It has been experimentally observed that the stress relaxation induced by dislocation generation in a strained epitaxial layer grown on a substrate is considerably less than expected from conventional equilibrium models[8]. Dodson[9] proposed that the discrepancy was the result of dislocation-dislocation interaction in which the early generated dislocations exert a back-stress on any subsequent dislocations. He drew an analogy between this type of process and work hardening in metals but provided no rigorous derivation of the magnitude of the effect. Willis and co-workers[6] made the first attempt to provide a theoretical basis for the calculation of the possible extent of work hardening. Their method was based on calculating the driving force for the introduction of the "last" dislocation to complete a periodic array of dislocations in the interface between a strained epitaxial layer and a substrate. Since their results predicted an attraction between the last dislocation and the incomplete array, they concluded that work hardening could not be used to explain the experimental observations.

The present work reports solutions of the energies of an array dislocations and of the dislocation-dislocation interaction for the subsequent generation of dislocations in a strained layer and applies the results to a consideration of potential work hardening.

THEORETICAL ANALYSIS

Energies of a dislocation array

Consider a periodic array of dislocations at an interface. All of the dislocations in the array have the same Burgers' vector $\mathbf{b}=[b_1,b_2,b_3]$ (Fig.1). The total energy for the array is expressed in terms of per unit area of interface. Thus, the energy per unit length of dislocation line is the product of the dislocation spacing, p, and the energy per unit area. The total energy and the formation energy are given by:

$$p\,E_t = p\,(E_s + E_{int} + E_m) \quad and \quad E_f = E_s + E_{int}, \qquad (1)$$

where E_t, E_f, E_s, E_{int} and E_m are, respectively the total, formation, self, interaction, and mismatch energies per unit area. The interaction energy defined here is the only driving force for dislocation generation and arises from the interaction between the mismatch stresses and the dislocations. The interaction energy among dislocations is included in the self energy at the moment. The appropriate energies per unit length of dislocation line can be evaluated by integrating the corresponding energy densities in a domain Σ, as shown in Fig.1. Consequently, the interaction and the mismatch energies are respectively evaluated as

$$pE_{int} = \frac{2\,\mu\,(1+\nu)}{1-\nu}\,\epsilon_m\,b_2\,h \quad and \quad E_m = \frac{2\mu\,(1+\nu)}{1-\nu}\,\epsilon_m^2\,h, \qquad (2)$$

where ϵ_m is the un-relaxed mismatch strain, h is the layer thickness, μ and ν are the shear modulus and Poisson's ratio respectively. Eq.(2) shows that the mismatch energy per unit area is independent of the dislocation spacing. The only driving force is ensured by a negative interaction energy, i.e. a negative product of the b_2 component of the Burgers' vector and the mismatch strain. If the dislocation core radius, r_0, is much smaller than either the layer thickness or the dislocation spacing, the self energy approximately equals:

$$pE_s = \frac{\mu}{4\pi\,(1-\nu)}\,(b_1^2+b_2^2+(1-\nu)\,b_3^2)\left[\ln\left(\frac{p\,(1-e^{-4Q})}{2\pi r_0}\right) + 2\,Q\right]$$

$$+\ \frac{2\mu\,(b_1^2+b_2^2)\,Q^2 e^{-4Q}}{\pi\,(1-\nu)\,(1-e^{-4Q})^2} + \frac{\mu\,(b_1^2-b_2^2)}{4\pi\,(1-\nu)}\left(1-2Q\frac{1+e^{-4Q}}{1-e^{-4Q}}\right) \quad with \quad Q=\frac{\pi h}{p}. \qquad (3)$$

The formation energy is obtained by the summation of the self energy, core energy, and interaction energy terms. Furthermore, the sum of the mismatch and the formation energies yields the total energy.

There are two limiting cases for the self energy of the dislocation array. The first is that the self energy distributed throughout the entire crystal reduces to that for an isolated dislocation as the dislocation spacing, p, approaches infinity. It can be shown that E_s does indeed approach:

$$pE_s = \frac{\mu}{4\pi(1-\nu)}(b_1^2 + b_2^2 + (1-\nu)b_3^2)\ln\left(\frac{2h}{r_0}\right) + \frac{\mu(b_1^2 + b_2^2)}{8\pi(1-\nu)}. \tag{4}$$

The other limiting case occurs as the layer thickness approaches infinity. Here, the array of dislocations becomes a dislocation wall in an infinite body. As a simple example, let us assume that the Burgers' vector of the dislocations in the array lies along the x_1 direction so that the dislocation wall is a pure tilt boundary. In this case, the self energy reduces to:

$$E_s = \frac{\mu b_1^2}{4\pi(1-\nu)p}\ln\left(\frac{e\,p}{2\pi r_0}\right). \tag{5}$$

This is identical to that given for a tilt boundary by Hirth and Lothe[7]. It should also be noted that, for the solution presented herein, if the Burgers' vector of the dislocations in the array lies along either the x_2 or x_3 direction, the self energy will approach infinity, as expected[7].

Work hardening

The work hardening is understood as an increase of the formation energy of a fresh dislocation. The fresh dislocation generates when other dislocations exist already. The formation energy of the fresh dislocation contains three terms: the self energy of the fresh dislocation which is defined as the self energy for an isolated dislocation given by Eq.(4), the interaction energy between the mismatch stresses and the fresh dislocation which is given by Eq.(2) and the interaction energy, E_{d-d}, between the fresh dislocation and the other pre-existing dislocations. Since the self energy and the interaction energy between the dislocation and the mismatch stresses are identical to each dislocation, the work hardening or the change of the formation energy must be caused by the dislocation-dislocation interaction. In this case, all energies have to be expressed in terms of per unit length of dislocation line.

In the present work, we assume that the fresh and pre-existing dislocations have the same Burgers' vector. First we consider the case of a single pre-existing dislocation. E_{d-d} is given by

$$E_{d-d} = \frac{\mu}{4\pi(1-\nu)}\left[(b_1^2 + b_2^2 + (1-\nu)b_3^2)\ln\left(\frac{4h^2 + x_{2d}^2}{x_{2d}^2}\right)\right.$$
$$\left. + 4h^2(b_1^2 + b_2^2)\frac{4h^2 - x_{2d}^2}{(4h^2 + x_{2d}^2)^2} + (b_1^2 - b_2^2)\frac{8h^2}{4h^2 + x_{2d}^2}\right], \tag{6}$$

where the x_1 axis is chosen to pass through the fresh dislocation and the location of the pre-existing dislocation is (h, x_{2d}). If there are N pre-existing dislocations, the interaction energy is directly obtained from Eq.(6) as a sum of the interactions between the individual dislocations

$$E_{d-d} = \frac{\mu}{4\pi(1-\nu)} \sum_{j=1}^{N} \left[(b_1^2 + b_2^2 + (1-\nu)b_3^2) \ln\left(\frac{4h^2 + x_{2d,j}^2}{x_{2d,j}^2}\right) \right.$$

$$\left. + 4h^2 (b_1^2 + b_2^2) \frac{4h^2 - x_{2d,j}^2}{(4h^2 + x_{2d,j}^2)^2} + (b_1^2 - b_2^2) \frac{8h^2}{4h^2 + x_{2d,j}^2} \right]. \qquad (7)$$

For the special case of a periodic array of pre-existing dislocations in the interface between the layer and the substrate the interaction energy for a fresh dislocation at the midpoint between any two pre-existing dislocations in the array is given by:

$$E_{d-d} = \frac{\mu}{4\pi(1-\nu)} \left[(b_1^2 + b_2^2 + (1-\nu)b_3^2) \ln\left[\cosh\left(\frac{2\pi h}{p}\right)\right] \right.$$

$$\left. - (b_1^2 + b_2^2) \frac{4\pi^2 h^2}{p^2 \cosh^2\left(\frac{2\pi h}{p}\right)} + (b_1^2 - b_2^2) \frac{4\pi h}{p} \tanh\left(\frac{2\pi h}{p}\right) \right]. \qquad (8)$$

RESULTS AND DISCUSSION

The critical thickness for dislocation generation can be obtained from the condition of a zero formation energy. Ignoring the dislocation core energy, one finds:

$$\frac{b_1^2 + b_2^2 + (1-\nu)b_3^2}{4\pi} \ln\left(\frac{2h_0}{r_0}\right) - 2(1+\nu)|\epsilon_m b_2|h_0 + \frac{b_1^2 + b_2^2}{8\pi} = 0. \qquad (9)$$

Eq.(9) determines the critical thickness, below which no dislocation can appear. A comparison of Eq.(9) with the results of Freund[1] reveals an extra term in Eq.(9) which yields a larger critical thickness. The reason for the discrepancy is that the earlier treatments assumed that the dislocation is already present in the strained layer, while, in the present work, the dislocation is assumed to generate from somewhere and subsequently moves to the interface. As an example, a Ge_xSi_{1-x} layer on a Si substrate with an interface, $x_1=h$, has a normal which coincides with the <100> direction for either material. For a 60^0 dislocation laying on a {111} plane, the dislocation line is along the <110> direction. A possible Burgers' vector in the coordinate system is $\mathbf{b}=b[-1/\sqrt{2}, -1/2, 1/2]$. The mismatch strain is $\epsilon_m=0.042x$ (Willis et al.[5]) and $\nu=0.22$[7]. For x=0.25, the critical thickness is approximately 39.3b or 133 Å. Without the extra term in Eq.(9), above data yield the critical thickness of 32.7b. The extra term makes the critical thickness larger by approximately 7b.

The total energy per unit area varies with the dislocation density. The data used to calculate the critical thickness above are adopted here again to demonstrate the total energy

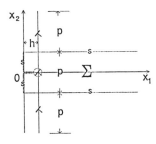

Fig.1. An array of dislocations with a period p are located in the interface between a thin layer and a substrate. The dashed lines represent the boundaries S of the domain Σ.

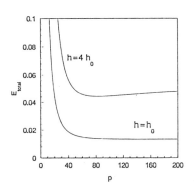

Fig.2. The variation of the total energy E_t with the dislocation spacing p, where p is in units of b and E_t is in units of $(\mu b)/(1-\nu)$, and h_0 stands for the critical thickness.

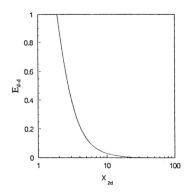

Fig.3. The variation of the interaction energy, E_{d-d} in units of $(\mu b^2)/[4 \pi(1-\nu)]$, between the fresh dislocation and a single pre-existing dislocation, with the distance between the two dislocations, x_{2d} in units of h.

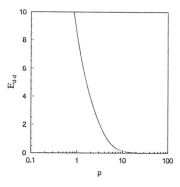

Fig.4. The variation of the interaction energy E_{d-d} in units of $(\mu b^2)/[4 \pi(1-\nu)]$, between the fresh dislocation and an array of existing dislocations, with the dislocation spacing, p in units of h.

423

change with dislocation density. Fig.2 shows that the total energy is independent of dislocation spacing at large p when the layer thickness is equal to the critical thickness. At very small dislocation spacings, however, the total energy rises sharply with decreasing dislocation spacing. For layer thicknesses larger than the critical thickness, the total energy gently decreases with decreasing dislocation spacing until it reaches a minimum value. As described above, the only energy term which is negative is the interaction energy. When the dislocation density in the array is relatively low, the increase in the interaction energy is more rapid with decreasing p than the self energy. Since the mismatch energy is independent of p, the total energy decreases. At dislocation spacings smaller than the minimum point, the total energy again increases very sharply with decreasing p. In this region the self energy becomes dominant as the dislocations begin to interact. Obviously, a critical dislocation density can be obtained by minimizing the total energy.

The dislocation-dislocation interaction energy as a function of the vertical component of the distance between the two dislocations is shown in Fig.3. As can be seen from the figure, the interaction energy is always positive, whether the fresh dislocation is above or below the pre-existing dislocation. This implies that extra work is always required for the generation of the second dislocation and supports the suggestion of work hardening. The interaction energy actually approaches infinity as the distance between the fresh and pre-existing dislocations decreases below about twenty layer thicknesses. At separations larger than twenty layer thicknesses, the interaction energy becomes negligible.

For the case of a pre-existing periodic dislocation array, Fig.4 shows the additional work required for the nucleation of a fresh dislocation at a position midway between any two pre-existing dislocations in the array. A comparison of Figs.3 and 4 shows that the interaction energy for an array of dislocations is similar to that for a single dislocation. Although it rises more rapidly with decreasing dislocation spacing in the array, it still becomes negligible above a spacing of approximately twenty layer thicknesses. If the layer thickness is assumed to be 30 b (approximately 100 Å), the self energy has a value of 5.5 in units of $(\mu b^2)/[4\pi(1-\nu)]$. As can be seen in Fig.4, the interaction energy will exceed the self energy when the dislocation spacing is smaller than twice the layer thickness.

REFERENCES

1.   L.B. Freund, *J. Mech. Phys. Solids*, **38**, 657 (1990).
2.   L.B. Freund, *Scripta Met. Mater.*, **27**, 669 (1992).
3.   T.J. Gosling, S.C. Jain, J.R. Willis, A. Atkinson and R. Bullough, *Phil. Mag. A*, **66**, 119 (1992).
4.   S.C. Jain, T.J. Gosling, J.R. Willis, D.H.J. Totterdell and R. Bullough, *Phil. Mag. A*, **65**, 1151 (1992).
5.   J.R. Willis, S.C. Jain and R. Bullough, R., *Phil. Mag. A*, **62**, 115 (1990).
6.   J.R. Willis, S.C. Jain and R. Bullough, R., *Appl. Phys. Lett.*, **59**, 920 (1991).
7.   J.P. Hirth, and J. Lothe, <u>Theory of Dislocations</u>, second ed. (John Wiley & Son, New York, 1982).
8.   R.M. Biefeld, C.R. Hills and S.R. Lee, *J. Cryst. Growth*, **91**, 515 (1988).
9.   B.W. Dodson, *Appl. Phys. Lett.*, **51**, 37 (1988).

DOMINANT MIDGAP LEVELS IN THE COMPENSATION MECHANISM IN GaAs

H. SHIRAKI, Y. TOKUDA*, E. TOHYAMA*, K. SASSA, AND N. TOYAMA
Central Research Institute, Mitsubishi Materials Corporation, 1-297 Kitabukuro-cho, Omiya, Saitama, Japan
*Department of Electronics, Aichi Institute of Technology, 1247 Yachigusa Yakusa, Toyota, Aichi, Japan

ABSTRACT

Properties of midgap levels in n-type GaAs crystals were studied by using trap density spectroscopy (TDS), capacitance–voltage and near infrared (NIR) absorption measurements. The TDS analysis for various n-type samples showed that the concentrations of the midgap level were from 2.0×10^{16} cm^{-3} to 3.4×10^{16} cm^{-3}. However, those of EL2 in the same crystals were found to be from 1.1×10^{16} cm^{-3} to 1.3×10^{16} cm^{-3} by NIR absorption measurements at room temperature. On the other hand, in undoped semi-insulating GaAs crystals with carbon concentrations ranging from 6×10^{14} cm^{-3} to 2.4×10^{16} cm^{-3}, the densities of the ionized EL2 determined by NIR absorption were only about 30 % of those of carbon acceptor determined by the localized vibrational mode absorption at room temperature. These differences suggest the presence of another midgap donor which does not give NIR absorption. The concentrations of this trap were estimated to be 0.8 to 1.6 times those of EL2 by decomposition of the TDS spectra into their components.

INTRODUCTION

Undoped semi-insulating (SI) characteristics in GaAs materials are generally described by the so-called three-level model that the net shallow acceptor compensates a part of the midgap donor EL2. The carbon is a dominant shallow acceptor in undoped SI materials. Electrical properties of these materials are explained by Shockley diagram based on this model qualitatively. The concentration of carbon in GaAs is determined by the localized vibrational mode (LVM) absorption, and those of the neutral and ionized EL2 can be determined by near infrared (NIR) absorption measurements [1,2]. According to this model, the concentration of the ionized EL2 must be nearly equal to that of carbon. Recent optical absorption measurements, however, suggested that the carbon density was about three times greater than the ionized EL2 concentration [2,3,4]. This discrepancy cannot be explained in the framework of the three-level model.

To investigate the compensation mechanism, properties of midgap levels in n-type GaAs crystals were studied by using trap density spectroscopy (TDS), capacitance–voltage (C-V) and near infrared (NIR) absorption measurements. For SI materials, the concentrations of the neutral and ionized EL2 were determined by NIR absorption measurements at room temperature. The electrical properties (resistivity and Hall mobility) of SI GaAs crystals were obtained from Hall measurements. A new model of the compensation mechanism is proposed to explain these results quantitatively.

EXPERIMENTAL

In order to study properties of midgap levels, n-type GaAs samples were prepared from crystals grown by HB, LEC methods, and arsenic-pressure

425

controlled CZ (PCZ) method [5] in which a GaAs crystal is grown under an arsenic pressure without any encapsulant in a hot wall chamber. Free carrier concentrations at 300 K in the samples were below 1×10^{16} cm^{-3}. To investigate the dependence of the electrical properties and the EL2 concentration on the carbon concentration, undoped SI crystals grown by the PCZ method were also prepared. The carbon concentrations were varied from 6×10^{14} cm^{-3} to 2.4×10^{16} cm^{-3}, which were determined by the LVM absorption at room temperature using a conversion factor of 11.8×10^{15} cm^{-1} [6].

The capacitance changes were measured isothermally with a 1 MHz capacitance meter on Au-Schottky diodes evaporated on chemically etched (100) n-type samples. Ohmic contacts were made by evaporating and alloying Au-Ge/Ni on the back surface of the samples. In order to determine the concentrations of midgap levels, the trap density spectroscopy (TDS) analysis [7] was carried out on the capacitance transients measured for long time duration at temperatures ranging from 280 K to 360 K. This method is capable of characterizing traps precisely, even if the trap concentration exceeds the net shallow donor concentration. However, the determination of shallow donor and free carrier densities is a serious problem for this analysis, if the thermally activated carrier density from other defect levels is not negligible in comparison with the free carrier density. For example, medium-deep traps (EL5, EL6 etc.) in n-type GaAs crystals with carrier concentrations below 1×10^{16} cm^{-3} can act as a dominant source of free carriers at about 300 K [8]. In this study, we used an improved analyzing method to precisely determine the amount of the space charge and the free carrier concentration in samples containing more than one deep trap, and thus to calculate the Fermi level and the so-called Lambda for the target trap [9].

The concentrations of the neutral and ionized EL2 were determined by NIR absorption measurements based on the photoionization cross sections of EL2 at room temperature [10]. The absorption coefficient α ($h\nu$) is given by

$$\alpha(h\nu) = \sigma_n^{\circ}(h\nu)[EL2^{\circ}] + \sigma_F^{\circ}(h\nu)[EL2^+], \qquad (1)$$

where $\sigma_n^{\circ}(h\nu)$ and $\sigma_F^{\circ}(h\nu)$ are the photoionization cross sections of the neutral and ionized EL2, respectively.

Hall measurements for n-type GaAs materials were performed by van der Pauw method on 7×7 mm samples at temperatures ranging from 80 K to 400 K. For SI materials, Hall measurements were carried out at 298 K.

RESULTS AND DISCUSSION

Figure 1 shows the depth profile of the space charge in the depletion layer obtained from C-V measurement at 80 K. This measurement was performed in the following way. First, the reverse bias ($V_R = -10$ V) was applied on the Au-Schottky diode at room temperature. After the capacitance reached a steady state, the sample was cooled down to 80 K with V_R applied. The C-V measurement was performed at 80 K by sweeping the applied bias from V_R to 0 V at a rate of 1 V/sec. In these conditions, C-V measurements could reveal the depth profile of the space charge in the depletion layer as formed at the reverse bias of V_R. The lowest and middle levels in the space charge profile correspond to the net shallow donor density (N_{SDA}) and the sum of N_{SDA} and the medium-deep trap density (N_{MD}), respectively. The highest plateau is due to the addition of the ionized midgap level. However, the concentration of the midgap level N_{DD} could not be determined correctly from the difference between the second and third plateaus, because the concentration of the ionized midgap level in the depletion layer decreased markedly during the C-V measurement due to electron capture. This speculation is supported by the following result. The dashed lines in Fig. 1 (a) and (b) represent the depth profiles of the space charge obtained from the C-V measurements in which the applied bias was swept from 0 V to V_R at 360 K. The concentrations of the space charge increase with depth. These variations are due to the effect of the Lambda for the midgap level. The midgap level densities in HB and PCZ samples are roughly estimated to be greater than 2×10^{16} cm^{-3}.

Figure 2 shows TDS spectra for midgap levels in n-type HB and PCZ samples at 360 K. The concentrations of the midgap level in HB and PCZ samples were 2.0×10^{16} cm^{-3} and 3.3×10^{16} cm^{-3}, respectively. These values were in good agreement with those obtained from the analysis of the depth profiles of the space charge at 360 K. The measured space charge density is given by,

$$N_{meas}=(N_{SDA}+N_{MD})+(1-\lambda/W)N_{DD}(W-\lambda) \qquad (2)$$

where W is the width of the depletion layer and λ is the distance between the depletion layer edge and the crossing point between the Fermi level and the midgap level [11]. The depth profiles of the midgap level calculated from Eq. (2) were represented by dotted lines in Fig. 1 (a) and (b). On the other hand, the EL2 concentrations determined by NIR absorption measurements at room temperature were 1.1×10^{16} cm^{-3} and 1.3×10^{16} cm^{-3} for the HB and PCZ samples, respectively. The result of the LEC sample was almost the same as that of the PCZ sample. The disagreement of midgap level density between the electrical and optical methods can not be ascribed to the used optical cross section values for EL2, because the

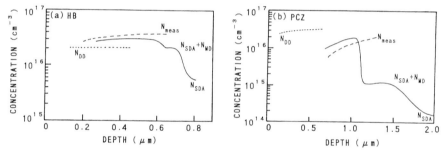

Fig. 1. Depth profiles of space charge for (a) HB and (b) PCZ samples. The solid lines are the results of the C-V measurement by sweeping the bias from V_R (-10 V) to 0 V at 80 K, after the samples were cooled from room temperature with the bias V_R applied. The dashed lines are results of the C-V measurement by sweeping the bias from 0 V to V_R at 360 K. The dotted lines represent the depth profiles of midgap levels calculated from N_{meas} obtained from the C-V measurement at 360 K.

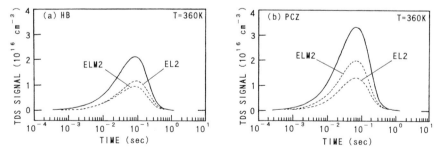

Fig. 2. TDS spectra of midgap levels in n-type (a) HB and (b) PCZ samples. The dashed lines are calculated components using the results of NIR absorption.

Table I. Properties of midgap levels in n-type GaAs crystals.

Sample	EL2			ELM2		
	$E_c - E_T$ (eV)	σ_∞ (cm^2)	N_T (cm^{-3})	$E_c - E_T$ (eV)	σ_∞ (cm^2)	N_T (cm^{-3})
HB	0.810	1.1×10^{-13}	1.1×10^{16}	0.800	8.1×10^{-14}	0.9×10^{16}
PCZ	0.805	1.0×10^{-13}	1.3×10^{16}	0.798	8.2×10^{-14}	2.0×10^{16}
LEC	0.818	1.4×10^{-13}	1.3×10^{16}	0.800	8.0×10^{-14}	2.1×10^{16}

concentration ratio N_{TDS}/N_{NIR} is not found to be constant. The ratio should be constant, even if $\sigma_n^0(h\nu)$ used in this study is not correct. We propose that an unknown midgap level other than EL2 should exist which does not give NIR absorption. This midgap level named ELM2 was estimated by decomposition of the TDS spectra into their components using the EL2 concentration determined by NIR absorption. The thermal emission activation energies, capture cross sections and concentrations of these midgap levels are listed in Table I.

We investigated SI materials in order to describe the compensation mechanism quantitatively. Figure 3 shows a typical NIR absorption spectrum due to EL2 in a SI PCZ sample at 298 K. The measured spectrum represented by solid line fitted well with the calculated curves based on Eq. (1). Figure 4 shows the dependence of the neutral, ionized and total EL2 concentrations on the carbon concentration. The total EL2 concentration [EL2] remained almost constant at about 1.25×10^{16} cm^{-3}. Although [EL2$^+$] increased with increasing carbon concentration [C$_{As}$], the ionized EL2 was only about 30 % of carbon acceptor in density. This result is in good agreement with those reported by other workers [3,4].

The electrical properties in SI PCZ materials are plotted as a function of the carbon concentration in Fig. 5. The maximum resistivity reaches approximately 1×10^9 ohm·cm at the carbon concentration of 2.4×10^{16} cm^{-3}. However, the material remains n-type and the Hall mobility is over 1000 cm^2/v·s. Otoki et.al. [4] reported that the material was n-type up to the same carbon concentration and became p-type beyond that. It was noticeable that the resistivity was higher than 1×10^6 ohm·cm at the carbon concentration of 3.3×10^{16} cm^{-3} in their data. These results

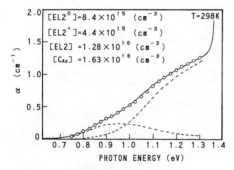

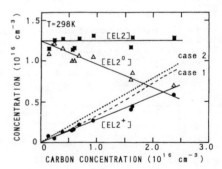

Fig. 3. A typical NIR absorption spectrum due to EL2 in a SI PCZ sample at 298 K. The solid line is an experimental spectrum. The open circles represent calculated absorption coefficients. The dashed lines are calculated components for the neutral and ionized EL2.

Fig. 4. EL2 concentrations as a function of carbon concentration. The triangles, circles and squares are the concentrations of the neutral, ionized and total EL2, respectively. The solid lines are fitted curves for experimental data. The dashed and dotted lines represent calculated ones for the ionized EL2 from case 1 and 2, respectively.

cannot be explained in terms of the three-level model, because the electrical properties must be converted to p-type in the carbon concentration range higher than the EL2 concentration.

In order to interpret the compensation mechanism, we attempt to calculate the charge balance based on a model including EL2 and ELM2 as midgap levels. Two energy levels of EL2 are given by [12,13]

$$E(0/+) = E_c - (0.759 - 2.37 \times 10^{-4} T) \text{ eV},\tag{3}$$

for the single donor state and

$$E(+/++) = E_v + 0.54 \text{ eV},\tag{4}$$

for the double donor state. However, the energy level of ELM2 is ambiguous, because the energy in Table I is not the net thermal depth but the activation energy for electron emission. Then, the energy level of ELM2 which was a single donor was assumed to be 10 meV shallower than that of EL2 (case 1) or the same as that of EL2 (case 2). The charge neutrality condition is given by the following expression, taking into account of the shallow donor N_{SD} and acceptor N_{SA}:

$$n + N_{SA} = p + N_{SD} + [EL2^+] + [ELM2^+] + 2 [EL2^{++}] ,\tag{5}$$

where n and p are the concentrations of free electron and hole, respectively. The resistivity ρ and Hall mobility μ_H are given by

$$\rho = \{ q (n \mu_n + p \mu_p) \}^{-1},\tag{6}$$

$$\mu_H = (n \mu_n^2 - p \mu_p^2) / (n \mu_n + p \mu_p),\tag{7}$$

where μ_n and μ_p are the electron and hole mobilities, their ratio being taken 17 in our calculation at 298 K. The calculated results are shown in Fig. 4 and 5. The measured electrical properties fitted well with the calculated curves shown in Fig. 5. The theoretical curve showed that the resistivity was higher than 1×10^6 ohm·cm at the carbon concentration of about 3×10^{16} cm^{-3}. However, the behavior of the ionized EL2 could not be reproduced well by the above model, especially in the carbon concentration range higher than 1×10^{16} cm^{-3}.

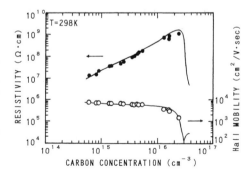

Fig. 5. Resistivity and Hall mobility as a function of carbon concentration at room temperature. The filled and open circles represent experimental data. The solid lines are calculated curves.

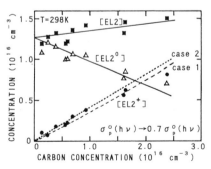

Fig. 6. EL2 concentrations as a function of carbon concentration. Fitted curves were calculated ones by our model using $0.7 \sigma_p^0 (h \nu)$.

Suemitsu et.al. [3,14] reported that the carbon-related donor existed in a constant ratio to the carbon acceptor. Their model could explain the dependence of the ionized EL2 concentration on the carbon concentration, but could not explain the change in the electrical properties.

What causes the disagreement between the experimental and calculated results?. We used two conversion factors in this study. Any or both of them may be not correct. However, the conversion factor for the carbon acceptor is not the case, because the change of the factor can not improve all fittings in Fig. 4 and 5 at the same time. It is possible that the photoionization cross sections of EL2, especially $\sigma_p^{\circ}(h\nu)$, used in NIR absorption analysis may be overestimated. Figure 6 shows the dependence of the neutral, ionized and total EL2 concentrations on the carbon concentration calculated by our model using a factor of 0.7 for $\sigma_p^{\circ}(h\nu)$. The calculated curves for the electrical properties were the almost same as those shown in Fig. 5. Precise re-examination of $\sigma_p^{\circ}(h\nu)$ is desired.

CONCLUSION

We investigated the properties of midgap levels in n-type GaAs crystals by using TDS, C-V and NIR absorption measurements. The TDS analysis and the analysis of the space charge profile for various samples showed that the concentrations of the midgap level were from 2.0×10^{16} cm^{-3} to 3.4×10^{16} cm^{-3}. However, those of EL2 in the same crystals were 1.1×10^{16} cm^{-3} to 1.3×10^{16} cm^{-3} by NIR absorption measurements at room temperature. On the other hand, for SI GaAs crystals with carbon concentrations from 6×10^{14} cm^{-3} to 2.4×10^{16} cm^{-3}, the ionized EL2 was only about 30 % of carbon acceptor in density. A new model was proposed that another midgap donor ELM2 existed which did not give NIR absorption, but played an important role in the compensation mechanism like EL2. The concentrations of this trap were estimated to be 0.8 to 1.6 times those of EL2.

REFERENCES

1. A. Winnacker, and F.X. Zach, J. Cryst. Growth, 103, 275 (1990)
2. J.S. Blakemore, L. Sargent, and R S. Tang, Appl. Phys. Lett., 54, 2106 (1989).
3. M. Suemitsu, M. Nishijima, and N. Miyamoto, J. Appl. Phys., 69, 7240 (1991).
4. Y. Otoki, M. Sahara, S. Shinzawa, and S. Kuma, Mater. Sci. Forum, 117–118, 405 (1993).
5. T. Atami, K. Shirata, H. Takahashi, K. Sassa, and K. Tomizawa, in GaAs and Related Compound, Karuizawa, edited by T. Ikegami, F. Hasegawa, and Y. Takeda (Inst. Phys. Conf. Ser., 129), p25.
6. T. Arai, T. Nozaki, J. Osaka, and M. Tajima, in Semi-insulating III-V Materials, Malmo, edited by G. Grossmann and L. Ledebo (Adam Hilger, Bristol, 1988),p.201.
7. T. Okumura, Jpn. J. Appl. Phys., 24, L437 (1985).
8. H. Shiraki, Y. Tokuda, and K. Sassa, in Defect Engineering in Semiconductor Growth, Processing and Device Technology, edited by S. Ashok, J. Chevallier, K. Sumino, and E. Weber (Mater. Res. Soc. Proc. 262, Pittsburgh, PA, 1992)p.105.
9. H. Shiraki, Y. Tokuda, K. Sassa, and N. Toyama, J. Appl. Phys., to be submitted.
10. P. Silverberg, P. Omling, and L. Samuelson, Appl. Phys. Lett., 52, 1689 (1988).
11. L.C. Kimering, J. Appl. Phys., 45, 1839 (1974).
12. G.M. Martin, J.P. Farges, G. Jacob, and J.P. Hallais, J.Appl.Phys., 51, 2840(1980).
13. J. Lagowski, D.G. Lin, T.P. Chen, M. Skowronski, and H.C. Gatos, Appl. Phys. Lett., 47, 929 (1985).
14. M. Suemitsu, K. Terada, M. Nishijima, and N. Miyamoto, Jpn. J. Appl. Phys., 31, L1654 (1992).

IDENTIFICATION OF THE 0.15 eV DONOR DEFECT IN BULK GaAs

Z-Q. FANG, J. W. HEMSKY, and D. C. LOOK
Physics Department, Wright State University, Dayton OH 45435

ABSTRACT

The well–known 0.15–eV Hall–effect center appearing in bulk, n–type GaAs quenches under IR illumination and recovers via an Auger–like process at a rate similar to the Auger rate of EL2. On the other hand, the 0.15–eV V_{As}–related center produced by 1–MeV electron irradiation does not quench at all. Based on these data and a detailed theoretical analysis by Baraff and Schluter, we argue that the bulk 0.15–eV center is related to the As_{Ga}–V_{As} defect or a related complex.

INTRODUCTION

Donor centers located at about 0.15 eV below the conduction band, and of concentration about 10^{15}–10^{16} cm^{-3}, have been commonly observed by temperature–dependent Hall–effect (TDH) and deep level transient spectroscopy (DLTS) measurements in as–grown [both horizontal Bridgman (HB) and liquid encapsulated Czochralski (LEC)], irradiated, and annealed GaAs over the last two decades [1–6]. In 1982, we showed that the 0.15–eV center in undoped bulk GaAs had a pure defect nature and was probably related to the As vacancy, based on comparison to a similar center in irradiated GaAs [1]. In this paper we give strong evidence that the 0.15–eV center in bulk GaAs is an As–vacancy/As–antisite complex, a form of which Baraff and Schluter [7] predicted should be abundant in GaAs, based on theoretical formation energies. The method we use relies on a comparison of the infrared (IR) quenching and thermal recovery characteristics of the 0.15–eV Hall–effect center in 1–MeV electron–irradiated MBE GaAs, which has very little EL2 (i.e. no apparent As antisite centers) and that in HB–grown bulk GaAs, which has an EL2 concentration of more than 10^{15} cm^{-3}. In dark–current measurements during a temperature sweep (82 K < T < 180 K) after IR illumination at 82 K, it is found that the irradiated material shows no quenching of the 0.15–eV center, whereas the bulk material shows strong quenching and a thermal recovery at about 130–140 K. The spectral dependence of the quenching of the 0.15–eV center is found to be similar to that of EL2 observed in undoped LEC–grown semi–insulating (SI) GaAs, but with some differences. A theoretical fit of the experimental data indicates that the quenched state thermally recovers by an Auger–like process at a rate $r = 2.3 \times 10^{-12} \, nv_n \exp(-0.18 \, eV/kT)$, which can be compared with that found for EL2, i.e., $r = 2 \times 10^{-14} \, nv_n \exp(-0.11 \, eV/kT)$ [8]. The experimental results support the assignment of a defect complex As_{Ga}–V_{As} to the bulk, 0.15–eV center.

EXPERIMENT

Two kinds of GaAs material, i.e., bulk, and irradiated molecular–beam–epitaxial (MBE) GaAs, were used in the study. The HB–grown bulk GaAs ingots, with electron concentrations in the mid–10^{14} cm^{-3} range, were doped with oxygen to suppress Si contamination. The dominant deep centers in various similar GaAs samples are located at E_c – (0.13–0.20) eV, as revealed by TDH studies, with a typical value of E_c – 0.15 eV (see Fig. 1 in ref. 1). For comparison, an undoped MBE GaAs layer with electron concentration of 1.5×10^{15} cm^{-3} and thickness of 15 μm was grown on an LEC SI GaAs substrate at a growth temperature of about 580 °C. To introduce deep centers into the epi–layer in a controlled manner, a series of consecutive 1–MeV electron irradiations (EIs) were performed at room temperature in air using a dose of 5×10^{14} cm^{-2} for each irradiation. As shown in Fig. 1, after the second EI the dark current of the epi–layer was dominated by a deep center at E_c – 0.15 eV, known as E2 and generally thought to be due to an As vacancy or possibly a vacancy–interstitial Frenkel pair [4].

For the IR quenching experiments, we measured current vs. temperature after IR light

Mat. Res. Soc. Symp. Proc. Vol. 325. ©1994 Materials Research Society

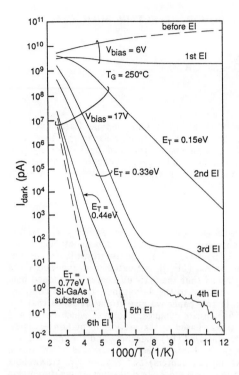

Fig. 1. Deep levels induced into a MBE GaAs ($N = 1.5 \times 10^{15}$ cm^{-3}) by consecutive electron irradiations, with a dose of 5×10^{14} cm^{-2} for each irradiation.

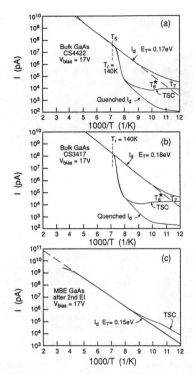

Fig. 2. A comparison of IR quenching behavior for the 0.15–eV center in bulk GaAs, (a) CS 4422 and (b) CS 3417 and the E2 center in irradiated MBE GaAs (c).

illumination at 82 K. A commercial DLTS system (Bio–Rad DL4600) was used for the temperature control and an electrometer (Keithley 617) was used to measure currents. The IR light was provided by a 25 W white light covered by a Si filter so that $h\nu \leq 1.12$ eV. The maximum IR light intensity was about 1×10^{16} phot./cm^2 s. For the current measurements, In contacts were first soldered on the two ends of 3 mm × 6 mm rectangular samples for the bulk GaAs and on the corners of a 6 mm × 6 mm square sample for the MBE GaAs, and then anneals were carried out at 420 °C for 4 min in forming gas. The EI was performed after the anneal to avoid the annihilation of the EI induced point defects at the annealing temperature. To measure the spectral dependence of the quenching of the 0.15–eV center in the bulk GaAs, a monochromator provided photon energies from 0.8 eV to 1.45 eV, at a constant intensity of 6×10^{14} phot./cm^2 s. Appropriate filters eliminated second–order light. The quenching parameter of the 0.15–eV center is defined as the ratio of the dark current before IR quenching to that after.

RESULTS AND DISCUSSION

Typical results for the dark current (I_d), the thermally stimulated current (TSC) and the quenched I_d are presented in Figs. 2a and 2b for two bulk GaAs samples (CS 4422 and CS 3417). In the experiment, the sample was cooled to 82 K in the dark, then IR light of intensity about 3×10^{15} phot./cm^2 s was turned on for 5 min, which caused partial quenching of the photocurrent (I_{ph}) at the beginning of the illumination. After turning the light off, the temperature was swept upward at a rate of 0.3 K/s. This is the typical procedure used for

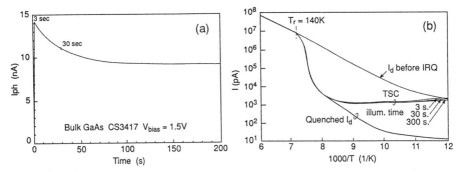

Fig. 3   a)   IR quenching of photocurrent at 82 K, using 1.10 eV light and
         b)   TSC signals and quenched I_d corresponding to different illumination times (bulk GaAs
               sample CS 3417, $V_{bias} = 1.5$ V).

TSC measurements in which the various traps are filled by the light–generated electrons and holes, and then the traps emit their carriers during the temperature sweep and produce TSC peaks at temperatures corresponding to the trap energy levels. A few traps, i.e., T_7 at 86 K, $T_6^\star$ at 96 K and even T_5 at 140 K, were found in these materials, as designated in the figures. As compared to the I_d curve, which was taken without IR illumination at 82 K, it can be seen that the I_d itself quenched due to the IR light illumination at 82 K and then thermally recovered at a temperature (T_r) of ~ 140 K. To observe the quenched I_d, the TSC signals had to be "thermally cleaned out." The cleaning procedure consisted simply of heating the sample to about 120 K, which is well below T_r, and then cooling it again to 82 K. In this way, the interfering traps were emptied and so the measured current in the subsequent temperature sweep was due only to the I_d contributed by the quenched 0.15–eV center. It is important to point out that at $T < 120$ K, the quenched system is basically in equilibrium, i.e., the quenched I_d curves in Fig. 2 are reproduced no matter which direction temperature is swept. In contrast to the bulk GaAs samples, the same IR light illumination at 82 K does not cause any quenching of the 0.15–eV–related I_d in the irradiated MBE GaAs sample, as shown in Fig. 2c. Instead, a few TSC signals are observed, some of which could be due to traps in the SI GaAs substrate.

A more detailed picture of the I_{ph} quenching upon illumination at 82 K and the corresponding TSC and quenched I_d curves, using 1.10 eV light with an intensity of about 2×10^{15} phot./cm^2 s and different illumination times, is shown in Fig. 3. The first thing to be noted (see Fig. 3a) is that I_{ph} quenching of only about 35% occurs (within about 50 s), which is very different from the observation of nearly complete I_{ph} quenching (close to 100%) occurring within the same period in bulk SI–GaAs [9]. The second thing (see Fig. 3b) is that only a short time illumination, e.g., 3 sec, is needed to quench the 0.15–eV center, as represented by the same quenched I_d and recovery behavior for three different illumination times, i.e., 3, 30, and 300 sec. The third thing is that, following the I_{ph} quenching the TSC signals are quenched too, which is commonly observed in the TSC spectra of SI–GaAs samples due to the IR quenching of EL2 [10]. From these results we see that the quenching of the 0.15–eV center is a fast process and is not identical to the IR quenching of EL2 in SI GaAs, such as that in photoconductivity [11], 1.1–μm absorption [12], and TSC (peak T_5) [13].

The recovery process of the quenched 0.15–eV center can be clearly observed in two different ways. In the first experiment, after IR quenching and thermally cleaning the TSC signals, the sample (CS 3417) was held at a particular recovery temperature, T_r, 82 K $< T_r <$ 130 K, for a time which depended on T_r. If T_r was chosen between 115 K and 130 K, then a normal recovery of the quenched I_d could be observed, i.e., the higher the T_r, the faster the recovery. However, if T_r was less than 110 K, then nearly no visible recovery could be found, which means that the system was still in the quenched or metastable state. Figure 4a shows the recovery of the quenched I_d as a function of the waiting time at $T_r = 120$ K. From the figure, we find that

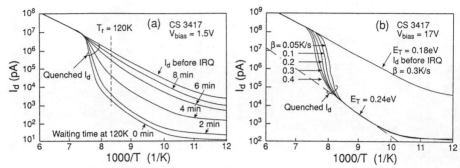

Fig. 4   a)   Thermal recovery of quenched I_d (for 0.15 eV center) as a function of the waiting time at 120 K and

b)   Thermal recovery of quenched I_d as a function of heating rate, after IR quenching using $h\nu \leq 1.12$ eV illumination and thermally cleaning out the TSC signals.

I_d is increased by more than two orders of magnitude as the waiting time is increased up to 8 min, resulting in a shift of the quenched I_d curve towards the initial I_d curve. In other words, the quenched fraction of the 0.15–eV center is reduced by thermal recovery at 120 K.

The thermal recovery process of the quenched I_d can also be demonstrated in a second way, i.e., by measuring the quenched I_d vs. T at different heating rates, as shown in Fig. 4b. From this figure, we find that the recovery temperature is increased from ~ 130 K to ~ 140 K, as the heating rate is increased from 0.05 to 0.4 K/s. This type of curve shift is typical of a thermally stimulated process, such as the thermal recovery of the photocurrent in bulk SI–GaAs due to the transformation of EL2 from its quenched (or metastable) state to the normal state [14] or the thermal recovery of the hopping conduction observed in low–temperature (LT) MBE GaAs due to the transformation of the As_{Ga}–related center from its quenched state to the normal state [15].   A theoretical fit to the experimental data in Fig. 4b [16], gives two important results: (1) the quenched (metastable) state has an electronic transition energy about 0.14 eV deeper than the ground (normal) state at $E_c - 0.15$ eV; and (2) the quenched state thermally recovers by an Auger–like process at a rate $r = \sigma_n\, n\, v_n\, \exp\,(-E_b/kT)$ with $\sigma_n = 2.3 \times 10^{-12}$ cm^2 and $E_b = 0.18$ eV, where n and v_n are the electron concentration and thermal velocity, respectively. The σ_n and E_b parameters can be compared with the same parameters found for EL2, i.e., $\sigma_n = 2 \times 10^{-14}$ cm^2 and $E_b = 0.11$ eV [8]. It may be significant that the Auger rates for EL2 and the 0.15–eV center are within an order of magnitude of each other over the entire transition range, 120–140 K.

Using long–term light at 82 K with different photon energies (0.8 eV < $h\nu$ < 1.45 eV), but the same light intensity (6 × 10^{14} phot./cm^2 s), the 0.15–eV center is found to be quenched in different amounts. Figure 5 shows the spectral dependence for the normalized quenching magnitude of the 0.15–eV center, which is determined by the ratio of the dark currents at 120 K (a temperature well below the full recovery temperature) before and after photoquenching. For comparison, we also show a spectral dependence for the normalized quenching magnitude of EL2 in SI–GaAs, which is determined by the ratio of the photocurrents before and after 5 min photoquenching at 82 K, using the same light as that in the 0.15 eV–center quenching experiments. It is found that the optimal quenching energies for the 0.15–eV center and EL2 are very close to each other, i.e., 1.15 eV vs. 1.10 eV, respectively. However, two additional features at higher and lower energies (about 1.40 eV and 0.95 eV, respectively) are observed for the 0.15–eV center. A similar result in the spectral dependence of absorption quenching for the As_{Ga}–related defect in LT MBE GaAs, with a peak efficiency at 1.40 eV, has been reported by Kaminska and Weber [17]. A comparison of the spectral dependences of the photoquenching for these various defects (0.15–eV center, EL2, and the As_{Ga}–related center in LT MBE GaAs) in different materials (bulk conducting GaAs, SI–GaAs, and LT MBE GaAs) indicates that the 0.15–eV center might be related to As_{Ga}.

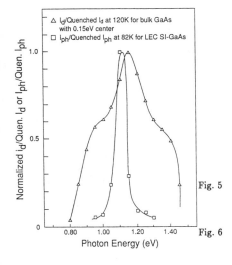

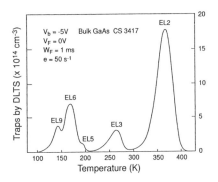

Fig. 5 Normalized spectral dependence of the photoquenching magnitude for the quenched I_d in bulk GaAs
with the 0.15–eV center ($\triangle$), and the quenched I_{ph}
in LEC SI–GaAs with the EL2 center ($\square$).

Fig. 6 Traps EL2, EL3 and EL6, observed by the DLTS
technique in bulk GaAs with the 0.15–eV center (n
@ 300 K = 2×10^{15} cm^{-3} and n @ 150 K = 2×10^{12}
cm^{-3} from TDH measurements).

To confirm the existence of EL2 in bulk GaAs containing the 0.15–eV center, DLTS measurements were performed using Schottky barrier diodes with evaporated Au contacts on the
top side of the materials. The DLTS spectrum for sample CS 3417 shown in Fig. 6 is typical of
that in bulk GaAs samples with electron concentrations at 300 K in the mid–10^{14} cm^{-3} range.
Similar to the case of common bulk HB GaAs wafers with shallow impurity levels and electron
concentrations in the low–10^{16} range, bulk GaAs with the 0.15–eV center also has several
well–known electron traps, such as EL2, EL3, and EL6. The EL2 concentration in CS 3417 is
higher than 1.5×10^{15} cm^{-3}, while the EL6 concentration seems at first glance to be lower than
1×10^{15} cm^{-3}. However, the lower apparent EL6 concentration is not accurate, since the
electron concentration at the EL6 peak temperature ($\sim$ 150 K) is only in the 10^{12}–cm^{-3} range,
which leads to n $\ll$ N_T (trap concentration), an improper condition for determining trap
concentration by DLTS measurements in the usual commercial instruments. Actually, according to Siegel et al., the concentration of EL6 in an undoped, medium–resistivity LEC GaAs
sample controlled by a center at E_c − 0.12 eV, can be as high as 1×10^{16} cm^{-3}, when C–V
measurements are corrected for the condition $N_T > n$ [2]. We did not conduct DLTS measurements on the undoped as–grown MBE GaAs or on the irradiated MBE GaAs samples.
However, studies of deep levels in undoped, as–grown MBE GaAs show that the dominant
electron traps are M1 (E_c − 0.21 eV), M3 (E_c − 0.33 eV) and M4 (E_c − 0.52 eV) with trap
concentrations in the 10^{13} cm^{-3} range or less, and the EL2 is usually not detected [18].
Low–dose EI studies on pure n–type MBE GaAs reveal the deep centers E1 (E_c − 0.045 eV), E2
(E_c − 0.15 eV) and E3 (E_c − 0.30 eV), but no EL2 [19]. Therefore, we believe that the main
difference between the bulk sample CS 3417 and the irradiated MBE sample is the existence of
EL2 (As$_{Ga}$) in the bulk sample, but not in the irradiated sample.

Based on the above experimental results, i.e., (1) the comparison of the IR quenching
between the 0.15–eV centers in two different materials (bulk GaAs with the 0.15–eV center and
irradiated MBE with the E2 center at 0.15 eV), and (2) the comparison of IR quenching and
thermal recovery characteristics between the 0.15–eV center in bulk n–GaAs and EL2 in SI
GaAs, we argue that the 0.15–eV center in bulk GaAs is not a simple As vacancy, but a combination of V$_{As}$ and As$_{Ga}$, such as the nearest–neighbor As$_{Ga}$ − V$_{As}$ center, which has been studied
theoretically by Baraff and Schluter (B–S) [7, 20]. From binding and formation energy
calculations, they show that "nearest–neighbor As$_{Ga}$ − V$_{As}$ pairs should be an abundant defect
in GaAs." Indeed, the only deep donors of high enough concentration to control E_F in bulk

GaAs are this one at $E_c - 0.15$ eV, another at $E_c - 0.43$ eV (unknown), and EL2 at $E_c - 0.75$ eV. All of our 0.15 eV samples also contain EL2, with concentrations about 10^{15} cm^{-3}, as determined by DLTS. Thus, As_{Ga} centers are available to form the $As_{Ga} - V_{As}$ complexes. According to B–S [20], nearest–neighbor ($As_{Ga} - V_{As}$) has three transitions within the bandgap: $(0/+)$, $(+/++)$, and $(++,+++)$, all of them associated with V_{As}–like (not As_{Ga}–like) wave-functions. Although the $(0/+)$ energy is shallower than 0.15 eV, and the $(+/++)$ energy deeper, still we believe that either is within theoretical uncertainty of 0.15 eV. They also find that the $(+)$ state has a metastable configuration, with the As atom moving to a position about 35% of the distance between the original As_{Ga} and V_{As} positions, and they show that IR illumination could induce this metastability by promoting an electron to an excited $(+)$ state. Such a transition is consistent with our IR quenching experiments. Also, they find that the metastable–state $(+/++)$ transition falls about 0.2–0.3 eV below the ground–state $(+/++)$ transition, which is certainly within error of our value of 0.14 eV [16]. Although some dif-ferences in the IR quenching characteristics for the 0.15–eV center and EL2 can be observed, the Auger–like thermal recovery rates for the two centers are very similar to each other. It is perhaps not unexpected that the recovery rate for $As_{Ga} - V_{As}$ should be close to that for As_{Ga} alone (for simplicity, assuming EL2 to be an isolated As_{Ga}), because the metastable states are basically $V_{Ga} - As_i - V_{As}$ and $V_{Ga} - As_i$ for ground states $As_{Ga} - V_{As}$ and As_{Ga}, respectively, and the $V_{Ga} - As_i$ distances are similar [7]. The only serious problem with a nearest–neighbor $As_{Ga} - V_{As}$ assignment for our 0.15–eV center is that in n–type material, the As atom should move completely over to the V_{As} position (forming V_{Ga}) and not stop at the metastable posi-tion. Thus, our center may not be nearest–neighbor $As_{Ga} - V_{As}$ although we still believe that it must involve both As_{Ga} and V_{As}.

In summary, we have directly observed a metastable state of the well–known $E_c - 0.15$ eV defect in bulk GaAs. Many of the properties of this defect are similar to these of the $As_{Ga} - V_{As}$ center, elucidated theoretically by Baraff and Schluter and predicted to be abundant in GaAs.

We would like to acknowledge T. A. Cooper for Hall–effect measurements and N. Blair for manuscript preparation. Z–Q. F. was supported under ONR Contract N00014–90–J–11847, D.C.L. was supported under USAF Contract F33615–91–C–1765.

REFERENCES

1. D. C. Look, D. C. Walters, and J. R. Meyer, Solid State Commun. 42, 745 (1982).
2. W. Siegel, G. Kühnel, H. A. Schneider, H. Witte, and T. Flade, J. Appl. Phys. 69, 2245 (1991).
3. W. C. Mitchel, G. J. Brown, and L. S. Rea, J. Appl. Phys. 71, 246 (1992).
4. D. Pons and J. C. Bourgoin, J. Phys. C 18, 3839 (1985).
5. D. C. Look, Appl. Phys. Lett. 51, 843 (1987).
6. D. C. Look, P. W. Yu, W. M. Theis, W. Ford, G. Mathur, J. R. Sizelove, D. H. Lee, and S. S. Li, Appl. Phys. Lett. 49, 1083 (1986).
7. G. A. Baraff and M. Schluter, Phys. Rev. Lett. 55, 2340 (1985).
8. A. Mitonneau and A. Mircea, Solid State. Commun. 30, 157 (1979).
9. Z–Q. Fang and D. C. Look, Appl. Phys. Lett. 59, 48 (1991).
10. Z–Q. Fang and D. C. Look, Mater. Sci. Forum Vol. 83–87, 991 (1992).
11. T. Hariu, T. Sato, H. Komori, and K. Matsushita, J. Appl. Phys. 61, 1068 (1987).
12. D. W. Fischer, Phys. Rev. B37, 2968 (1988).
13. Z–Q. Fang and D. C. Look, J. Appl. Phys. 73, 4971 (1993).
14. Y. N. Mohapatra and V. Kumar, J. Appl. Phys. 64, 956 (1988).
15. D. C. Look, Z–Q. Fang, and J. R. Sizelove, Phys. Rev. B47, 1441 (1993).
16. D. C. Look, Z–Q. Fang, and J. R. Sizelove, to be published.
17. M. Kaminska and E. R. Weber, Mater. Sci. Forum Vol. 83–87, 1033 (1992).
18. D. V. Lang, A. Y. Cho, A. C. Gossard, M. Ilegems, and W. Wiegmann, J. Appl. Phys. 47, 2558 (1976).
19. A. Mircea and D. Bois, Inst. Phys. Conf. Ser. 46, 82 (1979).
20. G. A. Baraff and M. Schluter, Phys. Rev. B33, 7346 (1986).

DEFECT REDUCTION BY THERMAL CYCLIC GROWTH IN GaAs GROWN ON Si BY MOVPE

W. KÜRNER, R. DIETER, K. ZIEGER, F. GORONCY, A. DÖRNEN, and F. SCHOLZ

4. Physikalisches Institut, Universität Stuttgart,
Pfaffenwaldring 57, D-70550 Stuttgart, Germany

ABSTRACT

The growth of GaAs epilayers on Si should combine the advantages of both materials. The lattice mismatch and the difference in thermal expansion coefficients, however, result in the yet unsolved problems of high dislocation density and thermal stress in the GaAs layer. Recently, considerable improvements have been achieved by a 'thermal cyclic growth' (TCG) process. In this study we focus on the reduction of high defect concentration and dislocation density. The improvement of the epilayer quality is verified by DLTS, PL and DCXD. Results of TEM and DLTS measurements lead to the identification of a dislocation related defect.

INTRODUCTION

The heteroepitaxial growth of GaAs on Si substrates has gathered a lot of interest in recent years [1]. The ability of combining the complementary properties of silicon and GaAs offered new ways to fabricate optoelectronic and high speed devices as well as high efficiency solar cells. But due to the lattice mismatch (4.1%) and the difference in the thermal expansion coefficient of around 60% high dislocation density in the order of 10^8 cm^{-2} and thermal strain of around 10^{-3} restricts the applicability of such structures. Several efforts were made to overcome these problems. A buffer layer between the substrate and the GaAs epilayer [2], strained layer superlattices [3], and in situ [4] or post growth annealing [5] were applied to reduce the dislocation density and the thereby resulting high defect concentration.

An alternative course is the so called *thermal cyclic growth* (TCG) process [6]. This is an alternating low temperature growth and a high temperature annealing step performed prior to the growth of the active epilayer. We analyse the particular steps of the TCG process and demonstrate the mechanism of the dislocation reduction as well as the decreasing defect concentration.

EXPERIMENTAL DETAILS

The samples used in this study have been grown by low pressure metal-organic vapor phase epitaxy (LP-MOVPE) in an *Aixtron* system *AIX 200* with TMGa and AsH$_3$ as precursors. For the heteroepitaxial growth of GaAs 2° misorientated (100)-Si wafers were used as substrates and Si-doped GaAs wafers were used as substrates for homoepitaxial reference samples. At a standard growth temperature of 700°C and a V/III ratio of 235 the growth rate was 2.3 μm/h.

All samples were n-type with an unintentional doping in the order of 10^{16} cm^{-3} in the case of Si substrates and 10^{15} cm^{-3} for GaAs substrates as determined by CV measurements. Deep defect centers were identified by deep level transient spectroscopy (DLTS) [7] with a *Polaron DL 4600 DLTS System*. Bandgap luminescence was recorded with a standard photoluminescence (PL) setup in order to get information about the thermal stress of the layers. The dislocation density was determined by the full width at half maximum (FWHM) of the (400) peak of the rocking curve measured by a double crystal X-ray diffraction (DCXD) setup. Transmission electron microscopy (TEM) images of several samples were made with a *JEOL 2000 FX*.

CONVENTIONAL GROWTH OF GaAs ON Si BY THE 2-STEP-GROWTH METHOD

In 1984 Akijama et al. suggested a method to grow GaAs directly on silicon substrates, the 2-step-growth method [8]. The main idea was to grow a thin nucleation layer at low temperatures (400°C) that will be the initial layer for the growth of the thicker GaAs layer at 700°C. Samples grown by this procedure feature a high dislocation density and thermal stress due to the incompatible properties of Si and GaAs. Figure 1 shows a comparison of DLTS measurements for GaAs grown on GaAs substrates (GaAs/GaAs) and on Si substrates (GaAs/Si). While the GaAs/GaAs sample shows only a single DLTS peak due to the well known EL2 center, the GaAs/Si sample exhibits additional defects hidden under a broad low temperature tail of the EL2 signal. The concentration of the EL2 defect in GaAs/Si is 20 times the concentration in the GaAs/GaAs sample. Also a broadening of the EL2 peak is observed due to the inhomogeneous stress distribution in the layer [9]. The PL spectrum is also affected by the biaxial tension of the GaAs epilayer. The hydrostatic part of the tensile stress shifts the spectrum to lower energies, the uniaxial part causes a splitting of the heavy-hole (hh) and the light-hole (lh) band. Hence, the normally sharp luminescence lines of the donor (DX) and acceptor (AX) bound exciton transitions remain unresolved. Only the Si-donor heavy hole (D_{Si}hh) transition as well as the Si-donor light hole (D_{Si}lh) transition can be observed [9]. The broadening and the intensity of the D_{Si}hh peak were taken as a measure of the quality of the layer.

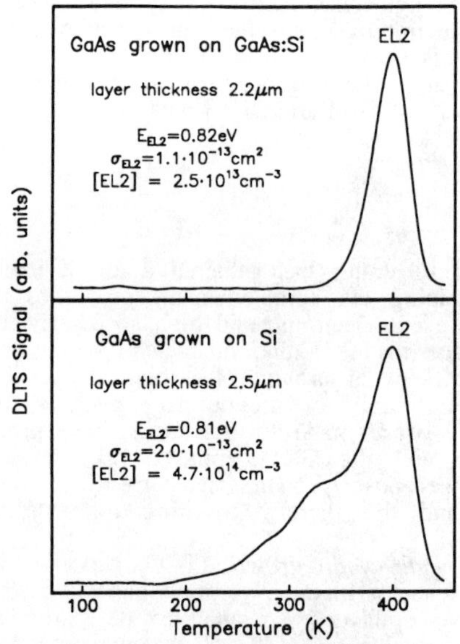

GaAs grown on GaAs:Si   EL2

layer thickness 2.2 μm

$E_{EL2}=0.82$eV
$\sigma_{EL2}=1.1\cdot10^{-13}$cm^2
[EL2] = $2.5\cdot10^{13}$cm^{-3}

GaAs grown on Si   EL2

layer thickness 2.5 μm

$E_{EL2}=0.81$eV
$\sigma_{EL2}=2.0\cdot10^{-13}$cm^2
[EL2] = $4.7\cdot10^{14}$cm^{-3}

DLTS Signal (arb. units)

100   200   300   400
Temperature (K)

Fig. 1: *Comparison of DLTS spectra of homo- and heteroepitaxially grown GaAs.*

THERMAL CYCLIC GROWTH

Depart from the 2-step-growth method [8] we developed a new method, where after 1.2 μm of conventional growth a TCG process is involved. As shown in Fig. 2 the TCG is initiated by a 50 nm layer grown at 400°C followed by a 900°C high temperature annealing step and a 50 nm growth step at 700°C. This procedure is repeated two times completed by a 200°C - 900°C - 200°C thermal cyclic annealing (TCA) step. Thereupon an additional 2.4 μm GaAs layer is grown. The improvement of the layer quality as observed by DLTS is presented in Fig. 3. The DLTS spectrum consists of only a weak EL2 signal, which is in accordance with a concentration as low as $3.5\cdot10^{13}$ cm^{-3} and is thus comparable to the GaAs/GaAs sample. The defect spectrum at the low temperature tail of EL2 has nearly vanished.

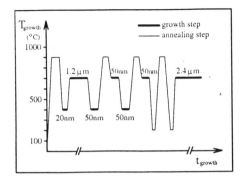

Fig. 2: *Growth procedure of a GaAs/Si sample incuding a TCG process.*

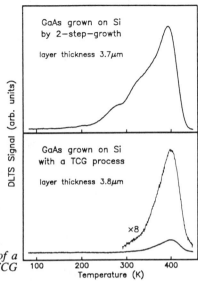

Fig. 3: *Comparison of DLTS measurements of a 2-step-grown sample and a sample with a TCG process.*

The PL intensity increases by a factor of 2 with respect to the 2-step-grown sample while the FWHM of the $D_{Si}hh$ peak decreases from 5.4 meV to 2.0 meV. The DCXD measurement yields a FWHM of the rocking curve of 98 ", which corresponds to a dislocation density of $1.6 \cdot 10^7$ cm^{-2} (211", $7.3 \cdot 10^7$ cm^{-2} for the 2-step-grown sample). The reason for this quality improvement can be seen in the TEM image of Fig. 4. After the TCG process at 1.2 μm layer thickness most of the dislocations have annihilated each other by forming loops whereas in an untreated sample the dislocations run to the surface. Thus in the upper 2 μm of the GaAs layer the quality approaches that of a homoepitaxially grown sample.

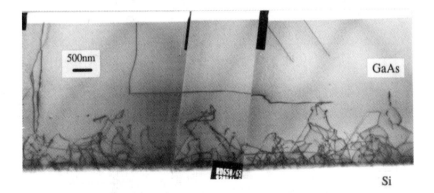

Fig. 4: *TEM image of a TCG sample. The formation of loops at the TCG reduces the dislocation density drastically.*

ANALYSIS OF THE TCG PROCESS

To get information on the effect of each particular step of the TCG process samples were grown where the TCG was slightly modified. In one sample the low temperature growth step was replaced by a layer grown at 700°C. In another sample instead of the 900°C annealing step the temperature was held at normal growth temperature of 700°C. In both cases the high quality of a complete TCG sample has not been achieved. For the sample with the omitted low temperature growth the EL2 concentration is reduced to a value of $7 \cdot 10^{13} \mathrm{cm}^{-3}$ but there is no suppression of the other defects. In the sample grown with no high temperature annealing step these defects have nearly disappeared but the EL2 concentration remains unaffected ($[EL2] = 1.2 \cdot 10^{14} \mathrm{cm}^{-3}$). The dislocation density in both cases is rather high with $3.5 \cdot 10^7 \mathrm{cm}^{-2}$ and the FWHM of the PL is 2.5 meV and 4.0 meV, respectively. Thus only the combination of both steps leads to a remarkable improvement of the layer quality.

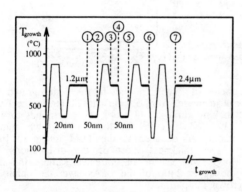

Fig. 5: *Growth termination at different steps ((1) to (7)) of a TCG process*

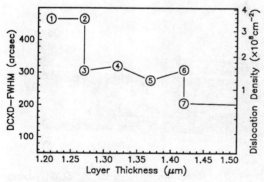

Fig. 6: *Dislocation density after each particular step of a TCG process*

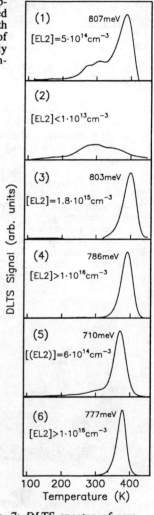

Fig. 7: *DLTS spectra of samples with terminated TCG*

More detailed information on what is happening during the TCG process is achieved by termination of growth after different steps of the TCG process as depicted in Fig. 5. The corresponding DLTS spectra are shown in Fig. 7 together with the concentration and the energy of the respective EL2 peak. The main features are as follows: (i) all samples exhibit the optical quenching of the photocapacitance that serves as the fingerprint for the EL2 center [11]. Also in sample (2) where the EL2 concentration determined by DLTS is below the detection limit of $1 \cdot 10^{13}$ cm^{-3} this quenching effect is observed. (ii) In sample (5) the peak is shifted to lower temperature compared to the other samples. Also the energy of 710 meV departs from the energy of EL2 of 815 meV [12]. Nevertheless, this defect exhibits the quenching effect so that we assume this defect belonging to the EL2 family [13]. This observation may also be the explanation of the observed quenching effect in sample (2) where at the position of this peak a small signal is observed which may be due to the same defect. (iii) The smaller low temperature shift in sample (4) and (6) is caused by the high concentration of EL2 with respect to the shallow donor concentration ($N_{EL2}/N_D \approx 0.3$). (iv) The effect of the low temperature growth ((1)→(2), (4)→(5)) is the reduction of EL2, whereas the following annealing step and the 50 nm 700°C growth ((2)→(3), (5)→(6)) favor the formation of EL2. This behavior is confirmed by data of homoepitaxial GaAs, where growth temperatures below 600°C inhibit EL2 formation and succeeding annealing at 700°C introduces this center [14]. (v) Figure 6 presents the effect of the single steps on the dislocation density. As clearly can be seen, directly after an annealing step ((2)→(3), (6)→(7)) the dislocation density decreases drastically. From all these observations we conclude that the low temperature layer is of minor quality with a lot of dislocations. By increasing the temperature to 900°C these dislocations interacts with those from the 1.2 µm layer and annihilate each other by forming loops. This effect is very efficient since repeating this procedure a second time has only little effect on the quality. The final TCA step reconstructs the surface of the layer and hence forms the basis of the following high quality growth of the thicker GaAs layer.

DISLOCATION RELATED DEFECT IN GaAs

The abrupt decreasing of the dislocation density at the TCG as shown in Fig. 4 was used to study the effect of dislocations on the defect spectrum at the low tem-
perature tail of EL2. For that purpose a sample was etched down to 2 µm thickness so that the space charge region of a Schottky contact reaches the TCG region with varying bias voltage. Since the Si concentration at the three annealing steps of the TCG is rather high, as confirmed by SIMS, a CV measurement will detect a strongly increasing donor concentration when the space charge region hits the TCG. Fig. 7 demonstrates the existence of a defect that is located directly at the TCG (curve (a)). 150 nm above the TCG this defect cannot be observed (curve (b)). This also ensures that this defect is not due to the etching process but results from the high dislocation density. The energy of the defect cannot be determined exactly by a DLTS measurement at dif-

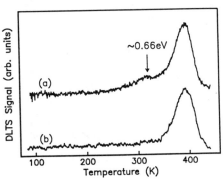

Fig. 8: DLTS measurement direct at the TCG (a) and 150nm above the TCG (b). The defect at 0.66eV can only be observed in measurement (a).

ferent ratewindows since the maximum of this small peak disappears in the background for smaller ratewindows. Hence a computer based analysis of the spectrum was applied to ascertain the energy of the defect level by subtracting the EL2 signal and the signal of other known defects [15] from the spectrum. The peak of the dislocation related defect has been reconstructed with the result of $E = (0.66 \pm 0.01)\,meV$ and $\sigma = (2.4 \pm 0.5) \cdot 10^{-13}cm^2\,cm^2$.

SUMMARY

In this study we report a detailed investigation of high quality GaAs layers on Si substrates grown with a TCG process. The mechanism of dislocation reduction and the decreasing of the defect concentration is studied by DLTS and DCXD. A detailed analysis of the particular steps of a TCG process together with TEM images leads to a better understanding of the mechanism of dislocation annihilation via formation of loops. DLTS measurements in the region with strongly increasing dislocation density reveals a dislocation related defect at 0.66 eV.

Acknowledgement

The authors wish to thank B. Roos (MPI-MF Stuttgart) for providing the TEM images and N. Draidia for valuable discussions. We also acknowledge M. Pilkuhn for his steady interest in this work. The financial support of the Bundesministerium für Forschung und Technologie (BMFT) under contract number 032 8861 A is gratefully acknowledged.

REFERENCES

[1]   *Heteroepitaxy on Silicon* (Mater. Res. Soc. Symp. Proc. Pittsburgh, PA, Vol. 67 (1986), Vol. 91 (1987), and Vol. 116 (1988))
[2]   R.D. Bringans, D.K. Biegelsen, L.-E. Swartz, F.A. Ponce, J.C. Tramonta Appl. Phys. Lett. **61**(2), 195 (1992)
[3]   References in P. Demeester, A. Ackaert, G. Coudenys, I. Moerman, L. Buydens, I. Pollentier, P. Van Daele, Prog. Crystal Growth and Charact. **22**, 53 (1991)
[4]   J.W. Lee, H. Shichijo, H. Tsai, R.J. Matyi, Appl. Phys. Lett. **50**(1), 31 (1987)
[5]   J.E. Ayers, L.J. Schowalter, S.K. Gandhi, J. Cryst. Growth **125**, 329 (1990)
[6]   R.J. Dieter, F. Goroncy, J.P. Lay, N. Draidia, K. Zieger, W. Kürner, B. Lu, F. Scholz, B. Roos, M. Braun, V. Frese, J. Hilgarth, Proc. 11[th]EC PVSEC, Montreux, Switzerland 1992, p.225
[7]   D.V. Lang, J. Appl. Phys. **45**(7), 3023 (1974)
[8]   M. Akijama, Y. Kawarada, K. Kaminishi, J. Cryst. Growth **68**, 21 (1984)
[9]   T. Soga, S. Sakai, M. Umeno, S. Hattori, Jap. J. Appl. Phys. **25**, 1510 (1986)
[10]   W. Stolz, F.E.G. Guimaraes, K. Ploog, J. Appl. Phys. **63**(2), 492 (1988)
[11]   G. Vincent, D. Bois, Solid State Communication **27**, 431 (1978)
[12]   J. Lagowski, D.G. Lin, T. Aoyama, H.C. Gatos, Appl. Phys. Lett. **44**, 336 (1984)
[13]   M. Taniguchi, T. Ikoma, J. Appl. Phys. **54**(11), 6448 (1983)
[14]   S.H. Xin, W.J. Schaff, C. Wood, L.F. Eastman, Appl. Phys. Lett. **41**, 742 (1982)
[15]   W. Kürner, unpublished

DEEP LEVELS INDUCED BY SiCl4 REACTIVE ION ETCHING IN GaAs

N.P. JOHNSON*, M. A. FOAD**, S. MURAD*, M. C. HOLLAND* AND C. D. W. WILKINSON*; *Department of Electronics and Electrical Engineering, The University, Glasgow G12 8QQ, UK; **Department of Electronic and Electrical Engineering, University of Salford, Salford M5 4WT, UK.

ABSTRACT

Deep Level Transient Spectroscopy (DLTS) is used to investigate the effect of SiCl4 Reactive Ion Etching (RIE) on GaAs. At high power (150-80 W) with high DC self bias (380-240 V), five deep levels trapping electrons are observed at energies of 0.30, 0.42, 0.64, 0.86 and ~0.8 eV below the conduction band edge. Depth profiling reveals an approximate exponential decay of the concentration of the deep levels. At low power the induced concentration falls, the small concentration of remaining deep levels is close to control (no etching) samples. The induced deep levels can account for reduced conductances in n+GaAs wires defined by RIE under similar experimental conditions.

INTRODUCTION

Reactive ion etching (RIE) is an important method of pattern transfer in semiconductors for both conventional devices and nanostructures[1]. Its advantage over wet etching is an ability to produce high aspect ratio near vertical structures. A common gas used in the RIE of GaAs is SiCl4. Under certain conditions the etching process can induce damage into the semiconductor[2]. Damage may be measured by a variety of techniques such as TEM[3] and Raman Scattering[4]. The effect on electrical properties has been measured by Foad et al[5] and Thoms et al[6] where the conductances of RIE etched wires were measured as a function of wire depth and width. Lootens et al[7] showed that RIE with SiCl4 produces an electron trap in GaAs which is not present in wet etched material. A model by Raman et al[2] based upon a phenomenological defect creation parameter with a characteristic length has been used to fit wire conductances. Also reported were five electron trap energies caused by the etching process.

In this paper we present the trap signatures for the induced traps measured by Deep Level Transient Spectroscopy[8] (DLTS). The ability of DLTS to depth profile is exploited to show differences in trap concentrations between high and low power etching and unetched samples. Under the low power conditions we show the damage as indicated by the presence of traps, is substantially reduced but the etching process still produces vertical side wall structures[9]. We also discuss the damage mechanism.

EXPERIMENT

To investigate the damage caused by SiCl4 RIE on GaAs, 0.5 μm of GaAs Si doped at 6 $\times 10^{16}$ cm^{-3} was grown on n+ substrates by MBE in a Varian Gen II modular system. Ohmic contacts were formed with Au/Ge/Ni/Au to the n$^+$ substrate. The wafer was cut into samples and etched at powers of 150 W, 80 W and 12-15 W, (corresponding to power density of 0.66, 0.35 and 0.053-0.066 Wcm^{-2}) unetched control samples were also prepared. The DC self bias was 380 V and 240 V in the high power samples and 48-60 V at low power. Ti/Au Schottky contacts, 200 μm diameter, were formed by photolithography and lift-off. The diodes were bonded to headers for measurement. I - V characteristics were measured to check reverse leakage current and barrier heights. The DLTS measurements were performed on a Biorad DLTS spectrometer in the temperature range 80 - 450 K. The positions of the peaks were determined by two temperature scans, first starting above the peak and scanning down, followed by a scan up to eliminate the small temperature hysteresis ~3 K. This system uses a boxcar signal averager to measure the capacitance transient. The fill pulse width was varied according to the rate window to give equal excitation and measurement times.

RESULTS

I High Power Etched GaAs

Arrhenius plots of deep levels induced by high power etching are shown in figure 1. The conventional correction for temperature dependence of the density of states and thermal velocity is made on the abscissa. Deep levels identified in the literature with similar Arrhenius plots are shown for comparison.

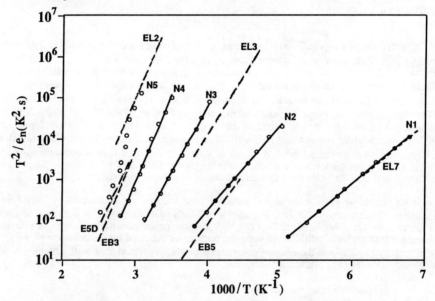

Figure 1 Arrhenius plots of deep levels N1 -N5 induced in GaAs by SiCl$_4$ RIE at high power 150 W (open circles and full lines). Deep levels in literature (broken lines).

The activation energies E_{na}, and apparent cross sections σ_a, of the deep levels the so called trap signatures are shown in table I. In some cases the discrepancy between the values quoted in the table from the literature and this work is large, nonetheless the best comparison is in the proximity of the Arrhenius plots as this allows comparison with directly measured peak position. The uncertainty in the values represent the variation over typically three different Schottky diodes. Both leakage current and field emission of carriers from traps are known to effect trap signatures. The leakage current from the diodes at 0.5 V reverse bias measured in the dark was in the range of 130-500 pA. These measurements were taken at low to medium reverse bias of 0.5 V over a depletion region of 70 nm to minimise field enhanced emission.

Table I Deep Level signatures measured from Arrhenius plots

Label	E_{na} (eV)	σ_a (cm^2)	Possible Comparison	E_{na} (eV)	σ_a (cm^2)	Ref
N1	0.296±0.003	4.9±0.8 x10^{-15}	EL7	0.30	7.2 x 10^{-15}	10
N2	0.416±0.006	6±2 x10^{-15}	EB5	0.48	2.6 x 10^{-13}	10
N3	0.641±0.009	6±3 x10^{-13}	EL3	0.575	(0.5-2.0) x 10^{-13}	10
N4	0.862±0.009	7±4 x 10^{-11}	EB3	0.90	3.0 x 10^{-11}	10
N5 low T	0.75±0.06	7 x 10^{-14}	EL2	0.825	(0.8-1.7) x 10^{-13}	10,11
N5 high T	0.85±0.07	6 x 10^{-12}	E5D	0.76	2.3 x 10^{-13}	11
N5 Low Power	0.69±0.03	4 x 10^{-14}	E5D	0.76	2.3 x 10^{-13}	11

Of the electron traps cited EL7, EB5 and E5D (the D suffix is used to distinguish the trap E5 measured by Day et al) are associated with as grown MBE material. EL3 and EL2 are associated with vapour phase epitaxy material, though it is well known that EL2 may be induced by thermal annealing above 500 K. The trap EB3 is caused by electron irradiation. The traps N1-N4 measured here all show normal exponential behaviour. The non-exponential behaviour of trap N5 is evident in figure 1, it follows EL2 at low measurement temperatures and E5D at high temperature. This is either caused by interference from another deep level, or transition of the material to another state which induces a different deep level. A further trap is observed but is out of range of measurement at temperatures above that for N5 and will not be discussed. The traps N1 - N5 are also present in material etched at medium power of 80 W.

2 Low Power Etched GaAs and control samples

Figure 2 shows an expanded portion of the Arrhenius plots shown in figure 1 overlaid with deep levels from low power etching and two control samples. The control (unetched) sample show only one trap present coincident with N4. The low power etched sample has two deep levels, one coincident with N4 present in both control samples and high power etched samples. The second trap follows the high temperature portion of N5 but displays normal exponential behaviour. This supports the hypothesis that N5 as measured on the high power samples is composed of two different deep levels.

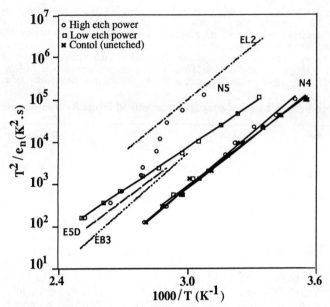

Figure 2 Arrhenius plot for various samples.

<u>3 Calculation of Depth Profiles</u>

By varying the pulse height and quiescent reverse bias and measuring the DLTS peak height, a depth profile can be constructed. The most important correction to the trap concentration N_T given in a corresponding expression by Sah,[12]

$$N_T = ((C_\infty / C_0)^2 - 1)N_D \qquad (1)$$

where C_0 is the capacitance change at t=0, C_∞ is the capacitance at infinite time after the fill pulse, N_D the donor concentration; is a correction factor given by Bleicher and Lange[13] which takes account of the depth over which deep levels take part in the measurement. The region is analogous to the depletion zone W, and is the distance from the edge of the depletion zone to were the Fermi level E_F crosses the energy of the deep level E_T under each bias condition V. The distance Λ, is given by,

$$\Lambda = \sqrt{\frac{2\varepsilon_r\varepsilon_0(E_F - E_T)}{q^2 N_D}} \qquad (2)$$

where q is the electronic charge, and ε_r and ε_0 are the relative permitivity, and permitivity of free space. The trap concentration including the correction factor is,

$$N_T = \frac{((C_\infty / C_0)^2 - 1)N_D}{(\sqrt{(E_F - E_T)}qV(C_\infty / C_0) - 1)^2} \qquad (3)$$

446

The distance plotted on in figures 3 (a) and (b) is the mid point of the $W_1 - \Lambda_1$ and $W_2 - \Lambda_2$ where the subscripts represent the upper and lower voltages of the fill pulse. A further correction for the abrupt depletion approximation was not applied, this may underestimate the concentration in the worse case of high temperature at low reverse bias[14].

The depth profile of the high power etched sample is shown in figure 3(a). The reproducibility of a single point of the depth profile point is typically 5%, though for different diodes on different samples this is typically 15% as shown by the two control samples in figure 3 (b).

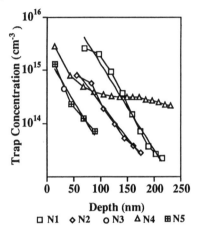

Figure 3(a) Trap concentration in high power etched sample.

□ N1 ◇ N2 ○ N3 △ N4 ⊞ N5

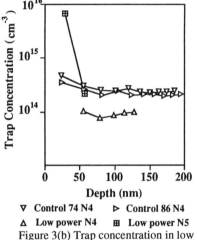

Figure 3(b) Trap concentration in low power and control (unetched) samples.

▽ Control 74 N4   ▷ Control 86 N4
△ Low power N4   ⊞ Low power N5

The straight lines in figure 3(a) give good least squares fits to an exponential function of the concentration for three of the five traps N1, N2 and N5 at 0.30 0.42 and 0.8 eV. The trap N3 at 0.64 has only one measurable point. The deep level at N4 at 0.86 eV shows an initial exponential decrease with depth and then a plateau at 2.3×10^{14} cm^{-3}. Referring to figure 3(b) control samples contain the trap N4 at this concentration and allows us to account for the non-exponential decrease in concentration as it approaches a common plateau. Broadly the data shows, the shallower the energy of the trap the deeper the physical penetration into the lattice.

The low power etched samples differ from the high power in three remarkable ways. First there is an absence of traps N1 N2 and N3, secondly the trap N5 coincident with N5 at high temperature in the high power etch sample has increased in concentration towards the surface. The trap N4 present in high and low power etched samples and control sample has apparently decreased in concentration from that present in the control samples.

<u>Discussion of damage mechanism</u>

The penetration of ions into the substrate by ion bombardment is small according to energy loss calculations. At low energies i.e. up to 150 eV the creation of vacancies and phonons all occur within 3 nm of the surface.
A possible explanation of the deep level distribution is that the surface is disordered by RIE (both high and low power etched samples show increased deep levels toward the surface). A common phenomena on the surface of metals[15] refactory materials[16] and semiconductors[17] is

the formation of ridges, steps and terraces when the material is heated. In some cases this is accompanied by adsorption of gases. These features observed by LEED, RHEED and STEM can range from single atom displacements to macroscopic features. The features form after an activation energy has been surmounted to produce a surface with a lower free energy. The etching process also provides energy to the surface in addition to chemical species. If the etched surface is in a different free energy state than the bulk of the lattice, or indeed an unetched surface, this transition from surface to bulk could be accommodated by point defects forming deep levels, therby relaxing the energy difference from surface to bulk material.

Relation to conductance cut off in dry etched wires.

It has been shown elsewhere[2] that a phenomenological model can account for the conductances in dry etched wires formed from selective masking of GaAs. Briefly, point defects are assumed to be generated from a source function and fall off over a characteristic length. For unmasked material the top surface is moving as it is etched. Similarly the side walls of wires formed by masking selective areas receive an integrated dose of ion bombardment (more at the side wall near the masked surface, less near the newly exposed side wall). Although the conductances of the wires can be successfully modeled with only two parameters, these represent the integrated contribution of the different traps.

Conclusions

Five (possibly six) deep levels have been measured and their signatures compared with traps in the literature. When the power density is reduced from 0.66 Wcm^{-2} to 0.053-0.066 Wcm^{-2} the traps N1, N2 and N3 disappear. At high power e.g. at the 3×10^{14} cm^{-3} concentration, N1 extends to ~140 nm, N2 to ~100 nm and N5 ~40 nm, N4 does not fall below 3×10^{14} cm^{-3} within the measured range. The low power etch shows an increase in concentration at the surface of the trap N5, however this falls off more quickly. Also the concentration of deep level N4 is apparently reduced from that of the control sample. All trap concentrations are below the 10^{14} cm^{-3} level at ~60 nm. These changes suggests a rearrangement of atoms near the surface under the different etch conditions which cause the point defects or deep levels.

The authors gratefully acknowledge R. Darkin and D. Clifton for assistance with RIE. This work is funded by SERC under Grant Nos. GR/F 31472 and GR/E 18186.

References

1   A R Long, M Rahman, I K MacDonald, M Kinsler, S P Beaumont, C D W Wilkinson and C R Stanley Semicod. Sci Technol. **8** 39-44 (1983)

2   M Rahman, N P Johnson, M A Foad, A R Long, M C Holland and C D W Wilkinson Appl. Phys. Lett. **61** (19) 2335-2337 (1992)

3   M.A. Foad, S. Hefferman, J.N. Chapman and C.D. W. Wilkinson 1991Gallium Arsenide and Related Compounds (IOP Conf. Ser 112) p293

4   R. Cheung, S. Thoms, M. Watt, M.A. Foad, C.M. Sotomayor-Torres, C.D.W. Wilkinson, U.J. Cox, R.A. Cowley, C. Dunscombe and R.H. Williams, Semicond. Sci. Technol. **7** 1189-1198 (1992)

5   M.A. Foad, S. Thoms and C.D.W. Wilkinson, J. Vac. Sci. Technol. B **11** (1) p20-25 (1993)

6   Thoms S, Beaumont S P, Wilkinson C D W, Frost J, Stanley C R 1986 Microelectr. Eng **5** p249

7   D Lootens, P. Van Daele, P Demeester, and P. Clauws J. Appl. Phys. **70**, 221 (1991)

8   D V Lang J. Appl. Phys., **45** 3023 - 3033 (1974)

9   Murad S.K., C.D.W. Wilkinson, P.D. Wang, W. Parkes, C.M. Sottomayer-Torres And N. Cameron. Very low damage etching of GaAs, Presented at the 37th. Int. Symp. on Electron, Ion and Photon Beams, 1-4 '93

10   G. M. Martin, A. Mitonneau and A. Mircea, Electronics Letters **13** (7) 191-192 (1977)

11   D S Day, J D Oberstar, T J Drummond, H Morkoc, A Y Cho, and B G Streetman, J Elect. Mater. **10**, 445-453 (1981)

12   C.T. Sah, L. Forbes, L.L. Rosier and A.F. Tasch, JR. Sol. Stat. Elect. **13** 759-788 (1970)

13   M. Bleicher and E. Lange, Solid State Electronics **16**, 375-380 (1973)

14   P.I. Rockett and A.R. Peaker, Appl. Phys. Lett. **40** (11), 957-959   (1982)

15   M. Flytzani-Stephanopoulos and L D Schmidt Prog. Surf. Sci. **9** p83 -111 (1979)

16   J. -K. Zuo and D. M. Zehner Phys. Rev. B **46**, 16122-16127 (1992)

17   G. A. Somorjai and M. A. Van Hove, Prog. Surf. Sci. **30** p201-231 (1989)

BAND EDGE OPTICAL ABSORPTION OF MOLECULAR BEAM EPITAXIAL GaSb GROWN ON SEMI-INSULATING GaAs SUBSTRATE

M. SHAH*, M.O. MANASREH**, R. KASPI**, M. Y. YEN**, B. A. PHILIPS***, M. SKOWRONSKI***, AND J. SHINAR‡.
*AIR FORCE INSTITUTE OF PHYSICS, AFIT/CTRD, WRIGHT-PATTERSON AFB, OH 45433
**Wright Laboratory, Wright-Patterson Air Force Base, Ohio 45433.
***Materilas Engineering Department, Carnegie-Mellon University, Pittsburgh, PA 15213
‡Ames Laboratory and Physcis laboratory, Iowa State University, Ames, IA 50011

ABSTRACT

The optical absorption of the band edge of GaSb layers grown on semi-insulating GaAs substrates by the molecular beam epitaxy (MBE) technique is studied as a function of temperature. A free exciton absorption peak at 0.807 eV was observed at 10 K. The free exciton line is observed in either thick samples (5 μm thick) or samples with ~0.1 μm thick AlSb buffer layers. The latter samples suggest that the AlSb buffer layer is very effective in preventing some of the dislocations from propagating into the MBE GaSb layers. The fitting of the band gap of the GaSb layers as a function of temperature gives a Debye temperature different than that of the bulk GaSb calculated from the elastic constants.

INTRODUCTION

There is an increasing interest in type II superlattices based on InAs/InGaSb for very long wavelength infrared detection.[1] The optical transitions in this class of materials are between the lowest miniband in the valence band well to the lowest miniband in the neighboring conduction band well. When the quantum confinement effects are small, band offset can produce band-to-band transition energies less than the bulk band gap of the constituent materials. However, there are a few problems preventing the full realization of the theoretical prediction of the above quantum structure. One of these problems is the choice of the substrate on which the superlattices are grown. In order to introduce the desired strain in InAs/InGaSb type II superlattices, the lattice constant of the substrate should be between those of InAs and InSb. GaSb substrate is one of these materials that satisfies the above condition. The substrate should also contain a low free carrier concentration in order to permit the interband optical absorption measurements in the far-infrared spectral region. Unfortunately, the techniques used to grow bulk GaSb substrates were unable to produce semi-insulating materials.

Mat. Res. Soc. Symp. Proc. Vol. 325. ©1994 Materials Research Society

In addition, the free carrier absorption in all GaSb substrates (typical thickness of 0.5 mm) tested by us was very high such that the transmission in the far-infrared region is almost zero. An alternative scheme[2] was developed to grow InAs/InGaSb type II superlattices on GaAs substrates with InAs or GaSb buffer layers.

In this article, we present optical absorption of molecular beam epitaxial (MBE) GaSb layers grown on semi-insulating GaAs substrates. The quality of the GaSb layers were greatly improved, as judged by the observation of free exciton absorption line when a thin AlSb buffer layer is grown between the GaSb layer and GaAs substrate. The band gap variation of GaSb layers as a function of temperature is studied.

EXPERIMENTAL TECHNIQUES

The MBE GaSb layers were nominally undoped 1.0 - 5.0 μm thick grown on (100) oriented semi-insulating GaAs substrates. The growth rate of GaSb was 0.93 monolayer/sec. A few GaSb samples were grown with a 0.10 μm thick AlSb buffer layer. The growth rate of the AlSb buffer layers was 0.71 monolayer/sec. The optical absorption spectra were recorded using Cary 05E spectrometer with a closed cycle refrigerator. The temperature was controlled between 10 and 300 K to within ±1 K.

RESULTS AND DISCUSSIONS

The optical absorption spectra recorded at 10 K for three different samples are shown in Fig. 1. Spectra (a) and (b) were taken for two samples with the AlSb buffer layer and spectrum (c) was taken for a sample without the AlSb buffer layer. The absorption peak at 0.807 eV is identified as the free exciton.[3] This exciton line is resolved up to 45 K. It is clear from Fig. 1 that the free exciton absorption is almost washed out in spectrum (c). This is a typical result for all GaSb samples grown on GaAs substrates without the AlSb buffer layer and with thickness less than 2 μm.

The quality of the GaSb layers is greatly improved by adding an AlSb buffer layer. The improvement of the quality of GaSb is judged by the observation of the sharp optical absorption line due to the free exciton as shown in Fig. 1. In addition, it was reported[4] that the hole mobility in p-type GaSb layers grown on GaAs substrates is improved by adding an AlSb buffer layer. It is tempting to assume that the improvement of the GaSb layer quality is due to the fact that AlSb buffer layer partially prevents the dislocation from propagating into the the GaSb layer. Thus, the band edge and free exciton absorption are smeared out in samples with high dislocation density or even high impurity concentrations. A complete study of impurity and dislocation effects on the band edge absorption is reviewed by Blakemore.[5]

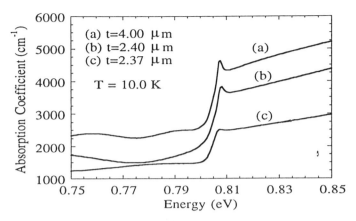

Fig. 1. *The optical absorption coefficient of three GaSb samples as a function of photon energy. Spectrum (a) was taken for a GaSb/AlSb/GaAs substrate sample with GaSb thickness (t) of 4.00μm. Spectrum (b) is the same as as spectrum (a), but t = 2.40μm. Spectrum (c) was taken for a GaSb/GaAs substrate sample with GaSb thickness of 2.37μm. All spectra were recorded at 10 K. The absorption peak at 0.807eV is due to the free exciton in GaSb layer. The structure observed below the band gap is interference fringes.*

Transmission electron microscopy (TEM) study shows that the dislocations density in GaSb layers was reduced by adding AlSb buffer layer between the GaSb layer and GaAs substrate. This is shown in Fig. 2. It is clear from this figure that the GaSb layer shown in Fig. 2(a) contains more dislocations than that with AlSb buffer layer as shown in Fig. 2(b).

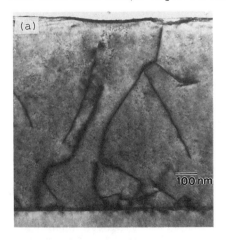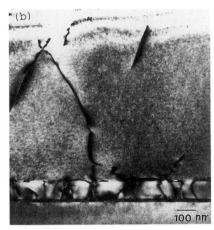

Fig. 2. TEM images of two GaSb layers grown on GaAs substrates. One layer was grown without (a) and with (b) AlSb buffer layer

A different scheme of improving the quality of GaSb epitaxial layers was reported.[6] In this scheme, a short period superlattice, (50 Å AlSb/50 Å GaSb)x10, was introduced which was found to be very effective in preventing the nonradiative recombination centers and carrier scattering centers from propagating in the GaSb layers grown on GaAs substrates. In addition, a free exciton absorption peak was observed[7] in 10 µm undoped MBE GaSb layer grown on GaAs substrate without the AlSb buffer layer. The observation of the free-exciton in thick epi-layers suggests that the dislocation density is decreased by increasing the layer thickness. Additional evidence of the latter assumption is provided in Fig. 3. In this figure we plotted three spectra òf three GaSb layers grown directly on GaAs substrates. Spectrum (a) was taken for 2.0 µm thick GaSb layer and spectra (b) and (c) were taken for 5.0 µm thick GaSb layers.

Spectrum (a) shows a weak free exciton line at ~1531.0 nm (~0.81 eV) which is the energy value of the free exciton in the bulk GaSb. Spectrum (b) shows a sharp and strong free exciton line at ~1535.0 nm (~0.808 eV). On the other hand, spectrum (c) shows a very week exciton line at ~1541.0 nm (~0.804 eV). Spectrum (c) indicates that the GaSb is highly strained probably due to the presence of excessive dislocations and defects that were generated during the MBE growth. However, it

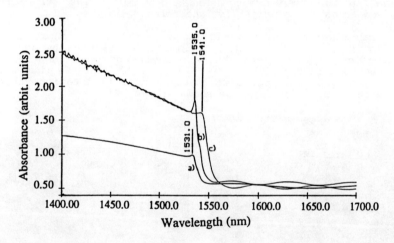

Fig. 3. Optical Absorption Spectra of three GaSb samples grown directly on GaAs substrates with different thicknesses (a) 2.0 µm thick, (b) 5.0 µm thick, and (c) 5.0 µm thick.

is clear that the quality of the layers is improved when the the thickness of the layer is increased. This is apparent in the case of spectrum (b) as compared to spectrum (a).

452

The band edge absorption of the GaSb layer is studied as a function of temperature. The result is shown in Fig. 4. It is well known that the band edge of any semiconductor becomes less steep (less abrupt) as the temperature is increased. This phenomenon causes a difficulty in determining the band gap energy as a function of temperature especially above 50 K where the free exciton peak is completely washed out. In the present study, the band gap is determined from the following relation: $E_g = (E_0+E_1)/2$ where E_g is the band gap, E_0 is the threshold energy at which the band edge absorption starts to rise, and E_1 is the energy at the top of the band edge absorption. Thus, E_1 is taken as the free exciton energy for $T \le 50$ K. The results are shown as open squares in Fig. 4. The experimental results are fitted with the following relationship.[9]

$$E_g = E_g(0) - \alpha \, T^2/(T+\beta), \qquad (1)$$

where Eg(0) is the band gap at 0 K, α is a constant, and β is approximately the 0 K Debye temperature [$\theta_D(0)$]. The fitting procedure gives $E_g(0) = 0.8056$ eV, $\alpha = 4.533 \times 10^{-4}$, and $\beta = 192$. The result is plotted as the solid line in Fig. 4.

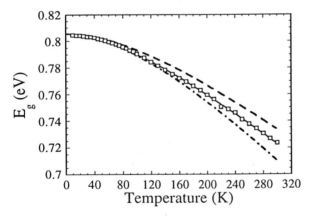

Fig. 4. The band gap (Eg) variation as a function of temperature. The open squares represent the experimental data, the solid line is the the results of fitting the experimental data using Eq. (1) with Eg, α, and β as fitting parameters. The dashed line is the same as the solid line, but replacing β with θD(0) = 265.5. The value of θD(0) was taken from reference 13. The dashed-dotted line is the same as the solid line, but with α = 6.0x10⁻⁴ (Ref. 8) and β = 265.5.

An expression of the 0 K Debye temperature, $\theta_D(0)$, is derived as a function of the elastic constants (C_{11}, C_{12}, and C_{44}) for semiconductors[10] which was then used to estimate $\theta_D(0)$ for many semiconductors.[11] Steigmeier estimated that $\theta_D(0) = 265.5$ for GaSb material.[11] The dashed line in Fig. 4 is the result of plotting Eq. (1) when $\theta_D(0) = 265.5$ is used in this equation instead of β. The values of E_g and α are the same for both the solid and dashed lines. On the other hand, the dashed-dotted line is the result of plotting Eq. (1) when α and β are taken for the bulk material as 6.0×10^{-4} and 265.5, respectively.[8,11] In both cases the results are different from the experimental data (open squares). If one assumes that Eq. (1) with $\beta \approx \theta_D(0) = 265.5$ is valid for the bulk GaSb materials[8,9], then it is tempting to explain the difference between the solid and dashed (or dashed-dotted) lines in Fig. 3 according the following proposition. The present GaSb layers are not completely relaxed and local (residual) strains are still present in these layers, which resulted in different elastic constants (C_{11}, C_{12}, and C_{44}) for the epitaxial layers as compared to the those of the bulk GaSb materials. Hence, the difference between β and $\theta_D(0)$, which is reflected in the solid and dashed lines in Fig. 3, can be used qualitatively to predict the presence of local strains. Additional support for the presence of residual strains in GaSb layers is that the free exciton absorption peak in spectra (a) and (b) in Fig. 1 is observed at ~0.807 eV. On the other hand, the free exciton absorption peak in 10 μm thick MBE GaSb sample[7] and bulk GaSb samples[3] is observed at 0.810 eV. The ~3 meV energy difference is most likely to be due to the presence of local strains in the present samples. Even a small energy difference (~0.5 meV) between the free exciton peak in Fig. 1 spectra (a) and (b) is detected. We also foud that the parameter β is sample dependent in agreement with the fact that each sample may contain different amount of strains depending on the dislocations and defects.

The assertion that $\beta \approx \theta_D(0) = 265.5$ for bulk GaSb is probably a very reasonable assumption. For example, recent measurements[12] on metal-organic vapor phase epitaxial GaSb show that $\beta \approx 265$ for 20 μm thick undoped samples. This parameter was found to be as low as $\beta \approx 94$ for liquid phase epitaxial Te-doped GaSb samples.[13] The variation of β values in different GaSb samples seems to provide additional support for the hypothesis that the difference between β and $\theta_D(0)$ [see Fig. 4] may be due to local strains in the GaSb layers.

CONCLUSIONS

In conclusion, we presented optical absorption measurements of molecular beam epitaxial GaSb layers grown on semi-insulating GaAs substrates. A free exciton absorption peak was observed in GaSb layers when an AlSb buffer layer was present. It is concluded that the improvement of the optical absorption spectra in thick GaSb layers and layers with AlSb buffer layers is due to the decrease of dislocation density and defect concentrations. TEM study also shows that the dislocations density in GaSb layers is decreased by adding AlSb buffer layers. The current GaSb/AlSb layers grown on semi-insulating GaAs substrates may present a very useful scheme as buffer layer for Sb-related strained-layer superlattices. The band gap energy of GaSb was studied as a function of temperature. It is found that the Debye temperature of GaSb layer estimated from the temperature dependence of the band gap to be smaller than the Debye temperature calculated from the bulk GaSb elastic constants. This discrepancy is attributed to the difference between the elastic constants of GaSb epi-layer and the bulk GaSb due to the presence of local (residual) strains in the MBE GaSb layers.

ACKNOWLEDGEMENT -- This work was partially supported by the Air Force Office of Scientific Research.

REFERENCES

1.   C. Mailhiot in *Semiconductor   Quantum   Wells   and Superlattices for Long Wavelength Infrared Detector*, edited by M. O. Manasreh (Artech House, Boston, 1993), chapt. 3. pp. 109 - 138.
2.   R. H. Miles, D. H. Chow, J. N. Schulman, and T. C. McGill, Appl. Phys. Lett. **57**, 801 (1990).
3.   E.J. Johnson and H. Y. Fan, Phys. Rev. **A 139**, 1991 (1965).
4.   H. Gotoh, K. Sasamoto, S. Kuroda, and M. Kimata, Phys. Stat. Sol. (a) **75**, 641 (1983).
5.   J. S. Blakemore, J. Appl. Phys. **53**, R123 (1982) and references therein.
6.   S. V. Ivanov et al., Semicond. Sci. Technol. **8**, 347 (1993).
7.   A.M. Fox, A.C. Maciel, J.F. Ryan, and T. Kerr, Appl. Phys. Lett. **51**, 430, (1987).
8.   H. C. Casey, *in Atomic Diffusion in Semiconductors*, edited by D. Shaw, (Plenum Press, New York, 1973), chapt.6, pp 351- 431.

9.   Y. P. Varshni, Physica **34**, 149 (1967).
10.   P. M. Marcus and A. J. Kennedy, Phys. Rev. **114**, 459 (1959).
11.   E. F. Steigmeier, Appl. Phys. Lett. **3**, 6 (1963).
12.   E. T. R. Chidley, S. K. Haywood, A. B. Henriques, N. J. Mason, R. J. Nicholas, and P. J. Walker, Semicond. Sci. Technol. **6**, 45 (1991).
13.   S. C. Chen and Y. K. Su, J. Appl. Phys. **66**, 350, (1989).

A TEM STUDY OF DEFECT STRUCTURE IN GaAs FILM ON Si SUBSTRATE

SAHN NAHM, HEE-TAE LEE, SANG-GI KIM, AND KYOUNG-IK CHO
Semiconductor Division, Electronics and Telecommunications Research Institute, Daedok
Science Town, P.O. Box 8, Daejeon, 305-606, Korea.

ABSTRACT

For the GaAs buffer layer deposited on Si substrate at 80 °C and annealed at 300 °C for 10 min, the size of most GaAs islands was observed as ~ 10 nm but large islands (~ 40 nm) were also seen. According to the calculation of spacing of moire fringes, large GaAs islands are considered to be rotated about 4 ° with respect to the Si substrate normal. However, for the main GaAs film overgrown on the GaAs buffer layer at 580 °C, moire fringes with the spacing of 5 nm (GaAs film without rotation) completely covered the surface of Si substrate. Misfit dislocations and stacking faults were already formed at the growth stage of buffer layer. Stacking faults and misfit dislocations consisting of Lomer and 60 ° dislocations were observed in GaAs films grown at 580 °C. However, after rapid thermal annealing at 900 °C for 10 sec, only Lomer dislocations with 1/2[110] and 1/2[-110] Burgers vectors were observed.

INTRODUCTION

There is a considerable interest in the heteroepitaxial growth of GaAs film on Si substrate because of the opportunity to combine the best properties of the two materials [1]. However, high densities of threading dislocations and stacking faults are formed due to the differences in lattice parameters and thermal expansion coefficients. The formation of crystal defects depends on the detail growth conditions at each growth stage. In this work, GaAs films were grown on tilted Si (001) substrate by a modified two-step molecular beam epitaxy (MBE) method. Two types of GaAs islands were found in the initial stage of growth and the interfacial defects developed at each growth stage were investigated using transmission electron microscopy (TEM).

EXPERIMENTAL DETAILS

In order to study the initial stage of growth, amorphous GaAs buffer layers 18 nm thick were deposited on Si substrates at 80 °C in a MBE system and they were furnace-annealed (solid phase epitaxy) at 300 °C for 10 min (sample 1) and 20 min (sample 2) in a N_2 atmosphere, respectively. For complete growth of the GaAs film, the substrate temperature was slowly increased to 580 °C and the main GaAs film of 0.5 μm thickness was grown after crystallization of the amorphous GaAs buffer layer in the MBE system (sample 3). Finally, to decrease the density of defects, rapid thermal annealing (RTA) treatment was carried out on sample 3 at 900 °C for 10 sec in N_2 atmosphere (sample 4). The Si substrate used in this work was tilted by 4 ° from [001] to [110] direction. The plane view TEM samples were prepared by mechanical grinding and subsequent ion milling at liquid nitrogen temperature. During the first stage of the ion milling, the energy of ions was 4 keV with glancing angle of 17 degrees and

457

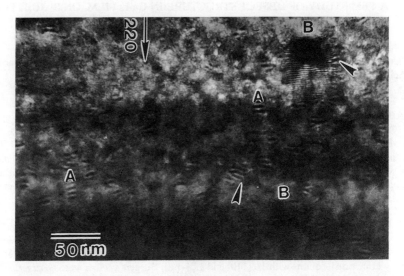

Figure 1.   Bright field image of GaAs film deposited at 80 °C and annealed at 300 °C for 10 minutes.

during the later stage of the ion milling, an even lower energy of 3 keV with 15 degrees of glancing angle was used to minimize the damage induced during the ion milling. A Philips CM20 T/STEM microscope was used to observe the samples.

RESULTS AND DISCUSSION

Figure 1 shows a plane view TEM bright field image taken under (220) two beam condition. Moire fringes due to the mismatch between Si and GaAs lattice parameter are shown in this figure and the regions of moire fringe indicate the GaAs islands on Si substrate. GaAs islands can be divided into two groups with respect to the size and spatial period of moire fringes. For most of the GaAs islands marked as A in fig. 1, the size is ~ 10 nm and the spacing of the moire pattern is about 5 nm. The spacing of the parallel moire pattern due to the (220) beam is calculated as 4.8 nm which agrees with our experimental value. The same type of moire pattern was observed in previous study [2].

For some of the GaAs islands marked as B in fig. 1, the size is about 40 nm and the spacing of the moire fringe is about 2.5 nm. The Si substrate used in this work was tilted by 4 ° from [001] to [110] direction and tilt of GaAs film was observed in previous study [3]. Even if [001] of GaAs film is tilted toward the [110] direction, parallel moire fringes with 5 nm spacing are expected. The spacing of moire fringes with 2.5 nm could be explained by taking into account the rotation of GaAs film about substrate surface normal. Using $d_m = (d_1 d_2)/(d_1^2 + d_2^2 - 2d_1 d_2 \cos\alpha)^{1/2}$ where d_1 is interplanar distance of Si (220), d_2 is interplanar distance of GaAs (220) and α is rotation angle [4], the amount of rotation is calculated as 4 °. In order to study the behavior of GaAs islands, one of the amorphous GaAs film on Si substrate has been

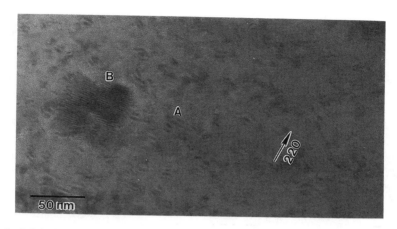

Figure 2. Bright field image of GaAs film deposited at 80 °C and annealed at 300 °C for 20 minutes.

annealed at 300 °C for 20 min (sample 2). Figure 2 shows plane view bright field image of sample 2 taken under (220) two beam condition. Two types of GaAs islands were also found and the number of small GaAs islands is increased. For sample 3 where main GaAs film was completely grown on Si substrate at 580 °C, moire pattern with 5 nm spacing completely covered the Si substrate as shown in fig. 3. Above results imply that at the initial stage of growth at 300 °C, both unrotated and rotated GaAs islands are formed but at later stage of growth at 580 °C, only unrotated GaAs islands are formed. Therefore, it can be concluded that the rotated GaAs islands stable at 300 °C become unstable as temperature increases and finally disappear at 580 °C. According to a previous study, for GaAs film grown on tilted Si substrate by migration-enhanced epitaxy exhibited not only tilt but also rotation [5]. It is suggested that

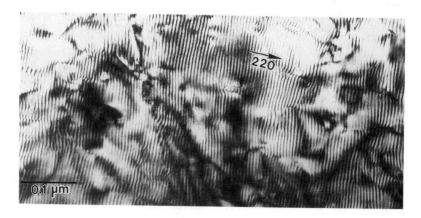

Figure 3. (220) dark field image of GaAs film overgrown on the GaAs buffer layer at 580 °C.

459

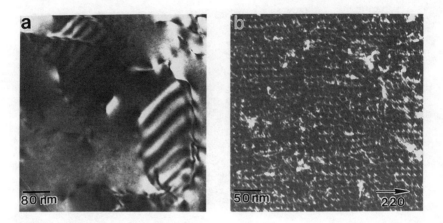

Figure 4. (a) (220) dark field image of the area away from the interface, and (b) $g_{(400)}$ and 3g weak beam image at the interface of GaAs/Si system grown at 580 °C.

the rotation of GaAs film is influenced by the initial growth condition [5]. However, for our system, more study is needed to explain the behavior of rotated GaAs islands.

Misfit dislocations are observed in both small and large GaAs islands as indicated by the arrow heads in fig. 1. Even if stacking faults were not found in fig. 1, they were observed in cross sectional high resolution TEM image [6]. Therefore, the lattice mismatch between GaAs film and Si substrate is accommodated by misfit dislocations and stacking faults in this early stage of growth.

Figure 4(a) is a (220) dark field image of sample 3 taken in the area away from the interface between GaAs film and Si substrate. Stacking faults and high density of threading dislocations can be observed in fig. 4(a). In order to investigate defect structure at the GaAs/Si interface,

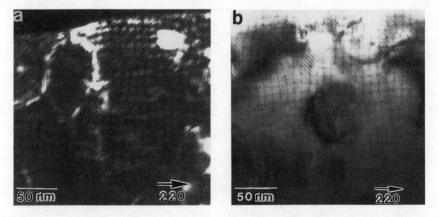

Figure 5. (a) $g_{(400)}$ and 3g weak beam image, and (b) bright field image taken under (400) two beam condition of GaAs/Si system with RTA treatment at 900 °C for 10 seconds.

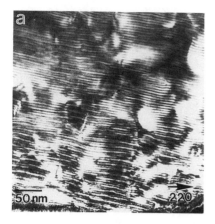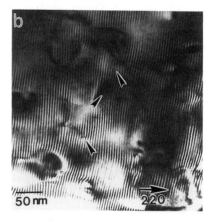

Figure 6. (a) (220) dark field image, and (b) (-220) dark field image of GaAs/Si system
with RTA treatment at 900 °C for 10 seconds.

ion milling was carried out for 15 minutes more on the GaAs surface. Figure 4(b) shows a
(400) weak beam image taken at the interface. Grid-like dislocation running along the [110]
and [-110] directions are shown in fig. 4(b). According to the previous study, both Lomer and
60 ° dislocation is present in this sample [7].

In order to decrease the defects in the system, one of sample 3 was annealed at 900 °C for 10
seconds. (400) weak beam and bright field images are shown in figs. 5(a) and 5(b),
respectively. Grid-like misfit dislocations with a distance of 10 nm are evenly developed in
both [110] and [-110] direction. In order to determine the Burgers vector of the dislocations
developed in this sample, pictures were taken under both (220) and (-220) two beam conditions
as shown in figs. 6(a) and 6(b), respectively. Misfit dislocations developed along [110]
direction are shown in fig. 6(a) taken using g=220 beam and those developed along [-110]
direction are shown in fig 6(b) (see arrow heads) taken using g=-220 beam. Therefore, Burgers
vectors of the misfit dislocations are 1/2[110] and 1/2[-110] which are in (001) plane,
indicating pure edge dislocation (Lomer dislocation). Since the distance between the
dislocations is 10 nm, the misfit (~ 4%) between GaAs and Si is completely accommodated by
pure edge dislocation. Above results indicate that rapid thermal anneling at 900 °C drastically
reduces the density of 60 ° dislocations and stacking faults, and misfit between GaAs film and
Si substrate is relieved by Lomer dislocations.

CONCLUSION

For GaAs buffer layers deposited at 80 °C and annealed at 300 °C for 10 min, the size of
most of the GaAs islands was observed as ~ 10 nm but large islands with ~ 40 nm were also
observed. According to the calculation of spacing of moire fringes, large GaAs islands are
considered to be rotated about 4 ° with respect to the Si substrate surface normal. Two types of
GaAs islands were also observed in GaAs film annealed at 300 °C for 20 min. However, for
main GaAs film overgrown on the GaAs buffer layer at 580 °C, GaAs film without rotation

completely covers the Si substrate. Therefore, it can be suggested that the rotated GaAs islands at the initial stage of growth (~ 300 °C) become unstable as temperature increases and finally disappear at the later stage of growth (~ 580 °C). Misfit dislocations were found in both small and large islands at the early stage of growth. Grid-like dislocations consisting of Lomer and 60 ° dislocations were observed at the interface of the GaAs/Si system grown at 580 °C. However, after RTA treatment at 900 °C for 10 sec, only Lomer dislocations with 1/2[110] and 1/2[-110] Burgers vectors were observed. The distance between the dislocations was measured as ~ 10 nm accommodating 4 % mismatch between GaAs film and Si substrate.

ACKNOWLEDGEMENT

This work was funded by Korea Telecommunication.

REFERENCES

[1] S. F. Fang, K. Adomi, S. Iyer, H. Morkoc, H. Zabel, C. Choi, and N. Otsuka, J. Appl. Phys. 68, R31(1990).
[2] S. M. Koch, S. J. Rosner, R. Hull, G. W. Yoffe, and J. S. Jr. Harris, J. Crystal Growth, 81, 205 (1987).
[3] I. Kim, M.S. Dissertation, Korea Advanced Institute of Science of Techonology (1992).
[4] Ludwig Reimer, Transmission Electron Microscopy (Springer-Verlag, Berlin, Heidelberg, New York, Tokyo, 1984).
[5] K. Nozawa, and Y. Horikoshi, Jpn. J. Appl. Phys. 32, 626 (1993).
[6] W. K. Choo, I. Kim, J. Y. Lee, K. I. Cho, J. L. Lee, and Y. J. Kim, Mat. Res. Soc. Symp. Proc. vol. 263, 131(1992).
[7] K. I. Cho, W. K. Choo, J. Y. Lee, S. C. Park, and T. Nishinaga, J. Appl. Phys. 69, 237 (1991).

TEMPERATURE AND POLARIZATION DEPENDENCE OF THE OPTICAL ABSORPTION IN ZnGeP$_2$ AT TWO MICROMETERS

M. SHAH*, M. C. OHMER*, D. W. FISCHER*, N. C. FERNELIUS*, M. O. MANASREH*, P. G. SCHUNEMANN** AND T. M. POLLAK**
*Wright Laboratory,WPAFB, OH 45433-7707
**Lockheed Sanders, Nashua, NH 03061-2035

ABSTRACT

The temperature and polarization dependence of the optical absorption in ZnGeP$_2$ at two micrometers is reported for the first time over the temperature range from 10K to 300K. The radiation was normally incident upon the face of a cubic sample which contained the c-axis. The absorption of o-rays (**E** parallel to **c**), and e-rays (**E** perpendicular to **c**) was determined. It was found that the e-ray absorption coefficient was always significantly larger than the o-ray absorption coefficient and that it had a less significant temperature dependence. For example, the ratio of e-ray to o-ray absorption coefficient was approximately two at 300K and five at 77K. Correspondingly the o-ray absorption coefficients were reduced upon cooling to 77K by a factor of 2.5, while the e-ray absorption coefficients were reduced only slightly (10%-20%). These results indicate that for Type I optical parametric oscillators (OPOs) which use an o-ray pump beam, that performance may be improved by cooling the crystal.

INTRODUCTION

The temperature and polarization dependence of the optical absorption in ZnGeP$_2$ at two micrometers is reported for the first time over the temperature range from 10K to 300K. This parameter is of interest because it is related to the performance [1] of two micrometer pumped OPOs which utilize ZnGeP$_2$ crystals. An OPO converts short wavelength laser output into longer wavelengths in a continuous fashion utilizing the non-linear optical (NLO) properties of the crystal. ZnGeP$_2$ is nearly the ideal candidate for this application because: its large second order NLO coefficient provides efficient energy conversion; it has the appropriate birefringence for phase

Mat. Res. Soc. Symp. Proc. Vol. 325. ©1994 Materials Research Society

matching from 2-11 micrometers; and it is extremely transparent in the mid-infrared. In addition, it has excellent mechanical and thermal properties particularily its large thermal conductivity which is 35 times larger than that of the state-of-the-art material $AgGaSe_2$ [2].

ZnGeP$_2$ is an ordered ternary diamond-like semiconductor with a pseudo-direct bandgap of 2.00 eV at 300K and has the chalcopyrite structure. In many aspects it behaves similarily to other diamond-like semiconductors such as Si and GaAs. It is the ternary analog of GaP as Zn and Ge are nearest neighbors to Ga in Period Four. The crystal is uniaxial, the c-axis is the optical axis and the birefringence is positive. ZnGeP$_2$ is lattice matched to both GaP and Si. Many of its intrinsic and extrinsic electrical and optical properties are orientation and polarization dependent. Significant near edge sub-bandgap absorption extending from .7 to 3 micrometers is observed in Bridgman and LEC grown crystals which are found to be universally p-type. It is due to a native defect which is a deep acceptor (or deep acceptors) designated as AL1. The concentration of AL1 can be minimized or maximized by appropiate growth conditions and post-growth processing.

The purpose of this paper is to report the effect cooling has on the e-ray and o-ray absorption coefficients for ZnGeP$_2$ in the sub-bandgap spectral region. Of special interest is a comparison between room temperature (300K) values and liquid nitrogen temperature (77K) values of the absorption coefficients. The previous relevant work is briefly reviewed. Most previous work was conducted at room temperature and ignored polarization effects. However, there are two interesting papers [3], [4] which report the temperature dependence of the absorption for energies near and greater than the bandgap (1.8 eV to 2.2 eV), one of which [4] considered substrate orientation and polarization effects.

Gorban et. al [3] in a study which did not consider orientation and polarization reported that the absorption coefficient spectrum shifts approximately 0.13 eV toward higher energies in cooling from 300K to 80K. As a result, near 2.0 eV (0.6 micrometers) the absorption coefficient was reduced by a factor of seven or greater via cooling.

Tregub [4] utilized the same geometry as in our study and reported that not only did the absorption spectrum shift to higher energies upon cooling but the highest absorbing polarity reversed comparing 300K to 4.2K. This data indicates that the birefringence of the crystal may have gone from positive to negative upon cooling to

4.2K in this spectral region. At 2.00 eV the ratio of e-ray to o-ray coefficient was approximately 1.5 at 300K and 0.65 at 4.2K. Correspondingly the o-ray absorption coefficient was reduced by a factor of 2.3 and the e-ray absorption coefficient was reduced by a factor of 5.3 by cooling to 4.2K. For simplicity, throughout the paper o-ray is defined as that polarization which is perpendicular to the c-axis at room temperature and correspondingly the e-ray polarization that which is parallel to the c-axis at room temperature. Tregub's result indicates that the polarizations of traditionally defined o-rays and e-rays may be temperature dependent.

The overall shift of the absorption spectrum to higher energies reported by Gorban et. al and Tregub is explained by the fact that the bandgap increases to 2.1 eV at low temperatures shifting everything by nearly 0.1 eV. We have observed this same effect for sub-bandgap absorption spectra in preliminary measurements which did not consider orientation and polarization effects.

EXPERIMENTAL TECHNIQUES

The crystals were grown using the horizontal freeze technique and annealed as previously described [5]. One of the two samples tested was as grown (42A) and the other (42B) was given a brightening anneal [6],[5] to reduce the concentration of AL1. The samples were nominally stoichiometric and the value of the two micrometer absorption although not equal to the best observed indicates that the samples were of good quality.

The samples were mounted in a closed cycle refrigerator which was then mounted in a CARY05E spectrophotometer. A calcite polarizer was used to select the desired polarization for the experiment. The CARY05E partially polarizes the light which is directed toward the sample. For this reason, the calcite polarizer was fixed in the orientation for maximum beam intensity and identical throughput while the sample was rotated to get the proper polarization of light through the sample for the two sample orientations.

The closed cycle refrigerator was cooled to 10K initially with the sample in place. Absorption scans were done at 10, 50, 77, 100, 150, 200, 250, and 300 degrees Kelvin. After scans for both orientations were completed, corresponding scans were made with the samples removed and the absolute transmission was obtained via the method of stored ratios. The absorption coefficient was calculated in the usual way assuming a index of refraction of 3.1.

RESULTS & DISCUSSION

The as-grown sample results listed in the Table and displayed in Figure (1) below show a smooth decrease in the absorption

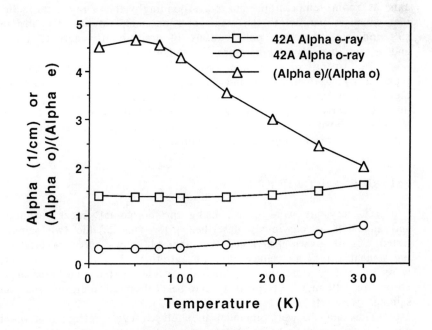

FIGURE (1) TEMPERATURE DEPENDENCE OF THE ABSORPTION
COEFFICIENTS OF AS-GROWN SAMPLE #42A

coefficient for the o-ray as temperature is decreased from room temperature down to about 50K, and then it increases slightly as the sample is cooled further to 10K. The e-ray absorption coefficient decreases slightly but noticeably. The annealed sample results listed in the Table above and displayed in Figure (2) below show generally the same trend. The ratio of the e-ray to the o-ray absorption coefficient vs temperature is also shown in the Table. It increases from 2.0 to 4.6 for the as grown sample and from 2.4 to 4.9 for the annealed sample upon cooling. The o-ray absorption for the as-grown sample was reduced by a factor of 2.6 and for the annealed sample it was reduced by a factor of 2.3 upon cooling.

Sample Temp Kelvin	42A alpha$_e$ 1/cm	42A alpha$_o$ 1/cm	42B alpha$_e$ 1/cm	42B alpha$_o$ 1/cm	42A/42B Ratios ------
10	1.42	.314	.845	.16	4.5/5.3
50	1.40	.300	.820	.155	4.7/5.2
77	1.38	.304	.786	.155	4.5/5.1
100	1.38	.321	.788	.161	4.3/4.9
150	1.39	.391	.782	.174	3.6/4.4
200	1.43	.474	.800	.200	3.0/4.1
250	1.52	.618	.798	.323	2.5/2.5
300	1.62	.800	.859	.356	2.0/2.4

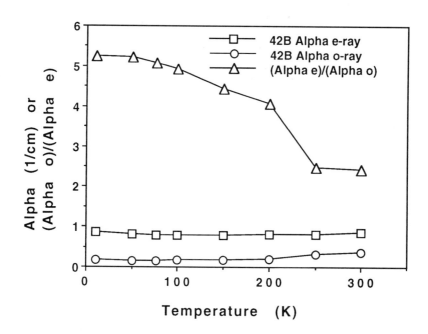

FIGURE (2) TEMPERATURE DEPENDENCE OF THE ABSORPTION
COEFFICIENTS OF ANNEALED SAMPLE #42B

The corresponding reduction factors for e-rays are 1.2 and 1.1. These results indicate that for Type I OPOs which use an o-ray pump beam, that performance may be improved by cooling the crystal.

ACKNOWLEDGEMENTS

This research was sponsored by Horst Wittmann and Gernot Pomrenke of the Air Force Office of Scientific Research. The crystal growth development effort conducted by Lockheed Sanders was partially supported by Wright Lab Contract F33615-88-C-5438.

REFERENCES

1.   P. A. Budni, P. G. Schunemann, M. G. Knight, T. M. Pollak, and E. P. Chicklis, OSA Proceedings on Advanced Solid State Lasers, Lloyd Chase and Albert Pinto, eds. OSA, DC, **13**, 380 (1992).
2.   G. C. Catella et al., Appl. Opt. **32**, 3948 (1993).
3.   I. S. Gorban, V. A. Gorynya, V. V. Lugovskii, and I. I. Tychina, Sov. Phys. Solid State, **16** ,1029 (1974).
4.   I. G. Tregub, Ukrainskiy Fizicheskiy Zhurnal, **25**, 1209 (1980).
5.   P. G. Schunemann, and T. M. Pollak, OSA Proceedings on Advanced Solid State Lasers, George Dube and Lloyd Chase, eds. **10**, 332 (1991).
6.   Y.V. Rud and R. Masagutova, Sov. Tech. Phys. Lett. **7**, 72 (1981).

# PART VIII

Superlattices and Quantum Wells

TUNING OF ENERGY LEVELS IN A SUPERLATTICE

FRANCOIS M. PEETERS
Departement Natuurkunde, Universiteit Antwerpen (UIA), Universiteitsplein 1,
B-2610 Antwerpen, Belgium; Email: peeters@nats.uia.ac.be

ABSTRACT

The gap between the minibands of a conventional superlattice (or between the subbands of a quantum well) can be controlled by introducing potential barriers in its wells. An appropriate choice of the position, the width d, and the height V_d of these barriers, achieved by standard methods, can reduce the energy minibands to the desired values. When these barriers are introduced at the center of the wells of the original structure, the position of the second miniband in energy space changes very little with d and/or V_d whereas that of the first miniband can change by one to two orders of magnitude. This leads to a tuning of the first miniband and of the energy gap between the first two minibands. Similar results are obtained for the case of wells in the barriers and for the tuning of impurity states in a superlattice. Possible applications include infrared photodetectors and tuning of the tunneling current.

1 Introduction

The success of semiconductors is based on the fact that semiconductor properties like the electrical carrier concentration, the mobility and the diffusion length can be varied by orders of magnitude through small changes in the chemical composition, defect structure or chemical and structural homogeneity of the semiconductor. There is much less flexibility in the choice of the bandgap of semiconductors which is fixed by the composition of the material. Changing external parameters like the temperature or the pressure will give some limited variability on the size of the band gap of the semiconductor.

Over the last two decades layered semiconductor structures, quantum wells and super-lattices, have emerged as a new class of materials structures with a wide range of interesting and novel optical and opto-electronic properties [1]. These properties have found numerous applications in devices, and an entire new technology, *photonics*, is based on such concepts. The rapid development of this field and its continued evolution is a direct result of the highly developed and sophisticated growth technologies, molecular beam epitaxy (MBE) and organometallic chemical vapor deposition (OMCVD), which permit control of composition and doping on the scale of a few monolayers in the growth direction. This unprecedented control offers the capability of tailoring or *quantum engineering* a desired set of electronic and optical properties for such layered heterostructures. This is typically achieved for simple quantum well stuctures by varying both the thickness and the composition of the well and barrier materials, but more sophisticated approaches are possible as discussed below.

The band gap is now determined by geometrical factors like the width of the quantum well. In fact the effective band gap becomes: $E_g^* = E_g + E_e + E_h$ where E_g is the band gap of the quantum well material with $E_e \sim 1/m_e W^2$ and $E_h \sim 1/m_h W^2$ the quantum

Mat. Res. Soc. Symp. Proc. Vol. 325. ©1994 Materials Research Society

confinement energies of the electron and hole respectively. Furthermore the quantum well can have several subbands with separation of the order $E_e(= 1. - 100.meV)$ which can be considered as little gaps in the electron spectrum. When such quantum wells are periodically repeated one obtains a superlattice in which the electron moves in minibands.

In semiconductor quantum wells the energy separation can be varied using different techniques. The energy between the ground and first excited subband can be tailored from a few meV to roughly 150 meV by changing both well and barrier parameters. In addition to tailoring the energies, there are two techniques that permit continuous tuning in a single quantum well structure. Application of an electric field provides the possibility of tuning over a limited range. The range is smaller the narrower the well. Another possibility is magnetic field tuning which is possible in II-VI based superlattices whose barriers contain magnetic ions (e.g. manganese). The problem of this system is that doping and transport properties are not as well known or controlled as in the III-V based systems. Also here the range of tunability is limited.

For certain infrared (IR) applications [2] it is desirable to have only two minibands with a small energy minigap. This could be achieved by varying, for example, the width and height of the potential barriers as well as the width of the wells. To obtain small transition energies this requires small barrier heights and large well widths. Furthermore, a precise control over *both* the barrier height and well width is essential. But, wide wells are much more sensitive to electric fields than narrow wells, and for small energy minigaps dark currents will be large for large well superlattices. This calls for narrow wells with small energy separation between the minibands.

In the present paper I give an overview of a method which is able to handle the above contradictory demands. Small energy separations (thus small energy minigaps) can be achieved in superlattices with narrow wells by introducing positive potential barriers in the middle of its wells or potential wells in the middle of its barriers. A similar technique is used to tune the energy of impurities in quantum wells and superlattices.

2 Quantum wells

Intersubband transitions can be used for the detection of infrared radiation. This requires quantum wells with appropriate widths and depths such that the first excited subband is either weakly bound [3] in the well or is slightly above the top of the well [4].

The position of the energy levels in a quantum well can be varied by changing the height of the barrier, but it is primarily more sensitive to the width of the quantum well (W), because the separation between the energy levels has approximately a W^{-2} dependence. Thus the tailoring of intersubband separation in a detector can be achieved by varying the width of the well. This, however, requires using fairly wide wells in order to obtain small transition energies. On the other hand, there are reasons to favor narrow quantum wells over broad wells. Wide wells are much more sensitive to an external (or internal) electric field then narrow wells. Such electric field bound states in a quantum well are resonant with the continuum and as a consequence acquire a certain width which is a measure for the probability of the electron to tunnel out of the quantum well. In a given electric field this width is much larger for a wider well and consequently it is much easier to tunnel out of the well. Therefore a much larger *dark current* is expected in wide wells than in narrow structures. Furthermore, the density of states in a narrow well is the same as in a wide well, and consequently we can gain more signal from a given width of the sample when it contains more wells.

In order to show how these contradictory demands can be satisfied, we will illustrate this using a simplified model of an infinitely deep potential well. In this quantum well we insert a barrier as illustrated in the inset of Fig. 1. The eigenvalue equation for the energy levels of this structure are given in the appendix of Ref. [5]. In Fig. 1(a) the first four energy levels are shown as function of the barrier width b for $s = 0$, i.e. the barrier is in the center of the well. For the quantum well we took the parameters for the material GaAs and for the barrier in the center those of Al_4Ga_6As. As b increases the levels 1 and 2 (or 3 and 4) move closer to each other and for $b > 50$Å the difference $E_2 - E_1$ is not resolved on the meV scale. Notice that it is mainly the n-odd levels that move whereas the n-even levels remain almost constant, at least for small b values. A similar behavior is obtained when b is fixed and V_0 varied. When the barrier moves away from the center of the well the separation between the $n = 1$ and the $n = 2$ energy levels increase.

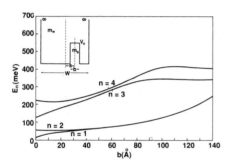

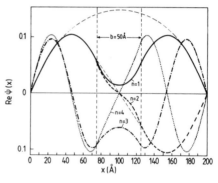

Fig. 1: The first four energy levels for a potential barrier in an infinitely deep potential well (see inset) as function of the width of the barrier.

Fig. 2: The real part of the wavefunction corresponding to the system of Fig. 1. The thin dashed curves correspond to the first two levels in the absence of the barrier.

We can understand this behavior with the help of Fig. 2 where the wave functions of the first four levels are plotted for $b = 50$Å and $s = 0$. The dashed lines are the wave functions of the first two levels in the absence of the barrier, i.e. $b = 0$. The wave function of the first level has a maximum at the center of the well, whereas that of the second level vanishes. Therefore, the wave function and the energy of the first level will be drastically affected by the introduction of the barrier whereas the corresponding quantities of the second level will not. This leads to an effective tuning of the first level and the gap $E_2 - E_1$.

The generalization of the above discussion to quantum wells with finite height can be found in Ref. [6]. Trzeciakowski and McCombe have shown theoretically how the intersubband separation can be varied over a wide energy range by depositing a thin barrier layer of AlGaAs in the middle of a GaAs well. They showed that in a sufficiently narrow well it is possible to push the subbands out of the well, i.e. the structure no longer has any bound states.

3   Tuning the minibands of a superlattice

In a real device several quantum wells are put in series in order to increase the detection area [2]. In such a multi-quantum well structure a gate voltage is applied over the structure. At low temperature and in the absence of any IR radiation electric conduction is via sequential resonant tunneling either between the ground states of the adjacent wells or between the ground state and the first excited state, depending on the voltage drop across the lattice period. This results in the so-called *dark current* which should be minimized by either increasing the thickness or the height of the barriers.

When IR-light hits the detector such that it is resonant with the intersubband transition $(E_2 - E_1)$ an electron will be excited from the doped ground state to the excited state where it can tunnel out of the well through the thin top of the barrier. This photogenerated hot electron produces a *photo-current* which can be detected. To increase the sensitivity of the device the ratio between the photo-current to the dark current should be maximized.

In order to maximize the absorption coefficient one should maximize the number of quantum wells in a given area. This calls for narrow quantum wells with narrow barriers. In this situation the multi-quantum well becomes a superlattice. Therefore tuning of minibands is needed.

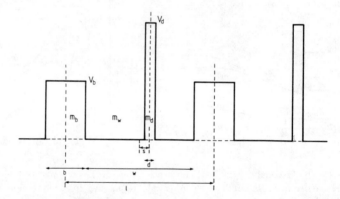

Fig. 3: Potential profile of the first two periods of the superlattice under investigation.

In order to realize tuning of the minibands in a superlattice a similar idea can be used as in previous section which is illustrated in Fig. 3. Barriers (called defect barriers) with width d and height V_d are placed in the quantum wells of a superlattice [7]. The resulting miniband structure of a $GaAs/Al_3Ga_7As$ superlattice with well width $w = 100\text{Å}$ and barrier width $b = 50\text{Å}$ is shown in Fig. 4 as function of the strength of the defect barrier. Here we choose to vary the width of the defect barrier d and keep the height V_d constant. The defect barriers, which are taken from the material Al_4Ga_6As, are placed in the center of the quantum wells. Notice that: 1) the width of the first miniband increases, 2) the two lowest minibands move closer to each other, and 3) for $d = 35.9\text{Å}$ the gap closes. At $d = 0$ the widths of the two lowest minibands are $\Delta E_1 = 1.8\text{meV}$ and $\Delta E_2 = 10.6\text{meV}$ respectively with a miniband gap of $\Delta E_g = 80.8\text{meV}$. At the closing of the miniband gap the resulting miniband has a width of 24meV. At the closing of the gap the group velocity $\partial E/\partial k$ of the electron at the point where the first and the second miniband touches is nonzero and thus a real closing of the gap is realized. This can be understood physically as follows: at the closing of the miniband gap the effective strength of the defect barrier becomes equal to the strength of the barrier of the host superlattice and the period of the superlattice becomes effectively half of its original value. Notice also that in contrast to the quantum well case where the separation between the energy levels is always nonzero, although it can be made very small, here we are able to make the separation between the two minibands exactly zero for some specific value of d for a given V_d.

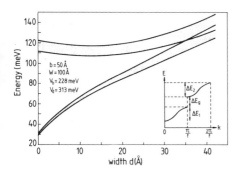

Fig. 4: Dependence of the extremal energy points of the first two minibands as a function of the width of the barrier which is centered in the middle of the quantum wells.

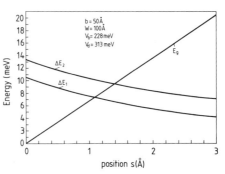

Fig. 5: Width of the first two minibands and the gap as a function of the position of the barrier in the quantum well measured from the center of the quantum well. The width of the barrier in the quantum well is such that the gap is zero when the barrier is centered in the middle of the quantum wells.

Beltram and Capasso [8] have shown that the introduction of a sheet of defects in the center of the barriers of a multi-quantum well structure: 1) enhances the width of the minibands, and 2) creates one extra miniband which is a consequence of the states which are introduced by the defects. They modelled this sheet of deep centers by a negative delta function potential. They found that the enhancement of the miniband widths is maximized when the energy level of the defect is matched to the ground state of the corresponding isolated wells. Because in a multi-quantum well structure the states can be localized in the direction perpendicular to the wells due to e.g. intralayer thickness fluctuations we have the interesting fact that the periodic introduction of a highly localized state in the barriers of a multi-quantum well system causes *delocalization* in the resulting states! This of course will have dramatic changes in the electronic properties of the system. When the position of the defect layer is moved away from the center of the barrier the lowest band splits into two narrow levels which are shifted in energy. The shift can be such that one of these levels is even below the bottom of the quantum well. The resulting superlattice bandgap is now *below* the bandgap of the bulk of both materials constituting the multi-quantum well! In fact this is similar to the formation of an impurity band below the conduction band in heavily doped semiconductors.

In Ref. [5] we generalized the above idea to the case in which 'defect' quantum wells of finite depth and nonzero width are introduced in the barriers of a superlattice. Similar results are obtained as before like the broadening of the minibands and closing of the gaps between the minibands. The difference is that now it is the second miniband (instead of the first miniband) which moves towards the first miniband and whose width is most strongly altered by the presence of the defect quantum well.

The closing of the miniband gap is very sensitive to the accuracy of centering the defect barrier in the quantum wells. This is illustrated in Fig. 5 where the miniband widths and the miniband gap is shown as function of the off-center position (s) of the defect barrier, where d is such that at $s = 0$ the gap between the two lowest minibands is closed. The strong sensitivity of the gap on the accuracy of centering the barrier in the quantum well may be a problem. This problem was also encountered in the case of positive defects in the barriers of a multi-quantum well as discussed by Beltram and Capasso [8]. Additional broadening of the energy levels, as a consequence of e.g. scattering with residual impurities and intralayer thickness fluctuations, may reduce this problem partially. Gumbs and Salam [9] have shown that specific choices of the shape of defect barriers may reduce this strong sensitivity on the centering.

The proposed superstructure, shown in Fig. 3, can be thought of as a superlattice with a complex basis and exhibits a much richer band structure than the conventional superlattices. There have been several works on superlattices with complex basis [10-13], some of which are generalizations of the present case.

4   Tuning of impurity levels

It has been demonstrated experimentally that the above discussed structures can be employed as IR-detectors [2]. In reverse mode those devices should act as emitters of IR-light. By applying a high gate voltage over the multi-quantum well structure electrons can be injected from the ground subband of a quantum well into the excited subband of the adjacent quantum well. This excited electron can relax to the ground state through emission of a photon. Unfortunately nonradiative phonon relaxation forms the major obstacle to produce efficient light emission out of such structure. The reason is that in a quantum well and in superlattices there are no true gaps in the electron spectrum due to the in-plane motion of the electron parallel to the interface. As a consequence electrons injected into the excited state of a quantum well will relax to the ground state through LO-phonon emission which is fast and nonradiative.

This problem can be avoided by constructing the quantum wells such that $E_2 - E_1 < \hbar\omega_{LO}$ which was realized experimentally by Helm $et\ al$ [14]. But it turned out that acoustic phonon emission now was the limiting factor for efficient IR emission.

Another way out of this problem is to have discrete energy levels, i.e. true energy gaps. This is realized in a quantum dot. In order to optimize the device in terms of minimal nonradiative phonon relaxation one should try to have $|E_i - E_j| > \hbar\omega_{LO}$ which requires confinement dimension such that $d < 150\text{Å}$. Present day technology is not advanced enough to produce reproducable and clean quantum dots of such small dimensions. The reason is that the strongest lateral confinement in a quantum dot is presently realized through the etching technique. A more clean way to realize confinement is e.g. through the growth of whiskers. But at present the controlled growth of whiskers is still in its infancy.

Most of the present day solid state lasers are based upon optical transitions between energy levels of impurities. Those impurities have a discrete spectrum and thus nonradiative phonon relaxation is eliminated. There are several drawbacks using such impurity systems: 1) the tuning of the energy levels is very limited (e.g. the energy levels of shallow donor impurities in GaAs are almost independent of the donor element); 2) the oscillator strength for bound-to-bound transitions is comparable to those of intersubband transitions in quantum wells. But free-to-bound transitions are required in order to convert current into radiation. For such transitions the optical cross section is very small and spread over a wide energy range limiting the energy sensitivity of the device.

One can try to combine the advantages of the discrete spectrum of impurities with the tunability and large oscillator strength of quantum wells by putting impurities in quantum wells. The wavefunction of the impurity states in the direction perpendicular to the interface is determined by the width of the quantum well structure while the wavefunction in the plane of the interface is determined by the Coulomb potential of the impurity which guarantees the discreteness of the energy spectrum.

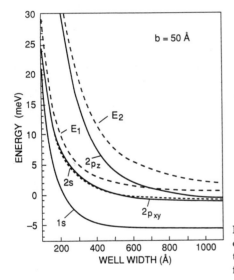

Fig. 6: The energy of the lowest levels of a donor in the center of the quantum well and the lowest minibands of the superlattice as a function of well width.

The tunability of the impurity energy levels is illustrated in Fig. 6 for a $GaAs/Al_{.3}Ga_{.3}As$ superlattice with barrier width of $b = 50$Å. This superlattice has a relative large barrier width such that its first two subbands, indicated by E_1 and E_2 respectively, have almost zero width. Notice that the $1s$, $2s$ and $2p_{x,y}$ impurity levels follow the lowest subband while the $2p_z$ is connected to the second subband. This can be understood by considering the wavefunction of the $2p_z$ state which is depicted in Fig. 7. This function has the same odd inversion symmetry in the z-direction as the second subband. This implies that tunability of E_2 will result in a tunability of the $2p_z$ level. Furthermore notice from Fig. 7 that the $2p_z$ state is spread out over the adjacent quantum wells of the superlattice which guarantees an appreciable overlap for tunneling of electrons from the adjacent wells into the $2p_z$ state of the central quantum well.

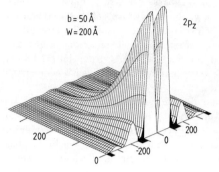

b = 50 Å
W = 200 Å

$2p_z$

200

200

0

-200

0

Fig. 7: The probability distribution of the $2p_z$ donor state in the ($\rho = \sqrt{x^2 + y^2}, z$) plane for an impurity located at the origin which is the middle of a quantum well. The position of the barriers is indicated by the solid areas.

In practice, transition energies are important. Therefore we show the allowed transition energies as function of the well width of the superlattice. We show the $1s \rightarrow 2p_{x,y}, 2p_z$ impurity transitions and the transitions $n \rightarrow n+1$ between the minibands of the superlattice, the width of it is illustrated by the thickness of the curves. The dots are the experimental results of Helm *et al* [15]. Notice the very limited tunability of the $1s \rightarrow 2p_{x,y}$ transition which is in contrast to the large variation of the transition energy $1s \rightarrow 2p_z$. The latter one has practically the same tunability as the miniband transition energies of the superlattice. Therefore the previous technique of tuning the minibands in a superlattice can also be applied to the one of the $2p_z$ impurity level. This was recently suggested by Parihar and Lyon [16].

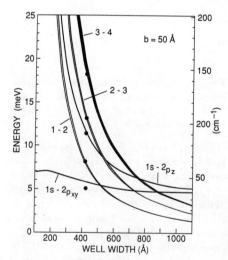

Fig. 8: Intersubband and donor transitions for a barrier width of 50Å as a function of well width. The symbols are the experimental results from a 400Å/50Å superlattice.

In previous example the nonradiative phonon relaxation is not completely quenched. The reason is that for $E_{2p_z} > E_1$ the impurity state can relax to a free electron state by emitting an acoustic phonon. Therefore in the case of Fig. 6 one has to take the well width $W > 700$Å. For impurities centered in the quantum well of a superlattice one always has $E_{2p_z} > E_{2p_{x,y}}$ and an electron in the $2p_z$-state can relax to the $2p_{x,y}$ state by acoustic phonon emission. As we showed in Fig. 8 the latter has a very limited tunability of its transition energy to the ground state and furthermore the $2p_{x,y}$ state is more localized in the z-direction and consequently much less effective in tunneling to the adjacent wells.

The above drawbacks can be circumvented by putting 'defect' barriers in the center of the quantum wells of the superlattice as illustrated in Fig. 3. Parihar and Lyon [16] recently showed theoretically that impurities localized in such defect barriers have the posibility to invert the position of the energy of the $2p_z$ and $2p_{x,y}$ states. Furthermore as demonstrated in previous section such a defect barrier is able to tune the energy of the second subband and consequently also the energy of the $2p_z$ state. This will decrease the width of the well of the superlattice at which $E_2 = E_{2p_{x,y}}$.

5   Conclusion

By putting barriers in the center of the well of a quantum well, or of a superlattice, allows one to tune the subbands and minibands to a desired value irrespective of the width of the quantum well. This allows one to grow superlattices or multi-quantum wells with narrow wells which have small energy separation between the energy levels. For applications in IR-devices such structures have the added flexibility required to overcome the disadvantages of having to build semiconductor structures with large well widths.

Furthermore such barriers have also the ability to tailor the energy of the impurity states in quantum wells and superlattices. This can have important implications for IR lasers.

Acknowlegments
    This work is supported by the Belgian National Science Foundation (NFWO). Part of this work was done in collaboration with P. Vasilopoulos and M. Helm.

REFERENCES

1. R. Dingle, in *Semiconductors and Semimetals* (Academic Press, New York, 1987), Vol. 24.

2. B.K. Levine, J. Appl. Phys. **74**, R1 (1993).

3. B.K. Levine, K.K. Choi, C.G. Bethea, J. Walker, and R.J. Malik, Appl. Phys. Lett. **50**, 1092 (1987).

4. B.K. Levine, C.G. Bethea, G. Hasnain, J. Walker, and R.J. Malik, Appl. Phys. Lett. **53**, 296 (1988).

5. P. Vasilopoulos, F.M. Peeters, and D. Aitelhabti, Phys. Rev. **B41**, 10021 (1990).

6. W. Trzeciakowski and B. D. McCombe, Appl. Phys. Lett. **55**, 891 (1989).

7. F.M. Peeters and P. Vasilopoulos, Appl. Phys. Lett. **55**, 1106 (1989).

8. F. Beltram and F. Capasso, Phys. Rev. **B38**, 3580 (1988).

9. G. Gumbs and A. Salam, Phys. Rev. **B41**, 10124 (1990).

10. S. Lin and J. Smit, Am. J. Phys. **48**, 193 (1980).

11. P. Yuh and K.L. Wang, Phys. Rev. **B38**, 13307 (1988).

12. S. Pan and S. Feng, Phys. Rev. **B44**, 5668 (1991).

13. J. Shi and S. Pan, Phys. Rev. **B48**, 8136 (1993).

14. M. Helm, E. Colas, P. England, F. DeRosa, and S.J. Allen, Jr., Appl. Phys. Lett. **53**, 1714 (1988).

15. M. Helm, F.M. Peeters, F. DeRosa, E. Colas, J.P. Harbison, and L.T. Florez, Phys. Rev. **B43**, 13983 (1991).

16. S.R. Parihar and S.A. Lyon, Appl. Phys. Lett. **63**, 2396 (1993).

OPTICAL AND ELECTRONIC PROPERTIES OF
GaAs/AlAs RANDOM SUPERLATTICES

E.G. WANG,* J.H. XU,** W.P. SU* ** AND C.S. TING,**
*Space Vacuum Epitaxy Center, University of Houston, Houston, TX 77204
**Texas Center for Superconductivity and Department of Physics, University of Houston, Houston, TX 77204

ABSTRACT

The optical and electronic properties of three-dimensional (3D) random GaAs/AlAs superlattices (SLs) has been studied by using a tight-binding Hamiltonian with second-neighbor interactions. We calculate three completely disordered sequences with the probability of GaAs layers being 30%, 50%, and 70%. The higher the GaAs composition, the narrower the indirect gap. An energy-level crossing is found at the bottom of conduction band, which originates from the M-state splitting induced by layer disorder. The localized states over two - four monolayers play an important role in the absorption edge of random SL. The highest absorption intensity of the band-edge transitions in our random models is about eight times stronger than that of short period ordered GaAs/AlAs SL. Our results are in good agreement with some recent photoluminescence measurements.

I. INTRODUCTION

Considerable effort has been devoted to the novel electro-optical devices fabricated from short period superlattice due to the ability to control the growth of each individual atomic layer. The best documented system is the $(GaAs)_m/(AlAs)_n$ $(n,m \leq 10)$ with the matched interfaces. However, as indicated by experimental and theoretical work[1], superlattices with m and n less than ten have indirect band gaps, which are inferior to the direct band semiconductor in optical absorption. Recent experiments[2] have demonstrated that short period random superlattices with n less than 4, exhibit remarkable intensity increase in photoluminescence. This is due to the quite different manner in description of the transition-matrix elements in an electromagnetic field. In ordered systems, the transition-matrix elements obey the k selection rule, while this rule is somewhat relaxed in disordered systems with their localized wave functions.

Although a theory capable of interpolating random media among the ordered super-lattices is desirable, no sufficiently detailed method to accomplish this end in a realistic three-dimensional systems currently exists. Many prior theories[3] have been designed to predict only trends of some specific quantities in one-dimensional models, such as optical transition and band-tail states in energy gap. As we know, in disordered SLs, the layers are randomly distributed along the growth direction, while the periodicity in the two other directions is well preserved. Both of these features are important in determining the electronic structure. Consequently, any reliable theoretical calculation of the electronic and optical properties of those random systems must rest on a realistic three-dimensional description.

We have proposed a periodic random superlattices $[(GaAs)_m/ (AlAs)_n]_l$[4], which are fabricated by repeating a large unit cell with a given period l along the growth direction, while the layers in the unit cell are randomly picked. For such realistic 3D random SLs, in this work, we introduce a tight-binding technique with stochastic functions that is designed to calculate the electronic structure and optical absorption of 3D random superlattices.

Mat. Res. Soc. Symp. Proc. Vol. 325. ©1994 Materials Research Society

II. TB HAMILTONIAN FOR 3D RANDOM SL

Our calculations are based on the 3D Hamiltonian

$$H_D = H_0 + \sum_i V(\mathbf{r} - \mathbf{r}_i) D(\delta_i) \tag{1}$$

where $D(\delta_i)$ is introduced to distinguish the random layers A:$(GaAs)_m$ and B:$(AlAs)_n$ picked along the growth direction. The traditional sp^3s^* [5] local orbitals are used as the basis functions in the tight-binding model. Therefore, we can expand the localized eigenfunctions $|\Psi_n\rangle$ in terms of the atomic orbitals $|\Phi_\alpha(\mathbf{k}, i)\rangle$,

$$|\Psi_n\rangle = \sum_{\alpha,i} \mathbf{C}_\alpha^n(\mathbf{k}, i)|\Phi_\alpha(\mathbf{k}, i)\rangle \tag{2}$$

where α is the orbital symmetry (here chosen among s, x, y, z, s^*), $\mathbf{k}$ is the momentum, and i labels the atomic position.

The Vogl's tight-binding parameters[5] are modified by using Yamaguchi formulae[6]. In addition, the second-neighbor interactions as given by Newman and Dow[7] have been adopted to allow an exact description of L-point energy of bulks. The calculated indirect band gap is 2.09 eV for the $m=n=1$ ordered SL by the above method, which is very close to the known experimental band gap 2.07 eV[8].

The intramaterial matrix elements are determined by using their bulk values. In the same way, we get the intermaterial nearest interactions as they share common anions at the interface. Some additional specifications should be noted for the interface second neighbor interactions of disordered SLs which consist of ultrathin GaAs and AlAs layers. In the case, we use an average of their corresponding bulk values to estimate the element $\langle\Phi_\alpha(i)|H_{AB}|\Phi_\beta(j)\rangle_2$ between two cations if i in GaAs (AlAs) and j in AlAs (GaAs), or between two anions at same interface or at different interfaces separated by a cation monolayer.

III. INFLUENCE OF RANDOM ON ELECTRONIC STRUCTURE

The band structures of the random superlattices $[(GaAs)_m/(AlAs)_n]_l$[4] have been carried out by using the tight-binding method described in Sec II. The disordered configurations with different compositions of GaAs and AlAs are produced by a random number generator. Three types of the sequences are chosen to study in more details here. They are:

1) d-SL_1: the sequence is constructed completely randomly with the probability of A-layers being 30% and B-layers being 70%.

2) d-SL_2: the sequence is constructed completely randomly by picking the A-layers and B-layers with equal probability.

1) d-SL_3: the sequence is constructed completely randomly with the probability of A-layers being 70% and B-layers being 30%.

The band structure calculations for periodic random SLs with $l = 10 - 20$ and $m=n=1$ are performed for various high symmetry points and lines in the Brillouin zone, taking into account the valence band offset $\Delta E_v = 0.5$ eV. For a direct comparison with the corresponding ordered SL, the band structure of $(GaAs)_1/(AlAs)_1$ with a supercell {ABABABABABABAB ABABAB} named o-SL is also calculated.

It is found that all the systems studied here have indirect band gaps. The introduction of the layer disorder induces a change in the relative order of the lower conduction bands. The lowest conduction band is at symmetry point M in a random SL, while the conduction band energies at X- and R- points are lower than that at M for an ordered SL, which is clearly shown in Table I. The indirect energy gap is very sensitive to the composition.

The higher the composition of the GaAs layers, the narrower the gap of the corresponding random SL.

Table I. Energies (in eV) of the lower conduction bands at different high symmetry points, where the energy zero is chosen at the top of the valence band.

	o-SL	d-SL$_1$	d-SL$_2$	d-SL$_3$
$\Gamma_c(0,0,0)/a$	2.131	2.109	2.023	1.837
$X_c(\sqrt{2}\pi,0,0)/a$	2.083	2.173	2.051	1.942
$M_c(\sqrt{2}\pi,\sqrt{2}\pi,0)/a$	2.104	2.049	1.937	1.835

Fig.1 shows the calculated band structures with the energy zero chosen at the top of valence band, where (a) corresponds to the ordered $(GaAs)_1/(AlAs)_1$ SL and (b) corresponds to the d-SL$_2$ disordered model, respectively. The lowering of the conduction band states at M originates from the splitting induced by layer disorder. Three energy levels with energies 1.937 eV, 2.054 eV, and 2.177 eV at the band edges of the point M of the random SL are diverged from the M state (E=2.104 eV) of the ordered SL. It can be seen from Fig.1 that the existence of disorder tends to push the energy levels to the band edge, and hence, enhance the densities of states near the band edge region.

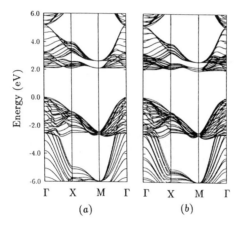

Fig.1 Calculated band structures of the $(GaAs)_1/(AlAs)_1$ ordered SL (a) with supercell {ABABABABABABABABABAB} for a direct comparison, and the d-SL$_2$ disordered SL (b).

IV. LOCALIZATION OF EIGENSTATE

To examine the localization of eigenstates in a 3D random SL, the electron distributions with atomic positions are calculated in real space. Fig.2 shows the planar average charge densities of o-SL ordered and d-SL$_2$ disordered SLs along the growth direction for the states at the bottom of the conduction band (a) and the top of the valence band (b). For the ordered system, the wave functions of the electron (hole) states are characterized by Bloch phase, which result in a periodic pattern of the electron (hole) distribution. The different amplitudes of the charge density in GaAs and AlAs layers are due to the potential difference of the cations at interface.

In a disordered SL, as expected, the electron states at the bottom of the conduction band are strongly localized over two to four monolayers (see Fig.2 (a)). However, it is rather surprising that the charge distributions corresponding to the states at the top of the valence band are rather extended (see Fig.2 (b)). It is certain that, at least based on our results, both localized and extended states exist at the band edges of a 3D random SL. The disorder affects mostly the conduction band states.

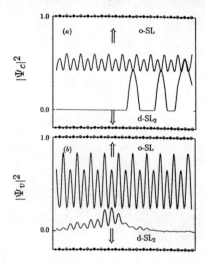 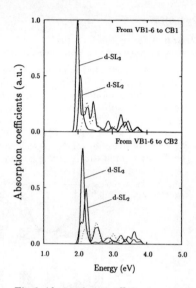

Fig.2 Plots of planar averaged charge densities of the ordered and the d-SL$_2$ disordered superlattices for the lowest M conduction state (a) and the highest Γ valence state (b) . • , As; ◆, Ga; ◇, Al.

Fig.3 Absorption coefficients corresponding to the optical transitions from total top six valence subbands VB1-6 to CB1 (top panel) and to CB2 (bottom panel) for ordered (dotted lines) and disordered SLs (solid lines).

V. OPTICAL ABSORPTION

The common form of the absorption coefficient is given by

$$\alpha(\hbar\omega) = \omega^{-1} \sum_{k} \sum_{m,n} |\hat{\epsilon} \cdot \mathbf{P}_{nm}(\mathbf{k})|^2 \delta(E_m(\mathbf{k}) - E_n(\mathbf{k}) - \hbar\omega) \tag{3}$$

where $\hat{\epsilon}$ is the direction of polarization and $\mathbf{P}_{nm}(\mathbf{k})$ is the momentum matrix element between superlattice eigenstates

$$\mathbf{P}_{nm}(\mathbf{k}) = \sum_{\alpha\beta} \sum_{i,j} \mathbf{C}_{\alpha}^{n^*}(\mathbf{k}, i) \mathbf{C}_{\beta}^{m}(\mathbf{k}, j) \langle \Phi_{\alpha}(\mathbf{k}, i) | \mathbf{P} | \Phi_{\beta}(\mathbf{k}, j) \rangle \tag{4}$$

All the optical matrix elements $\langle \Phi_{\alpha}(\mathbf{k}, i) | \mathbf{P} | \Phi_{\beta}(\mathbf{k}, j) \rangle$ for the SLs are chosen from the parameters of the bulk materials, which were given by Chang et al.[9] . In figure 3, we present the absorption coefficients for the transitions between the total top six valence subbands (VB1-6) and the separate bottom two conduction subbands (CB1 and CB2), where the dotted lines correspond to that of the o-SL ordered SL. Here we consider only the (x,y) polarization, supposing the incident light is propagation along the SL growth direction. It is obvious that the localized states induced by disorder play an important role in the band edge absorption of a random SL. They enhance the optical transitions in this region considerably. The first absorption peaks for the transitions from VB1-6 to CB1 are shifted towards the lower energy region by 0.26 eV and 0.19 eV, respectively, for d-SL$_3$ and d-SL$_2$, compared with that of ordered SL.

Finally, the calculated total absorption coefficients of the ordered and disordered SLs are plotted in Fig.4, where (a), (b), (c), and (d) correspond to o-SL, d-SL$_1$, d-SL$_3$, and d-SL$_3$ respectively. For comparison, the joint densities of states for these materials are also shown by the dotted curves. For an ordered short-period SL $(m=n=1)$, two main peaks are found at 3.7 eV and 4.8 eV which correspond one by one to those of bulk GaAs.[10] The dramatic changes between the ordered and the disordered systems occur near the absorption edges. This can be explained by the conduction band tails formed by strongly localized states in a random SL, which make the optical transitions no longer forbidden in this region . The absorption intensities of the band edge transitions for d-SL$_3$ and d-SL$_2$ models are about 8 and 5 times stronger than that of the o-SL, respectively. Our results are in fairly good agreement with the photoluminescence (PL) measurements at 4.2 K, where Yamamoto et al[2] has pointed out the short period SL radiate PL less than one tenth of the disordered SL.

Away from the absorption edge, the main features of the band center optical transitions are still retained as seen in Fig.4. For an ordered SL, the highest absorption peak is located at 4.8 eV, but it is shifted below 4.0 eV in a disordered systems. The different effect of the disorder on the conduction and valence band states leads the coexistence of localized and extended band-edge states in a 3D random SL.

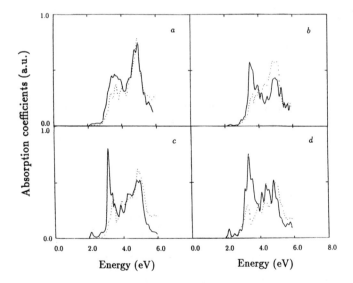

Fig.4 Absorption coefficients of the o-SL (a) ordered, d-SL$_1$ (b), d-SL$_2$ (c), and d-SL$_3$ (d) disordered superlattices, where the dotted lines stand for the joint densities of states.

VI. CONCLUSION

We have studied the electronic structures and optical properties of 3D disordered SL, based on a tight-binding Hamiltonian with stochastic functions. The second-neighbor interactions included in the theory are essential for obtaining the correct band-edge energies for the bulks and the short-period GaAs/AlAs superlattices at different high symmetry points. For three kinds of disordered SLs with the probability of GaAs layers being 30%, 50%, and 70%, our band structure calculations show that the random systems have indirect gaps, which is very sensitive to the composition. The higher the GaAs composition, the narrower the indirect gap. An energy-level crossing is found at the bottom of conduction band, which originates from the M-state splitting induced by disorder. The different effect of the disorder on conduction and valence band states leads to the coexistence of localized and extended band-edge states in a 3D random SL. We find that disorder affects mostly conduction bands. It is noticed that the localized states play an important role in the edge absorption of a random SL. The absorption intensities of the band-edge transitions for d-SL$_3$ and d-SL$_2$ random models are about eight and five times stronger than that of the short period SL, respectively. A contrary case occurs in d-SL$_1$ disordered SL with more AlAs layers, where the band-edge absorption intensity is weaker than that of the ordered one. Thus, increasing of the GaAs composition in disordered SLs is useful to enhance the optical absorption near the band edge. Away from the absorption edge, the main features of the band center optical transitions are still retained.

ACKNOWLEDGMENTS

EGW and JHX would like to thank Y. A. Zhang in SVEC at University of Houston for her helpful discussions during the course of this work. WPS thanks Y.C.Chang for bringing Ref.13 to our attention. The work was supported by the Texas Advanced Technology Program under Grant No. 003652-228, Texas Center for Superconductivity at University of Houston, and a grant from the Robert A. Welch foundation.

References

[1] A. Chomette, B. Devesud, A. Regreny, and G. Bastard, Phys. Rev. Lett. **57**, 1464 (1986).

[2] T. Yamamoto, M. Kasu, , S. Noda, and A. Sasaki, J. Appl. Phys. **68**, 5318 (1990).

[3] J. Strozier, Y. A. Zhang, C. Horton, A. Ignatiev, and H. D. Shih, Appl. Phys. Lett. **62**, 3426(1993); W. P. Su, and H. D. Shih, J. Appl. Phys. **72**, 2080 (1992).

[4] E. G. Wang, J. H. Xu, W. P. Su, and C. S. Ting, Appl. Phys. Lett. **63**, 1411 (1993).

[5] P. Vogl, H. P. Hjalmarson, and J. D. Dow, J. Phys. Chem. Solids **44**, 365 (1983).

[6] E. Yamaguchi, J. Phys. Soc. Jpn. **56**, 2835 (1987).

[7] K. E. Newman, and J. D. Dow, Phys. Rev. **B30**, 1929 (1984).

[8] Weikun Ge, M. D. Sturge, W. D. Schmidt, L. N. Pfeiffer, and K. W. West, Appl. Phys. Lett. **57**, 55 (1990).

[9] Y. C. Chang, and J. N. Schulman, Phys. Rev. **B31**, 2069 (1985).

[10] Zhizhong Xu, Solid State Com. **75**, 1143 (1990).

A TIGHT-BINDING THEORY OF THE ELECTRONIC STRUCTURES FOR RHOMBOHEDRAL SEMIMETALS AND Sb/GaSb, Sb/AlSb SUPERLATTICES

J.H. XU,* E.G. WANG,** C.S. TING,* AND W.P. SU* **
*Texas Center for Superconductivity and Department of Physics, University of Houston, Houston, TX 77204
**Space Vacuum Epitaxy Center, University of Houston, Houston, TX 77204

ABSTRACT

The band structures of three group-V semimetals As, Sb, and Bi with rhombohedral $A7$ symmetry are studied using a second-neighbor tight-binding model including spin-orbit interaction with an sp^3s^* basis. Then the bulk tight-binding parameters are used to investigate the electronic properties of semimetal-semiconductor superlattices made of alternating (111) layers of Sb and GaSb or AlSb. It is found that the band gap can be adjustable depending primarily on the thickness of the Sb layers. An interface state is observed in the region of the gap.

I. INTRODUCTION

Recent experiments have demonstrated that semimetallic antimony layers can be grown on a semiconductor GaSb substrate grown along the (111) direction by molecular beam epitaxy (MBE) technique.[1] Bi/CdTe (111) heterostructures have also been successfully fabricated.[2] Bulk Sb and Bi are group-V semimetals with equal numbers of electrons and holes. Their conduction band minima (at the L-point) lie at lower energy than the valence band maxima (at the H-point in Sb, at the T-point in Bi). Thus, an interesting quantum size effect is expected if the carriers in these semimetals are spacially confined. Such confinement can be achieved by sandwiching the semimetal film between layers of a suitable non-matallic barrier material. When the thickness of the semimetal film decreases, the electron subbands should move up in energy while the hole subbands move down. At a certain thickness, the electron and hole subbands will cross and a semimetal-semiconductor transition occurs. It has recently been suggested that a narrow-gap semiconductor whose band alignment is indirect in momentum space would have highly attractive properties in optical and electro-optical device application.[3] Indirect narrow-gap heterostructures such as Sb/GaSb (111), Sb/AlSb (111) and Bi/CdTe (111) superlattices could potentially open new possibilities for devices.

In the present work, we first present a second-neighbor tight-binding theory for bulk rhombohedral crystals. We show that a tight-binding method using a few interaction parameters can give a reasonable band structure for group-V semimetals. Then the bulk parameters of antimony are used to calculate the electronic properties of the semimetal-semiconductor Sb/GaSb and Sb/AlSb superlattices.

II. TIGHT-BINDING THEORY FOR BULK SEMIMETALS

The tight-binding method for bulk As, Sb, and Bi we use here is equivalent to that of Slater and Koster.[4] Our model has the following properties: (i) the chemistry of s^2p^3 bonding is manifestly preserved. (ii) The spin-orbit interaction is included in the theory. And (iii) the theory successfully reproduces not only the valence bands but also the lower conduction bands, even in indirect (negative) gap semimetals.

Mat. Res. Soc. Symp. Proc. Vol. 325. ©1994 Materials Research Society

In rhombohedral A7 structure crystals, there are two atoms in the primitive cell. For each tight-binding basis function centered on these atoms, two Bloch functions can be constructed:

$$|n, \alpha, \mathbf{k}\rangle = \frac{1}{\sqrt{N}} \sum_i e^{i\mathbf{k}\cdot\mathbf{R}_i + i\mathbf{k}\cdot\mathbf{r}_\alpha} |n, \alpha, \mathbf{R}_i\rangle. \tag{1}$$

The quantum number n runs over the $s\beta$, $p_x\beta$, $p_y\beta$, $p_z\beta$, and $s^*\beta$ orbitals with β ($= \uparrow$ and $\downarrow$) being spin index. The N wavevectors $\mathbf{k}$ lie in the first Brillouin zone; the site index α is either 1 or 2. The atom 1 is located at $\mathbf{R}_i$. The Hamiltonian matrix in the $|n, \alpha, \mathbf{k}\rangle$ basis can be calculated from the Slater and Koster approach[4]

$$H_{\alpha\alpha'}(\mathbf{k}) = \sum_i e^{i\mathbf{k}\cdot\mathbf{r}_i} \langle 0, \alpha | H | i, \alpha' \rangle, \tag{2}$$

We have obtained these matrix elements for rhombohedral semimetals and their explicit expressions will be given elsewhere.[5] The interaction parameters can be obtained by fitting *ab initio* results.[6] The resulting band structure of Sb is shown in Fig.1 where the Fermi level is chosen as energy zero. One can see from this figure that the minimum of the lowest conduction band corresponds to L, while the maximum of the valence band is at H, which agrees well with the *ab initio* calculation.[6]

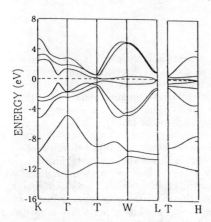

Fig.1. Band structure of Sb. The usual notation for symmetry points has been used.

III. BULK GaSb and AlSb PARAMETERS

The tight-binding parameters for GaSb are determined by fitting the pseudopotential calculation.[7] While for AlSb there has been no available pseudopotential or first principle band structure results in the literature that take into account the spin-orbit interaction. We determine the tight-binding parameters for AlSb based on the sp^3s^* basis by the following two steps: first the parameters are determined by fitting the pseudopotential results without spin-orbit effect;[8] then the parameters are adjusted to include the experimental values of the spin-orbit splitting.[9] Our parameters for AlSb give an indirect gap of 1.657 eV and a correct order of conduction band minima L-X-Γ, as well as the correct spin-orbit splitting energies $[\Delta_0(\Gamma_{15v}) = 0.75 \text{ eV and } \Delta_0'(\Gamma_{15c}) = 0.3 \text{ eV}]$. The parameters for bulk Sb determined in the last section give an overlap between conduction and valence bands about 120 meV. The values of the various interaction parameters are given in Table I.

Table 1: Tight-binding parameters (in eV) for Sb, GaSb, and AlSb. The off-site matrix elements (V's) are in the standard notation of Slater and Koster. Here V and V' stand for nearest- and next nearest-neighbor interactions in Sb respectively, and λ's are the spin-orbit couplings.

Sb			GaSb	AlSb
E_s	-7.7085	E_{sa}	-7.2820	-6.3720
E_{s^*}	20.1440	E_{sc}	-3.8580	-2.2680
E_p	-0.1337	E_{s^*a}	8.2000	8.4600
$V_{ss\sigma}$	-0.9973	E_{s^*c}	8.8000	9.0700
$V_{sp\sigma}$	2.2740	E_{pa}	0.6743	1.1580
$V_{s^*p\sigma}$	3.1952	E_{pc}	2.7323	3.1920
$V_{pp\pi}$	-0.6900	$V_{ss\sigma}$	-1.5495	-1.5682
$V_{pp\sigma}$	2.2776	$V_{sa,pc\sigma}$	2.0342	1.8942
$V'_{ss\sigma}$	-0.3880	$V_{s^*a,pc\sigma}$	2.0517	2.2184
$V'_{sp\sigma}$	0.3217	$V_{sc,pa\sigma}$	1.8676	1.5711
$V'_{s^*p\sigma}$	1.0144	$V_{s^*c,pa\sigma}$	1.6961	1.8954
$V'_{pp\pi}$	-0.3760	$V_{pp\pi}$	-0.6282	-0.4681
$V'_{pp\sigma}$	1.7880	$V_{pp\sigma}$	2.4992	2.5796
λ	0.2480	λ_a	0.3122	0.3428
		λ_c	0.0512	0.0072

IV. BAND STRUCTURE OF Sb/GaSb AND Sb/AlSb SUPERLATTICES

In this section, we study the electronic properties of Sb/GaSb and Sb/AlSb superlattices using the bulk tight-binding parameters obtained above. These superlattices could be interesting narrow-gap materials for applications in infrared devices. A typical $(Sb_2)_2/(GaSb)_2$ or $(Sb_2)_2/(AlSb)_2$ superlattice structure is depicted in Fig.2.

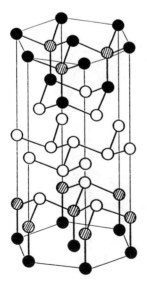

● Ga (Al) ◍ Sb ○ Sb

Fig.2. The crystal structure of $(Sb_2)_2/(GaSb)_2$ or $(Sb_2)_2/(AlSb)_2$ (111) superlattices.

It can be seen from Fig.2 that there are two kinds of interfaces in these superlattices: the Ga-Sb (Al-Sb) interface and the Sb-Sb interface. Fortunately both interface bonds exist already in bulk GaSb (AlSb) and Sb. So it is reasonable to assume that the Ga-Sb (Al-Sb) and Sb-Sb interface bonds remain the same lengths as in bulk GaSb (AlSb) and Sb respectively.

In the absence of experiments on the band line-up of Sb/GaSb and Sb/AlSb superlattices, the band-offset is undetermined. We assume the band-offset so that the Fermi energy of Sb lies at the effective middle gap positions:[10] i.e., 0.07 eV and 0.45 eV above the valence band maximum of GaSb and AlSb respectively. For Sb/GaSb this assumption is in coincidence approximately with the prediction of the common anion rule.[11] This rule, however, is believed to be inapplicable to the Al-compounds.[11] So we determine the band line-up for Sb/AlSb superlattice using the middle gap theory in our calculation.

The band structures of the $(Sb_2)_2/(GaSb)_8$ and $(Sb_2)_2/(AlSb)_8$ superlattices are shown in Fig.3. Eight GaSb (AlSb) layers are chosen so that the superlattice behaves like a quantum well as far as the near Fermi-level Sb states are concerned. All energies are measured with respect to the valence band maximum of bulk GaSb (AlSb). We see from the figure that the results for $(Sb_2)_2/(GaSb)_8$ and those for $(Sb_2)_2/(AlSb)_8$ superlattices are very similar, and they are all semiconductors with an indirect band gap since the valence-band maximum (H'_v point) and the conduction-band minimum (Γ_c point) are at different points in momentum space. In addition, we find an interface state (denoted by IS in Fig.3) which lies in the gap region. Fig.4 shows the charge density of this interface state at the M and X points of the Brillouin zone. It is obvious that this interface state has a sharp peak at the interface, indicating that the strong localization of this state appears around the M and especially the X points. But this state still has some extended components in the Sb layers. We have also calculated the charge density of the band edge states (not shown in the figure) and found that all band edge states are confined two dimensionally in the Sb layers.

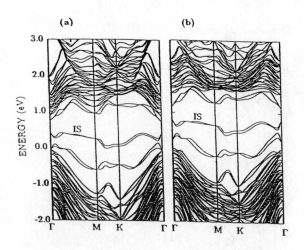

Fig.3. Band structures for (a) $(Sb_2)_2/(GaSb)_8$ and (b) $(Sb_2)_2/(AlSb)_8$ superlattices.

The energy differences between the conduction band edge states and the valence band maximum (at the H′ point) have been calculated as a function of m, the number of the Sb layers, with fixed semiconductor thickness $n = 8$ layers. As shown in Fig.5, the $(Sb_2)_m/(GaSb)_8$ and $(Sb_2)_m/(AlSb)_8$ superlattices for $m \leq 10$ are semiconductors with an adjustable band gap.

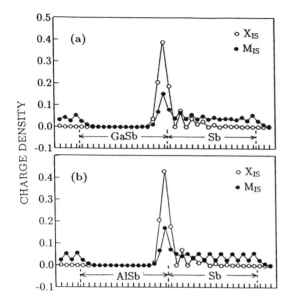

Fig.4. Charge densities of interface states for (a) $(Sb_2)_8/(GaSb)_8$ and (b) $(Sb_2)_8/(AlSb)_8$ superlattices.

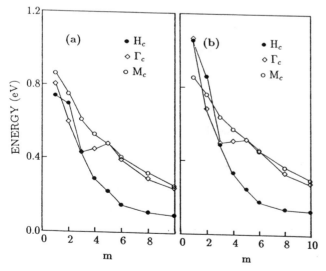

Fig.5. The energy differences between the conduction band edge states and the valence band maximum of (a) $(Sb_2)_m/(GaSb)_8$ and (b) $(Sb_2)_m/(AlSb)_8$ superlattices as a function of m.

Both of them exhibit a transition from indirect gap to direct gap elctronic structure around $m = 3$. The energies of interface state depend on both the interface length and the band offset. We have tried several different values of the band offset and found that the interface state does not disappear from the gap for any reasonable offset.

V. CONCLUSIONS

In conclusion, We have developed a tight-binding theory for group-V semimetals. The bulk parameters have been used to calculate the electronic structures of Sb/GaSb and Sb/AlSb superlattices. Our results show that these superlattice are good candidates for new narrow gap materials and may have potential applications in optoelectronic devices. Furthermore, we have found an interface state located at the gap region which may affect the determination of the intrinsic band gap. We are continuing our theoretical study of the properties of these superlattices and interface.

ACKNOWLEDGMENTS

This research was supported by Texas Center for Superconductivity at University of Houston, a grant from the Robert A. Welch foundation, and the Advanced Technology Program of the Texas High Education Coordinating Board under Grant No. 003652-228.

References

[1] T. J. Golding, J. A. Dura, W. C. Wang, J. T. Zborowski, A. Vigliante, J. H. Miller, and J. R. Meyer, Semicond. Sci. Technol. 8, S117 (1993).

[2] A. Divenere, X. J. Xi, C. L. Hou, J. B. Ketterson, G. K. Wong, and I. K. Sou, Appl. Phys. Lett. (in press).

[3] J. R. Meyer, F. J. Bartoli, E. R. Youngdale, and C. A. Hoffman, J. Appl. Phys. 70, 4317 (1991).

[4] J. C. Slater and G. F. Koster, Phys. Rev. 94, 1498 (1954).

[5] J. H. Xu, E. G. Wang, C. S. Ting, and W. P. Su, Phys. Rev. B48, December, 1993, in press.

[6] X. Gonze, J. P. Michenaud, and J. P. Vigneron, Phys. Rev. B41, 11827 (1990).

[7] J. R. Chelikowsky and M. L. Cohen, Phys. Rev. B14, 556 (1976).

[8] H. Hathieu, D. Anvergne, P. Marle, and K. C. Rustagi, Phys. Rev. B12, 5846 (1975); A. B. Chen and A. Sher, Phys. Rev. B23, 5360 (1981).

[9] A. Joullie, B. Girault, A. M. Joullie, and A. Zien-Eddine, Phys. Rev. B25, 7830 (1982).

[10] J. Tersoff, Phys. Rev. Lett. 56, 2755 (1986).

[11] J. O. McCaldin, T. C. McGill, and C. A. Mead, Phys. Rev. Lett. 36, 56 (1976).

THE GROWTH OF InAsSb/InGaAs STRAINED-LAYER SUPERLATTICES BY METAL-ORGANIC CHEMICAL VAPOR DEPOSITION

R. M. Biefeld, K. C. Baucom, S. R. Kurtz, and D. M. Follstaedt
Sandia National Laboratory, Albuquerque, NM

ABSTRACT

We have grown $InAs_{1-x}Sb_x/In_{1-y}Ga_yAs$ strained-layer superlattice (SLS) semiconductors lattice matched to InAs using a variety of conditions by metal-organic chemical vapor deposition. The V/III ratio was varied from 2.5 to 10 at a temperature of 475 °C, at pressures of 200 to 660 torr and growth rates of 3 - 5 Å/s and layer thicknesses ranging from 55 to 152 Å. The composition of the InAsSb ternary can be predicted from the input gas molar flow rates using a thermodynamic model. At lower temperatures, the thermodynamic model must be modified to take account of the incomplete decomposition of arsine and trimethylantimony. Diodes have been prepared using Zn as the p-type dopant and undoped SLS as the n-type material. The diode was found to emit at 3.56 µm. These layers have been characterized by optical microscopy, SIMS, x-ray diffraction, and transmission electron diffraction. The optical properties of these SLS's were determined by infrared photoluminescence and absorption measurements.

INTRODUCTION

The growth of $InAs_{1-x}Sb_x/In_{1-y}Ga_yAs$ SLS's is being explored by us for their possible use in mid-wave, 2-5 µm infrared optoelectronic and heterojunction devices. This system was chosen because the compositions span the 2-5 µm wavelength range and they can be grown lattice matched to InAs. Recent results by Menna et al. on a metal-organic chemical vapor deposition (MOCVD) grown 3.06 µm diode laser with a maximum operating temperature of 35 K and threshold current densities of 200 - 330 A/cm² indicate the potential and the need for devices operating in this wavelength range.[1] Our previous studies in the Sb rich end of the InAsSb ternary system have demonstrated accurate composition control through the use of a thermodynamic model and high quality infrared detectors have been made in our laboratory from doped strained-layer superlattices (SLS's) grown by MOCVD in the Sb rich end of this ternary.[2,3]

The initial studies reported on here focused on the growth of $InAs_{1-x}Sb_x/In_{1-y}Ga_yAs$ heterostructures on InAs. The studies concentrated on the growth and characterization of the InAsSb layers due to the importance of this material in the active devices. The thermodynamic model used to describe composition control in the Sb rich compositions can also be used to predict the composition of As rich materials. The details of the growth conditions for $InAs_{1-x}Sb_x/In_{1-y}Ga_yAs$ SLS's and $InAs_{1-x}Sb_x$ alloys, and the growth and characterization of an infrared photodiode and the applicability to infrared devices of As rich $InAs_{1-x}Sb_x$ materials are also discussed.

EXPERIMENTAL

This investigation was carried out in a previously described horizontal MOCVD system.[4] The sources of In, Sb and As were trimethylindium (TMIn), trimethylantimony (TMSb), trimethylgallium (TMG), and 100 percent arsine (AsH₃), respectively. Hydrogen was used as the carrier gas at a total flow of 6 SLM. The III/V ratio was varied from 2.7 to 10.3 over a temperature range of 475-525 °C, at total growth pressures of 200 to 660 torr and growth rates

Mat. Res. Soc. Symp. Proc. Vol. 325. ©1994 Materials Research Society

of 0.75 to 3.0 μm/h. The growth was performed on (001) InAs substrates. Diodes have been prepared using Zn as the p-type dopant with diethylzinc as the precursor and undoped SLS as the n-type material. InAs was cleaned by degreasing in solvents and deionized water. It was then etched for one minute in a 10:1 mixture of sulfuric acid and hydrogen peroxide, rinsed with deionized water and blown dry with nitrogen.

Sb compositions, x, reported for the $InAs_{1-x}Sb_x$ layers were determined by x-ray diffraction using a Cu x-ray source and a four crystal Si monochromator. The (004) reflection was used to measure the lattice constant normal to the growth plane and the (115) or (335) reflections were used to determine the in-plane lattice constant.[4] In this way the composition determination is corrected for partial strain relaxation by misfit dislocations. Some samples were also checked for impurities using secondary ion mass spectroscopy (SIMS).

Infrared photoluminescence was measured at 14 K using a double-modulation technique which provides high sensitivity, reduces sample heating, and eliminates the blackbody background from infrared emission spectra.[5] The absorption measurements were carried out on polished samples using a Fourier transform infrared spectrometer.

Cross-sectional specimens of some samples were prepared by cleaving along (110) and epoxying two alloy surfaces together. A cylinder centered on the epoxied interface was ultrasonically cored and epoxied into a brass tube for mechanical strength. Disks were sliced from the tube and were mechanically thinned at their centers with a rotating wheel "dimpler" using 1 μm diamond paste. Final thinning was done by ion milling with Ar, followed by a short milling time in I_2 vapor to remove In residue on the surface. Transmission electron microscopy (TEM) examination was done using a Philips CM20T (200 keV) microscope. The specimens were examined in a <110> direction perpendicular to the [001] growth direction. Both transmission electron diffraction (TED) patterns and TEM images were obtained. Surface optical photomicrographs were obtained with a Zeiss Ultraphot microscope equipped with Nomarski interference contrast objectives.

RESULTS AND DISCUSSION

The growth of $InAs_{1-x}Sb_x/In_{1-y}Ga_yAs$ SLS's was investigated using III/V ratios between 2.7 and 10.3 at a temperature of 475 °C, at total growth pressures of 200 to 660 torr and growth rates of 1-1.8 μm/h. The group V molar fraction of TMSb in the vapor phase $[n_{TMSb}/(n_{TMSb} + n_{AsH3})]$ was varied from 0.06 to 0.09. The growth results for the As rich end of the $InAs_{1-x}Sb_x$ ternary are similar to those described previously for the Sb rich end of the ternary.[6] The growth rate was found to be proportional to the TMIn and the TMGa flow into the reaction chamber and independent of the TMSb and AsH3 flow.

The observed trends for the effects of input vapor concentrations on the resulting solid composition can be completely described by a thermodynamic model.[6] The model predicts that the thermodynamically more stable III/V compound will control the composition. For the InAsSb system when III/V < 1, As is preferentially incorporated into the solid and the solid-vapor distribution coefficient of Sb (k_{Sb}) is < 1. This is because InAs is more stable, has a lower free energy of formation, than InSb at 475-525 °C. For III/V ratios close to one, k_{Sb} approaches one and for III/V ≥ 1, k_{Sb} = 1. When III/V ≥ 1, all of the group V materials, As and Sb, will be incorporated into the solid at their vapor concentrations. These trends are also found for this data for As rich InAsSb. However, some slight deviations from the predicted behavior of the thermodynamic model are observed for the present results. For some of the samples grown at 475 °C, k_{Sb} appears to be ≥ 1 . This can be explained by assuming that not all of the AsH3 is decomposed at this temperature. This assumption is consistent with the reported incomplete decomposition of AsH3 at temperatures below 500 °C.[7]

An x-ray diffraction pattern of the (004) reflection of a selected $InAs_{1-x}Sb_x/In_{1-y}Ga_yAs$ SLS is shown in Figure 1. The pattern shown is for sample CVD1242 which was also fabricated into a light emitting diode (LED) structure. The calculated pattern is for an $InAs_{0.91}Sb_{0.09}/In_{0.94}Ga_{0.06}As$ SLS with 110 Å/ 116 Å layer thicknesses. The slight asymmetry

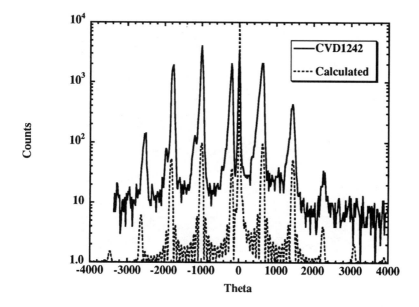

Figure 1. The x-ray diffraction pattern of the (004) reflection of CVD1242 indicates a lattice mismatch with InAs equal to -200 arc seconds which corresponds to a $\Delta a/a_0$ of 1.6×10^{-3}.

and line widths of the peaks can be explained by assuming either that there are some dislocations in the structure or that layers of $In_{1-x}Ga_xAs_{1-y}Sb_y$ occur at interfaces in the SLS.

The bandgap of the unstrained, $InAs_{0.9}Sb_{0.1}$ alloy was determined to be approximately 270 meV from optical studies of bulk ternary alloy and subsequent studies of SLSs and quantum wells. The accepted value for $InAs_{0.9}Sb_{0.1}$ is ≈ 330 meV.[8] Throughout our studies of As-rich, InAsSb (5-50% Sb), the optically determined bandgap of InAsSb alloys was smaller than accepted values.[8] This InAsSb bandgap anomaly was observed in both MOCVD and MBE grown samples.

Figure 2 shows the transmission electron diffraction (TED) pattern for an $nAs_{0.86}Sb_{0.14}$ alloy. The TED shows bright reflections of the <110> zone pattern of the zinc blend lattice as well as non-zinc-blende reflections at half the distance between (000) and {111} spots. These weak reflections indicate that compositional ordering of the Cu-Pt type is occurring in $InAs_{0.86}Sb_{0.14}$. The type of ordering which occurs is the same as that which was previously observed for $InAs_{0.6}Sb_{0.4}$ alloys and SLSs and is on the {111} planes of the group-V sublattice (CuPt-type).[9] Compositional ordering of InAsSb should result in bandgap reduction.[9,10] Details about ordering and the optical characterization of these SLSs will be discussed in a later publication.

Initial attempts to dope these SLSs produced infrared LEDs. The LED consisted of 1 μm of p-type SLS (Zn $\approx 5 \times 10^{18}$ cm^{-3}) grown on 2 μm of undoped SLS (n-type background doping) on an n-type InAs substrate. The structure of the SLS determined by x-ray analysis was

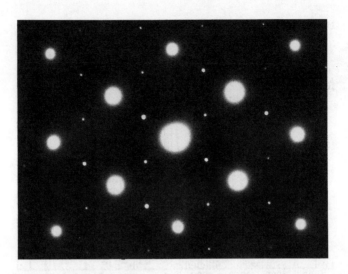

Figure 2. Transmission electron diffraction pattern from a <110> cross section of an InAs$_{0.86}$Sb$_{0.14}$ alloy grown by MOCVD on (001)-oriented InAs. The strong reflections are due to the zinc-blende lattice and the weak ones to CuPt-like ordering.

InAs$_{0.91}$Sb$_{0.09}$/In$_{0.94}$Ga$_{0.06}$As (110 Å/ 116 Å layer thicknesses). The LEDs were fabricated into unpassivated, mesa structures. The forward-bias LED emission, photoluminescence spectrum, and photovoltaic spectral responsivity (all at 77K) for the LED are shown in Figure 3. All spectra were obtained with an FTIR spectrometer equipped with an InSb detector. A Nd:YAG laser (1.06 μm) was used for excitation in the photoluminescence experiment. Photoluminescence and LED emission were measured by operating the FTIR in a double-modulation mode. The peak of the photoluminescence and the edge of the spectral response both occur at approximately the same energy, 340 meV. The LED emission occurs at slightly higher energy, with a peak of 348 meV (3.56 μm) at high injection levels. With the shift of the LED emission to higher energy than the PL peak, we conclude that the band-filling increase in the emission energy dominates the injected carrier-induced decrease in the bandgap.

At 77K, the LED output power was approximately 0.02 W/cm^2 for a current density of 100 A/cm^2. This results in an efficiency of 0.06%. The LED emission intensity falls-off rapidly above 100 K, and we could not observe LED emission at 300 K. Also, the LED I-V characteristic was soft, and the photoluminescence efficiency of the LED sample was less than that observed in our best SLSs. Overall, the LED emission efficiency and I-V characteristic of this initial SLS device were poor compared to that demonstrated for thick, LPE-grown InAsSbP LEDs emitting at 4 μm and operating at 300 K.[11,12] There was significant residual lattice mismatch between the SLS and the substrate (Δa / a = .0016) and we speculate that the emission efficiency and the I-V characteristic of the SLS device were degraded by dislocations forming as a result of this mismatch.

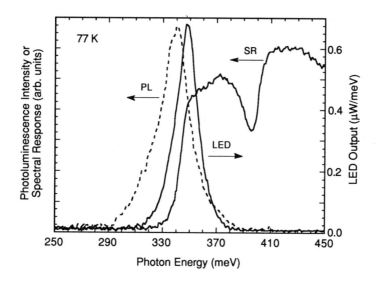

Figure 3. Forward-bias LED emission (LED), photoluminescence spectrum (PL), and photovoltaic spectral responsivity (SR) of the LED sample. All spectra were taken at 77 K, and the injection current used for the LED emission measurement was 100 mA.

In conclusion, we have established the growth conditions for $InAs_{1-x}Sb_x/In_{1-y}Ga_yAs$ SLS's. The vapor-solid distribution coefficient for Sb can be described by a thermodynamic model. The PL peak energies of the SLS's and alloys grown under the conditions of this study are lower than those previously reported. The lower energy anomaly in these SLSs is probably due to ordering which is observed in $InAs_{0.86}Sb_{0.14}$. The first midwave infrared emitters that utilize SLSs with biaxially compressed InAsSb have been demonstrated. An SLS LED was constructed which emitted at 3.56 μm with 0.06% efficiency at 77 K. Demonstration of this device should allow us to construct electrically injected lasers. We anticipate that with improvements in material quality, we may soon demonstrate the reduction of Auger rates through higher temperature operation of midwave emitters with biaxially compressed InAsSb.

ACKNOWLEDGMENTS

The authors wish to acknowledge the assistance of J. A. Bur and M. P. Moran with the infrared measurements and the TEM sample preparation, respectively. This work was supported by the US DOE under Contract No. DE-AC04-94AL85000.

REFERENCES

[1] R. J. Menna, D. R. Capewell, R. U. Martinelli, P. K. York, and R. E. Enstrom, Appl. Phys. Lett. 59, (1991) 2127 .

[2] R. M. Biefeld, J. Crystal Growth 75 (1986) 255.

[3] R. M. Biefeld, S. R. Kurtz, and S. A. Casalnuovo, J. Crystal Growth 124 (1992) 401.

[4] R. M. Biefeld, C. R. Hills and S. R. Lee, J. Crystal Growth 91 (1988) 515.

[5] S. R. Kurtz and R. M. Biefeld, Phys. Rev. B 44, 1143 (1991).

[6] R. M. Biefeld, J. Crystal Growth 75 (1986) 255.

[7] C. A. Larsen, S. H. Li, N. I. Buchan, and G. B. Stringfellow, J. Crystal Growth, 102 (1990) 126.

[8] Z. M. Fang, K. Y. Ma, D. H. Jaw, R. M. Cohen, and G. B. Stringfellow, J. Appl. Phys. 67, 7034 (1990).

[9] S. R. Kurtz, L. R. Dawson, R. M. Biefeld, D. M. Follstaedt, and B. L. Doyle, Phys. Rev. B 46, 1909 (1992).

[10] Su-Huai Wei and Alex Zunger, Appl. Phys. Lett. 58, 2684 (1991).

[11] N. P. Esina, N. V. Zotova, B. A. Matveev, N. M. Stus', G. N. Talalakin, and T. D. Abishev, Sov. Tech. Phys. Lett. 9, 167 (1983).

[12] A. Krier, Appl. Phys. Lett. 56, 2428 (1990).

OPTICAL PROPERTIES AND ELECTROLUMINESCENCE OF ORDERED AND DISORDERED ALAS/GAAS SUPERLATTICES

D. J. ARENT, R. G. ALONSO, G. S. HORNER, A. E. KIBBLER, J. M. OLSON, X. YIN[#],
M. C. DELONG[#], A. J. SPRINGTHORPE[†], A. MAJEED[†], D. J. MOWBRAY[±], AND
M. S. SKOLNICK[±]

National Renewable Energy Laboratory, 1617 Cole Blvd. Golden, CO 80401
[#] *Department of Physics, University of Utah, Salt Lake City, UT 84112*
[†] *Bell Northern Research, Ltd. P.O. Box 3511 Station C, Ottawa, Ontario, Canada K1Y 4H7*
[±] *Department of Physics, University of Sheffield, Sheffield , U.K. S3 7RH*

ABSTRACT

In this study, we report the optical properties of a variety of compositionally equivalent disordered superlattices in the AlAs/GaAs system. Markedly different signatures are seen in the steady-state optical signatures of a random pseudobinary $Al_{0.5}Ga_{0.5}As$ alloy and $(AlAs)_n(GaAs)_{4-n}$ superlattices (SLs) where n = 2 (ordered) or n is randomly chosen (disordered). Relative to the properties of the pseudobinary alloy or ordered SL, intense, red-shifted, photoluminescence (PL) peaks and broad non-excitonic absorption are observed from disordered $(AlAs)_n(GaAs)_{4-n}$ SLs where n is randomly chosen from the sets [0, 1, 2, 3, 4] or [1, 2, 3]. Our observations suggest that a large density of states at energies lower than the compositionally equivalent pseudobinary band gap exist, and including wider wells and barriers may lead strongly localized regions and the onset of quantum effects. Room temperature electroluminescence is also observed from disordered SL *pn* samples, and the energy, intensity, and efficiency are shown to vary with deposition sequence.

Introduction

The observation of carrier localization in one-dimensional random systems was first studied in purposely disordered GaAs/AlGaAs superlattices where the authors were able to demonstrate that vertical transport in superlattices decreased with increasing disorder.[1] Additionally, it was shown that the vertical transport is thermally activated from localized levels with trap energies of 5 - 66 meV, depending on the degree of disorder and that the transport displays behavior characteristic of an Anderson transition as the disorder exceeds a value which leads to the localization of all eigenfunctions (2 monolayers for the SLs studied of nominal width 30Å).[1,2]

In 1989, new, severely disordered, materials requiring monolayer (ML) precision were proposed.[3] These materials, called disordered monolayer superlattices (d-SLs), are composed of randomly distributed layers with thicknesses L_z no greater than 6 Å and average disorder of $\frac{L_z}{2}$.

As shown in Fig. 1, a disordered monolayer superlattice, for example, composed of an alternating sequence of 1 ML of GaAs followed by 3 ML of AlAs, then 2 ML GaAs, 2 ML AlAs, 3 ML GaAs, 1 ML AlAs, etc., would produce a random superlattice with an overall periodicity of 4 ML and average composition of $Al_{0.5}Ga_{0.5}As$. This sequencing is in contrast to a normal, ordered superlattice of the same dimension where 2 ML of GaAs and 2 ML of AlAs are repeatedly deposited in alternating sequence, or a bulk alloy where both Ga and Al atoms impinge on the growing surface at the same time (no intentional spatial separation).

Mat. Res. Soc. Symp. Proc. Vol. 325. ©1994 Materials Research Society

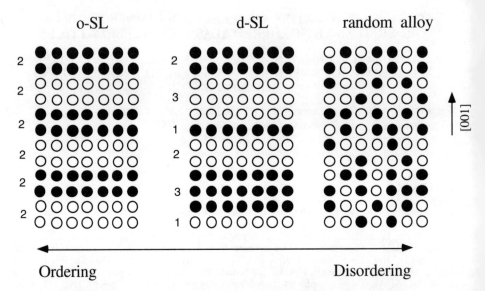

Figure 1. Various deposition sequences of AlAs/GaAs monolayer SLs with equivalent compositions of $Al_{0.5}Ga_{0.5}As$.

Early reports have shown that d-SL layers exhibit (1) increased PL and electroluminescence (EL) intensity compared to ordered and bulk alloy indirect band-gap-equivalent materials, especially near 300 K; (2) reduced band-gap energies relative to compositionally equivalent random alloys and ordered superlattices of the same, or even longer, periodicity; (3) a temperature dependence of PL intensity similar to that of amorphous materials, which is described by $I \propto \left[1 + A \exp\left(T/T_0 \right) \right]^{-1}$, where T is absolute temperature, T_0 is the characteristic temperature corresponding to the energy depth of localized sates from a mobility edge, and A is a tunneling factor; and (4) a shorter PL decay lifetime than equivalent indirect-gap alloys or ordered SLs.[4-10] The luminescent behavior was observed to depend on the probability distribution of the layer thicknesses and was attributed to different localized states created by the disordering in atomic arrangement.[8] The results of these experiments and early theoretical treatments[11-13] support the assignment of the observed properties to localization effects similar to those ascribed to amorphous materials and Anderson localization.[14]

In this paper, we report on the influence of the (dis)ordering deposition sequence (randomization) on the optical properties of d-SL materials. We also show that the emission energy, intensity, and efficiency of room temperature EL vary with deposition sequence.

Experimental

The following undoped samples were grown by solid source MBE on exact [001] GaAs substrates: (i) a 3-D random binary alloy of $Al_{0.5}Ga_{0.5}As$ (RBA), (ii) a $(AlAs)_2(GaAs)_2$ [001] ordered superlattice [to be denoted as o-(2,2) SL], (iii) a $(AlAs)_n(GaAs)_{4-n}$ d-SL where n = 1, 2, or 3 is a random number [(d-(1,3)-A], (iv) $(AlAs)_n(GaAs)_{4-n}$ d-SL where n is varied randomly from the set [0, 1, 2, 3, 4], denoted as d-(0,4) and the n=0 component indicates that the layer is skipped. The undoped d-SL layers were 0.3 μm thick and capped by 100 Å of GaAs. For the EL samples, 1 μm thick d-SL layers Si doped 1 x 10^{17} cm^{-3} were followed by a 0.2 μm d-SL region

500

Be doped at 1 x 10^{19} cm^{-3} and a 0.2 μm GaAs contact layer. The d-SL layers in the EL samples consisted of (i) a n-type d-(1,3)-A layer, (ii) a (AlAs)$_n$(GaAs)$_m$ d-SL where n, m = 1, 2, or 3 are random numbers [(d-(1,3)-B], and (iii) a (AlAs)$_n$(GaAs)$_{8-n}$ d-SL where n = 2, 4, or 6 is a random number [(d-(2,4,6)-A]. A significant variation in the sequencing of the GaAs and AlAs layer was used: layers defined as (GaAs)$_n$(AlAs)$_{4-n}$ or (AlAs)$_n$(GaAs)$_{8-n}$ require that the subunit thickness = $L_{z, AlAs}$ + $L_{z, GaAs}$ be constant whereas those formulated as (GaAs)$_n$(AlAs)$_m$ have no short range correlation and are completely random. In each d-SL we have studied, the probability of occurrence for each n,m was equal.

The SLs were grown without the use of growth interrupts at a low growth rate of 0.25 ML/s and a low growth temperature of 560°C to maximize interface abruptness and limit cation diffusion.[15] These conditions contrast those used in previous studies: 600°C and 3-5 s growth interrupts with a growth rate of 0.5 ML/s.[4] To fabricate isolated *pn* detectors and EL samples, large area mesas (0.1 cm^2) were etched , Au contact fingers electroplated, and the contact layer was selectively removed. Aluminum concentration, sample thickness, and periodicity were confirmed by double-crystal X-Ray diffraction and transmission electron microscopy (TEM).

Results

PL (solid lines) and photoluminscence excitation spectroscopy (PLE) (dashed lines) spectra from four samples are plotted in Fig. 2. The arrow above each PL spectrum indicate the energy at which the PLE spectrum was recorded. Note that all the structures have the same average composition. However, by controlling the deposition sequence of the group III atoms (Ga and Al), we have been able to change the peak emission energy by more than 0.2 eV, as well as numerous other spectral details.

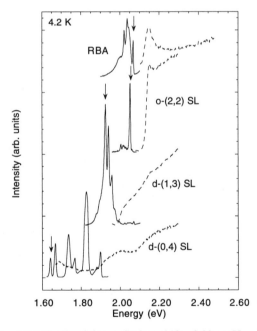

Figure 2. Photoluminescence (full lines) and photoluminescence excitation emission (dashed lines) as a function of photon energy taken at 4.2 K for Al$_{0.5}$Ga$_{0.5}$As RBA, (GaAs)$_2$(AlAs)$_2$ ordered SL, d-(1,3)-A SL and d-(0,4) SL. Excitation intensity was 5 mW/cm^2 at 5400 Å. The arrows indicate the detection energies used to collect the PLE spectra.

The PL and PLE spectra obtained from the RBA and o-(2,2) SL (Fig. 2) show relatively weak PL emission associated with the lowest energy indirect transition E$_{X-\Gamma}$ at ~ 2.05 eV and phonon replica similar to that of earlier findings.[16] For the RBA sample, the PL is a broad manifold upon which sharp peaks are superimposed, with full width at half maximum (FWHM) as small as 5.1 meV. The PLE spectrum (detected at 2.066 eV) shows well a resolved free excitonic absorption at 2.148 eV, giving a Stokes shift of 82 meV. The maximum emission intensity is at 2.037 eV, and the peak separations are consistent with the LO phonon energies of GaAs (*X*) (31 meV) and AlAs (*X, Γ*) (48 meV).[16]

In comparison, the optical signatures of the o-(2,2) SL show that PL is dominated by a single emission at 2.052 eV, 15 meV lower than the high-energy PL peak detected from the RBA sample, and phonon replica. PLE detected at the PL peak energy indicates a weakly excitonic absorption with a band-gap energy of 2.150 eV. The measured Stokes shift of 98 meV and reduced PL peak energy relative to the RBA are consistent with the predicted Type II pseudoindirect nature of the lowest energy transition [17] and earlier observations. [16]

The PL spectrum of d-(1,3)-A SL exhibits a strong peak with up to five shoulders, indicating a complicated convolution of emission processes. At 4.2 K under 5 mW/cm^2, 5400 Å excitation, 1.926 eV emission (with FWHM = 9.4 meV) dominates from the d-(1,3)-A SL. Photoluminescence from the two highest energy transitions was observed only with selective excitation at 2.296 eV, suggesting that these peaks are deep-level recombination associated with the buried $Al_{0.5}Ga_{0.5}As$. Also, PL obtained with lower energy excitation (1.941 eV) is nearly half as intense as that obtained at higher excitation energy (2.032 eV), suggesting that a significant density of states exists at the lower energies. Using the PL peak energy as representative of the band-gap energy, the introduction of the randomization sequence has red-shifted the effective band gap by more than 120 meV.

The PLE spectrum for the d-(1,3)-A SL displays no indication of the presence of free excitonic absorption. The broad absorption profile is consistent with band tailing or disorder-induced-localization, where the required nonperiodic potential has been synthetically produced by using a random deposition sequence for the Group III atoms. [5,18] Also consistent with the existence of a large density of states at lower energy is the fact that emission from the three lowest energy bands could be excited at an energy of 2.032 eV (not shown).

The PL spectrum of the d-(0,4) SL is qualitatively different from that of the d-(1,3)-A sample in that seven reasonably well-defined PL peaks are observed between 1.625 eV and 1.898 eV. The two highest in energy (1.880 and 1.898 eV) are detected only with excitation energies greater than 2.2 eV (i.e., larger than the band-gap energy of the $Al_{0.5}Ga_{0.5}As$ layer). Additionally, PLE detected on these two highest energy emissions show essentially the same sharp, well-defined band edge at ~2.17 eV. These two peaks are therefore associated with the buried $Al_{0.5}Ga_{0.5}As$ layer, which does not absorb the lower energy (2.0 eV) excitation beam. These high-energy emissions, observed in both d-SL PL spectra, may be associated with regions of the d-SL that are close to the RBA buffer layer and separated from the "bulk" of the d-SL by a thin, minimally absorbing AlAs layer.

The PLE spectra detected on the five lower energy peaks of the d-(0,4) SL (a representative spectrum is shown in Fig. 2) exhibit extremely broad absorption tails with a stair case-like profile which is normally associated with 2-dimensional confined quantized levels. In this case, however, the absorption is extremely weak and excitonic effects are not observed. In the extreme, these transitions would be associated with isolated single quantum wells with relatively ill-defined interfaces, as evidenced by FWHM values $\geq$19 meV in the 4.1 K PL spectra. Inspection of the actual computer-generated layer sequence reveals GaAs and AlAs layers with thicknesses up to 11 and 14 ML, respectively. We analyzed the computer generated growth sequences, calculated transition energies, and found no correlation to the observed PL peak energies and PLE absorption edges. This discrepancy could arise from incomplete analysis of the wavefunction overlap between adjacent areas of the d-SL structure as well as ambiguous termination of the well and barrier regions in the complex sequence. Interestingly, we observed very little spectral dependence of the PL from the d-(0,4) SL with either temperature or excitation intensity. These observations suggest that the radiative recombination transitions may be fundamentally different from those of the d-(1,3)-A SL ; they also suggest that carrier localization inhibits migration to a free surface or other nonradiative recombination site.

We observed additional evidence for the complex nature of the d-(1,3)-A SL sample in the changes in peak energy and spectral line shape of PL with excitation intensity (Fig. 3). The PL spectra of d-(1,3)-A initially (low excitation intensity) shows emission at low energy (1.925 eV). A tenfold increase in excitation power reveals two higher energy transitions at ~1.945 and 1.960 eV and an additional lower energy peak at 1.915 eV. At an excitation intensity of ~13 W/cm^2 (400 mW, unfocused), the PL spectrum is dominated by two transitions, ~1.915 and 1.946 eV with only a small shoulder visible at ~1.98 eV. As the incident intensity increases, energy levels with a low density of states become saturated, and stronger emission occurs from states with shorter lifetimes or higher densities. As higher energy states become populated, new recombination paths become available (for example, electrons may spatially diffuse to nearby lower energy states) and new emission processes at lower energies may turn-on. In this case, time-resolved photoluminescence may indicate an extra delay time in the turn on of the 1.915 eV emission with respect to the 1.945 eV emission, since spatial electron diffusion may be involved. In marked contrast, the profile of PL spectra of the RBA and o-(2,2) and d-(0,4) (not shown) change very little with increasing excitation intensity.

The enhanced PL properties of the d-SL structures suggested that active devices made from these materials may outperform those composed of the indirect-band-gap equivalent pseudobinary alloy. Hence, we fabricated *pn* junction devices to investigate the absorption[19] and EL properties. Shown in Fig. 4, the room temperature EL peak energies vary more than 100 meV for structures with the same average composition . Specifically, we observe EL at 1.5V from the $(AlAs)_n(GaAs)_{8-n}$ [n = 2, 4, 6] d-SL with a peak energy of 1.81 eV that is 35-fold more intense than EL from either the d(1,3)-A or d(1,3)-B samples, with respective peak energies of 1.85 and 1.97 eV. These energy differences are due to disorder-induced localized states where the effective peak and width of the density of states varies with the randomization sequence.[13] Additionally, doubling the layer thickness while maintaining the *exact* layer sequence seems to have a rather dramatic effect on the extent of localization. Comparing observations from the d-(2, 4, 6)-A and d(1, 2, 3)-A samples, we note that the peak EL energy is red shifted ~ 50 meV, the turn on voltage for emission visible to the eye (a few μW) is reduced 0.15 eV, and the intensity is 35-fold greater at 1.5V bias. Interestingly, the d-SL structures with some short range correlation have not been considered theoretically, though our observations suggest that this structure may provide a feasible means to obtaining high quality optical material with band-gap energies previously unattainable in many materials systems.

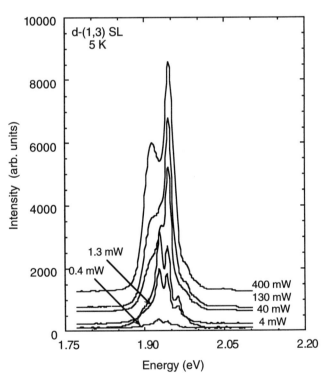

Figure 3. PL as a function of photon energy and excitation power for the d-(1,3)-A SL. The sample temperature was 4.2 K and the excitation was at 5400 Å. The spectra are offset for clarity.

Conclusions

In summary, optical spectroscopy has revealed that randomization of the layer thickness in short period AlAs/GaAs superlattices induces significant changes in the band structure. We observed optical signatures from d-SL structures indicating the presence of a significant density of localized or band tail states at energies more than 200 meV below the band gap of the random $Al_{0.5}Ga_{0.5}As$ alloy, as well as bright EL intensities at room temperature, suggesting that these novel material structures may be suitable for advanced optoelectronic applications.

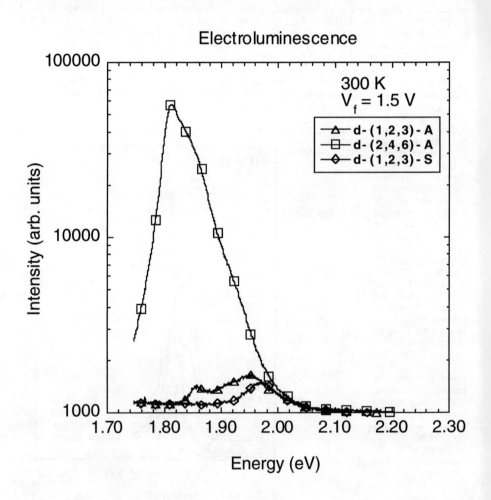

Figure 4. Room temperature EL at 1.5 V bias from variously disordered SL sequences. The integrated intensity from the d(2,4,6)-A d-SL is 35-fold greater than that measured from either d(1,3) d-SL structure.

Acknowledgments

Part of this work was supported by the US. Department of Energy, Office of Energy Research, Division of Basic Energy Sciences and by contract No. DE-AC02-83CH10093 and NREL contract XR-2-12121-1.

References

1. A. Chomette, B. Deveaud, A.Regreny, and G. Bastard, Phys. Rev. Lett. **57**, 1464, (1986).

2. See E. Tuncel and L. Pavesi, Phil. Mag. B. **65**, 213 (1992) and references therein.

3. A. Sasaki, M. Kasu, T. Yamamoto, and S. Noda, Jap. J. Appl. Phys. **28**, L1249, (1989).

4. M. Kasu, T. Yamamoto, S. Noda and A. Sasaki, Jap. J. Appl. Phys. **29** , 828 (1990).

5. M. Kasu, T. Yamamoto, S. Noda and A. Sasaki, Jap. J. Appl. Phys. **29**, L1055, (1990).

6. M. Kasu, T. Yamamoto, S. Noda, and A. Sasaki, Appl. Phys. Lett. **59**, 800 (1991).

7. M. Kasu, T. Yamamoto, S. Noda and A. Sasaki, Jap. J. Appl. Phys. **29**, L1588, (1990)

8. A. Sasaki, J. Cryst. Growth **115**, 490, (1991)

9. X-L. Wang, A. Wakahara, and A. Sasaki, Appl. Phys. Lett. **62**, 888 (1993).

10. X. Chen, B. Henderson, and K. P. O'Donnell, Appl. Phys. Lett. **60**, 2672 (1992).

11. K. F. Brennan, Appl. Phys. Lett. **57**, 1114 (1990).

12. W. P. Su and H.D. Shih, J. Appl. Phys. **72**, 2080 (1992).

13. J. Stozier, Y.A. Zhang, C. Horton, A. Ignatiev, and H.D. Shih, J. Vac. Sci. Tech. A, **11**, 923 (1993).

14. P.W. Anderson, Phys. Rev. **109**, 1492 (1958).

15. J. Grant, J. Menéndez, L. N. Peiffer, K. W. West, E. Molinari, and S. Baroni, Appl. Phys. Lett. **59**, 2859 (1991).

16. W. Ge, M. D. Sturge, W. D. Schmidt, L. N. Pfeiffer, and K. W. West, Appl. Phys. Lett. **57**, 55, (1990).

17. S.-H. Wei and A. Zunger, J. Appl. Phys. **63**, 5794 (1988).

18. N. Mott, and E.A. Davis, Electronic Processes in Non-Crystalline Materials, (Clarendon Press, Oxford, 1979).

19. D. J. Arent, A. Kibbler, Sarah R. Kurtz, J. M. Olson, A. J. SpringThorpe, and A. Majeed, unpublished.

OPTICAL INVESTIGATION OF STRAINED-LAYER
GaInAs/GaInAsP HETEROSTRUCTURES

I. QUEISSER, V. HÄRLE, A. DÖRNEN, AND F. SCHOLZ
4. Physikalisches Institut, Universität Stuttgart
Pfaffenwaldring 57, D-70550 Stuttgart, Germany

ABSTRACT

We performed low temperature photocurrent and photoluminescence excitation spectroscopy on tensile and compressively strained $Ga_x In_{1-x}As/GaInAsP$ quantum well layers to determine the band offset of the heterojunction ($0.3 < x_{Ga} < 0.7$). The ratio of the conduction band discontinuity to the heavy hole discontinuity has been obtained from well to barrier transitions and is found to be about 35/65 for gallium contents between 0.4 and 0.6. We obtained the effective heavy hole mass by comparison of PLE transition energies with calculations of the subband levels. We observe that the effective heavy hole mass increases with the gallium content from $0.3\,m_0$ for $x_{Ga} = 0.31$ to about $0.45\,m_0$ for $x_{Ga} = 0.55$.

INTRODUCTION

Strained quantum well systems have very interesting properties for the design of semiconductor devices. For example due to the modified valence band structure strained-layer GaInAs quantum well devices show an improved laser characteristic [1], and an improved modulator performance [2].

The band offset of GaInAs/GaInAsP is scarcely investigated although it determines the potential barriers of the heterostructures and thereby transport characteristics of carriers. The size of the band offset and the effective masses are important quantities for the calculation of energy levels in quantum wells (QW). For unstrained $Ga_{0.47}In_{0.53}As/GaInAsP$-layers the conduction band discontinuity was estimated from photoluminescence excitation spectroscopy (PLE) to be 38% of the bandgap difference [3].

SAMPLES

We studied the band offset and the effective heavy hole masses of $Ga_x In_{1-x}As/GaInAsP$ strained-layer heterostructures by photocurrent (PC) and PLE measurements. These single quantum well (SQW) structures were grown by low-pressure metal-organic vapor phase epitaxy (LP-MOVPE) on (100) oriented semi-insulating InP:Fe substrate. The GaInAs SQWs have a well width of 20 nm and gallium content in the range of 0.3 ... 0.7. The 100 nm wide GaInAsP-barriers (low temperature bandgap of 1.08 eV) were grown lattice-matched to InP. The gallium content and the well width of the SQWs were cross-checked using high resolution X-ray diffractometer measurements on multi quantum-well structures.

EXPERIMENT

The photocurrent measurements were carried out with a Fourier spectrometer (BOMEM DA3.01). The samples were contacted with standard ohmic contacts (alloyed Au/Sn/Ag/Au). The photoluminescence (PL) signal was excited with the 514 nm line of an Ar⁺ laser. For PLE measurements we used the light of a 50 Watt halogen lamp dispersed by a 1 m Spex monochromator. The luminescence of the lowest intersubband transition was detected by a liquid nitrogen cooled germanium detector which followed a 0.25 m monochromator. Standard lock-in technique was used to amplify the signal. The spectra were corrected for the system response.

THEORY

In order to deduce the band discontinuities and the effective masses from a comparison of experimental data with theory we calculated the energy levels of the confined carriers. We performed these calculations with a finite-quantum-well model including effects specific for strained material [4]. We included : (i) the composition dependence of the unstrained bandgap $E_g^{s.f.}$ and (ii) the influence of strain on the bandgap of the bulk material.

For the composition dependence of the strain free $Ga_xIn_{1-x}As$ bandgap we used the relation: $E_g^{s.f.}(x) = 0.4105 + 0.6337x + 0.475x^2$ [5]. Lattice mismatch between the epitaxial GaInAs layer and the substrate leads to compositional strain in the layer. This biaxial strain causes a change in the fundamental bandgap and a splitting of the light and heavy hole band edges E_{lh} and E_{hh} which are degenerate at the lattice matched composition $x_{Ga} = 0.47$. With increasing gallium content the heavy hole bandgap E_g^{hh} increases stronger than the light hole gap E_g^{lh}. The effects on the energy bands have been described by the deformation potential theory [6, 7]. The lattice constant of unstrained $Ga_xIn_{1-x}As$, the elastic constants, the deformation potentials, and the effective electron masses have been estimated by interpolating the parameters of the binaries [7, 8]. With the bulk parameters of strained GaInAs obtained in the way described above the energy levels of the vertically confined carriers have been calculated within the finite potential well model.

RESULTS FROM PHOTOCURRENT MEASUREMENTS

Figure 1a shows photocurrent spectra of strained $Ga_xIn_{1-x}As/GaInAsP$ SQWs measured at low temperatures (5K) for gallium contents of 0.41, 0.48, and 0.55. For all compositions we observe thresholds in the photocurrent at photon energies between the bandgap energies of the quantum well and the barrier. The $Ga_xIn_{1-x}As$ bandgap is indicated with E_g^{hh} and E_g^{lh}. At photon energies of $\sim 1.08 eV$ the bandgap of the confinement material $E_g(GaInAsP)$ appears. The threshold energy between the bandgap energies of well and barrier, marked as E_{th}^{hh} (E_{th}^{lh}), corresponds to the lowest transition from the heavy (light) hole subband to the conduction band of the barrier. For the threshold energy of the heavy hole case the following equation holds:

$$E_{th}^{hh} = 1hh + E_g^{hh} + \Delta E_c = E_g^{GaInAsP} + 1hh - \Delta E_{hh}.$$

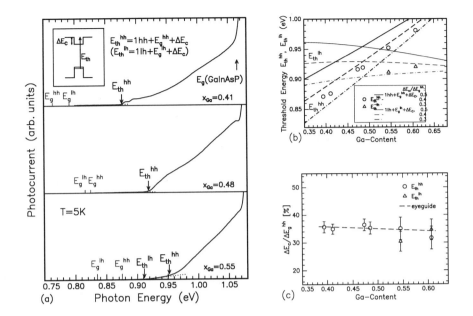

Figure 1: (a) Low-temperature PC-spectra of $Ga_xIn_{1-x}As/GaInAsP$ SQW samples with various gallium contents. Thresholds in the photocurrent marked with E_{th}^{hh} (E_{th}^{lh}) result from transitions from the lowest heavy (light) hole state into the conduction band of the barrier. These experimental threshold energies are compared in (b) with calculated values. The only adjustable parameter is $\Delta E_c/\Delta E_g^{hh}$. The resulting relative conduction band discontinuity is shown in (c).

An equivalent relation exists for E_{th}^{lh}. These thresholds can be used to determine the conduction band discontinuity ΔE_c and the heavy (light) hole band discontinuity ΔE_g^{hh} (ΔE_g^{lh}) very precisely since the first hole quantization energy is small (for the heavy hole case $1hh \sim 2$ meV and for the light hole case $1lh \sim 11$ meV) compared to ΔE_c. For biaxial compression ($x_{Ga} = 0.41$), where the heavy hole bandgap is smaller than the light hole bandgap, the transition energy E_{th}^{hh} is smaller than the transition energy E_{th}^{lh}. Thus the observed threshold energy is E_{th}^{hh}. For $x_{Ga} = 0.48$, a composition which is close to the unstrained case ($x_{Ga} = 0.47$), heavy and light hole threshold energies coincide. In the case of biaxial tension ($x_{Ga} = 0.55$) the first onset of photocurrent is at the light hole threshold energy E_{th}^{lh}. In the photocurrent spectrum one can observe also a second onset which corresponds to the heavy hole threshold energy.

From a comparison of the experimental data with theoretical calculations of the threshold energy (see Fig. 1b) we can determine the ratio of the conduction band discontinuity to the difference in heavy hole bandgaps $\Delta E_c/\Delta E_g^{hh}$ to be $(35 \pm 3)\%$. This value is almost constant in the investigated range of gallium contents from 0.4 to 0.6 as shown in Fig. 1c. Our value for the conduction band discontinuity of the unstrained material is in good agreement with earlier measurements on the unstrained GaInAs on GaInAsP [3].

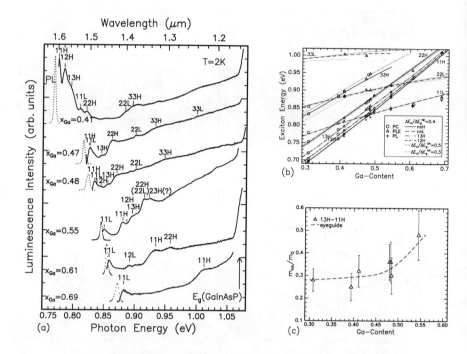

Figure 2: (a) Low-temperature PLE (solid line) and PL (dotted) spectra of $Ga_xIn_{1-x}As/GaInAsP/InP$ SQW samples with different gallium content x_{Ga} and a well width of 20 nm. The excitonic transitions which are used for the comparison with theory (b) are marked with arrows. For all transitions a good agreement between experiment and theory is observed for $\Delta E_c/\Delta E_g^{hh} \approx 0.4$. The 13H transition which only slightly depends on the band discontinuity distribution has been used to calculate the gallium dependent effective masses shown in (c).

PLE MEASUREMENTS

We performed photoluminescence measurements in order to check our results for the band discontinuity with the energetic position of higher transitions. Furthermore we deduced the effective heavy hole masses from the observed transition energies. Figure 2a shows low-temperature (2K) PLE spectra of strained $Ga_xIn_{1-x}As/GaInAsP$ samples. The detection wavelength was set to the peak energy of the corresponding PL spectrum included as dotted lines in this picture. The peak halfwidth vary in the range of 3.3 - 6.5 meV. These narrow PL lines show the good quality of the samples.

Several excitonic transitions have been observed. Lines mark positions of identified transitions which have been compared with theoretical calculations. Some selection rule forbidden transitions (12H and 13H) can be observed. Especially the relatively strong occurrence of the 12H transition (1e-2hh transition) is presently not understood. The intersubband transition, which is most sensitive to the value of ΔE_c is the 3e-3hh transition

(33H). The energy difference between the 11H and 13H transition is most sensitive to the effective quantization mass of the heavy hole.

Figure 2b shows a comparison of optical transition energies measured by PLE (see Fig. 2a) with calculated excitonic energies. The effective electron and light hole masses were linearly interpolated between the values of the binaries (InAs: $m_e/m_0 = 0.023$, $m_{lh}/m_0 = 0.026$, GaAs: $m_e/m_0 = 0.0665$, $m_{lh}/m_0 = 0.082$, m_0: vacuum electron mass [8]). The exciton binding energy is assumed to be ~ 6 meV for 11H transition and ~ 7 meV for 11L transition [9]. Higher transitions have slightly lower exciton binding energies [10]. The energetic position of the lowest optical transition 11H (11L) is caused by the large well width determined by the gallium content. The higher transitions are most sensitive to a variation of the band discontinuity, especially the 22H and 33H heavy hole transitions. We observe a good agreement between the experimental peak positions and the results of our calculations using the previously determined band discontinuity.

To separate the influences of the band offset and the effective mass on higher transitions we investigated the splitting of the 11H (12H) and 13H transition, which is (besides the influence of well width) determined by the effective heavy hole mass.

We can now calculate the effective heavy hole mass in growth direction from the 11H, (12H), 13H splitting. A careful analysis shows that the effective heavy hole mass of GaInAs cannot be described by a linear interpolation between the standard bulk values of the effective masses of the binaries InAs ($m_{hh}/m_0 = 0.40$ [8]) and GaAs ($m_{hh}/m_0 = 0.45$ [8]). We determined the heavy hole mass in the compressively and the nearly unstrained case to be about $0.3\,m_0$ For biaxial tension the effective heavy hole mass appears greater. For $x_{Ga} = 0.55$ we deduced for the heavy hole mass $m_{hh}/m_0 = 0.45 \pm 0.09$ (see Fig. 2c) An uncertainty of 1 nm in well width is taken into account. An analysis of the 22H and 33H transitions leads to the same result (see Fig. 2b). The deviation from linearly interpolated values of the heavy hole mass is not fully understood and may arise from a more complicated valence band structure in the (strained) quantum wells.

SUMMARY

In conclusion photocurrent spectra show threshold energies E_{th}^{hh} (E_{th}^{lh}) which we attribute to transitions from the lowest heavy (light) hole subband to the conduction band of the confinement material. From these threshold energies we determined the band offsets. The ratio of conduction band discontinuity to heavy hole discontinuity is found to be almost constant ($\approx 35/65$) for gallium compositions between $0.4 < x_{ga} < 0.6$.

The large well widths allows us to identify several excitonic transitions in the PLE spectra between the confined electron and hole levels. From the higher heavy hole transitions especially the splitting of the 11H and 13H transition we determined the effective heavy hole mass in growth direction. We observe for the compressively and unstrained GaInAs layer a heavy hole mass of about $0.3\,m_0$ which increases up to about $0.45\,m_0$ for $x_{ga} = 0.55$.

We would like to thank M. H. Pilkuhn for his steady interest in our work and K. Pressel for stimulating discussions.

REFERENCES

1. M. P. C. M. Krijn, G. W. 't Hooft, M. J. B. Boermans, P. J. A. Thijs, T. van Dongen, J. J. M. Binsma, L. F. Tiemeijer, and C. J. van der Poel, Appl. Phys. Lett. **61**, 1772 (1992).

2. Y. Chen, J. E. Zucker, N. J. Sauer, and T. Y. Chang, IEEE Photonics Technol. Lett. **4**, 1120 (1992).

3. S. L. Wong, R. J. Nicholas, C. G. Cureton, J. M. Jowett, and E. J. Thrush, Semicond. Sci. Technol. **7**, 493 (1992).

4. D. Gershoni, and H. Temkin, J. Luminescence **44**, 381 (1989).

5. K. H. Goetz, D. Bimberg, H. Jürgensen, J. Selders, A. V. Solomonov, G. F. Glinski, and M. Razeghi, J. Appl. Phys. **54**, 4543 (1983).

6. C. G. Van de Walle, Phys. Rev. B **39**, 1871 (1989).

7. M. P. C. M. Krijn, Semicond. Sci. Technol. **6**, 27 (1991).

8. 'Numerical Data and Functional Relationship in Science and Technology', Landolt-Börnstein, New Series, Group III, Vol. 17, (Springer-Verlag, Berlin, 1982)

9. D. Campi and C. Villavecchia, IEEE J. Quantum Electron. **QE-28**, 1765 (1992).

10. G. Bastard, J. A. Brum, and R. Ferreira, Solid State Phys. **44**, 229 (1991).

CATHODOLUMINESCENCE SPECTROSCOPY STUDIES OF GROWTH INDUCED DEEP LEVELS AT GaInP.

R. ENRIQUE VITURRO*, JOHN D. VARRIANO**, and GARY W. WICKS**
*Xerox Webster Research Center, 114-41D, 800 Phillips Rd., Webster, NY 14580.
**Institute of Optics, University of Rochester, Rochester, NY 14620.

ABSTRACT

We report a cathodoluminescence spectroscopy study of growth-induced deep levels at GaInP epilayers grown by Molecular Beam Epitaxy under various conditions. This approach allows the identification of deep levels which appear to play an important role in the band to band radiative recombination efficiency of these GaInP films. Control of these electronic defects is crucial for the performance of visible optoelectronic devices.

INTRODUCTION

III-V semiconductors of the quaternary system $Al_xGa_yIn_zP$, with x+y+z=1, are among the selected materials for fabrication of visible laser diodes and light emitting devices.[1, 2] A relevant scientific and technological issue is the understanding and control of the electronic properties which determine the quality of those materials. Here we report results of cathodoluminescence and photoluminescence spectroscopy (CLS and PLS) performed on Molecular Beam Epitaxy (MBE) grown GaInP films[3, 4] and correlate them with growth conditions. This approach allows the identification of deep levels that appear to play an important role in the band gap radiative recombination efficiency of GaInP epilayers.

Cathodoluminescence spectroscopy is used because of the several advantages this technique has over photoluminescence spectroscopy.[5, 6] The determination of defect states, their physical location-depth distribution, energies and relative intensities by using CLS has been previously shown.[5-8] The properties of the low energy CLS technique result from the dependence of electron energy loss to the solid on incident electron beam energy, which differs from photon absorption.[6] Figure 1 shows energy loss profiles for 1.0, 3.0, and 5.0 keV electron beam energies as a function of the depth into the solid. As the electron beam energy increases so does the excitation volume, which then extends between a few nanometers to about 300 nm. Thus, the CL technique provides control of the excitation depth. Its importance in studying deep levels at semiconductor films will transpire when the luminescence results are presented.

CL excitation was produced by an electron beam impinging on the sample surface at 30^0. The electron beam energy can be varied between 500 - 5000 eV. We carried out room temperature (RT) and low temperature (LT) luminescence measurements. The temperature was determined by the shift in band gap emission of the GaAs substrate and using the known dependence of band gap energy on temperature.[9] PL excitation was produced by a 6 mW HeNe laser or a 5

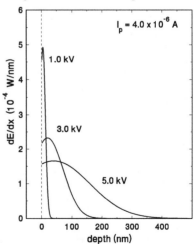

Figure 1. Electron energy loss profiles as a function of depth into the solid.

Mat. Res. Soc. Symp. Proc. Vol. 325. ©1994 Materials Research Society

mW HeCd laser. The resulting luminescence was focused into a Single Leiss monochromator. Transmitted signals were phase detected by means of either a Ge detector (North Coast) or a S-1 type photomultiplier, and a lock-in amplifier. The system response was deconvolved by comparison of the black body spectrum with the emission spectrum of a 3000 K carbon filament.

The samples were grown by MBE in an UHV Riber system. Details of the growth conditions can be found elsewhere.[3] Briefly, 500 nm thick GaInP epilayers were grown on GaAs substrates (n-type, Si, <111> misoriented 4°), using P_2 beam equivalent pressures (BEPs) between 1×10^{-5} Torr and 6×10^{-4} Torr and substrate temperatures (Ts) between 475 °C and 535 °C. Ga and In BEPs were kept constant at about 5×10^{-7} Torr and 7.6×10^{-7} Torr, respectively. All the epilayers were grown under phosphorus stabilized 2x1 surface reconstructions. Sample compositions were determined by PLS, Raman spectroscopy, and double crystal x-ray diffraction measurements.[3] The compositions of the samples were found to be $Ga_{0.51}In_{0.49}P$ within 1%. Phosphine gas has been the standard source of phosphorus for the growth of phosphides by both organometallic vapor phase epitaxy and MBE. For the most part, this type of phosphorus source produces high quality material in both growth techniques. The major disadvantages of phosphine are the high toxicity and great expense associated with its use. The use of a valved, solid red phosphorus source for MBE growth avoids these problems.[3]

RESULTS AND DISCUSSION

The dependence of the RT band gap (BG) luminescence intensity on P_2 BEP and growth temperature of GaInP epilayers is shown in Figure 2.[3] At LT these intensity differences are much smaller. The optimum P_2 BEP at Ts = 475 °C is about 2×10^{-4} Torr. At higher growth temperatures the values of P_2 BEP which may maximize the luminescence intensity are excessive for our purposes and remained undetermined.[3] The optimum value of P_2 BEP is about 10–20 times larger than As_4 BEPs typically used in the growth of GaAs/AlAs structures.[3] Three regions, of high, low, and extremely poor BG luminescence can be distinguished. We would like to point out that samples with the highest luminescence compares well to those of the best reported values of organometallic vapor phase epitaxy grown samples.

All the samples were exposed to same ambient conditions when transported between the growth chamber and the analysis chamber. Then, we expect that detrimental surface effects on BG luminescence, i.e., surface recombination velocity and band bending, are similar for all these samples. Because of the thickness of the GaInP epilayers, of about 500 nm, contributions to the spectral shape from the GaAs substrate are always observed. Even though for 5.0 kV electron beam energies the electron range is about 300 nm, diffusion enables excited carriers to reach the substrate and recombine there. For low excitation energies, these substrate contributions diminish, thus, they can be identified and separated. Luminescence contributions from n–type (Si) GaAs substrates are intense and strongly dependent on temperature. For illustration, Figure 3

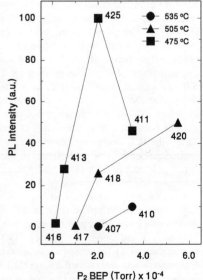

Figure 2. Band Gap PL intensity of MBE grown GaInP for several P_2 BEP and Ts.

shows the PL spectra of the GaAs substrate as a function of temperature. The RT band gap emission at 1.425 eV (BG_S) shift to 1.5 eV at 100 K. There is a strong RT deep level emission at 0.97 eV (DL_S), which shift to 1.03 eV at 100 K. There is a deep level emission at 1.2 eV (CB_S-VGa_S), not observed at RT, whose intensity is strongly dependent on temperature. For temperatures below 180 K, its intensity dramatically increases and dominates the spectral shape. This emission is usually identified as a transition between the conduction band (CB_S) and Ga vacancies (VGa_S).[7] Thus, it is imperative to record both a wide spectral range and the depth dependent spectra when thin epilayers are studied to separate spectral contributions from the substrate.[8]

Figure 4 shows CL spectra of sample #413 as a function of photon energy and electron beam energy. Five emission bands whose relative intensity change with excitation depth are observed. The 1.97 eV emission is assigned to a GaInP BG transition. Most remarkable is the relative intensity change among the 1.4 eV emission (labeled CB-Vc) and the 1.5 eV, 1.2 eV

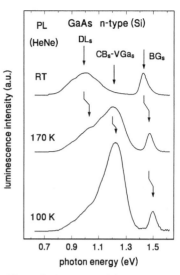

Figure 3. PL spectra of GaAs substrate as function of temperature.

and 1.03 eV emissions (labeled BG_S, CB_S-VGa_S, and DL_S) with increasing excitation depth. Comparing these results with those of Figure 3 indicates that the BG_S, CB_S-VGa_S, and DL_S transitions have their origin in the GaAs substrate. On the other hand, the CB-Vc transition has its origin in the GaInP epilayer, and its intensity is strongly dependent upon growth conditions. Figure 5 shows CL spectra of sample #411 as a function of photon energy, temperature and

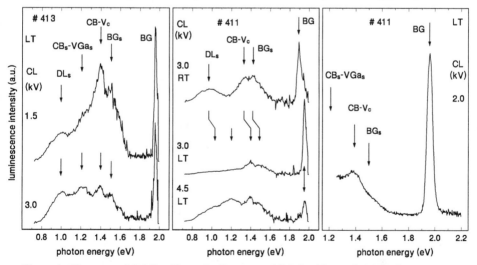

Figure 4. CL spectra of GaInP #413 as function of excitation depth.

Figure 5. CL spectra of GaInP #411 as function of temperature and excitation depth.

Figure 6. 2.0 kV CL spectrum of GaInP #411. Detector: S-1 photomultiplier.

electron beam energy. The RT spectrum shows the GaInP BG at 1.89 eV , the GaAs BG at 1.425 eV, a deep level emission at 1.34 eV, associated with the GaInP epilayer, and DL_S at 0.97 eV. The 3.0 kV LT spectrum shows the CB-Vc and BG_S, and BG emissions shifted to 1.4 eV, 1.5 eV, and 1.97 eV, respectively. Also, the relative intensity of the DL_S emission diminishes and it broadens to higher energies. The 1.2 eV (CB_S-VGa_S) transition is now clearly detected. Also, the substrate spectral features are enhanced in the 4.5 kV CL spectrum. Figure 6 shows in detail a LT 2.0 kV CL spectrum of sample #411 as a function of photon energy using a S-1 photomultiplier. It shows the same spectral shape but better signal to noise ratio than that of Figure 5. The full width at half maximum of the BG emission is about 40 meV. Notice the absence of emission in the 1.6-1.9 eV range. Overall, these spectra indicate the presence of a GaInP deep level associated with the 1.34 eV emission at RT, which shifts to 1.4 eV at low temperature. This feature is common to samples in the low BG luminescence region, see Figure 2, grown with relatively large P_2 BEPs and low Ts. Also, this feature is weak in sample #425.

The CL spectra of samples from the poor luminescence region show the dramatic changes produced in the electronic structure of GaInP with varying growth conditions, i.e., decreasing P_2 BEPs and increasing growth temperature. Figure 7 shows CL spectra of GaInP (sample # 416) as a function of photon energy, excitation depth and temperature. Figure 7(a) shows CL spectra at RT. Optical emission from the GaInP and the GaAs band gaps are observed. No deep level emissions are detected at RT. This observation is distinct from that of samples from the low luminescence region, Figures 5 and 6, which show a 1.34 eV deep level emission at RT. Figure 7(b) shows the dependence of the CL spectral shape on temperature. With decreasing temperature a deep level emission at about 1.7 eV (labeled V_P-VB) appears. As shown in Figure 7(c), which depicts the CLS dependence on excitation depth, this V_P-VB emission has its origin in the GaInP epilayer, and its intensity is strongly dependent on temperature.

Samples grown at higher Ts and P_2 BEPs show CL spectra similar to those of Figure 7. Figure 8 shows CL spectra of GaInP sample # 407. The V_P-VB emission broadens and shifts to higher energies. The overall luminescence intensity is extremely low. The BG emission is broad. No BG emission from the GaAs is observed, and the CB_S-VGa_S emission is weak

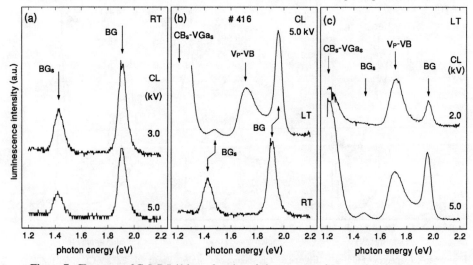

Figure 7. CL spectra of GaInP # 416 as a function of photon energy for several excitation depths and temperatures. Detector: S-1 photomultiplier.

(compare with Figure 7(c)). These observations suggest a high level of non radiative recombination processes and low carrier mobilities. Overall Figures 7 and 8 indicate that the presence of the deep level associated with the 1.7 eV emission is highly detrimental for the efficiency of the BG emission of GaInP. Whereas the higher growth temperature can also affect the incorporation of In,[3] the deficit in phosphorus incorporation appears to be more detrimental for the luminescence efficiency. These results strongly suggest that samples grown at low effective P_2 BEPs have the poorest BG luminescence efficiencies.

We now discuss the assignment of the deep level emissions to specific point defects. First, the 1.34 eV (1.4 eV at LT) emission is assigned to transitions involving the CB and cation vacancy levels, Ga and or In. Our analysis is based on the correlation with the growth conditions and the results of the CL measurements of GaInP epilayers carried out in the range 300–100 K. These measurements indicate that both BG and deep level emissions shift from their RT

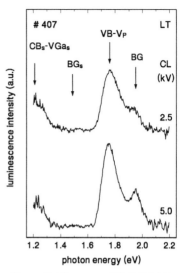

Figure 8. CL spectra of GaInP # 407.

value by approximately 65 meV with decreasing temperature, see Figure 5. The BG shift to higher energies with decreasing temperature follows the general behavior observed in III-V semiconductors, i.e., the temperature dependence of the band gap is described by a Varshni type equation.[9] The energy shift in the 1.34 eV deep level emission with temperature tracks the BG shift. In order to characterize this deep level transition it is necessary to know the change of both the valence and conduction bands with temperature. These changes are difficult to measure, because most experiments relate to band gap changes. Intuitively, the valence band is more rigid than the conduction band, because of its larger bond strength. Upon any perturbation that causes changes in volume, i.e., pressure, temperature, etc., changes in conduction band energy are expected to be larger than changes in valence band energy.[10, 11] Theoretical and experimental work confirm this picture: the change in band gap energy is mostly due to changes in conduction band energy. Similar concepts apply to shallow donor and acceptor states, where the impurity potential is dominated by the long-range (Coulomb) term. On the other hand, a deep level is a level that originates from the short-range central-cell defect potential, and this potential is very strong within the effective radius of the defect. It follows that deep levels are less sensitive to changes in temperature or pressure than the semiconductor bands.[12, 13, 14] Both deep level and VB energies remain roughly unaltered in the temperature range considered, and CB–VB and CB–DL transitions have similar shifts, whereas DL-VB transitions have small or no shift with temperature. Thus, the 1.34 eV emission band comprises a transition between the CB(donor) and DL states localized about 0.5 eV above VBM. This property characterizes Column III vacancies, Ga and In, acceptor states.[13, 14] These states are

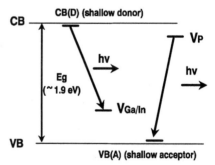

Figure 9. Schematic diagram of probable transitions in GaInP.

closer to the valence band, as cation vacancies are, because they are formed by anion wavefunctions, which have VB character.[13, 14] On the other hand, the 1.7 eV emission shows, within the small temperature range in which it is detected, no temperature dependence of the peak position. Its strong intensity dependence on temperature is due to the fact that this deep level is close to a band (similar to VGa in GaAs), and thermalization is likely to occur. Then, we associate the 1.7 eV emission with a transition between a deep level localized about 200 meV below the conduction band minimum and the valence band, V_P-VB. These properties characterize anion vacancies, V_P.[13, 14] Figure 9 shows schematically the suggested transitions and deep level energies for GaInP. This V_P assignment correlates well with recent results from thermal-electric effect spectroscopy measurements performed on similar samples.[15]

The 1.34 eV emssion may also be interpreted as involving antisite defects. These defects are common in GaAs, the main reason being that the atomic size of Ga and As are similar, and little energy is required to swap these two atoms. For example, the 0.97 eV emission, see Figure 3, may be related to transitions involving some excited state of As antisite defects.[8] For GaInP the formation of antisite defects may be difficult, because of the larger difference in atomic radius of the involved atoms. However, we cannot conclusively rule out the absence of antisite defects in these GaInP samples, and additional experiments should be performed to clarify this point.

CONCLUSIONS

The efficiency of the BG luminescence for GaInP samples grown either at high temperatures and or low P_2 BEPs appears associated with the presence of point defects, most likely cation and anion vacancies. The detriment in BG luminescence is larger for phosphorus deficient growth conditions, when optical emission involving point defects which can be associated with anion vacancies are found. The reported correlation between the CL results and the growth conditions allows the determination of defects which appear to play an important role in controling the BG radiative recombination efficiency of GaInP epilayers.

REFERENCES

[1] M. B. Panish, J. Elechtrochem. Soc. **127**, 2731 (1980).
[2] R. P. Schneider, R. P. Brian, J. A. Lott, and G. R. Olbright, Appl. Phys. Lett. **60**, 1830 (1992).
[3] G. W. Wicks et al., Appl. Phys. Lett. **59**, 342 (1991).
[4] J. A. Varriano, M. W. Koch, F. G. Johnson, and G. W. Wicks et al., J. Electronic Materials. **21**, 195 (1991).
[5] R. E. Viturro et al., J. Vac. Sci. Technol. **B6**, 1397 (1988).
[6] B. G. Yacobi and O. B. Holt, *Cathodoluminescence Spectroscopy of Inorganic Solids,* (Plenum, New York, 1990), and references therein.
[7] R. E. Viturro et al., J. Vac. Sci. Technol. **B9**, 2244 (1991).
[8] R. E. Viturro, M. R. Melloch, and J. M. Woodall, Appl. Phys. Lett. **60**, 3007 (1992).
[9] Y. P. Varshni, Physica 34, 149 (1967).
[10] C G. Van de Walle, Phys. Rev. B **39**, 1871 (1989).
[11] M. Cardoan and N. E. Christensen, Phys. Rev. **B 36**, 2906 (1987).
[12] M. Jaros, *Deep Levels in Semiconductors* (Adam Hilger Ltd, Bristol, 1982), page 191.
[13] H. P. Hjalmarson, P. Vogl, D. J. Wolford, and J. D. Dow, Phys. Rev. Lett. **44**, 810 (1980).
[14] John. D. Dow et al., J. Electronic Materials **19**, 829 (1990), and references therein.
[15] C. R. Wie and Z. C. Huang, private communication.

Author Index

Abernathy, C.R., 261, 273
Ager III, J.W., 347
Alonso, R.G., 499
Arent, D.J., 499
Aziz, M.J., 189

Bachmann, T., 119
Baeumler, M., 409
Baribeau, J.-M., 217
Baucom, K.C., 493
Bello, A.F., 389
Bennahmias, M.J., 389
Bergman, J.P., 19
Bert, N.A., 401
Bhat, R., 113
Bhattacharya, Pallab, 147
Biefeld, R.M., 493
Bimberg, D., 329
Bishop, S.G., 131
Boon, C.L., 217
Brandt, M.S., 341
Brown, Steven W., 177
Bryant, Garnett W., 49
Bube, Richard, 335
Bullough, T.J., 241

Calamiotou, M., 383
Calleja, E., 223
Chaldyshev, V.V., 401
Chatterjee, B., 125
Chen, A.C., 131
Chen, P., 37
Cheng, J.-P., 79
Cheng, K.Y., 131
Cheng, Y.M., 273
Cho, Kyoung-Ik, 457
Christou, A., 383
Colomb, C.M., 197
Cox, H.M., 113
Cunningham, J.E., 247

Dadgar, A., 329
Davidson, B.R., 241
Delerue, C., 317
Delong, M.C., 499
Deshpande, Sadanand V., 177
Di Giuseppe, M.A., 113
Dieter, R., 437
Dinh, L.N., 389
Ditmire, T.R., 389
Dorman, D.R., 347
Dörnen, A., 437, 507
Dow, John D., 3
Ducroquet, F., 235
Dumas, K.A., 43
Duxstad, K.J., 347

Erskine, D.J., 389
Erzgräber, H.B., 171

Evwaraye, A.O., 353

Fang, Z-Q., 431
Fernelius, N.C., 463
Ferreira da Silva, A., 93, 183
Ferreira, A.C., 19
Fischer, D.W., 463
Fitzgerald, E.A., 159
Foad, M.A., 443
Follstaedt, D.M., 493
Fresina, M.T., 197
Fujioka, H., 377

García, R., 223
Gauneau, M., 235
Gaworzewski, P., 171
Goloshchapov, S.I., 401
Goroncy, F., 437
Gossard, A.C., 19, 73
Grillot, P.N., 159
Gu, S.Q., 131
Guillot, G., 235
Gulari, Erdogan, 177
Gutiérrez, G., 223

Haller, E.E., 347
Hanson, A.W., 197
Härle, V., 507
Harris, C.I., 19
Harris, T.D., 113
Hatzopoulos, Z., 383
Heitz, R., 329
Hemsky, J.W., 431
Hipólito, O., 93, 183
Hirt, G., 101
Hobson, W.S., 261
Hoffman, R., 125
Holland, M.C., 443
Holmes, S., 79
Holtz, P.O., 19, 73
Hong, C.H., 235
Hong, K., 235
Hongbin, Huang, 165
Horner, G.S., 499
Horng, S.F., 85
Hsieh, K.C., 131
Hsieh, K.Y., 395
Húang, J.W., 305
Huang, Z.C., 137, 209
Hwang, H.L., 85
Hwang, Y.L., 395
Hyeon, J.Y., 329

Ichimura, S., 67
Ivonin, I.V., 401

Jantz, W., 409
Johnson, N.M., 341
Johnson, N.P., 443

Jones, R., 241
Joyce, T.B., 241

Kai, Yang, 165
Kainosho, K., 101
Kanayama, T., 67
Kaspi, R., 449
Kayiambaki, M., 383
Kibbler, A.E., 499
Kim, Sang-Gi, 279, 457
Knecht, A., 329
Koch, M.W., 137
Kolbas, R.M., 395
Komuro, M., 67
Konagai, Makoto, 229
Kozyrev, S.V., 401
Krüger, D., 171
Kuech, T.F., 305
Kunitsyn, A.E., 401
Kürner, W., 437
Kurtz, S.R., 493
Kuttler, M., 329

Lagadas, M., 383
Lannoo, M., 61, 317
Lavrentieva, L.G., 401
Leão, Salviano A., 93, 183
Lee, El-Hang, 279
Lee, Hee-Tae, 457
Lee, Jinju, 147
Li, M. Siu, 285
Li, Shin-Hwa, 147
Li, W.J., 79
Liliental Weber, Zuzanna, 377
Liqun, Hu, 165
Liu, X.C., 131
Look, David C., 361, 431
Lubyshev, D.I., 401
Luo, Jinsheng, 229

Maigne, P., 217
Majeed, A., 499
Makita, Yunosuke, 267
Manasreh, M.O., 293, 449, 463
Mariella Jr., R.P., 389
McCombe, B.D., 79
Merz, J.L., 19, 73
Missous, M., 371
Mitchel, W.C., 353
Mitha, S., 189
Molina, S.I., 223
Molnar, R.J., 341
Monemar, B., 19, 73
Morgenstern, T., 171
Moustakas, T.D., 341
Mowbray, D.J., 499
Moy, A.M., 131
Müller, G., 101
Müller, P., 119
Murad, S., 443

Nahm, Sahn, 457
Nee, T.W., 43

Newman, R.C., 241
Nozaki, Shinji, 229

O'Hagan, S., 371
Öberg, S., 241
Oda, O., 101
Ohmer, M.C., 463
Olego, D.J., 347
Oliveira, L., 285
Olson, J.M., 499
Osinsky, A., 171

Packard, William E., 3
Park, Y., 293
Pasquarello, A., 73
Pavlidis, D., 235
Pearton, S.J., 113, 261, 273
Peeters, Francois M., 471
Perry, M.D., 389
Philips, B.A., 449
Ping, Han, 165
Poker, D.B., 189
Pollak, T.M., 463
Preobrazhenskii, V.V., 401
Pritchard, R.E., 241

Qi, Li, 165
Qi, Ming, 229
Queisser, I., 507

Radousky, H.B., 389
Rand, S.C., 177
Redfield, David, 335
Rehn, V., 43
Ren, F., 261
Ren, Shang Yuan, 3
Reuter, E.E., 131
Richter, U., 119
Ringel, S.A., 125, 159
Rong, Zhang, 165
Ronghua, Wang, 165
Rosseel, T.S., 293
Roth, A.P., 217
Russo, O.L., 43
Rybicki, George C., 311

Sacedón, A., 223
Sassa, K., 425
Scalvi, Luis V.A., 285
Schaff, W., 79
Scheffler, H., 329
Schiferl, D., 189
Schmalz, K., 171
Schnoes, M.L., 113
Scholz, F., 437, 507
Schumann, H., 329
Schunemann, P.G., 463
Semyagin, B.R., 401
Shah, M., 449, 463
Shen, Jun, 3
Shibata, Hajime, 267
Shinar, J., 449
Shirakashi, Junichi, 229

Shiraki, H., 425
Shulin, Gu, 165
Sieg, R., 125
Singh, R., 341
Skolnick, M.S., 499
Skowronski, M., 293, 449
Smith, Steve R., 147, 353
Sohn, H., 377
Springthorpe, A.J., 499
Srocka, B., 329
Stavola, M., 273
Steckl, A.J., 37
Stievenard, D., 61
Stillman, G.E., 197
Stockman, S.A., 197
Su, W.P., 481, 487
Sugiyama, Y., 67
Sundaram, M., 19, 73

Takahashi, Kiyoshi, 229
Taskar, N.R., 347
Theiss, S.D., 189
Ting, C.S., 481, 487
Tohyama, E., 425
Tokuda, Y., 425
Tokumitsu, Eisuke, 229
Toyama, N., 425
Tret'yakov, V.V., 401
Tsang, W.T., 247

Varriano, J.A., 137
Varriano, John D., 513
Vatnik, M., 171
Veinger, A.I., 401
Vilisova, M.D., 401

Viturro, R. Enrique, 31, 513

Wada, T., 67
Wang, E.G., 481, 487
Wang, J.L., 79
Wang, Y.J., 79
Wang, Z.M., 409
Watson, G.P., 159
Weber, E.R., 377
Weinberg, I., 125
Wendler, E., 119
Wesch, W., 119
Wicks, Gary W., 31, 137, 513
Wie, C.R., 137, 209
Wilkinson, C.D.W., 443
Williams, Wendell S., 311
Windscheif, J., 409
Wolk, J.A., 347

Xie, Y.H., 159
Xu, J.H., 481, 487

Yakubenya, M.P., 401
Yamada, Akimasa, 267
Yang, H.W., 85
Yen, M.Y., 449
Yi, Shi, 165
Yin, X., 499
Yoo, Byueng-Su, 279
Youdou, Zheng, 165

Zhang, T., 395
Zhang, Tong-Yi, 419
Zhao, Q.X., 19, 73
Zieger, K., 437

Subject Index

activation
    energy, 247
    volume, 189
AlGaAs, 285
AlInAs, 235
alloy composition, 197
annealing, 119, 197, 311
    dynamics, 361
    high pressure, 101
    isochronal, 67
arsenic antisite, 361

band offset, 61
bound exciton, 73
buffer layers, 223
    strain free, 217

carbon(-), 197, 425
    doped GaAs, 273
    pair defects, 273
carrier trapping, 159
cathodoluminescence, 31
compensation, 293, 425

defect(s), 19, 61, 341
    acceptor(s)
        identification, 113
        in GaAs/AlGaAs quantum wells, 73
    assisted tunneling effect, 61
    deep levels, 137, 147, 209, 235, 353, 443
    distribution of deep levels, 165
    donor, 431
        identification, 113
    introduction rate, 311
    level(s), 425
        in InGaAs, 329
        in InP, 329
    metastable, 335
    mid-gap
    point, 43
    residual donors, 101
degradation of 2DEG in AlGaAs/GaAs
    heterostructure, 67
dislocation(s), 159
    array, 419
    distribution, 223
    misfit, 279, 457
DLTS, 125, 159, 165, 305, 311, 329, 383
doping engineering, 247
DX-center, 285

electrical properties, 119, 171
electron Hall mobility, 279
excitons, 49

filling factors, 79

GaAs, 305, 383, 409, 431, 443, 457
    layer on Si substrate, 377, 383, 437

GaInP, 209
GaN, 341
GaSb, 449

heteroepitaxy, 279, 437
hydrogen, 341
    atomic hydrogen in III-V compounds, 241
    hydrogenation, 125, 279
    of nitrides, 261

implantation temperature, 119
impurities, 19
    Be-doped GaAs, 73
    D^- impurities, 79
    erbium, 171
    Fe deep acceptors, 101
    impurity
        bands, 93, 183
        complexes, 293
        identification, 113
    oxygen
        contamination, 209
        incorporation, 293, 305
    rare-earth impurities in semiconductors, 317
    Rh-doping, 329
    ytterbium, 317
InGaAlN, 261
InGaAs, 229
InN, 261
InP, 113, 119, 125, 311, 409
InP/InGaAs HBT, 197
InSb, 279

local
    density functional theory, 341
    vibrational modes, 347
localized states, 481
low
    energy electron irradiation, 67
    temperature grown semiconductors
        200°C, 395
        250°C, 371, 383
        300°C, 389
        AlGaAs, 395
        arsenic-precipitate, 395
        defects in low temperature grown
            GaAs, 361
        GaAs, 361, 371, 401
        Lt-grown GaAs, 383, 393
        superconductivity phase, 401

metal-organic vapor phase epitaxy (MOCVD), 235
molecular beam epitaxy (MBE), 371, 383, 401
    MOMBE, 229, 273
    MOVPE, 305

nanostructures, 49

optical absorption, 463
  infrared absorption, 241, 293, 347, 449
  optical
    admittance spectroscopy, 353
    band gap, 177
    properties of 3D superlattices, 481
    spectroscopy, 19, 347

passivation, 19, 197, 279, 347
  of defects, 341
  of shallow acceptors, 241
phosphorus vacancies, 137
photoluminescence, 131, 171, 177, 267, 481
  excitation, 85
  microscopy, 409
  topography, 409
polarizations, 463
proton irradiation, 311

quantum
  dots, 49, 85
  wires, 49, 85

Raman spectroscopy, 73, 229, 347
recombination, 19, 159
  nonradiative, 19
  radiative, 49

relaxation time distribution, 285
resistivity, 305

semi-insulating
  behavior of undoped InP, 101
  oxygen doped GaAs, 305
Si inversion layers, 183
SiC, 6H-SiC:N, 353
SiGe, 147, 159
  alloy, 171
solubility limit, 247
stretch exponentials, 335
superlattices, 3, 37
  $(GaP)_2/(InP)_2$ short period, 131
  tuning of energy levels, 471
  type II superlattices, 3

TEM, 223, 377

vacancy injection, 37

x-ray, 217, 383

Zeeman splitting, 73
$ZnGeP_2$, 463
ZnSe, 217, 347, 383